Manufacturing Processes

Automation, Materials, and Packaging

by

J. Barry DuVall
Professor,
College of Technology & Computer Science
Director, Technology Advancement Center
Division of Academic Outreach
East Carolina University
Greenville, NC

and

David R. Hillis
Associate Professor Emeritus,
College of Technology & Computer Science
Department of Technology Systems
Co-Director, Center for Wireless
and Mobile Computing
Division of Academic Outreach
East Carolina University
Greenville, NC

Publisher
The Goodheart-Willcox Company, Inc.
Tinley Park, Illinois
www.g-w.com

The Goodheart-Willcox Company, Inc. Brand Disclaimer: Brand names, company names, and illustrations for products and services included in this text are provided for educational purposes only and do not represent or imply endorsement or recommendation by the author or the publisher.

The Goodheart-Willcox Company, Inc. Safety Notice: The reader is expressly advised to carefully read, understand, and apply all safety precautions and warnings described in this book or that might also be indicated in undertaking the activities and exercises described herein to minimize risk of personal injury or injury to others. Common sense and good judgment should also be exercised and applied to help avoid all potential hazards. The reader should always refer to the appropriate manufacturer's technical information, directions, and recommendations; then proceed with care to follow specific equipment operating instructions. The reader should understand these notices and cautions are not exhaustive.

The publisher makes no warranty or representation whatsoever, either expressed or implied, including but not limited to equipment, procedures, and applications described or referred to herein, their quality, performance, merchantability, or fitness for a particular purpose. The publisher assumes no responsibility for any changes, errors, or omissions in this book. The publisher specifically disclaims any liability whatsoever, including any direct, indirect, incidental, consequential, special, or exemplary damages resulting, in whole or in part, from the reader's use or reliance upon the information, instructions, procedures, warnings, cautions, applications, or other matter contained in this book. The publisher assumes no responsibility for the activities of the reader.

Library of Congress Cataloging-in-Publication Data

DuVall, J. Barry
    Manufacturing Processes - Automation, Materials, and
Packaging / by J. Barry Duvall, David R. Hillis
    p. cm.
    Includes index
    ISBN 1-59070-780-X

    1. Manufacturing processes. I. Title.

TS183.D88   2008
670—dc20

About the Authors

Dr. J. Barry DuVall is Director of the East Carolina University Technology Advancement Center (TAC), a research and development incubator and test bed that conducts research on technology for learning and organizational effectiveness. Dr. DuVall was previously Codirector of the Center for Wireless and Mobile Computing (2002–2004) and OWLS (Online Wireless Learning Solutions), a project funded by the U. S. Department of Education; Ericsson, Incorporated and East Carolina University (ECU) (1999–2003). He was also the Director of the Defense Industry Partnership Project, a project entitled *The Factory as a Learning Laboratory*. This was a project that provided education and training to Black & Decker (U.S.) associates and defense industry scientists and engineers in six locations using interactive video and Internet transmission (ARPA/TRP/NSF 1994–1997). This resulted in the first Internet-based programs for ECU and the first Internet programs in industrial technology in the nation. Prior to accepting his current assignment as Director of the TAC, DuVall enjoyed teaching courses in manufacturing processes, industrial materials, productivity improvement, and digital communications.

Dr. DuVall received his Bachelor of Science and Master of Science in Industrial Technology from Indiana State University and his PhD in Industrial Education and Technology from the University of Maryland. He has more than thirty years experience teaching in industry and at various universities, and has worked in industry in the fields of electronics, design, and manufacturing.

Dr. David R. Hillis taught courses in manufacturing and quality in the Department of Industrial Technology at East Carolina University for almost 16 years. In 1994, he taught his first course using distance education technology. His experience in distance education ranges from computer-based bulletin boards to closed-circuit television and interactive Internet-based methods. In 1997, Dr. Hillis received a grant from East Carolina University to develop a "hands-on" Internet-accessible laboratory facility. This work created a CNC milling machine with design software that students were able to access, operate, and view entirely over the Internet.

Dr. Hillis received his Bachelor of Science in Industrial Engineering and Master of Science in Engineering Administration from Bradley University, Peoria, Illinois. He has an EdD in Training and Development from North Carolina State University. Hillis had over 20 years of industrial experience in engineering and management with an electrical equipment manufacturer before beginning his teaching career at East Carolina University. His industrial experience has continued to grow through his active consulting practice. David R. Hillis specializes in manufacturing operations and systems.

Introduction

Manufacturing Processes is a book based on practical experiences your authors have had working in manufacturing and with thousands of students who were designers, machine operators, technicians, engineers, and managers, all wanting to make their companies as competitive as possible. There are many exciting stories to be told about the magic that occurs when products are made profitably and on schedule. When a winning product emerges, one never ceases to feel the excitement of what has been accomplished. It is a magical process. It is this magic that motivated us over the span of five years to develop this new edition. This would not have been possible without being able to draw on the work of so many student innovation teams working on cost-saving projects in manufacturing.

This is a book about doing things better, outpacing the competition, and surviving in an area where complacency often results in plant closures or even the movement of entire industries to other countries. Some managers believe all that has to be done is to reduce costs. Too often, skilled and highly motivated, hard working people are asked to sacrifice pay, benefits, and pensions only to see their plant still unable to compete. Certainly, controlling costs are important, but the key element in creating a winning manufacturing plant is selecting and matching appropriate manufacturing processes, materials, and systems to meet both the demands of the product design and the customer. This book describes the knowledge, principles, and techniques that can meld manufacturing processes, materials, systems, and people to create a competitive factory. It is a book that will help you find a better way to make quality products faster, better, and cheaper.

Manufacturing Processes should be viewed as a resource text for use in many different fields of study and for better understanding manufacturing in many different types of industries. This book provides a comprehensive survey of hundreds of materials and processes, and can be used at both introductory and advanced levels in courses preparing learners for careers in business and industry. Because of the magnitude of coverage, the text will find a home in several courses. It will be particularly useful in fields and courses in technology systems, industrial and manufacturing technology, engineering technology, engineering, operations management, and business management.

After exposing you to the structure of manufacturing, our discussion quickly turns to "Cost-Saving Approaches" (Section 2). We have learned that when looking at costs in manufacturing, many firms are quick to concentrate on cutting back on personnel when things need to be tightened up. Too often, managers view automation as the next step in cost cutting. This is not always the best approach. Section 3 is devoted to "The Decision to Automate."

Now, we need to think about the product, and make good choices about selecting the right material for the job at hand. "Selecting Appropriate Materials" (Section 4) exposes the reader to materials used by hard-good manufacturing firms to make products from industrial stock. There are thousands of materials that may be studied, and new materials are created every day. We have reduced this list to a more useful form by organizing the chapters according to the major types of industrial materials: metals, ceramics, plastics, wood, and composites.

Selecting the most appropriate process is based on the material, what will be done with it, the life cycle of the product, and other concerns. Section 5 presents the "Manufacturing Process Database." We have tried to make things easier to understand by grouping manufacturing processes according to the types of process action that occurs. These process actions are forming, separating, fabricating, conditioning, and finishing.

The North American Industry Classification Index (NAICS) is the government body that organizes and classifies industry in the U.S. NAICS identifies more than 700 different types of manufacturing industries, and each of these uses many hundreds of different processes. There is a lot to cover.

In *Manufacturing Processes*, we reduce all of this to the most important processes used by twenty major industries producing industrial products from metal, ceramic, plastic, wood, and composite materials. Some of these processes are used with multiple materials; others can be used efficiently with only one.

When the product is completed, it needs to be shipped out to the buyer. Section 6 covers "Packing Products for Distribution." Packaging is an interesting field in and of itself. We cover the major processes in use today. There is also information about the future of packing processes.

Manufacturing is truly the driving force behind any industrialized nation, having a direct relationship to the quality of life and standard of living of the nation's people. However, keeping businesses running and competitive is not an easy job. Global competition is greater than ever before. It is our hope that this book will be viewed as a type of "survivor's guide" that will keep the reader interested in continued exploration and discovery. What lies ahead is an exciting world where there is more to learn and change. Continuous improvement is a way of life for a manufacturing workforce that wants to advance their company and themselves into a productive and healthy future.

Acknowledgments

The authors would like to thank the following individuals, companies, and organizations for their assistance and contributions:

Our wives, Jean DuVall and Carol Hillis, for their encouragement and for giving up many hours of time we might have spent with them over the past five years. And, special thanks to A. and all the others that offered their support and time.

All of our friends and Associates at Black & Decker in Tarboro and Fayetteville, NC and Easton, MD, and engineers in the Coastal Plains Chapter of the Society of Manufacturing Engineers for the opportunity to learn with and from you.

Hundreds of our students and learning partners in the Department of Industrial Technology at East Carolina University for tolerating our hair-brained schemes and enthusiasm and for helping us with this project.

Special appreciation is also due to the following individuals who assisted with the Product Showcases in Chapter two. Their support, guidance, and knowledge made it possible to show some of the exciting processes on the factory floor in these case studies. Thanks to: Gary L. Butts, Production Manager and Kelly Kurz, Production Planner, The Hammock Source, Greenville, NC; Mike Watson, Manufacturing Engineer, Kinston Neuse Corporation (KNC/Crown Equipment Corporation), Kinston, NC; Paul S. Gleim, Operations Manager, Phoenix Fabrication,

Inc., Ayden, NC; Joe Banks, Manufacturing Engineer and James Suggs, Manufacturing Trainer, NACCO Materials Handling Group, Inc., Greenville, NC; and Wayne E. Price, Plant Manager, Louisiana Pacific Company, Wilmington, NC.

Thanks are also due to Jerry Rose, Marketing Manager, Maxi-Blast, Inc., South Bend, Indiana, for the photographs and information provided on blast cleaning media and deflashing.

The following companies and agencies generously provided photographs and other illustrations:
3D Systems, Inc.
Abbey Etna Machine Co.
ABS Packaging
ACIPCO Steel Products
Aircraft Spruce and Specialty Company
Alibaba.com Corporation
The American Ceramic Society
American Machine & Tool Co.
American Plywood Association
AMP Inc.
Anocut, Inc.
Apple Rubber Products, Inc.
Arbonite Corporation
Aristech Chemical Corporation
Ball Corporation
Bauteck Marine Corporation
Benjamin Moore
Betar, Inc.

Black & Decker
Branson Ultrasonics Corporation
Caleca USA Corp.
Ceradyne, Inc.
Ceramic-Refractory Corp.
Clarks Wood Company Ltd., UK
Coherent General
Consumer Protection Safety Commission
Corning Incorporated
Creative Pultrusions, Inc.
Crown Equipment Corporation, Kinston, North Carolina
The Cyril Bath Co.
Delta International Machinery Corporation
DoAll Company
Dreis & Krump
Drierite, Inc.
Duo-Fast Corporation
DuPont
Elox Corporation, an AGIE Group Company
Foodservice and Packaging Institute, Inc.
Forest Products Laboratory, USDA Forest Service
Garret Engine Division, Allied Signal Aerospace
GE Plastics
Gebruder Netzsch Maschinenfabrik GmbH
General Electric
Hammermill Paper Group
HDE Systems, Inc.
Herman Miller, Inc.
Hitachi Seki USA, Inc.
Hoge Lumber Company
Hudson Machinery Worldwide
Ingersoll-Rand Waterjet Cutting Systems
Kayex
Kelly James
Kennametal, Inc.
Lake Geneva Spindustries, Inc.
Laramy Products Co., Inc.
Laser Technology, Inc.
Lenox China
Macco Adhesives
MagneTek
Malco Products, Inc.
Manufacturing Technology, Inc.
Masonite
Maxi-Blast, Inc.
Metal Improvement Company
Milwaukee Electric Tool Corporation
The Minster Machine Co.
Mitsubishi Carbide

NACCO Materials Handling Group
NASA/JPL-Caltech
NASA/Marshal Space Flight Center
National Academy of Engineering
National Broach & Machine Co.
National Machinery Co.
National Twist Drill
NatureWorks, LLC
NIOSH
OTEA
PHI-Tulip
Phoenix Fabrication, Inc.
Pitt Community College Manufacturing Engineering Program, Greenville, NC
PMW Products, Inc.
Positrol, Inc.
Powermatic Houdaille, Inc.
PPG Industries
Reed Rolled Thread Co.
Robersonville Products
Robert Larson Company
Rocon Equipment Company
Rofin-Sinar, Inc.
Rohm and Haas Co.
Roto Finish Company, Inc.
S.S. White Technologies, Inc.
Sandia National Laboratories
Schuler Incorporated
Scott Gauthier
Seawolf Design, LLC
Shibuya Kogyo Co.
Sonic-Mill, Albuquerque Division, Rio Grande Albuquerque, Inc.
Soudronic, Ltd.
Stanton Manufacturing Co., Inc.
Matt Bentz
Stapla Ultrasonics Corporation
SUMCO USA
Tantec
Technicut
Timesavers, Inc.
Toyoda Machinery Co.
U.S. Amada, Ltd.
U.S. Bureau of Census
USG Interiors, Inc.
Wagner Electronic Products, Inc.
Weyerhaeuser Company
Workrite, Inc.
Wright Coating Company
Z Corporation, Inc.

Contents

Section 6—Packing Products for Distribution

Section 1
Manufacturing Today

Introduction to Manufacturing

Chapter Highlights

- Selecting appropriate manufacturing processes.
- Strategies for establishing market superiority among global competitors.
- Strengthening alliances with the *maquiladoras* under NAFTA.
- Manufacturing processes—tools, techniques, and technical systems for converting raw materials into hard good manufactured products.
- Using NAICS to classify manufacturing industries.

Magic on the Factory Floor

Manufacturing Processes is a book about carefully guarded knowledge, principles, and techniques used with manufacturing processes and materials to make better products. Your authors have experienced the excitement and adventure from the plant floor by working as machine operators, technicians, and managers, and through hundreds of visits to factories that manufacture hard good products (durable goods) in this country and abroad. Much has been learned, and many exciting stories could be told about the magic that occurs when products are made within specifications and on schedule. Only the right combination of materials and processes will produce products that look great and perform well. Much of the magic that motivated us to develop this edition of *Manufacturing Processes* was a result of the fun we have had working with student innovation teams on cost-saving projects in manufacturing. We have been fortunate to live in an area where there is a great diversity of manufacturing. Manufacturing companies cover the gamut from aerospace, pharmaceuticals, metal fabrication, and racing catamarans to textiles, automated food production, composites, and industrial ceramics.

It is our hope that you will use this book as a resource tool that will help you find answers when alternative solutions may be needed. If you are already employed in manufacturing or soon will be, you know that change is a necessary condition of all that you do. Your search for a better way, for processes and materials that can be used to make a quality product faster, better, and cheaper than your competition will result in profit for your employer, and job security and satisfaction for you and many others. It is serious business, but you are in a position to help make a difference. Take advantage of this opportunity, and make it happen!

Not all of the processes out there are new and different. Some have been around for a long time, and have contributed to the manufacture of many products useful to society. Others are modifications of old processes, but are used with new materials or to perform new applications. Do not worry though; some of the processes presented in this book are brand new, and are just beginning to demonstrate their usefulness in industry. You may be the first to take advantage of these in your area.

Historians are quick to point out that it is often possible to anticipate the future by studying the past and present. However, in the field of technology, the best forecasts are often sketchy and incomplete. Throughout history, breakthrough innovations have occurred with little foundation in the past or present. Nothing would have led to the prediction of the birth of flight at Kitty Hawk, or to the invention of the lightbulb, personal computer, and cell phone.

We will start by thinking a bit about the environment for manufacturing in the decades of the 1980s and 1990s. What was important to manufacturing back then is to some extent still important today. At least, it provides the foundation that supports what is going on today. In the 1980s and 1990s, the hot topics in manufacturing were labor content, lean manufacturing, design for assembly, workplace safety, and staying ahead of regional competitors. Most of the technical literature provided articles on these topics, and training workshops were offered worldwide to address issues in these areas. In some areas, these are still taking place.

However, there was something very different when comparing today's world of manufacturing with what we saw back then. Back then, the "dot.com" industry (computers/telecommunication/communications) was in its infancy, and people still viewed the world largely as a collection of independent countries and cultures, each competing for access to limited natural resources. Beating out your competitor to get a contract for making metal parts to be sold to GM was enough to keep everyone happy.

Then change came rapidly with the rise and fall of the digital telecom industry, an industry that was by many, viewed as a major driver and stimulus to manufacturing and economic development in the United States and developing countries. In the late 1990s and early 2000s, both the telecom and manufacturing industries had to contend with changing conditions such as governmental regulation concerning monopolies, increased global competition, corporate mergers, and downsizing. The doors of many companies, both large and small, were closed forever because of their inability to restructure themselves quickly enough to cope with changing market conditions.

Renewed optimism about the future came with the advent of a new century. However, lackluster sales and market instability in many sectors led to declining employment in many regions of the country. Short bursts of renewed performance occurred with substantial growth in construction, real estate sales, insurance and services, and unprecedented growth of the Internet and e-commerce. All indications were that an economic rebound and a brighter future for manufacturing were on the horizon.

Then came the terrorist attacks of September 11, 2001. Increased media attention to a wave of terror and efforts throughout the world to cope with this problem changed the way people viewed themselves and their world. Terrorism was understood to be a worldwide problem and nations banned together to develop strategies to eliminate this threat.

A positive outcome from this was that many people came to realize that the world is a finite place and that activities influencing one region and country will influence many others. Global concern and efforts to strengthen homeland security while still ensuring the rights of people in a free society have started to influence the way we (nations across the world) conduct business and live our lives today and into the future.

The way we view our world has a direct relationship on how we think about growth and competition in manufacturing. With the rapid growth of the Internet, e-commerce, and our digital economy, it is difficult to imagine operating a growing company with a geographically isolated focus on production, sales, and the market. An isolationist perspective would mean that the company would only be concerned about operating in a limited sales region and the major competitor would be down the street or in the same locale. Materials, services, and manufactured products can easily be bought and sold worldwide. Directly opposite from the isolationist perspective would be companies that view their own growth with a global perspective. It will be shown later in this chapter that many American manufacturing companies have operations offshore (outside of the United States). The same can be said for manufacturing companies from other developed countries. Many firms that originate in countries such as Canada, Japan, Germany, Mexico, China, and Australia have installations in the United States as well.

Even though we understand that much of the future is unknown, we do have indications from the events of today about many of the things that will be likely to occur in the future. We know that growth in manufacturing will continue to require a global perspective. This will continue to intensify as we continue development of the International Space Station and look to outer space for new environments for manufacturing. We also know that manufacturing will continue to be the driving force that leads to economic prosperity and global well-being.

The opportunity to carefully select the best processes to form, separate, fabricate, condition, or finish materials to make a product is a rewarding experience. If the correct combination of materials and processes is made, and if the product is made in an efficient and timely manner, it is likely that the firm will be successful. If the wrong combination is selected, it is likely that the firm will not be able to survive for long in a highly competitive market. Part of this magic can be realized by drawing an analogy between manufacturing and the game of football. On the old 1950s record, *What It Was, Was Football*, Andy Griffith described football as s*ome kindly of a game where you have to carry the ball from one end of a cow pasture to the other without either getting knocked down or steppin in something*. In the game of manufacturing, one company has to take the product (the ball) to the marketplace (the field). Then marketing or e-commerce (the players) have to generate interest in the product while it still has potential sales value. Things have to occur quickly because each product will appeal to the consumer for a limited amount of time. Like players with the ball on the field, marketing is with its product and has to fend off all competitors. If the product is poorly

constructed, is introduced at the wrong time, or if the wrong combination of processes and materials are selected, then it will have limited sales appeal. If this happens, to return to the vernacular of Andy's football, the company and product will have missed the mark, and everyone will have *gotten knocked down and stepped in something*!

The magic of manufacturing processes should be understood and appreciated. It is serious business. The life or death of a manufacturing firm and impact on members of society is dependent on how well engineers, technologists, and technicians select and apply manufacturing processes and materials. If products are well designed, and are made effectively and efficiently, they will sell. The profit that will be generated will not only pay the company's bills, but it may even lead to the growth of nations, improvement of quality of life, and prosperity. On the plant floor, results evidenced by making products faster, better, and cheaper is what separates the winners from the losers. And the winners will have to be making and assembling parts in many locations, and shipping them throughout the world.

New materials are being discovered each day, and processes are being created to reshape these raw materials into useful products. The National Aeronautics and Space Administration (NASA), has been a major leader in this initiative. On April 3, 1996, an experimental Starfire rocket launched by NASA, Aerojet Corporation, and Lawrence Berkeley National Labs carried its payload 13 miles higher than the typical orbit of the Space Shuttle, and then returned to the earth. One of the experiments on board dealt with manufacturing *aerogel*, a rigid material that is extremely light and remarkably strong. Aerogel had previously been produced on the ground, but one of the objectives of this NASA research was to determine whether a more uniform and transparent gel could be made in space. Many other efforts to manufacture aerogel in space took place over the next few years. In May of 2002, the *Guinness World's Records*, awarded Dr. Steven Jones and NASA's Jet Propulsion Laboratory the record for the world's lightest solid (the space version of aerogel). It weighed in at .00011 pounds per cubic inch. See **Figure 1-1**.

This is how the material is constructed. Silica aerogel begins as a liquid, consisting primarily of water, alcohol, and silica. It then gels into something that looks like gelatin. The challenge is to dry the aerogel without it collapsing into a dense slab. This is accomplished by exchanging the alcohol with

Figure 1-1. Aerogel is a material with thermal insulating properties so efficient that the heat of a flame is unable to melt the crayons on top. (NASA/JPL-Caltech)

liquid carbon dioxide and then removing the carbon dioxide at high pressure.

Aerogel was used for thermal insulation on the Mars Pathfinder Mission, as a particle-collecting material on NASA's Stardust spacecraft and the Mars Exploration Rover program in 2003. In 2004, aerogel was used to capture particles from the comet Wild 2. Ten years later, aerogel will be a major contributor on the Mars Scout program.

Aerogel has the equivalent thermal insulating capability of up to 20 glass windowpanes. On a per weight basis, aerogel is the strongest and lightest transparent building material. It is only three times the weight of air. A single inch thickness of this silica-based material has the internal surface area as large as a basketball court. This gives it an insulating quality so efficient that it can protect the human hand from the heat of a cutting torch. It is anticipated that in the house of the future, aerogel will be used for clear, highly insulating windows, skylights, oven door panels, space-saving refrigerators, and many other applications. In the *Technology to Watch* section of Fortune Magazine (June 27, 1997), the use of aerogel was cited for more than 800 different product applications.

The International Space Station will create many more opportunities for conducting experiments with materials and processes. Expedition Five, ISS Flight UF2, STS-111 Space Shuttle Flight (Summer 2002) contained the first two materials science experiments conducted on the International Space Station. Both experiments were conducted inside of the *Microgravity Science Glovebox*, a sealed work environment used for melting semiconductor crystals. Semiconductor crystals are used in products such as integrated circuits used in computers and controllers, sensors, medical imaging devices, and devices for detecting nuclear radiation. See **Figure 1-2**.

Processes such as convection and sedimentation are reduced in a microgravity environment. This means that fluids do not move and will not deform. This makes space an ideal place to test melt motion in the production of semiconductor crystals. However, there are challenges that must be addressed with manufacturing in space.

When metals are melted on earth, air bubbles are formed during the melting process. These bubbles just escape into the air and the material is left with few air bubbles. In space, the lighter bubbles do not escape, but stay inside of the material. When the metal is cooled and returns to its solid form, these bubbles result in porosity, which reduces the material's strength and usefulness.

The second experiment that was conducted on the International Space Station is called the *Pore Formation and Mobility During Controlled Directional Solidification in a Microgravity Environment Investigation (PFMI)*. With this experiment, a melting furnace will

Figure 1-2. The first Materials Science experiment on the International Space Station *–Solidification Using a Baffle in Sealed Ampoules (SUBSA).* Astronauts will conduct experiments by inserting their hands into a pair of gloves reaching inside the sealed *Microgravity Science Glovebox.* (NASA/Marshal Space Flight Center)

be installed inside of the *Microgravity Science Glovebox* and used to melt various chemical compounds.

Today it is much more difficult for companies to establish and maintain leadership in manufacturing than it was just a decade ago. Some firms are still rooted in the past, stuck with doing things as they have always been done. Others have an introspective vision of what is possible; an attitude that what occurs at their plant is all that is important. Some have little interest in competing in global markets, or from creating joint ventures with others.

Global competition in manufacturing will continue to challenge the positions of superiority in many market sectors long held by U.S. firms. In areas where inability to keep pace with improved materials, processes, and techniques, entire industries (steel, shipbuilding, furniture making, and textiles) have been lost to global competition. At the same time, one should be cautioned not to judge the growth and importance of American manufacturing solely by the presence of companies on American soil. It should be realized that flexibility is necessary if firms are to remain competitive in the world of the future. Cost savings are often realized when manufacturers are located closer to the materials used to produce their products. An increasing number of manufacturing firms have factories located worldwide to reduce the cost of transporting raw material to the factory.

No matter where these companies are located, they will need aggressive leadership to remain competitive, and to create new ventures and opportunities. The hope for the future of American manufacturing lies with employees (associates) who are willing to work harder than ever before. They will also have to be able to work in teams and to get along with others. They will need to have a good attitude toward work, and to appreciate the need for happy, positive attitudes around them. They will need to embrace change, and will appreciate the need for life-long learning. It will be these new types of employees that will lead their companies into the future.

These associates will be working for companies that are *learning organizations* (Warren Bennis, *The Learning Organization*). Many companies are already in place today, functioning like thinking machines, thriving on change, and constantly reinventing the way they do business. What these firms have in common is the realization that their most precious commodities are their employees, not their facilities, materials, or technology.

The Global Challenge

There are 473 different types of manufacturing industries. These range from guided missile and space vehicle manufacturing, polystyrene foam manufacturing, tortilla manufacturing, and surgical and medical equipment manufacturing, to semiconductor and electronic component manufacturing, audio and video equipment manufacturing, glass container manufacturing, and textile carpet manufacturing. Manufacturing covers literally everything from soup to nuts, pretzels, and chewing gum, to jawbreakers, DVDs, shipbuilding, motor homes, machine tools and much more.

Most analysts recognize that the United States holds an overall lead throughout the world in science and technology. This is quite an accomplishment when considering the number of manufacturing industries listed by NAICS. Of course, manufacturing only scratches the surface when it comes to the total universe of science and technology.

According to the U.S. Department of Commerce, the gross domestic product (GDP) by industry in current dollars increased from 1,223.2 (in billions of dollars) in 1994 to 1,566.6 in 2000. Throughout the decade of the 1990s, output of manufactured products contributed 29 percent of the growth in the GDP, adjusted for inflation. In the 1980s, this was only 21.5 percent (National Association of Manufacturers). Manufacturing's importance to the economy of the world can be fully appreciated when considering that in the 1990s, the growth shown in manufacturing was greater than in any other industrial group. In this same time period, the service industries contributed 19 percent; transportation and utilities, 10 percent; and finance, insurance and real estate, 13 percent.

The 2005 Report on GDP by Industry produced by the Commerce Department indicated that manufacturing GDP increased 4 percent over the previous year in inflation-adjusted terms, or half a percentage point faster than the 3.5 percent pace of the overall economy. The good news for 2006 was that the economy grew at an annual rate of 2.9 percent in the second quarter, compared with 5.6 percent in the first. (Bureau of Economic Analysis, U.S. Department of Commerce)

Growth in manufacturing is occurring in most regions throughout the world. In some instances, the technology gap between the U.S. and other countries is closing as other nations gain experience in specialized fields. In the past decade, one of the most progressive countries in terms of technological leadership was Japan. For example, consider what they have accomplished in a short amount of time.

Due largely to government-industry cooperation, a strong work ethic, and exploitation of high technology, Japan has advanced to the rank of the second most technologically powerful economy in the world. China is third. One reason for this is the effective integration of manufacturers, suppliers, and distributors into closely-knit groups called *keiretsu*. Another reason for this success is due to the fact that historically, Japan has guaranteed lifetime employment to a large percentage of the workforce. Robotics is one of Japan's major strengths, with Japan owning 410,000 of the world's 720,000 operational robots. (CIA Factbook)

In the steel industry, Japan is one of the world's leading exporters. It has accomplished this despite having to contend with importing the basic raw materials for steel production. Japan currently has ten steel plants that are larger than any in the United States. Japan uses steel in the manufacture and sale of automobiles, consumer electronics, and other products to the U.S. and other countries.

The automotive industry has a major impact on many different types of manufacturing. When sales are strong, manufacturing thrives and there is a positive impact on the employees and suppliers. When a plant no longer remains competitive, sales may drop and the plant may be closed. The impact can be devastating for many thousands of people in a long chain of supportive industries that depend on the company for their own survival.

Between 1961 and 1991, one of the classic models of *mass production* was the General Motors assembly plant in Framingham, Massachusetts. When the plant closed in 1991, GM sold the buildings to Adsea Corporation, a large car auction house from Indianapolis, Indiana. With the plant closing, GM shifted most of the manufacturing of the Buick Century and Chevrolet Celebrity to Mexico and Canada. In its heyday, the Framingham plant employed 4,000 people, many making as much as $21 an hour. Today, some of the previous GM employees are still working there, but not in manufacturing. Some are driving used cars around in circles for auction houses, showing the same cars they once built, to dealers for $7–$12 an hour.

The Framingham plant was one of the most productive automotive plants in the United States during the heyday of mass production. But it was not productive enough to remain competitive. To better understand this, let us contrast data from this plant against what was once one of its major competitors, Toyota Motors, of Takaoka, Japan. According to the International Motor Vehicle Program Assembly Plant Survey conducted in 1986 by the Massachusetts Institute of Technology, the gross assembly time to produce one vehicle at GM Framingham was 40.7 hours. At Toyota, the time to assemble a comparable vehicle was 18.0 hours. There was also a great disparity in the number of assembly defects per 100 cars: GM had 130; Toyota, only 45. In terms of the inventory of parts kept on hand to support assembly, Framingham had a two-week supply. Takaoka had only two *hours*.

It should not be difficult to see why Framingham is not operational today. It was an impressive accomplishment for the Japanese to beat this industrial giant, and to beat it so soundly. They beat GM/Framingham based on significant advancements in management, productivity, and quality. These advancements had an impact reaching far beyond the automotive industry. All types of production are being influenced by the new dynamics of manufacturing.

In the future, Japan will continue to try to capture many new markets through extensive investments in research and development. Some areas where Japan's efforts will be concentrated include semiconductors, factory automation, telecommunications, lasers, fiber optics, pharmaceuticals, and industrial ceramics. But Japan no longer holds the lead as a major competitor in many of today's markets.

Lean Production vs. Mass Production

James P. Womack, Daniel T. Jones, and Daniel Roos popularized the term, *lean production* in their book, *The Machine That Changed the World*. The authors referred to the system implemented at Toyota as lean production. Lean production incorporates some of the best thinking from continuous (mass) production and craft (one of a kind) production. It also involves major changes from what was typical in the 1980s and 1990s in terms of product design, parts inventories, quality control, product rework, and reducing *work in process (WIP)* (components that are partially processed, waiting for the time when they will be used to complete an assembly).

Lean manufacturing is a popular term today among consultants and engineering professionals. The Society of Manufacturing Engineers has established a *Lean Manufacturing* e-newsletter. The term is often used in conjunction with other quality management concepts such as *Six Sigma*, *Total Quality Management (TQM)*, and *Waste Free Manufacturing (WFM)*. But thinking *lean* involves more than just improving quality. A major change in thinking is also made when it comes to the relationships that must be created with parts suppliers. The goal of lean manufacturing is to reduce the number of parts warehoused by the manufacturing plant for immediate use, thereby cutting costs to maintain this inventory. Lean manufacturing involves a whole new way of *thinking about the process of making products faster, better, and cheaper.* You will learn more about lean manufacturing later in Chapter 3.

Up until the 1950s, most automotive assembly plants used a moving production line, not much different than the one designed by Henry Ford to produce Model Ts at his Highland Park, Michigan location. In 1927, Ford shifted production from Highland Park to his new 2,000-acre site on the Rouge River. During World War II, the Rouge was the largest single manufacturing complex in the United States, with peak employment of close to 120,000 employees. Henry Ford achieved self-sufficiency and vertical integration in production at the Rouge plant, with continuous flow from iron ore and raw materials to finished automobiles. The complex was extensive, with dock facilities, blast furnaces, open-hearth steel mills, foundries, a rolling mill, metal stamping facilities, an engine plant, a glass manufacturing facility, a tire plant, and its own power station supplying both steam and electricity.

The River Rouge site was recreated by Ford Motor Company to reflect an environmentally conscious production, and reopened in 2003. The new Rouge facility is one of the most environmentally conscious automotive plants in the world. Some of the highlights of this new future-focused factory are listed below:

- It has the world's largest living roof on any industrial building. Approximately 454,000 square feet of assembly plant roofing are covered with sedum, a succulent groundcover, and other plants. One advantage of this type of roof is that it will reduce storm water runoff by holding an inch of rainfall. The living plants will absorb carbon dioxide, so oxygen is emitted and greenhouse gases are reduced.

- Shallow ditches seeded with indigenous plants are used to regulate water flow and evaporation and improve storm water management on site.
- Michigan State University is partnering with Ford on a Phytoremediation project, using natural plants to rid the soil of contaminants.
- All of the pavement is porous, permitting filtration of water through retention beds with 2–3 feet of compacted stones.
- Trellises for flowering vines and other plants provide shade and help cool the Rouge Office Building.
- Solar cells and fuel cells are used to provide energy-efficient power.
- More than 1,500 trees and thousands of other plantings were grown to provide a habitat for songbirds and other wildlife.

While the assembly line and conventional mass production has long been the backbone of American industry and productivity, many of today's companies use different types of approaches to manufacture their products. Increased flexibility is often provided through systems such as cellular manufacturing, demand flow technology, waste free manufacturing, cells in a line and empowerment teams that follow the product down the line from start to finish, from one business unit (production center) to another.

In the aerospace industry, manufacturing has traditionally involved producing high quality precision parts in low volume. Normally, parts are made of high strength, lightweight, hard-to-machine metals such as titanium, stainless steel, and nickel steel alloys. The machine tools that are used to cut, shape, and fabricate these materials must be able to handle large parts at high speed, in high-temperature cutting conditions. One way that the U.S. is able to keep this industry from moving offshore to areas with lower wages is through improved productivity utilizing faster machines and improved materials. Machine tool manufacturers are constantly looking for improved technology to automatically handle tasks that would have previously been conducted by a machine operator. As an example, in the area of computer-assisted numerical control (CNC), five years ago the operator would have had to set the machine tool at the origin, or starting point on the workpiece. Today this is often done using low-cost automation.

In the days of conventional mass production, many thousands of "look-alike" products were made in what was referred to as a **batch**, or **job lot**. For many generations, printing companies have manufactured letterheads and business cards based on the size of the job. The larger the batch, the less it costs the consumer for each individual sheet or card. In many other types of manufacturing, a larger batch size also means lower cost for each product. Examples can be seen in look-alike products where variation is undesirable, e.g. golf tees, metal washers, plastic bottles, tractor tires, and soft drink cans.

Continuous production is often used when there are a limited number of variables that have to be controlled in the process of manufacture, and when few changes in the product are necessary. Manufacturers often produce *plain vanilla machines.* That is, the basic foundation of the machine remains constant, but low-cost automation options are available. In today's world, the consumer is able to shop around for tailor-made products to meet their needs. In many cases, products can be ordered directly from the sales floor allowing information about the design features for an individual's car, boat, or motor home to be transmitted directly to the factory. Each product is then constructed so it appears to be somewhat different from the products made before and after it in the flow of production. In reality, the special features are usually just added as components or subassemblies to the *plain vanilla* basic product. In fewer cases, products are truly tailor-made to meet consumer needs from the ground up.

Today, consumer expectations for improved quality and product variety place great demands on the manufacturer. With increasing global competition, manufacturers must often respond quickly to demand with smaller batch sizes. In some cases, the size of these batches may be as small as one. A single unit of one product may be produced, followed by a single unit or small batch of a different product. Production may still be *continuous,* but major adjustments have to be made to reduce changeover time for each new product. Changes taking place in continuous production are evident in the implementation and evolution of production concepts such as intermittent manufacturing, cellular manufacturing, demand flow technology (DFT), waste free manufacturing (WFM), six sigma, total quality management (TQM), just-in-time manufacturing (JIT), and computer-integrated manufacturing (CIM). These will be discussed more completely later on in the textbook.

International Manufacturing

As recently as the 1960s, companies located their manufacturing facilities within the same country. Since then, cheap labor, air transportation, and government incentives have given companies reasons to construct plants in several different countries. The Tax Reform Act of 1986 (TRA) influenced the development of international joint ventures by American firms by creating tax advantages that encouraged exports. The TRA requires companies to use separate tax credits on income received from foreign corporations with 50 percent or less American ownership. Prior to the act, it was permissible for companies to average income. With the advent of the TRA in 1986, the number of international joint ventures to U.S. investors has declined. This is particularly apparent when it comes to joint ventures in low-tax countries.

In spite of this legislation, there are still many highly successful international joint ventures that have been established. NUMMI (New United Motor Manufacturing, Inc.), a Japanese/American automotive assembly plant in Fremont, California, was one of the first of these successful joint ventures (1984). The plant is a 50/50 joint venture between General Motors (GM) and Toyota. NUMMI manufactures the Toyota Corolla, Tacoma truck, and Pontiac Vibe, and in 2002 started producing the right-hand drive Toyota Voltz for export to Japan. The firm can produce 220,000 cars and 150,000 pickups per year. NUMMI represents a successful marriage between the classic mass producer, GM, and the classic lean producer, Toyota. The company says that the net result is a car or truck that is made faster, cheaper, and with fewer defects.

GM is involved in other offshore ventures. Major brands include Buick, Cadillac, Chevrolet, GMC, Pontiac, Saab, and Saturn. GM also produces cars for Holden, Opel, and Vauxhall. General Motors has a 49 percent ownership of Isuzu Motors, 20 percent of Fuji Heavy Industries, Suzuki Motor, and Fiat Auto (manufacturer of Alfa Romeo and Lancia), and 42 percent ownership of South Korea's Daewoo Motor. (Hoover's Online; Company Capsule)

Ford Motor Company now makes vehicles for Aston Martin, Jaguar, Lincoln, Mercury, and Volvo. It has 33 percent controlling interest in Mazda and has purchased BMW's Land Rover sport utility vehicle (SUV) operations.

DaimlerChrysler was created in 1998 through a $37 billion acquisition of Chrysler by Germany's Daimler-Benz. The company manufactures 4.7 million vehicles per year. Chrysler brands include Dodge, Eagle, Jeep, and Plymouth. DaimlerChrysler also produces Mercedes luxury sedans, commercial vehicles, and SUVs. The company also has a 10 percent interest in Hyundai Motor and 37 percent in Mitsubishi Motors. (Hoover's Online; Company Capsule) One of the company's subsidiaries, the Beijing Jeep Corporation (BJC) is a joint venture with the Beijing Automotive Works of China.

Covisint, Inc. is a company that is promoting global Internet exchange for improving supply chain efficiency within the automotive industry. GM, Ford, and DaimlerChrysler created the concept for Covisint. The venture was then purchased by Renault/Nissan and Delphi, the world's largest manufacturer of automotive parts. Covisint, Inc. is a holding company operating through Covisint LLC, with offices in Michigan, Amsterdam, and Tokyo. (Hoover's Online; Company Capsule)

There are many other active markets and joint ventures outside of the automotive community. Names such as Hitachi, Toshiba, Sony, Tonka Toys, Ciba-Geigy/Chiron, Courtaulds-Nippon, and Fujitsu are recognized for the production of high-quality products.

UTC Fuel Cells, a subsidiary of United Technologies, is a leader in the manufacture of fuel cells. UTC has provided hydrogen-powered fuel cells for all manned space flights since 1966. It also makes fuel cell power plants for commercial, transportation, and residential applications. It has signed a development contract with Nissan to develop fuel cells for homes.

ABB, Inc. is a company that has expanded into dozens of industries. Its Zurich-based parent company, ABB Ltd., has operations in more than 100 countries. ABB firms in the U.S. are engaged in automation, energy, engineering, financial services, and industrial processes.

An interesting company that has been around a long time is the Diamond Chain subsidiary of Armsted Industries. Diamond Chain produced the propeller chain for the Wright brothers' aircraft. Customers today include industrial distributors, locomotive and railcar manufacturers, and automotive original equipment manufacturers (OEMs). Amsted has roughly 50 plants worldwide.

The extent to which many global corporations have diversified makes it increasingly difficult to determine which country can be given credit for leadership in a particular technology. As an example, Sumitomo Heavy Industries, Ltd. (SHI) is a world leader in steel, heavy machinery, shipbuilding, and bridge manufacturing industries. It also produces recycling and power transmission equipment, lasers,

medical equipment, material handling systems, research equipment such as particle accelerators, and moving sidewalks. SHI has more than 30 major subsidiaries and affiliates operating in Japan and other Asian countries, Europe, and the U.S. While SHI has a global customer base, Japan accounts for about 90 percent of its sales. (Hoover's Online; Company Capsule)

In fields such as space technology, aerospace, microprocessors, medical technology, food technology, and bioengineering, the United States maintains a competitive advantage. However, in high-growth areas, such as fiber optics, composites, security, and superconductivity, the race is still on. Only time will tell which nation will be able to establish superiority in these fields. The trickledown effect of success in these areas will have an impact on many different types of manufacturing and service industries. Any nation that achieves superiority in these fields will have a tremendous economic advantage.

Other Competitors for Manufacturing Superiority

Capturing and maintaining leadership in any hard goods manufacturing sector would be much simpler if there were only a handful of countries engaged in manufacturing. However, such a thought is no longer realistic; many countries throughout the world are gaining rapidly in specialized high-growth technology fields. The U.S. Department of Commerce's Office of Trade and Economic Analysis (OTEA) has established twelve trade zones. See **Figure 1-3**.

In 2001, the fourth largest trading partner with the U.S. was China. China is the country to watch in terms of major competition with the U.S. China has a large population, low wage rate, and people who like to work in teams. This means that they have a winning combination when it comes to the principles of lean manufacturing. China is now working hard to capture the high-volume mass production markets. In 2000, China exported $232 billion in products in commodity areas including machines and equipment, textiles, footwear, toys, sporting goods, and mineral fuels. According to the National Association of Manufacturers, actual two-way trade with the U.S. in 2000 was $116 billion. China has major strengths in iron and steel, coal, machine building, armaments, textiles, petroleum products,

Figure 1-3. International trade zones established by the U.S. Department of Commerce's Office of Trade and Economic Analysis (OTEA).

Trade Zones	Countries
Apec	Asian Pacific Economic Cooperation (APEC) markets are Australia, Brunei, Canada, Chile, China, Hong Kong, Indonesia, Japan, Malaysia, Mexico, New Zealand, Papua New Guinea, Peru, Philippines, Russia, Singapore, South Korea, Taiwan, Thailand, and Vietnam.
Asean	The ten Association of Southeast Asian Nations (ASEAN) members are Brunei, Burma, Cambodia, Indonesia, Laos, Malaysia, Philippines, Singapore, Thailand, and Vietnam.
Caribbean	The Caribbean countries include Anguilla, Antigua and Barbuda, Aruba, the Bahamas, Barbados, Belize, British Virgin Islands, Cayman Islands, Dominica, Dominican Republic, Grenada, Guyana, Haiti, Jamaica, Montserrat, Netherlands Antilles, St. Kitts and Nevis, St. Lucia, St. Vincent and the Grenadines, Suriname, Trinidad and Tobago, and Turks and Caicos Islands.
Central America	The Central American countries consist of Costa Rica, El Salvador, Guatemala, Honduras, Nicaragua, and Panama.
Eastern Europe	Eastern Europe includes Albania, Armenia, Azerbaijan, Belarus, Bulgaria, Czech Republic, Estonia, Georgia, Hungary, Kazakhstan, Kyrgyz Stan, Latvia, Lithuania, Moldova, Poland, Romania, Russia, Slovakia, Tajikistan, Turkmenistan, Ukraine, and Uzbekistan.
European Union	The European Union includes Austria, Belgium/Luxembourg, Denmark, Finland, France, Germany, Greece, Ireland, Italy, Netherlands, Portugal (including Azores and the Madeira Islands), Spain (including Spanish Africa and the Canary Islands), Sweden, and the United Kingdom.
Former Soviet Republics	The Former Soviet Republics consist of Armenia, Azerbaijan, Belarus, Estonia, Georgia, Kazakhstan, Kyrgyz Stan, Latvia, Lithuania, Moldova, Russia, Tajikistan, Turkmenistan, Ukraine, and Uzbekistan.
Middle East	The Middle East includes Bahrain, Iran, Iraq, Israel (including the Gaza Strip and the West Bank), Jordan, Kuwait, Lebanon, the Neutral Zone, Oman, Qatar, Saudi Arabia, Syria, United Arab Emirates, and Yemen.
Nafta	The NAFTA Markets are Canada and Mexico.
Opec	The Organization of Petroleum Exporting Countries (OPEC) consists of eleven countries: Algeria, Indonesia, Iran, Iraq, Kuwait, Libya, Nigeria, Qatar, Saudi Arabia, United Arab Emirates, and Venezuela.
South America	The countries of South America are Argentina, Bolivia, Brazil, Chile, Colombia, Ecuador, Paraguay, Peru, Uruguay, and Venezuela.
Sub-Saharan Africa	Sub-Saharan Africa consists of Angola, Benin, Botswana, Burkina, Burundi, Cameroon, Cape Verde, Central African Republic, Chad, Comoros, Congo, Djibouti, Equatorial Guinea, Eritrea, Ethiopia, Gabon, Gambia, Ghana, Guinea-Bissau, Guinea, Ivory Coast, Kenya, Lesotho, Liberia, Madagascar, Malawi, Mali, Mauritania, Mauritius, Mozambique, Namibia, Niger, Nigeria, South Africa, Rwanda, Sao Tome and Principe, Senegal, Seychelles, Sierra Leone, Somalia, Sudan, Swaziland, Tanzania, Togo, Uganda, Zaire, Zambia, and Zimbabwe.

cement, chemical fertilizer, footwear, toys, automotives, consumer electronics, and telecommunications. (CIA Factbook) Direct investment in China by U.S. firms increased from $200 million in 1989 to more than $7.8 billion in 2000. (*Could your job go to China*, September 7, 2001, CNN.com)

In April of 2001, the U.S.-China Security Review Commission and U.S. Trade Deficit Review Commission funded a pilot study on the impact of Washington-Beijing trade relations on workers, wages, and employment. The Cornell University study found that in the period from October 1, 2000 to April 30, 2001, 26,267 U.S. jobs moved to Mexico and 9,061 to Asian countries other than China. In the same period, 80 U.S. corporations announced their intentions to move their production to China. Data from the study suggests that U.S.-based multinational corporations shifting production to China intend to serve both a global and U.S. market. Examples include LaCrosse Footwear (winter boots), Lexmark (printers), Rubbermaid (cookware and kitchen products), Motorola (cell phones), Raleigh (bicycles), Cooper Tools (wrenches), Mattel Murray (Barbie doll playhouses), and Samsonite (luggage). (*Could your job go to China*, September 7, 2001, CNN.com)

There are still many unfair policies that regulate international trade. In some instances, governments charge as much as 30 percent tariff on machine tools imported to their country, while the U.S. charges no tariff at all on machine tools imported to our country. China has been a consistent example of this imbalance.

As a result of this, China has agreed to reduce its industrial tariffs on goods brought into their country from an average (1997) of 25 to 8.9 percent for most goods by January 1, 2005. China has agreed to completely eliminate its tariffs on beer, furniture, and toys. The tariffs charged for these products in 1997 were 70, 22, and 23 percent respectively. China also joined the Information Technology Agreement (ITA). A strong compliance plan has been developed by the U.S. Department of Commerce to help China meet WTO obligations. (*China's Entry into the WTO—What it Means for U.S. Industry*, NAM Online, July 5, 2002)

China was approved as a member of the World Trade Organization (WTO) on December 11, 2001. In spite of these efforts, it is taking a very long time to see significant results. In May, 2005, China still had not joined the World Trade Organization Government Procurement Agreement so that U.S. firms could compete in China's market on the same basis as local units.

The United States has expressed concern about China's efforts to implement regulations for procurement of computer software by government offices. (Charles Freeman, testimony to the Committee on Government Reform of the House of Representatives, May 13, 2005). According to Mr. Freeman, proposals made by China would "put U.S. firms at a significant disadvantage in the Chinese market; qualified products would have to be made in China, intellectual property rights would have to be held by a Chinese citizen, and China-based development costs would have to comprise at least half of total development costs." (*China to adhere WTO procurement rules*, The Economic Times, May 14, 2005.) The software sector is the first area for which implementing language has been written.

Automotive assembly plants in other countries such as Brazil, Korea, Mexico, and Taiwan are showing extraordinary performance. In terms of manufacturing revitalization in the developing countries, the Ford plant in Hermosillo, Mexico, has surfaced as a model of innovation for others to follow. This plant is based on conventional mass production, updated to include the new thinking of lean production.

Other countries, particularly those in Western Europe, are now working hard to surpass the United States in terms of annual growth in *manufacturing productivity*. However, contrary to popular belief, U.S. productivity in manufacturing is increasing, even at a rate that exceeds annual increases in Japanese and German manufacturing. This is in contrast to what is happening in other developed countries, where the productivity revolution in manufacturing is largely over. Let us take a moment to look more carefully at how the U.S. measures productivity.

There are two programs charged with developing Productivity and Costs data for elements of the U.S. economy. The *Major Sector Productivity and Costs Program* produces quarterly and annual output per hour and unit labor costs for the U.S. business, nonfarm business, and manufacturing sectors.

The *Industry Productivity Program* publishes annual measures for output per hour and unit labor costs for over 500, three and four-digit Standard Industrial Classification (SIC) industries (forerunner of NAICS) in the U.S. This covers manufacturing, retail trade, and some other sectors. In the first quarter of 2002, there was an 8.4 percent increase in growth per hour of all persons in the nonfarm business sector compared with the previous quarter. (Bureau of Labor Statistics; U.S. Department of Labor)

According to the U.S. Census Bureau, the nation's international deficit in goods and services decreased to $31.6 billion in March of 2002, from $31.8 billion in February of 2002. The February to March change in exports of goods reflected increases in capital goods; automotive vehicles, parts, and engines; and other goods. Decreases occurred in industrial supplies and materials, foods, feeds, and beverages. Consumer goods were virtually unchanged.

The goods deficit with Mexico increased from $2.7 billion in February to $3.5 billion in March. Exports increased $0.3 billion (primarily automobile parts) to $7.6 billion, while imports increased $1.1 billion (primarily automobile and automobile parts; electrical machinery; and petroleum and petroleum products) to $11.1 billion. The goods deficit with China decreased from $6.5 billion in February to $5.6 billion in March. Exports increased $0.1 billion (primarily transport equipment) to $1.6 billion, while imports decreased $0.8 billion (primarily toys, games, sporting goods, footwear, and apparel) to $7.2 billion. (U.S. Census Bureau)

The goods deficit with Western Europe increased from $4.8 billion in February to $5.5 billion in March. Exports increased $1.5 billion (primarily transport equipment and computers and computer products) to $14.7 billion, while imports increased $2.2 billion (primarily petroleum and petroleum products; transport equipment; and organic chemicals) to $20.2 billion. (U.S. Census Bureau)

Exporting goods overseas is an essential part of doing business for many small and medium-sized manufacturers. Of all the companies that export products from their home locations in the U.S., 93 percent are small and medium manufacturers (SMMs). These firms employ from 10 to 2,000 employees, and employ nearly 9.5 million people. SMMs that export their products add jobs 20 percent faster than firms that remain totally domestic and are 9 percent less likely to go out of business. (National Association of Manufacturers Web site, June, 2002; http://www.nam.org) There is also a tendency for these firms to become more diffused, and to farm out business (subcontract) with other firms in their region.

The top 15 countries with which the U.S. trades represent 74.22 percent of U.S. imports and 73.52 percent of U.S. exports in goods. (Month of April 2006) See **Figure 1-4**.

In *Managing for the Future—the 1990s and Beyond*, (1993) Peter Drucker, pointed out that about one-fifth of the total capital invested by U.S. manufacturing firms is in *offshore* facilities (operated outside the United States). This is interesting to note when considering that nearly one-third of the world's trade in manufactured products takes place between plants owned by one company but located in two different countries. Such trade might involve materials transported from a Sony plant in Mexico to a Sony plant just across the border in San Diego, California. Today, Mexico is America's second-largest trading partner, ranking behind Canada and followed by Japan and China.

Production Moves to Mexico

One of the great steps in revitalizing the world economy following World War II was the General Agreement on Tariffs and Trade (GATT). In November of 1993, the U.S., Mexico, and Canada negotiated the

North American Free Trade Agreement (NAFTA). This agreement has helped to encourage free trade between the three countries. It is anticipated that the agreement will eventually be extended to include eight more South American countries as well: Argentina, Brazil, Chile, Colombia, Paraguay, Peru, Uruguay, and Venezuela. The NAFTA Secretariat was established pursuant to Article 2002 of the North American Free Trade Agreement (NAFTA). The role of the Secretariat is to resolve trade disputes between national industries and/or governments in a fair, timely and impartial manner.

Prior to creation of NAFTA, Mexico was the third most significant trading partner with the U.S. After NAFTA, it moved into second place. NAFTA has already led to the creation of a *North American Economic Community (NEC)* consisting of Canada, the U.S., and Mexico. Such a community may eventually be more populous than the *European Economic Community (EEC)*.

The Maquiladoras

The passage of NAFTA is only one factor that has contributed to Mexico becoming such an active

Figure 1-4. Top 15 U.S. trading partners. (U.S. Bureau of Census)

The Top 15 Countries with Which the U.S. Trades (April, 2006)		
Rank	**Country**	**Total Trade—Imports and Exports (billions)**
1	Canada	178.1
2	Mexico	106.6
3	China	98.1
4	Japan	66.5
5	Germany	42.3
6	United Kingdom	31.4
7	Korea, South	25.4
8	France	20.1
9	Taiwan 6.7	18.6
10	Malaysia	15.4
11	Netherlands	14.8
12	Venezuela	14.7
13	Italy	14.4
14	Brazil	13.7
15	Singapore	12.9

trading partner with the U.S. Another reason is due to Mexico's greatest economic initiative, the *maquiladora*. Maquiladoras are industrial parks that in most cases are owned by Mexican or American corporations. There are thousands of maquiladoras that were first located near the Mexico-U.S. border from Tijuana to Reynosa. Over the past several years, they have expanded further down into Mexico.

The first maquiladoras were built in 1966 in Baja, California and Ciudad Juarez, Mexico, under what was then referred to as the *Border Industrialization Program*. Initially, the maquiladoras could only be built along a strip of land 12.5 miles (20 km) wide along the Mexico-U.S. border, and in the Baja, California free trade zone. The goal of these maquiladoras was to utilize excess labor in the border areas and to encourage Mexican exports. It was also hoped that maquiladoras would help develop the Mexican manufacturing base and lead to the transfer of technology to Mexico. This enabled the sale of a portion of the maquiladoras' production in the Mexican domestic market.

The factories that are located in these maquiladoras are called *maquilas*. The maquilas can temporarily import into Mexico, on a duty-free basis, machinery, equipment, materials, parts and components and other items needed for the assembly or manufacture of finished goods intended for export. Corporations outside of Mexico own many of the maquilas. In June of 2002, ownership of the top 100 maquilas (based on overall employment) was as follows: United States (79), Japan (8), Mexico (6), Korea (2), Germany (2), Netherlands (1), Holland (1), Canada (1). (www.maquilaportal.com)

The terms "maquiladora," and "maquila" were derived from the Spanish word "maquilar" which historically referred to the milling of wheat into flour. The farmer would compensate the miller with a portion of the wheat and the miller's compensation was referred to as "maquila". Today, the terms refer to any partial activity in the process of manufacturing, such as assembly or packaging that is carried out by someone other than the original manufacturer.

The primary reason for foreign investors establishing assembly or manufacturing operations in Mexico is to take advantage of relatively inexpensive Mexican labor and the close proximity to the United States market. Even before Mexico's Foreign Investment Law was liberalized in 1993, in anticipation of the implementation of NAFTA, there were few restrictions on foreign investment.

The development of the maquiladora industry in Mexico was assisted by a U.S. customs program, which allows goods "assembled" outside the United States from U.S components to be imported into the United States without payment of U.S. import duties on the value of the U.S. components. Import duties are paid only on value-added when their products are moved back across the border. Plants that are not located in a maquiladora must pay costly import duties and deal with bureaucratic red tape at the border.

Today, the maquiladora industry may be the most important sector of the Mexican economy, with annual growth remaining constant at nearly 10 percent for the past 10 years. One out of five manufacturing jobs in Mexico are in maquila plants. According to SECOFI, Mexico's exports have increased 163.7 percent from 1994 to 1999 and 70.7 billion in U.S. dollars has been invested in Foreign Direct Investment. (June 2002).

During the 1980s, approximately one-half million new jobs were created in the maquiladoras. Today, there are thousands of maquilas, and millions of jobs created by the program. The largest maquila is American-owned Delphi Automotive Systems, with 47 different plants in Mexico and 49,586 employees. The second largest maquila is Yazaki Corporation, a Japanese-owned firm with 27,500 employees and 28 Mexican factories. Today, the largest maquiladora regions are El-Paso-Ciudad Juarez (132,000 workers in 254 plants), San Diego-Tijuana (77,000 workers in 554 plants), and Brownsville-Matamoros (38,000 workers in 96 plants).

The Ford Motor Company built one of the largest plants at Hermosillo, about 175 miles south of the U.S. border. Today, Ford has 10,587 employees in seven Mexican locations. The Hermisillo plant is still recognized as one of the most efficient automobile plants in the world.

Most analysts realize that the maquiladoras are taking jobs away from workers in the United States. These are high-volume mass production and low-skilled assembly jobs. Many proponents contend, however, that without the maquiladora program, some American-owned companies would be unable to compete with products produced in other countries with lower labor costs. By United States standards, workers at plants in the maquiladoras are paid very low wages. See **Figure 1-5**. They also note that by shifting low-skilled, lower-paying jobs to the maquiladoras, companies have often been able to retain more highly skilled jobs in their own countries, and

Figure 1-5. Labor Rate in Maquilas. (Bancomext, Inegi, and State Government Statistics, from www.maquilaportal.com)

Location	Wages	Classification
Matamoras	$2.57	Direct labor
Tijuana	$1.90	Direct labor
CD Juarez	$1.75	Direct labor
Mexacalli	$1.90	Direct labor
Matamoros	$6.27	Technician
Tijuana	$5.62	Technician
CD Juarez	$4.66	Technician
Mexacalli	$5.39	Technician
Average (All maquilas)	$72,800	Quality supervisor
Average (All maquilas)	$56,000	Production engineer
Average (All maquilas)	$75,040	Production manager

expand their own manufacturing operations in new high-growth markets.

Labor's True Impact on Product Cost

It is important to understand that many of today's most progressive finance managers are realizing that traditional blue-collar wages (direct labor costs) are becoming less significant as the major driver when calculating the total cost of a company's products. Certainly, labor costs are important to the determination of the cost of a product, but often more significant cost savings can be realized by concentrating on indirect labor costs (support, rework, maintenance) or low-cost automation. Cost savings are often realized by reducing the number of parts in a product (design for assembly or manufacture), making more informed choices regarding selecting of materials, reducing transportation costs for materials by selecting vendors closer to the plant, and for improving methods for transporting products to the market. Management strategies such as waste-free manufacturing, lean manufacturing, and cellular manufacturing have also proven their worth as cost-saving strategies.

A rule of thumb used by many companies with multinational operations is that offshore production must be five to eight percent lower in cost than domestic production to compensate for the higher costs attributed to distance. Offshore production incurs higher costs for communication, financing, travel, transportation of goods, and insurance. At the same time, corporations must compete in a situation where the rules of the game are not always fair and equitable. For example, a mold maker in China may quote a U.S. customer a finished price that is cheaper than U.S. companies would have to pay for the raw material. Sometimes there are governmental subsidies provided to specific industries in order to build up the infrastructure to support that industry.

U.S. automakers such as Ford and GM have worked hard to integrate some of the concepts typified by Toyota and NUMMI in their plants. It is likely that Ford is now just about as lean in its North American assembly plants as the average Japanese transplant to the United States. Many automotive component manufacturers are also incorporating new methods of production. These efforts will require time to be implemented and tested. A catch-up game will have to be played; continuous improvement will be necessary to beat aggressive competitors in global markets.

A new breed of engineer and technical manager is needed to keep pace with technology and provide leadership in the factories of the present and future. These individuals will have to perform the roles of technical strategists, mentors, and executive coaches more than managers. They will have to know where to go to find answers, how to solve problems, and what to do to motivate others to be more productive. These people may hold a variety of different job titles: managers, supervisors, engineers, technologists, associates, or facilitators. But none of these traditional job titles will really describe what they will have to do. They will have to increase profits for their employers, and they will have to work quickly to do it!

Manufacturing managers must be able to work with people and technology. Merely understanding sophisticated technical systems will no longer be enough to stay ahead of the competition. Successful companies will have to do things differently to stay ahead of the competition. Futurists have written books about the role of technology and automation in the factory of the future, but there are still very few totally automated, workerless factories. Technology has little value without some human intervention. Warren Bennis, a well-known author from the University of Southern California, jokingly referred to the process of manufacturing in the factory of the future like this: *There will be two employees in the factory of the future—a man and a dog. The man will be there to feed the dog, and the dog will be there to keep the man from touching the equipment.*

Business Ethics in a Global Marketplace

Before the 1990s, many small manufacturers were satisfied with developing and maintaining a strong market for their products in their home country, geographical region, or sales territory. Only the corporate giants were strong enough to compete in larger markets. Now, with firms competing on a worldwide scale, such a limited vision is self-crippling. In the years ahead, this trend toward global manufacturing will continue as an increasing number of mid- and small-sized companies begin to actively participate in worldwide markets. In order to achieve and maintain a leadership position in any one developed country, it will soon be necessary for manufacturing firms to be able to research, design, develop, engineer, and manufacture their products in any part of the developed world. These companies will have to become *transnational*.

The approach to competing on a global basis that is being used by many companies is what the Germans call a community of interest. In the U.S., this arrangement might be called a joint venture or cross-licensing agreement. The concept is one of "leveraging knowledge."

No matter what it is called, partnerships are not limited to small or medium-sized companies. For example, ASEA of Sweden, the world's largest heavy-engineering company, has joined forces with equipment manufacturer Brown Boveri of Switzerland. Already world leaders in their own right, the two firms feel that forming a *community of interest* will make them even more competitive in the North American and Far Eastern markets.

Performance also seems to have a direct relationship on how well a firm contributes to society and uses limited natural resources. Research based on *Business Ethics* magazine's *100 Best Corporate Citizens* list concluded that financial performance of the top 100 companies was *significantly better* than others in the *S&P 500*. (*Good Guys Are Prospering: 100 Best Corporate Citizens Outperform S&P 500 Peers*, April 22, 2002, DePaul University, Chicago) The ranking is based on quantitative measures of corporate service to stockholders, employees, customers, the community, environment, offshore stakeholders, and women and minorities. These 100 companies are applying innovative ideas in ways that benefit people, our limited natural resources, and the company's bottom line.

The top four companies for 2002, in rank order are IBM, Hewlett-Packard, Fannie Mae, and St. Paul Companies. Let us take a closer look at the list for companies that are engaged in manufacturing. The manufacturing top four are IBM, followed by Hewlett-Packard, Procter and Gamble, and Motorola. Other manufacturing companies in the top 100 are shown in **Figure 1-6**.

There are many other companies that provided support to the manufacturing industries that were also on this list. Lets take another look at the manufacturing top four. It is almost overwhelming to see just how involved these companies are throughout the world. See **Figure 1-7**.

In February 2002, The ***KLD-Nasdaq Social Index (KLD-NS Index)*** was introduced as the first benchmark for socially screened securities traded on The Nasdaq Stock Market. Based on the Nasdaq Composite, the KLD-NS Index is a market-value-weighted index reflecting the performance of some 280 of the largest U.S. corporations in technology, financial, and telecommunications sectors.

Figure 1-6. *Business Ethics*, Other Socially Responsible Firms Engaged in Manufacturing in 2002.

Other Socially Responsible Firms Business Ethics Study
Cummins Engine
Corning, Inc.
3Com Corp.
Network Appliances
Arrow Electronics
Charles Schwab
General Mills
Cisco Systems
Texas Instruments
Nucor
Delphi Automotive
Oracle Devices
Qualcomm
3M
Eastman Kodak
Clorox
Nordson Corp.
Sun Microsystems

Figure 1-7. *Business Ethics* Top 4 Manufacturing Firms for 2002, Global Presence.

Manufacturing Top 4	Global Presence (Number of Countries)
IBM	166
Hewlett-Packard	46
Procter and Gamble	94
Motorola	71

The KLD-NS Index is the first index to merge the Nasdaq indexes with the growing demand for socially responsible investment products.

To construct the KLD-NS Index, KLD applied social responsibility criteria to Nasdaq Composite domestic companies with over $1 billion in market capitalization. KLD's measurement of social performance evaluates the company's environmental stewardship, employee relations, non-U.S. operations, its involvement in the tobacco industry, and other issues. (www.kld.com)

The global nature of many markets means that producers have to think differently about the way they design, market, and distribute their products. If they are effective in meeting the needs of consumers from different lands and cultures, an increasing number of customized products will be needed. In a global marketplace, the traditional concept of mass production, making thousands of identical products for consumers with essentially identical needs, will have limited value. In addition to producing more localized products, manufacturing firms will have to produce better-quality products at a lower cost to consumers. They will also have to be more careful about how they use limited natural resources and the company's overall impact on the environment. The *Business Ethics* top 100 are making effective strides in this direction. For example, furniture maker Herman Miller won the Waste Wise Program Award from the EPA in 2001 for recycling 23 million pounds of waste in one year, and then reusing another 21 million pounds of waste for fuel to heat their main site.

Future of Industrial Production

What does the future of manufacturing hold? With aggressive global competition and the rise and fall of industrial leadership in many different markets, it may seem at first glance that the future of manufacturing is plagued with problems and difficulties. There is no doubt that change is needed in order to stay ahead. Doing things as they have always been done will not fill the bill. However, there is a great future out there for those who can think differently. Let us take a closer look at what is going on.

The major goal of most accounting offices in manufacturing is to find ways to cut costs. Often the first thing to come under scrutiny is the cost for direct labor (machine operators). In some cases the appeal of low cost wage rates in Mexico or China is irresistible, and a move is made to shift production offshore. Suddenly, the toaster oven business is in China. There are many layoffs back in the U.S., and plants may even be closed. Suppliers lose business, and more people are laid off. The community feels the brunt of this action when people have to move to other locations. More local businesses lose business because there are fewer people in the area to buy their products.

Then the time comes for the toaster ovens produced offshore to be shipped back to the U.S., marketed, and sold. In some cases, the accounting or finance office did not take the long and careful look at what might eventually happen. To their surprise, the American people, who have the desire to buy American toaster ovens, may not desire the products. Everyone may have lost, because accounting had a short-term solution. The shift offshore needs to be carefully evaluated to be certain that the best assessment is done of all factors that influence the cost of the product.

In many industries, companies have realized that complex work will stay here in this country, as companies become more diverse. High-volume mass production work may go wherever the wage rate is lowest (often offshore). Complex high volume machines are becoming less popular in the U.S. There are several reasons for this. First, quantities desired are becoming increasingly smaller. Larger companies have capital but they are interested in reducing capital in areas that cannot be justified. Small and mid-sized companies do not have the finances to purchase large costly equipment. Transfer lines are being replaced with computer numerical control (CNC), cellular manufacturing, and multitasking equipment (i.e., a lathe equipped with a turning center). Products have very short life cycles, so flexibility and changeability of equipment is important. Changeover of complex equipment to produce different types of products takes time, and time is money.

Many of today's processes do not require skilled hand labor and require the operator to spend more time on programming than pushing buttons. Today's manufacturing firms must be able to provide options for making parts in the best way possible. In many cases this means that they must be able to produce sophisticated, close tolerance parts that cannot be moved from one machine to another without destroying the integrity of the part. This means that they are produced in cell or on a flexible machining center. Sometimes this requires the concept of a cell in a line. Often, cost savings is accomplished through low cost automation.

One way to develop perspective on the bright future of manufacturing is to study the performance of Hass Automation, Inc. of Oxnard, California. Hass is one of the largest manufacturers of machine tools in the world. Since the 1990s, Hass has been able to reduce the cost of their machine tools as much as 50 percent. At the same time their equipment is much more efficient because of computer technology, improved design features, and automated systems. Current equipment is able to maintain better surface finishes and closer tolerances, and is able to run at faster speeds. A vertical machining center was able to run at up to 40 inches per minute, with a spindle speed of somewhere around 4,000 rpm. Today, their fastest machines are able to run at spindle speeds up to 30,000 rpm and operate at cutting speeds up to 400 inches per minute. Bigger, better, faster and cheaper is what results in sales and leadership in this industry. Hass is also a leader in the field of laser cutting tools used in the medical, electronics, fabrication and manufacturing industries. (Interview with Jeffrey's Manufacturing Solutions, Greensboro, NC)

A vital consideration today for any company competing in the world marketplace, and especially, in the European Economic Community, is *ISO certification*. The *ISO 9000 standard*, developed by the *International Standards Organization*, is a recognized and agreed-upon method of determining quality. *ISO 9000* is primarily concerned with *quality management*. The standardized definition of *quality* in ISO 9000 refers to all those features of a product (or service) that are required by the customer. *Quality management* refers to what the organization does to ensure that its products conform to the customer's requirements. The ISO 14000 standard was created to address *environmental management*, what the organization does to minimize harmful effects on the environment caused by its activities. To date,

ISO's work has resulted in some 12,000 International Standards, representing more than 300,000 pages in English and French (terminology is often provided in other languages as well).

Certification is a process that involves meeting strict requirements that demonstrate the company's ability to control all the processes that affect the acceptability of its product by buyer. Certification can take a year or more, since it requires considerable work and strong commitment at all levels of an organization, from top management throughout the company to production, clerical, and support personnel.

In June 2002, some 170 representatives from international organizations, business, industry and standards bodies from throughout the world, met in Trinidad to examine the feasibility and needs for ISO to develop internationally accepted standards for corporate social responsibility (CSR). The ISO committee for consumer policy, ISO/COPOLCO, recommended the establishment of a strategic advisory group to explore further the issue of whether or not the organization should launch the development of standards for CSR.

It will be difficult for some companies to adjust to many challenges and opportunities that now face manufacturing. This is where the technology strategist must provide enthusiastic leadership, energy, and guidance. The list of plant closures should serve as an early warning for what will come. It is no longer adequate to depend solely on the habits of the past to run a manufacturing enterprise. There has to be a bridge to the future. This bridge involves continuous improvement and the search for doing things faster and better.

The primary purpose of this book is to serve as a resource tool for evaluating and using manufacturing processes. It is hoped that this material will stimulate continued exploration and discovery. What lies ahead is an exciting world in which new materials and processes are being developed each day, so there is always something new to learn. We have tried to present the most important processes and emphasize their applications in industry.

What Is Manufacturing?

Manufacturing is the application of knowledge demonstrated through the use of tools, processes, machines and systems to transform raw materials or substances into new products. Repair or rework of old products is not classified as manufacturing. However, manufacturing does include companies that assemble component parts into a product.

Manufacturing is really the driving force behind any industrialized nation, since it has a direct relationship to the quality of life and standard of living of the nation's people. Nearly two-thirds of all of the wealth-producing activities in the U.S. come from manufacturing.

The **Standard Industrial Classification (SIC) Index**, prepared by the U.S. Office of Management and Budget, has for many years been the primary method for determining what industries and types of companies could be classified as manufacturing. The *SIC Manual* lists classifications and subclassifications of industries with their numerical codes. SIC identified twenty major industries engaged in manufacturing. These covered everything from *Food and Kindred Products*, *Paper and Allied Products*, and *Primary Metal Products* to *Industrial and Commercial Machinery and Computer Equipment*, *Apparel and Other Finished Products Made from Fabrics*, and *Furniture and Fixtures*. With five to ten subclassifications under each of the twenty major classifications, several hundred different types of manufacturing industries are listed in the *SIC Manual*. According to the SIC, industries involved in the production of liquors and wines, and products of agriculture such as mining, fishing, and quarrying, are classified as manufacturing. Other industries that you might not normally associate with manufacturing are also included: oyster shucking, apparel jobbing, publishing, logging, and ready-mixed concrete production.

The SIC classification scheme had other areas that made the process of classification complicated. Mining of copper is classified as manufacturing, while the extraction of coal or other nonmetallic materials is classified as mining. Tobacco products and milk bottling and pasteurizing are classified as manufacturing, but milk processing on farms and stemming leaf tobacco in the fields are not. In spite of the complexity of this, the three and four-digit SIC classification scheme is still used by some agencies and organizations to classify manufacturing.

In 1997, statistical bodies representing the U.S., Canada, and Mexico developed another classification system called the **North American Industry Classification System (NAICS)**. NAICS is currently replacing SIC as a major classification system. NAICS places additional emphasis on classification of the construction, wholesale trades, and service industries. An annual survey of manufacturers (2005) report entitled *Value of Shipments, Manufacturing, Mining and Construction Statistics* from the U.S. Census Bureau points out that there are 473 six-digit NAICS product groups and approximately 1,500 seven-digit NAICS product classes.

Another report, *The Manufacturers' Shipments, Inventories and Orders* released in August of 2006 from the U. S. Department of Commerce, provides more evidence of the importance of manufacturing to our economy. Let us take a look at some of the highlights of this report.

- Sales for hard good manufactured products dropped in July of 2006 following two consecutive monthly increases, decreasing $2.4 billion to $405.1 billion. This followed a 1.5 percent June increase.
- Shipments, increased in April and May, from $0.1 billion to $403.9 billion.
- Unfilled manufacturing orders in fourteen of the last fifteen months covered by the survey increased from $8.2 billion to $630.2 billion. This was the highest level since 1992.
- The report also indicated that inventories were up nine of the last ten months, increasing from $3.0 billion to $473.0 billion. See: http://www.census.gov/indicator/www/m3/

There are now 473 major product groups listed as manufacturing by NAICS. Nearly 350 new industries were recognized for the first time with NAICS. Several of these industries represent "high tech" industries such as fiber optic cable manufacturing, satellite communications, and the reproduction of computer software.

Two years later, another system was introduced by Canada, Mexico, and the United States. (Federal Register Notice, April 16, 1999) This system was called the **North American Product Classification System (NAPCS)**. The long-term objective of NAPCS is to develop a market-oriented classification system for products that (a) is not industry-of-origin based but can be linked to the NAICS industry structure, (b) is consistent across the three NAICS countries, and (c) promotes improvements in the identification and classification of service products across international classification systems, such as the *Central Product Classification System of the United Nations*. NAPCS will give special attention to service products, new products, and advanced technology products.

During 2001 and 2002 the primary efforts of NAPCS were on the service industries. Phase III really began during the days of May 5–9 of 2003, when the Trilateral Steering Committee on Economic Classification began consolidating committee assignments to address NAPCS goods-producing industries. Results from Phases I, II and III will be incorporated incrementally into services, annual surveys and inclusively into the 2007 *Economic Census*.

The Manufacturing Process

Now that we know how manufacturing is classified, let us take a closer look at the process of manufacturing. It was said that manufacturing involves transforming or changing materials into products. As an example, a raw material, such as zinc or magnesium, might be melted in a die-casting machine and forced under pressure into a closed mold to produce the casing for a power drill. In the process of die-casting, the mold normally opens and a robot reaches in and picks out one casing or several connected casings. The raw material (metal ingots) was changed into another shape (the casing) that would eventually become part of the finished product (a power drill). The primary manufacturing process used to do this is called *die-casting*.

Manufacturing can be done through mechanical or chemical means. Manufacturing is normally done in a factory, mill, or plant, but can take place just about anywhere—on this earth and even in other galaxies. Today mobile factories in trailers perform manufacturing operations every day. The *factory* is just moved to the job site or point-of-use. Examples would be mobile locksmith shops, mobile document shredders, mobile sawmills, and mobile welding shops.

In order to better understand what manufacturing is, it may be helpful to spend some time clarifying terminology. There are many different types of manufacturing firms, or companies. Each of these produces and/or assembles a specific line of products, such as motorcycles or cosmetics or woodworking tools or computers. Typically, a company will manufacture a variety of models or styles in its particular product line. For example, an automobile manufacturer would offer two-door, four-door, or convertibles; different "packages" or features; and several power train options.

If you were to consider all of the companies manufacturing a particular product (say fishhooks), and a significant quantity of that product was sold each year, then chances are that it would be classified as an industry—in this case, the fishhook industry. Currently, there are not too many fishhook manufacturers around, nor are vast numbers of fishhooks being sold, so none of the three classification systems

lists this as a separate industry. However, both SIC and NAICS cover fishhook production as a subclassification under a major type of manufacturing industry, *other miscellaneous manufacturing*. Other products, such as cars and trucks, have established their identity as a separate industry (the *automotive industry*).

The reader should have a better understanding now of the great diversity of manufacturing that occurs every day throughout the world. The different types of industries classified as *manufacturing* range from companies that make hard good products, such as earthmoving equipment and laser cutting machinery, to others that blend plastics, resins, and oils. Even textile, tobacco, pretzel, and chewing gum producers are classified under manufacturing. There are 473 product groups based on material and product types listed in the National American Industry Classification system. It would be next to impossible to study all of the processes important to 473 different types of industries, with its 1,500 different product classes and millions of specialized companies. NAICS has grouped all of these industries into 21 major types. See **Figure 1-8**.

Often, the terms *industry* and *company* are used interchangeably and incorrectly. In each of the individual *industries* listed in the NAICS classification scheme there are many thousands of different *companies*. For example, the automotive industry has millions of separate companies responsible for manufacture of automobiles; for production of subassemblies supporting the industry; for manufacture of devices, systems, and components to be assembled in automobiles; for production of materials used in making automobiles; for transportation and distribution of automobiles, and more.

Important Terms

aerogel
batch
Border Industrialization Program
European Economic Community (EEC)
Industry Productivity Program
International Standards Organization
ISO Certification
ISO 9000 Standard
job lot
KLD-Nasdaq Social Index
lean production
Major Sector Productivity and Costs Program
manufacturing

Figure 1-8. NAICS manufacturing industries.

NAICS Classification Number	Number of Establishments	Sales ($1,000)	Paid Employees	Annual Payroll ($1,000)
31-33 Manufacturing	363,753	3,842,061,405	16,888,016	572,101,070
311 Food Mfg	26,361	423,978,723	1,471,050	38,532,086
312 Beverage & Tobacco Product Mfg	2,729	97,124,576	175,996	6,746,774
313 Textile Mills	4,706	58,804,269	393,914	10,099,969
314 Textile Product Mills	7,906	31,107,992	236,170	5,120,015
315 Apparel Mfg	17,065	68,428,564	719,269	12,748,228
316 Leather & Allied Product Mfg	1,870	10,899,471	84,822	1,835,675
321 Wood Product Mfg	17,411	89,211,563	574,426	14,401,357
322 Paper Mfg	5,896	150,635,435	576,920	22,271,191
323 Printing & Related Support Activities	42,916	97,944,985	838,240	26,109,332
324 Petroleum & Coal Products Mfg	2,155	176,217,259	107,878	5,554,842
325 Chemical Mfg	13,513	419,617,444	884,321	39,887,185
326 Plastics & Rubber Products Mfg	16,876	160,317,732	1,029,976	30,028,561
327 Nonmetallic Mineral Product Mfg	16,385	87,010,210	504,443	16,271,427
331 Primary Metal Mfg	5,095	170,188,704	611,714	24,069,436
332 Fabricated Metal Product Mfg	62,501	243,254,492	1,774,874	57,040,954
333 Machinery Mfg	30,665	270,357,157	1,421,820	53,059,543
334 Computer & Electronic Product Mfg	17,465	438,209,195	1,698,529	72,717,428
335 Electrical Equipment, Appliance, & Component Mfg	6,946	111,809,707	594,914	18,978,904
336 Transportation Equipment Mfg	12,980	571,979,634	1,848,558	79,649,337
337 Furniture & Related Product Mfg	20,758	63,939,540	604,845	14,977,736
339 Miscellaneous Mfg	31,554	101,024,753	735,337	22,001,090

manufacturing productivity
maquiladora
maquilas
mass production
North American Economic Community (NEC)
North American Industry Classification System (NAICS)
North American Free Trade Agreement (NAFTA)
North American Product Classification System (NAPCS)
Standard Industrial Classification (SIC)
work in process (WIP)

Questions for Review and Discussion

1. With the passage of NAFTA, free trade agreements between the U.S., Mexico, and Canada, and the rapid growth of maquiladoras just across the U.S.-Mexico border, what do you expect will be the impact on low-skilled workers who are presently assembling household products in U.S. plants? What type of manufacturing do you think will be most significantly impacted? How will this change the future of manufacturing in the U.S.?

2. From which direction would you expect the real challenge to United States manufacturing to emerge in the next decade: Canada, Mexico, Japan, China, or the European Economic Community? Provide a rationale to support your answer.

3. List the U.S. manufacturing industries that you believe could be labeled *major growth industries.* Explain why you feel these industries are growing faster than others. Do you think this growth trend will continue?

4. Describe the manufacturing process starting with raw materials and ending with the final product. Use a consumer product you are familiar with to get things started.

5. How important is it to think about manufacturing and sales of industrial products with a global perspective? Provide examples to support your position.

6. Pick a country that you are familiar with, other than the United States, and then discuss the impact of manufacturing on its development.

The transition from the simple spinning wheel to this large-scale thread-manufacturing machine is a good example of the rapidly changing world of manufacturing.

2 Material and Process Classifications

Chapter Highlights

- The North American Industry Classification System (NAICS) is used to identify hundreds of manufacturing industries.
- There are five major material families, or types of materials used in the manufacture of hard good manufactured products.
- Differentiation between primary and secondary material processing industries.
- The major processing actions—forming, separating, fabricating, conditioning, and finishing—can be used to classify types of manufacturing processes.
- Case studies of processes used in industry, associating process actions with material families.

Introduction

There are 473 six-digit NAICS product groups and approximately 1,500 seven-digit product classes listed in the North American Industry Classification System (NAICS) for the twenty-one different industrial classifications. These are based on the type of product and/or material. Let us take a closer look at twenty-one major types of industries listed.

Manufactured Consumer Products

Two NAIC primary products codes are concerned with the general classifications of food, beverages, and tobacco products. Three codes cover the general classification of textile products. Three codes cover the general classification of wood, paper, and printing products. Two codes address the general classification of chemical products. Three codes are provided for metal products. Other codes cover major product types such as: furniture, computers and electrical products, plastics and rubber products, and nonmetallic mineral products.

Consumer Product Types

Many of these industries are concerned with the manufacture of consumer products that are *consumable*. Some of these are digestible and are eaten. Other products may be chewed, smoked, or absorbed. Many other industries are concerned with the manufacture of products that are *nonconsumable*. These include *hard good consumer products* (durable goods) such as automobiles, machine tools, chemicals, solvents, adhesives, and finishes. Many industries are responsible for the manufacture of *nonconsumable soft good consumer products*. These include paper products, textile products, and computer software.

Most of our emphasis in this book will be on the processes and materials—metallics, ceramics, polymeric/plastics, polymeric/woods, and composites —that are used to make *nonconsumable, hard good consumer products*. See **Figure 2-1**.

Major Material Families

The major materials associated with the manufacture of nonconsumable, hard good consumer products can be grouped into five major material families: metals, plastics, woods, ceramics, and composites. Most of the manufacturing activities conducted today involve the use of these materials. Many different types of industries, products, and materials are associated with each of these material families. For example, the glass industry is part of the ceramic material family.

Figure 2-1. Manufactured consumer products.

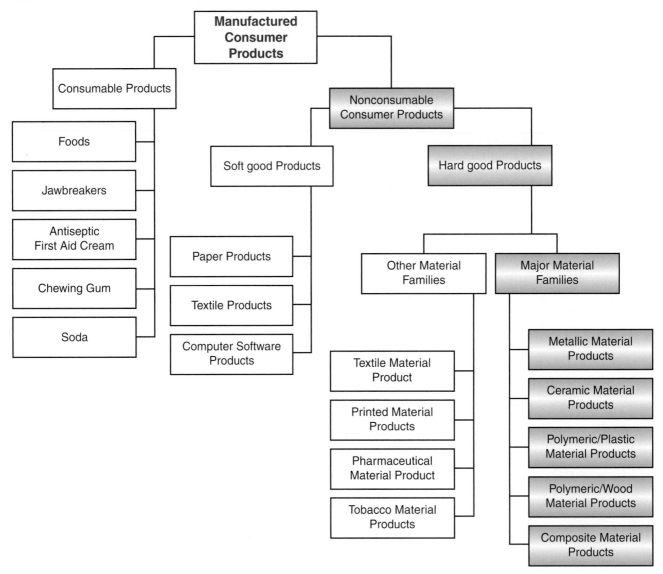

There are many different sectors in the glass industry, and many different types of companies making glass or glass products. Examples are plate glass, fiber optic rods, bottles, electrical insulators, and scientific glassware, to list only a few.

Even though there are many different types of products and materials involved, *all* manufacturing firms either: (1) make industrial stock from raw materials, (2) make new products from industrial stock, or (3) assemble parts into new products.

Raw Materials

The online database, MatWeb (www.matweb.com), lists data sheets for 59,000 metal, plastic, ceramic, and composite materials. Raw materials can be classified in general terms as solids, liquids, or gases.

There are *crystalline* solids, such as sodium chloride; *amorphous* solids, such as glass; *covalent* solids, such as germanium; and *molecular* solids, such as solid argon. The properties of solid materials depend on the properties of atoms in the material, on the position of these atoms in the solid, and on temperature and pressure conditions present when the solid was formed.

Raw materials in liquid state can only be distinguished between other materials in solid and gaseous form when the liquid's molecules are composed of less than twenty atoms. When the number exceeds this limit the material may be cooled below its true melting point to create materials such as glass, an *amorphous solid* with many of the properties of a solid, but without crystalline order. Materials such as nylon and rubber do not have fixed melting points and behave in this way.

Raw materials in gaseous form range from simple gases, whose particles are single atoms (neon or helium), to complex multi-atom gases such as hydrocarbons produced by the petroleum refining industry. Atoms and molecules in gases move about freely in three dimensions. This results in continuous collision and bombardment of atoms and molecules against each other and generates pressure.

Extractive Industries

Manufacturing processes are used by **extractive industries**—industries that collect raw materials, such as oil, gas, trees, and minerals—to convert the basic raw materials into refined materials. These refined materials are then purchased as *industrial stock* and used to make finished products by hard good manufacturing firms. The companies using this stock use manufacturing processes to change the shape or composition of industrial stock through mechanical, chemical, thermal, or electrical means.

Semifinished goods move to the final consumer through various distribution channels. The products made by these industries are then moved through their distribution channels. Sometimes they are sent directly to the retailer, or consumer. Usually they are transported to wholesalers before they reach the hard goods manufacturing firm, the retail market, and the consumer. In less-developed countries the channels of distribution are usually shorter and simpler than in more highly industrialized nations.

To get a better perspective on the magnitude of different types of raw materials used by industry, we will look at some of the materials produced by a leading manufacturer of raw materials used in industry, E.I. DuPont deNemours and Company. See **Figure 2-2**.

Firms such as DuPont take raw materials and manufacture other products that are used as the raw material or industrial stock by other manufacturing companies to make hard good products. On these factory floors is where the magic really comes alive and stock is transformed into products that are useful to the consumer.

Figure 2-3 shows what happens within a typical manufacturing process. Here, a process called *turning* is used to remove metal from titanium stock that is mounted between centers in a machining center. Since the titanium metal is difficult to cut because of its extreme hardness, a ceramic carbide insert is used as the cutting tool. The figure provides a close-up view of the titanium chip that is being removed as the tool cuts into the *workpiece* (metal stock). The titanium bar stock that is used by this manufacturer is the finished product from another manufacturing industry that produces titanium bar stock.

Every manufacturing process involves following a planned sequence of operations or steps to create the desired product. If the same sequence of operations is followed each time the process is used and if the conditions in the work environment do not change, products of consistent quality can be produced.

Material Processing in Manufacturing

While all manufacturing companies are concerned with transforming materials into products,

Figure 2-2. Diversity of raw materials created by DuPont.

DuPont products used as Raw Materials by Other Manufacturers	Description
Zonyl® Fluoroadditives	White, free-flowing Teflon® PTFE powders designed for use as additives in other materials or systems
1,4-Butanediol	A clear, odorless liquid or white solid, depending on ambient temperature Very pure, urethane grade, liquid glycol (diol)
Elvamide® nylon multipolymer resins	Thermoplastic, alcohol-soluble resins
Sorona®	A family of polymers made from 1,3 Propanediol (PDO) that in fiber form have unique properties such as resilience, stretch with recovery, easy dye ability
Ditac®	Integrated circuit adhesives
Vexar®	Plastic netting made from polyethylene, polypropylene, and other plastic resins Available in tube or sheet form in a wide variety of sizes, textures;
Ink Jet Barrier Films	Barrier film to make ink jet printer pen heads
OmniDex®	Holographic photopolymer materials used to produce high-quality holograms in three operations: imaging, an ultraviolet cure and a brief heating step
Zodiaq®	A new category of surfacing material for commercial and residential interiors; It is composed of 93% quartz and has exceptional toughness and hardness
OmniDex®	Holographic photopolymer
Dacron® Polyester Fiberfill	Sleeping bag filler
KEVLAR®	Sporting goods, helmets, fishing boats

Figure 2-3. Turning is a typical manufacturing process in which industrial stock is transformed into a product. Turning consists of holding a cutting tool against a rotating workpiece to remove material. In this example, a replaceable cutter made of ceramic carbide is being used to shape hard titanium metal. (Kennametal, Inc.)

not all of them make or assemble products that are ready for use or consumption by the consumer. In some cases, a product from a manufacturing firm is ready for use or consumption when it leaves the manufacturing plant. In other cases, the product from one plant becomes the raw material for use by another, thus undergoing successive transformations. **Figure 2-4** shows a microwave amplifier package that was hermetically sealed (made airtight) with the heat generated by a process called laser welding. With this product, the cover was welded using a Nd: YAG (neodymium yttrium aluminum garnet) laser moving at a rate of 8 to 10 inches per minute. The advantage of using a laser for this application is that it can apply pulsed or intermittent blasts of energy to the workpiece and therefore transmit less heat to the surface being joined. The individual parts making up the amplifier package might have been produced by several different manufacturers and then sold to the amplifier manufacturing firm as components to be assembled in the product shown. The amplifier, in turn, might become a component in a more complex product.

The process of *successive transformation* (production in stages) can extend across a number

Figure 2-4. A laser beam was used to weld the cover of this microwave amplifier in place, sealing it airtight. The attached tag provides an idea of the size of this small component. (HDE Systems, Inc.)

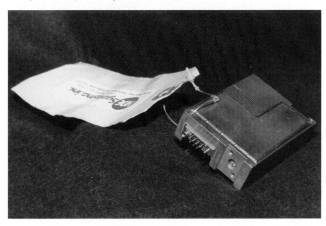

of industries. Consider the manufacture of electric motors. First, the copper smelting plant produces copper for use by electrolytic refineries. The refinery then manufactures a product called refined copper. This is purchased as the raw material for use by the copper wire mill. The copper wire it produces is purchased by the electrical equipment manufacturer as the raw material for use in making electric motors. The electric motor may then be purchased as a windshield wiper subassembly by other manufacturers for use in their cars or trucks.

Material Processing Categories

There are many ways to organize the way we think about the various stages of transformation, or phases of material processing. To simplify the

process, we will classify the different types of processing into two categories:

- *Primary processing*, manufacturing carried out by firms engaged in the preparation of standard industrial stock.
- *Secondary processing*, manufacturing done by firms that manufacture hard good consumer products.

The relationship of primary and secondary processing is shown in **Figure 2-5**. Remember that our emphasis here in this book is on secondary processing.

There are many thousands of different manufacturing processes used by primary and secondary manufacturing firms. Primary processors start with raw materials, like trees or minerals, and then use manufacturing processes to change these materials into industrial stock: sheets, rods, pellets, marbles, powder, or liquids. As noted earlier, products often go through many different stages in production. For example, some primary processing firms deal with smelting and separating metal from the impurities. Others distill petroleum to manufacture jet fuel or make plastic pellets.

There are many types of industrial stock produced by primary manufacturing firms, ranging from perfumes, ceramic particles, and sheets of composite laminate to powdered metal and containers of lubricating oil. The chemical refining, filtering, mixing, and separating processes used by primary processing firms are outside the scope of this book.

Secondary processing firms are those companies that purchase industrial stock, such as powdered clay, particles of glass, or plastic pellets, and then

Figure 2-5. Primary and secondary processing industries.

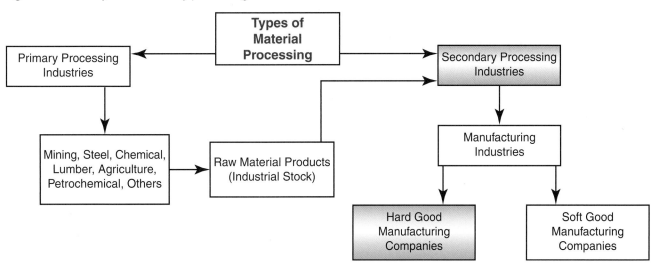

use manufacturing processes to transform this stock into a hard good product for sale to the consumer. Hard good, nonconsumable products are durable goods that are used over substantial periods of time. Examples include boats, computers, cars, and earth moving equipment. These products are purchased less frequently and require more thought before purchase than fast-moving consumer goods like toothpaste, stationary, and food items.

By turning stock into useful products, secondary processing firms add more value to the basic raw material. The focus of this book is on the *manufacturing processes* used by *secondary processing* firms to produce nonconsumable, *hard good consumer products*. Even with this attempt to narrow the field of study, there are still thousands of manufacturing processes used by these manufacturing firms. Further delimitation is still necessary.

Major Material Families

There are many types of materials used by the hard good manufacturing industries. However, manufacturers concerned with producing hard good consumer products use some of these materials more often than others. **Figure 2-6** shows the major material families: metallics, ceramics, polymeric/plastics, polymeric/woods, and composites. These five are identified as major material families because most processes used in the manufacture of hard good consumer products involve one or more of these materials. The five major material families constitute the foundation for the study of manufacturing processes and materials. Most manufacturing sector employment involves making products from these materials.

You may recall that, in **Figure 2-1**, a major grouping labeled as *Other Material Families* was shown under the major heading of *hard good products*. This grouping included printed, electronic, tobacco, and textile products. These material classifications are responsible for many different types of manufactured products, but are not within the scope of this book. The rationale for assigning a secondary priority to these is because the manufacturing processes used in these areas are often associated with industries classified outside of manufacturing by the Standard Industrial Classification or North American Industrial Classification indices. Examples would be electronic products (communications or telecommunications industry), printed products (printing industry), and tobacco products (tobacco industry). While these industries *do* manufacture products, they represent specialized groupings and unique processes that are generally outside the common body of knowledge in manufacturing.

Figure 2-6. Major material families and associated industries.

Metallic Family	Ceramic Family	Polymeric/ Plastics Family	Polymeric/Woods Family	Composites Family
Primary Metal Mfg.	Nonmetallic Mineral Product Mfg.	Plastics and Rubber Products Mfg.	Furniture and Related Products Mfg.	Computer and electronic Product Mfg.
Fabricated Metal Products Mfg.	Misc. Mfg.	Chemical Mfg.	Wood Product Mfg.	Transportation Equipment Mfg.
Machinery Mfg.	Clay Product and Refractory Mfg.	Petroleum and Coal Products Mfg.	Paper Mfg.	Chemical Mfg.
Computer and Electronic Product Mfg.	Pottery, Ceramics, and Plumbing Fixture Mfg.	Plastics, foil, and Coated Paper Bag Mfg.	Printing and Related Support Activities	Textile Mfg.
Steel Wire Drawing Mfg.	Glass and Glass Product Mfg.	Polystyrene Foam Product Mfg.	Engineered Wood Mfg.	Abrasive Product Mfg.
Aluminum Sheet, Plate, and Foil Mfg.		Plastic Bottle Mfg.	Reconstituted Wood Mfg.	Semiconductor and Other Electronic Component Mfg.
Transportation Equipment Mfg.		Doll, toy, and Game Mfg.		

Refer to **Figure 2-1** again. Note that there are two types of manufactured products, consumable and nonconsumable. *Consumable products* include foods, chewing gum, soft drinks, and even some pharmaceutical products (medicines). *Nonconsumable products* are hard goods, such as plastic recycling containers, ceramic grinding wheels, and metal lawn furniture.

There is nothing finite about the use of classification schemes to organize thinking. Graphic models can be used to group information according to the best judgment of the designer.

Although we have delimited the scope of our study a great deal, there are still thousands of manufacturing processes that are used to produce hard good consumer products using materials from the five major families (metallics, ceramics, polymeric/plastics, polymeric/woods, and composites). For each of these material families, there are major industries that can be identified.

Material Family	Examples of Major Industries
Metallic	Chrome plating, copper foundries
Ceramic	Clay and glass
Polymeric/plastics	Plastic and elastomeric
Polymeric/woods	Wood and paper products
Composite	Aerospace and automotive

Composite materials are used by all of the manufacturing industries. However, the metallic, ceramic, polymeric/plastics, and polymeric/woods material families have their own unique types of products. As human or animal family members have similar characteristics and behaviors, so do the members of material families. All of the materials in a particular family behave in a similar way when they are transformed into products using a manufacturing process. We will look more closely at the structure of each of these industries.

Metal Manufacturing Industries

There are 94 different types of *metal* product manufacturing industries recognized by NAICS. Reduction of these into the major groupings is a subjective process. **Figure 2-7** illustrates the ten significant metal manufacturing industries.

Ceramic Manufacturing Industries

The American Ceramics Society (www.acers.org) classifies the content of ceramics into seven major

Figure 2-7. Major types of metal manufacturing industries.

Metal Manufacturing Industries	
Primary Metal Manufacturing Industries	**Ferrous and Nonferrous Metal Manufacturing Industries**
Metal containers	Hardware, spring, and wire
Industrial and commercial machinery	Metallic coating and finishing
Transportation equipment	Electronics and computers
Computer	Fabricated metal products
Measuring, analyzing, and controlling instrument	Metal production, smelting, and refining

types of industries. See **Figure 2-8**. At first glance, most people think that the term *industrial ceramics* would be concerned only with clay products. However, as shown in the figure, there is much more involved here than just making structural clay products. It may be useful here to list some of the many different types of products manufactured by these seven major ceramic industries. See **Figure 2-9**.

Polymeric Manufacturing Industries

The polymeric industry can be studied in terms of two primary groupings: plastics and woods. **Figure 2-10** shows ten of the major industries that use *polymeric* materials in woods and plastics.

Material Processing Families

Over the past quarter century, new polymeric, ceramic, and composite materials have brought revolutionary changes in the way we design and manufacture

Figure 2-8. Major types of ceramic manufacturing industries.

Ceramic Manufacturing Industries
Structural clay products
Abrasives
Whiteware
Cement
Refractory
Advanced ceramics
Glass

Figure 2-9. Ceramic industry classification and products. (The American Ceramic Society)

Major Industrial Classification	Products	NAICS Classification Numbers
Structural Clay Products	Brick, sewer pipe, roofing tile, clay floor and wall tile, flue linings	327111, 327121, 327122, 327123
Whitewares	Dinnerware, floor and wall tile, sanitaryware, electrical porcelain, decorative ceramics	327112, 327113
Refractories	Brick and other monolithic products used in iron and steel, non-ferrous metals, energy conversion, glass, cements, petroleum, and chemical industries	327124, 327125
Glasses	Window glass (flat glass), container glass (bottles), pressed and blown glass (fine dinnerware), glass fibers (insulation), and advanced glass (optical fibers)	327210, 327211, 327212, 327213, 327215
Abrasives	Natural materials (diamond, garnet) and synthetic materials (silicon carbide, fused alumina, diamonds); used for grinding, cutting, polishing, lapping, or blasting of materials	32799, 327910, 327991, 327992
Cements	Materials used in construction of buildings, roads, bridges, water and sewer systems, and dams	327310, 327320, 327331, 327332, 327390,
Advanced Ceramics	Wear parts (bioceramics, cutting tool inserts, engine components), electrical components (capacitors, insulators, substrates, IC packages, piezoelectrics, magnets and superconductors), coatings (engine components, cutting tools, industrial wear parts), chemical and environmental parts (filters, catalysts, catalyst support, membranes)	327999

products. Never before have we had the opportunity to not only design and build products, but also to create the materials used in their manufacture. Sophisticated new materials are now available that can withstand the abuse of harsh environments, including the rigors of space exploration and manufacturing. These *engineered materials* can withstand exposure to chemicals, pollution, contamination, and even collision with floating debris in space. Such human-engineered materials offer great promise for manufacturing; they also provide new challenges for the technologist.

Figure 2-10. Major types of polymeric/plastics and polymeric/wood manufacturing industries.

Plastic and Wood Manufacturing Industries
Plastics and rubber
Polystyrene foam
Rubber and elastomeric products
Resin compounding
Plastic film, sheet, and bag
Veneer, plywood, engineered wood, and reconstituted wood
Plastic bottles
Paper products

Each material has its own set of behavioral characteristics; not all manufacturing processes work equally well with all materials. Important production decisions often have to be made in selecting the correct process for a particular material and unique set of conditions. Manufacturing strategists (manufacturing supervisors, managers, engineers, technologists, and technicians) have to be able to make informed judgments to choose the process that will accomplish the most for the least (provide the highest quality with the least expenditure of resources).

On the factory floor, good decisionmaking is judged in terms of profit. Profit results from making the correct number of quality products at a cost that makes them more attractive than those offered by competitors. Poor judgment in selecting and using manufacturing processes will result in poor quality, lost time, and lost profits.

To identify manufacturing processes for study in this book, it was first necessary to determine which processes are used most frequently in each of the four major material processing families. Some processes, of course, are used primarily with metals. Others are used mostly with plastics or woods. Still other processes are common to the field of industrial ceramics. However, many of the processes used with ceramics are also used with plastics, metals, or woods.

Since composites involve several different materials integrated in a matrix as one material, the processes used with them are often the same as they would be when used with the other major materials in the composite structure. In other cases, the process must be modified so that it is more suitable for use with the combination of materials that has been created.

Some processes are used with a number of different materials; others can be used efficiently with only one. Throughout this text, efforts have been made to present each process in conjunction with the material area where the process is most frequently used.

Process Action

Because of the many manufacturing processes used with the five material families, it would be difficult to develop perspective on how all of these processes work if they were presented and studied at random. In order to help you understand the similarities and differences between processes, they will be presented according to the type of process action that they perform. There are five major process actions:
- Forming.
- Separating.
- Fabricating.
- Conditioning.
- Finishing.

A *process action* is the description of what happens when a process changes the internal structure or the outward appearance of a material (ceramic, composite, metallic, plastic, or wood). In the future, when you are thinking about processes used in industry, it will be helpful to have an organizational scheme for organizing your thinking. Grouping processes according to process actions will help you quickly organize a large amount of content according to the action the process performs when used with industrial stock. Remember, a process action is not the same thing as a manufacturing process.

In the section that follows, five case studies on cutting-edge manufacturing companies studied by the authors will be provided to better illustrate the nature of processes used to form, separate, fabricate, condition, and finish materials. Please read over the characteristics of each of the manufacturing process actions, and then think carefully about how the firm that is showcased uses manufacturing processes to create their products.

Forming

Forming processes are used to change the size or shape of industrial stock. With forming processes, hydraulic and mechanical pressure is normally employed to squeeze or shape the material. Often, there is no loss of weight or volume when material is changed using a forming process. Bending, stretching, casting, pressing, and molding processes are all classified under the heading of forming. These are processes used to form metal, polymeric, ceramic, and composite materials.

Now in order to better understand what happens when processes are used to form materials, we will take a moment to look at several of the processes used by Kinston Neuse Corporation (KNC), a subsidiary of Crown Equipment Corporation with headquarters in New Bremen, Ohio. The Crown/KNC facility is located in Kinston, North Carolina.

Company Overview

Crown is a family-owned, privately held company that manufactures heavy-duty electric lift trucks used in transporting materials and goods in warehouses and distribution centers throughout the world. Crown has domestic manufacturing facilities for fork lift trucks in Kinston, North Carolina; New Bremen, Ohio; and Greencastle, Indiana. Electric motors and plastic components are manufactured at a Crown Equipment facility in New Knoxville, Ohio. Crown also operates manufacturing facilities in Sidney, Australia; Basingsoke, England; Galway, Ireland; Roding, Germany; and Queretaro, Mexico.

The Crown product line includes narrow-aisle stacking equipment, powered pallet trucks, order picking equipment, and fork lift trucks capable of

Product Showcase #1				
Company	**Location**	**Internet URL**	**Process Action**	**Product to be Showcased**
KNC (Kinston Neuse Corporation)	Kinston, North Carolina	www.crown.com	Forming	Forks for Lift Trucks

moving loads as heavy as 8,000 pounds and lifting these loads into the air as high as 45 feet. Crown makes a wide selection of lift trucks ranging from hand pallet trucks to V.N.A. (Very Narrow-Aisle) turret trucks, material handling products, and storage solutions. The KNC facility makes electric pallet trucks and heavy-duty walkie stackers, electric pallet trucks that can lift loads several feet up.

Forming Process Overview

The forming process that we want to show here is a flange forming and bending process completed on a 350-ton Cincinnati Computer-Assisted Numerical Control (CNC) hydraulic press brake. See **Figure 2-11**.

Several different operations are performed to form metal on the forks for lift trucks. Before the forming process takes place, the metal plate is flame cut using plasma torches. It is then formed and deep drawn using a TMMT 500-ton hydraulic press. Once these operations have been completed, the workpiece is ready for the forming process emphasized in this showcase.

First, .312″ thick metal stock is moved to the press brake by the press operator using an overhead crane with a magnetic lift. See **Figure 2-12**.

The stock is then inserted between the working jaws on the press brake. A computer software program is written and used to direct the action of the ram and back gages on the press brake. When the operator calls for the CNC program, information is transmitted from the program and controller to the press brake. The controller tells the brake how to adjust the back gages and ram for the next bend or

Figure 2-12. Crown/KNC uses an overhead lift to support the workpiece and assist the operator in transporting it to the press brake.

fold. The operator then moves the stock and the operation is completed. The process is semiautomatic. It is necessary for the operator to obtain the part, and then insert it into the brake after each operation. See **Figure 2-13**. When all of the bends and folds are completed, the operator removes the workpiece using the overhead crane. See **Figure 2-14**. There are several different bends and folds that must be completed to form the frame for lift truck forks. See **Figure 2-15**.

There is much more to the process than we have been able to show you here. Our goal was to provide an overview of the forming process. Much more detail on how computer software, systems, manufacturing processes, and hardware are used to form metals, plastics, woods, ceramic, and composite materials will be provided later on.

Figure 2-11. 350 ton Cincinnati brake used by Crown Equipment Corporation, Kinston, North Carolina.

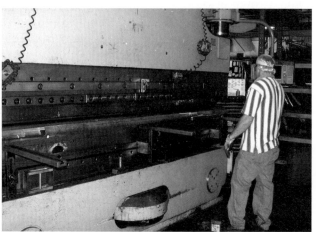

Figure 2-13. Operator inserting workpiece into press brake.

Figure 2-14. Removal of workpiece from press brake.

Figure 2-15. Forks after forming by press brake.

Separating Processes

Separating processes are used to remove material or volume. They can be classified according to the type of chip produced as a cutting tool strikes the material. The force involved is also a determining factor in classifying separating processes. There are three different types of separating processes:

- *Mechanical chip-producing separating.* This process involves wedge-type cutting action with loss of material or volume (Example: sawing).
- *Mechanical non-chip-producing separating.* In this process, there is wedge cutting action with no chip being produced and no loss of volume of the material (Example: shearing).
- *Non-mechanical separating.* This process involves cutting action that may or may not generate a chip, with no force involved (Example: photochemical milling).

In order to better understand what happens when processes are used to separate materials, let us take a moment to showcase a product being made by Phoenix Fabrication in Ayden, North Carolina. We will be focusing on a *mechanical non-chip-producing separating* process.

Company Overview

Phoenix Fabrication, Incorporated in Ayden, North Carolina is a large job shop known for its ability to produce very high tolerance work from the ground up, from design engineering, manufacturing engineering, programming, computer assisted design and manufacturing to automated processing on the plant floor. Phoenix offers a high level of automation to perform laser cutting, punching, machining, welding, and assembly. Phoenix also provides a complete array of processes for machining, welding, and cutting.

Phoenix is a recognized leader in laser cutting of thin and thick materials up to 3/4″ mild steel and 1/2″ stainless steel using 2,500; 3,000; or 4,000 watt lasers. Phoenix also uses three TRUMPF CNC press brakes to perform ACB (Automated Controlled Bending). Phoenix also offers cutting edge processes equipped with quick change tooling systems for punching of 5′ x 10′ sheets of mild steel up to 5/16″ thick. One of the unique features about Phoenix Fabrication is their ability to interface laser cutting with automated punching and bending.

Product Showcase #2				
Company	**Location**	**Internet URL**	**Process Action**	**Product to be Showcased**
Phoenix Fabrication, Inc.	Ayden, North Carolina	www.phoenixfabrication.com	Separating	Stainless steel parts

Figure 2-16. 4′ × 8′ sheet of mild steel being fed automatically onto a conveyor and into an automated punch equipped with automated tool changer.

Separating Process Overview

One of the most impressive systems used by Phoenix is a fully automated TRUMPF punch (Trumatic 500) operating in conjunction with a TRUMPF feeder (Trumalift Sheetmaster) and roller conveying system. This is a good example of the types of process that will be found in factories of the future. When we visited, the machine was running unattended, feeding sheets, punching out parts, removing parts, and changing punches. See **Figure 2-16** and **Figure 2-17**.

This is a separating process that requires mechanical force to drive the punch down into the sheet metal stock. The punch cuts (stamps) out the part. The parts can be entirely separated from the sheet of stock. When the machine picks up the parts and drops them into the appropriate loading bins, the skeleton is removed. What is important to understand about this punching system is that through the use of CAD/CAM (computer assisted design and computer assisted manufacturing) software such as TRUMPF CNC TOPS, maximal part utilization can be calculated on each sheet. Phoenix is normally able to use at least 90 percent of a sheet for parts using state of the art *part nesting* practices. This is a great example of how a company such as Phoenix can outpace the competition. See **Figure 2-18**.

Fabricating

Fabricating processes are those used to join and fasten materials together. There are three major types of fabricating processes: adhesion, cohesion, and mechanical joining.

Figure 2-17. TRUMPF Trumatic 500 press punching out parts automatically. After punching, the machine automatically picks up all the parts and deposits them in proper loading bins. The skeleton that remains after the parts are picked up is removed. What is particularly impressive is the efficient way that Phoenix takes advantage of the maximum amount of stock that is available.

Soldering, brazing, and gluing are fabricating processes involving adhesion. Adhesive joining processes normally take place with two materials being joined through the use of another material. The material that joins the others together creates a bond at the point of attachment. Adhesion processes do not provide the strength attained through cohesive bonding or mechanical joining.

Fabricating processes based on the principle of cohesion create a molecular bond that unites the atoms of the materials being joined. Fusion welding and resistance welding are examples.

In order to better understand what happens when processes are used to fabricate materials, let us take a moment to study a process used by NACCO

Figure 2-18. Stock after parts have been punched out. Note efficient nesting of parts on the sheet, resulting from effective use of CAD/CAM nesting software.

Product Showcase #3				
Company	**Location**	**Internet URL**	**Process Action**	**Product to Be Showcased**
NACCO Materials Handling Group, Incorporated	Greenville, North Carolina	N/A www.nmhg.com	Fabricating	Frames and bodies for lift trucks

Materials Handling Group in Greenville, North Carolina. The process we have selected is used to join two materials together through adhesion.

Company Overview

NACCO Materials Handling Group (NMHG), with corporate headquarters in Portland, Oregon, is one of three core businesses in the diversified holdings of NACCO Industries, Inc. NACCO Materials Handling Group manufacturers Yale and Hyster lift trucks and material handling products at its manufacturing facilities in Greenville, North Carolina. Other manufacturing facilities are located in Berea, Kentucky; Danville, Illinois; Sulligent, Alabama; and Lenoir, North Carolina. NMHG also has manufacturing facilities in Sao Paulo, Brazil; Ramos Arizpe, Mexico; Craigavon, N. Ireland; Irvine, Scotland; Nijmegen, The Netherlands; Masate, Italy; Obu, Japan; Manila, Philippines; and Pudong, China.

NACCO/Greenville is the location for this case study. The Greenville facility is the home of the Americas Division Headquarters staff, manufacturing plant and warehouse engineering group, and marketing. NACCO/Greenville is responsible for the manufacture of all sizes of Hyster and Yale trucks, ranging in capacity from 2,000 to 36,000 pounds. Sitdown rider models propelled by electric, gas, LP-gas, diesel or compressed natural gas are available.

Fabricating Process Overview

We have chosen a fabrication process called MIG (Metal Inert Gas) welding for this case study because of its great popularity in metal product manufacturing. MIG welding is also known in the industry as GMAW (Gas Metal Arc Welding). MIG uses argon, helium, or carbon dioxide gas to reduce contamination and oxidation in the weld area. Then the welder feeds a consumable wire electrode through the welding nozzle. The wire carries an electrical charge and the workpiece is grounded. When the wire is held close to the workpiece, an arc is created and welding begins.

The wire is fed when the operator presses a trigger on the electrode holder or nozzle. A bead of molten metal is created. See **Figure 2-19**.

The fabrication department at NACCO is organized into individual work areas so their welders can work on large pieces without interruption, sometimes in an open area and at other times inside their welding booth. See **Figure 2-20**. Curtains surround the welding area to protect people from being exposed to the intense light created by the spark and flashing of the electrode. Large parts are supported on pallets and carried by overhead conveyors to the welding area.

Fixtures and clamps are used to hold parts together. It is often necessary for the welder to work in confined spaces. Since high-quality weldments are essential, a great deal of skill, patience, and experience is required. Much more detail will be provided on MIG welding in the chapter on fabrication of metal materials. See **Figure 2-21**.

There are often many different welds that must be made on large workpieces such as the components of a lift truck. The completed frame from the welding department is then transported through other departments in the factory. See **Figure 2-22**.

Figure 2-19. MIG welding at NACCO Materials Handling Group.

Figure 2-20. Welder fabricating shell for lift truck at NACCO Materials Handling Group.

Figure 2-21. Fixtures and clamps are used to hold large workpieces together for welding

Conditioning

Conditioning processes are used to change the internal or external properties of materials. Conditioning processes often involve heat, shock, electrical impulses, magnetic methods, chemical action, or mechanical means to change the structural characteristics of the material. Many times, the result of conditioning is not visually apparent. Often, conditioning processes change the arrangement of atoms within the material. Conditioning processes are used to influence the hardness, strength, rigidity, resistance to wear, and fatigue (product life) of a material.

In order to better understand what happens when products are conditioned, we will take a moment to showcase one of the products made at an engineered lumber facility of Louisiana Pacific Company, in Wilmington, North Carolina. The two major products made at LP-Wilmington are *I-Joists* and *Laminated Veneer Lumber (LVL)*. Samples of LVL are shown being transported by quality control. See **Figure 2-23**. We will be showcasing one of their products called Gang-Lam® Laminated Veneer Lumber (LVL) to illustrate a conditioning process used with wood materials.

Company Overview

Louisiana Pacific Company (LP) was founded in 1973 and is headquartered in Portland, Oregon. LP is a leading manufacturer of building materials in North America, with facilities in the United States, Canada, and Chile. LP owns almost a million acres

Figure 2-22. Fabricated part ready to leave welding and move to other work centers at NACCO.

Product Showcase #4				
Company	**Location**	**Internet URL**	**Process Action**	**Product to Be Showcased**
Louisiana Pacific Company	Wilmington, North Carolina	www.lpcorp.com	Conditioning	Engineered Wood/Laminated Veneer Lumber (LVL)

of timberland, predominantly in the southern United States, and has more than 55 manufacturing facilities in North America.

LP manufactures specialty framing products that are distributed to retail, wholesale, homebuilding, and industrial customers.

LP's Gang-Lam® Laminated Veneer Lumber (LVL) is designed to outperform traditional solid wood beams. The products are made from ultrasonically and visually graded veneers patterned to prevent naturally occurring defects from affecting the beams' performance. They are then bonded together under pressure and heat with waterproof adhesives. The process ensures that Gang-Lam® LVL products are exceptionally strong, solid, and straight, making them excellent for most primary load-carrying beam applications. Compared to wood milled directly from trees, Gang-Lam® LVL is available in longer and thicker dimensions, yet it is not made from old-growth trees. (Louisiana Pacific Company; www.lpcorp.com)

Conditioning Process Overview

The raw material for Gang-Lam® LVL is Southern yellow pine, which arrives at the plant in 1/8" thick sheets. After a number of grading processes, the sheets are moved on continuously-moving roller conveyors through many different manufacturing processes. One of the first processes for LVL is called *scarfing*. This is a planing process that produces a true edge on the sheet (separating process). Other processes such as the *clipper composer* (also a separating process) are used offline to recover undersized or cracked sheets so they can be resized and used later for core material. Sheets are then conveyed to the "beam and header" line. A *curtain coater* machine is responsible for edge gluing sheets (finishing and conditioning process). Let us take a closer look at how this product is manufactured.

After the sheets are glued they are transported to the layup station. **Figure 2-24** shows the sheets in place on top of each other at the *layup station*. This is where the process of creating a cohesive bond between the sheets (forming process) takes place. The laminated sandwich of materials is then conveyed to the *prepress* area where they are compressed at 3,500 pounds per square inch (psi).

At this point the product has been laminated, pressed, and sprayed over its entire surface. Next comes the *hotpress*. Here the sandwich is heated and

Figure 2-23. LVL (laminated veneer lumber) being inspected by quality control.

Figure 2-24. Sheets sandwiched together at the layup station.

Figure 2-25. Hot pressing LVL engineered wood products at a pressure of 3,500 psi.

Figure 2-26. The Gang-Lam® Laminated Veneer Lumber (LVL) which has been formed, being conveyed to trimming, QA, and final finishing.

compressed further. At this point, a typical Gang-Lam® LVL product would be 1.75" thick and up to 72' long. The hotpress process takes about twenty minutes. See **Figure 2-25**.

Now that the LVL product has been heated and compressed, the completed engineered wood product is stronger than its natural counterpart and resistant to chemicals, pests, and environmental contaminants. **Figure 2-26** shows the Gang-Lam® LVL, that has been formed, being conveyed to other processes for trimming, edge sealing, quality inspection, testing, and packaging.

There are several conditioning processes that are involved in the manufacture of this product. After the glue was applied and sheets were joined together, they were compressed in the prepress stage and conveyed to the hot press. Both of these processes caused the glue to be distributed throughout the sandwich, which lowered the water content of the product and resulted in a material that was stronger than the original raw materials. The combination of pressure, heat, and cohesive bonding agents (resins) created a new material with changed characteristics.

Finishing Processes

Finishing processes are often used to improve the outward appearance or protect the exterior surface of

a part. In other cases, finishing processes are used to prepare the surface of a workpiece prior to the application of a coating.

There are hundreds of coating processes used in manufacturing. Many of these are unique, ranging from fluidized bed coating of ceramics to spray metallizing of plastic to curtain coating of wood products.

In order to better understand what happens when processes are used to finish materials, we will take a moment to showcase a product being made by The Hammock Source in Greenville, North Carolina.

Company Overview

The Hammock Source, with corporate offices in Greenville, North Carolina, is the parent company and the design and manufacturing center for two companies, Hatteras Hammocks® and The Original Pawleys Island Rope Hammock®. The Hammock Source makes, distributes, and is the world's leading

Product Showcase #5				
Company	**Location**	**Internet URL**	**Process Action**	**Product to Be Showcased**
The Hammock Source	Greenville, North Carolina	www.thehammocksource.com	Finishing	Metal hammock stand

manufacturer of rope hammocks. All of their hammocks are 100 percent inspected, made by hand, and hand assembled.

The Original Pawleys Island Rope Hammock® was created in 1889 at Pawleys Island, one of the oldest summer resorts on the South Carolina coast. Captain Joshua Ward, a riverboat pilot transporting rice and supplies by barge between Georgetown and the great rice plantations of the South Carolina low country, wanted a replacement for the hot grass-filled mattresses on his boat.

He decided to make a cool and comfortable cotton rope hammock for use on his boat. His creation was a hammock made without knots and employed the use of wooden spreaders. This original design proved to be so comfortable that it is still used in the Original Pawleys Island Rope Hammocks® today.

Hatteras Hammocks® were first made in 1971, when the companies founder, Walter Perkins Jr., began selling hammocks that he had purchased on business trips out of the trunk of his car. Mr. Perkins disassembled one of these sought after rope hammocks in his garage and set about improving it.

In 1971, Hatteras Hammocks® was incorporated. By 1987, Hatteras Hammocks® had grown at such a tremendous rate that it became the largest rope hammock manufacturer in the world. Today, Hatteras Hammocks® continues to be the industry leader with continuous additions to its innovative product line that include hammocks, stands, swings, and hammock accessories, as well as state of the art improvements to its manufacturing capabilities.

The Hammock Source uses nearly 31 miles of cotton yarn to make each rope hammock. See **Figure 2-27**. Each strand of rope is made from 35

Figure 2-27. Bobbins supplying yarn to the Twister.

Figure 2-28. Yarn being fed through a template into the Twister.

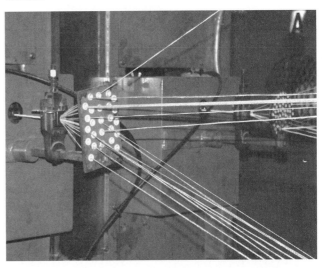

individual strands carried on bobbins to a machine called a *twister*. See **Figure 2-28**. After the 35 strands are twisted into three strands, the three strands are taken to a *layer* machine that makes the rope. See **Figure 2-29**. The layer machine is responsible for wrapping the strands of rope together and winding them onto a roll. See **Figure 2-30**. The rope is then woven into hammocks. An experienced hammock weaver can produce approximately 80 hammocks in a 40-hour workweek.

Finishing Process Overview

The finishing process that we have chosen for this showcase is not involved with rope making. The process we selected is called powder coating. This is

Figure 2-29. Layer machine responsible for making the rope.

Figure 2-30. Winding rope on rolls for use by weavers in many types of hammocks.

an exciting process that is completely automated. It is used by the Hammock Source to coat 12 gage steel hammock stands. See **Figure 2-31**. Power coating is a process that is important to many manufacturers of metal products. According to The Hammock

Figure 2-31. Hammock stand powder coated by The Hammock Source.

Source, in 1999 they used more than 750 miles of steel tubing in the construction of stands for Hatteras hammocks.

The Hammock Source uses a fully automated Nordson powder coating system. There are seven major components in this system: (1) overhead part conveyor, (2) wash and sealing unit, (3) electrostatic control module, (4) powder feeder and delivery pump, (5) spray guns, (6) powder recovery system, and (7) curing oven.

Powder coating is a finishing process where powdered paint is electrostatically charged, converted to a fluid-like state using compressed air, and sprayed onto a grounded part. Paint particles are attracted to the surface of the part through electrostatic attraction. Then the part is conveyed to an oven where heat is used to liquefy and cure the coating on the part. Powder resulting from overspray is recovered through a filtering system in the recovery system and reused.

The major advantage of powder coating over other finishing processes is that approximately 99 percent of the powder that is sprayed is actually deposited on the parts or recovered and reused. There is very little material that is wasted and virtually no contamination or pollution of the environment. Another advantage is that color changes can be made rapidly by changing the cartridge in the powder-feeding unit.

The process begins with the parts of the hammock stand being transported by an overhead conveyor track through a multistage wash, rinse, iron phosphate dip, and sealing unit. When the cleaning and sealing process is completed, the part is then automatically conveyed to the powder coating system. See **Figure 2-32**.

Figure 2-32. Nordson powder coating system.

Before the part enters the powder spray booth, the operator must program the guns so they will spray the correct amount and consistency of powder on the parts. This information is transmitted to the guns through the electrostatic control module (controller). See **Figure 2-33**.

The guns are staggered on the right and left inside the spray booth. In the system used at The Hammock Source, eight guns spray from the right and eight from the left. See **Figure 2-34**.

A critical part of the automated powder coat spray system is the conveying track and mechanism used to carry parts to and from the powder coating booth. An overhead conveyor continuously transports parts through the spray booth and past the guns. The two sets of guns, staggered on both sides of the booth, spray fluidized powder onto the parts. After the parts are sprayed, they are transported on the conveyor to the curing oven.

Many companies are using powder coating systems to provide finishes on their metal parts. Often these systems are not automated, and workers spray the power coat inside of a spray booth. What is most impressive about the operation seen at The Hammock Source is that their entire process is automated. Machines perform low-skilled tasks where the operation could be hazardous to human operators.

This is one example of how a firm can outthink and outpace the competition. The goal is to keep the machine running all the time, performing tasks where a minimal amount of thinking is required to perform the operation. This means that people can then be used in other ways for jobs such as design, programming, and control of the machines.

Figure 2-34. Right-hand bank of eight spray guns inside powder coating booth.

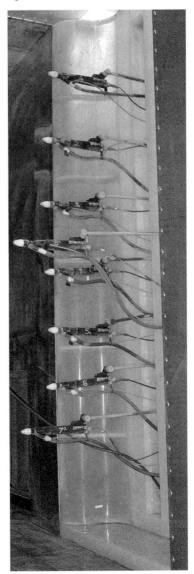

Figure 2-33. Electrostatic spray module (controller), which transmits information to the spray guns located inside the spray booth. Sixteen different guns are controlled by this controller.

Important Terms

conditioning
consumable products
extractive industries
engineered materials
fabricating
finishing
forming
nonconsumable products
primary processing
process action
secondary processing
separating
successive transformation

Questions for Review and Discussion

1. List the five major material families.
2. Would you say that *nonconsumable hard good consumer products* are produced by primary or secondary manufacturing industries? What would be examples of nonconsumable hard good consumer products?
3. What is the difference between primary and secondary manufacturing industries? Pick a raw material you are familiar with and explain where primary manufacturing of the stock ends, and secondary manufacture of the product begins.
4. What are *Process Actions*? List the five process actions, and be able to correctly classify processes someone might select with these actions.
5. What is manufacturing? What is the difference between assembly and manufacturing?

Sheets of .312″ thick steel are formed and welded to produce the fork blades of this lift truck. (Crown Equipment Corporation)

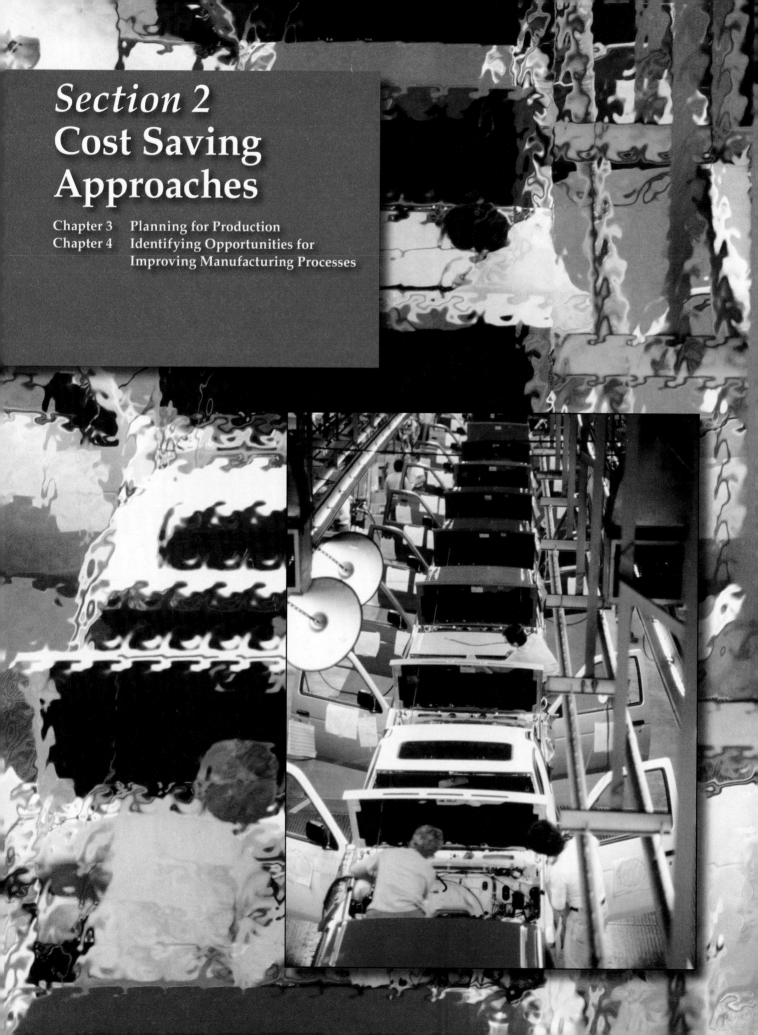

Section 2
Cost Saving Approaches

3 Planning for Production

Chapter Highlights

- The factory system developed in the nineteenth century did not rely on skilled labor.
- The production line created by Henry Ford is one of the most effective manufacturing systems in use.
- Product-oriented manufacturing systems are replacing people-oriented systems.
- Production managers are changing manufacturing systems to manufacture smaller lot sizes with less lead time.
- Push, pull, and kanban are names associated with responsive manufacturing systems.

Introduction

More than 200 years ago, economist Adam Smith in his book *Wealth of Nations* wrote about the possibility for improving manufacturing output through the division of labor. He believed that by breaking up the work to make a product into many simple tasks a company would see an overall increase in productivity.

A few years later (1798), Eli Whitney, the inventor of the cotton gin, demonstrated the value of interchangeable parts in the production of rifles. Today, the interchangeability of parts is an essential, if not obvious, requirement for all manufacturing. However, Whitney's most noteworthy accomplishment was the creation of a basic production system that applied Smith's concept of the division of labor coupled with manufacturing processes that were able to make interchangeable parts.

Whitney put this approach into action when he accepted a contract to manufacture 10,000 muskets. First, he set about to design machinery to produce gun parts that were identical and therefore interchangeable. Next, he organized production so that workers were able to specialize in producing just one part while others assembled the finished product. Whitney made the muskets and demonstrated the worth of a manufacturing system that would be used by industry throughout the twentieth century.

Much of the manufacturing in the nineteenth century, however, did not follow Whitney's approach. It is true that industry recognized the need for interchangeable parts but not the concept of task specialization. During this time, manufacturing was carried out in factories that were organized around trades or crafts. The exceptions were in the textile industry where there had been a breakthrough in mechanizing the manufacture of yarns and cloth.

The textile plants used sophisticated machinery, which dictated how the plant should be laid out. The equipment also changed the nature of the work being done. In general, the capability to do the work existed in the design of equipment meaning that the operators could be unskilled.

Outside the textile industry, most factories were operated as job shops doing a variety of low volume work. Consequently, manufacturing management believed they needed skilled workers who could handle a wide range of jobs. Because of the shortage of skilled workers, output was very limited. Sometimes a company could hire a person that was self-taught but more often employees learned their trade through a long period of on-the-job training. In some trades, there was an apprentice/journeymen system or a trade school, which provided a more comprehensive approach for gaining the necessary job skills. The limited supply of skilled people meant that the nineteenth century approach to manufacturing was not capable of mass production or creating products with consistent quality.

However, at the end of the century, the United States was expanding rapidly and the demand for manufactured products was enormous. The uncertain productivity of the craft-based system of manufacturing together with the shortage of skilled workers meant something had to change.

The Unskilled Factory System

The change in manufacturing was brought about in part by an engineer, Frederick W. Taylor. Taylor was born in 1856 (died 1915) into a wealthy Philadelphia family. He disappointed his parents by pursuing a career in manufacturing. His first factory job was as a machinist in a metal working company, where he later became a foreman. At work, he was astonished by the factory's inefficiency and chaotic method of operation. Later as a mechanical engineer, Taylor wrote in his book (published in 1911) about solutions that he believed would solve both problems. Taylor put Adam Smith's concept on the division of labor together with some of his own ideas and created an approach to manufacturing that he called scientific management. Taylor believed that complicated jobs had to be broken down into a series of simple tasks. Then he would have a person learn just one of these tasks. An inexperienced person could quickly learn how to do the work, which eliminated the need for skilled workers. Taylor, however, wanted to ensure that the task would be done the same way each time. To do this he would study the task to eliminate unnecessary motions or steps. Once the task was defined and the worker was following the correct procedure, Taylor would time it. This procedure became known as *time-and-motion study*. Knowing the time it takes to do a task enabled managers to predict how many parts could be made in a time period and how many people would be needed to make the products at the rate required. Taylor believed this practice would achieve an optimum method of manufacturing and management. Over the next few years, scientific management was adapted to some extent in nearly every industry in the United States. Throughout the twentieth century, manufacturing management used time-and-motion analysis to establish and control the amount of labor needed to make a product. Taylor's approach created a tremendous number of productive jobs that could be filled with unskilled workers, enabling American manufacturers to make the equipment and materials needed by the country.

The Production Line Manufacturing System

One of the most popular and important products of the last century is the automobile. At the beginning of the twentieth century, automobiles were not readily available and the few that were being made were very expensive. The construction of an automobile required scores of skilled people and a great deal of time. Nevertheless, there were hundreds of automobile manufacturers, although most were only assemblers that put together cars using purchased parts. The benefits of being an assembler meant that the company did not have to obtain and manage the skilled people and technologies needed to manufacture the parts. Henry Ford, however, realized that if he were to be just an assembler, most of the costs of manufacturing would be outside his control.

There were other problems, too. Ford found that these suppliers were not very productive and their product's quality was poor. His solution was to develop a manufacturing approach, known as *vertical integration*, where Ford created the parts instead of relying on outside suppliers. This ambitious approach meant that the Ford Company made everything from the steel to the paint used in their automobiles. This created a manufacturing facility that was enormous and nearly impossible to control and coordinate unless some overall manufacturing system could be developed to manage it.

The manufacturing system that Ford set out to put together was unique in that it was naturally organized. His engineers used Taylor's approach to define an individual's work, but they added an important new element. After the tasks needed to make an automobile were established, the engineers then arranged those tasks into a sequence similar to a "bucket brigade" to manufacture the automobile. A further enhancement was to use a conveyor to move the car from task to task. The conveyor served three purposes. First, it was a material handler eliminating the need for a person to push the chassis to the next task or workstation. Second, it established a specific physical destination for components and materials. However, the real benefit was the third purpose— the establishment of a fixed manufacturing rate or *cycle time* to complete a task. The rate of production and the cycle time to complete a task are the basic measurements for determining and maintaining the output of the manufacturing system. This element was missing from manufacturing systems a hundred years ago.

Ford's manufacturing strategy or paradigm focused on productivity and minimizing or eliminating non-value added activities. To achieve the goals of this strategy, Ford started with Whitney's manufacturing system based on the division of labor and the consistent replication of parts and materials—the foundation for every successful manufacturing system. Ford's system, however, was distinctive in the way it controlled the routing and pace of work by using a production line. Although using a production line as a way to manage product flow seems as obvious as interchangeable parts, it was at the time a brilliantly effective system for managing and controlling a factory even by present day standards. Ford's production system produced the Model T automobile—the first affordable mass produced automobile. Many other automobile companies quickly adopted Ford's system and succeeded while companies that did not disappeared.

People-Oriented versus Product-Oriented Manufacturing Systems

Taylor's concepts also enjoyed a great deal of success in other industries. Companies that adopted Taylor's scientific management were now able to expand operations using less costly unskilled workers. Maintaining product flow, however, was not recognized as essential. Manufacturing management in most factories seemed to focus only on the labor cost in the product. The objective was to minimize the time it took to complete the task, which is a good start. But, the time and motion study people did not consider any other aspect of manufacturing beyond reducing the cycle time for the task. When workers completed the task, they placed their completed work aside in a box (tote pan) or on a pallet. Keeping workers busy was the focus of management. As a result, companies that emphasized reducing cycle times and keeping workers busy created a buildup of materials on the factory floor. This material is called **Work In Process (WIP)**. If you visited companies that followed this approach, you would see factory aisles and work areas that would be filled with tote pans or pallets of WIP. Every manufacturing system has to have some WIP, but if the part or assembly is not being worked on, having value added, then it becomes *excess WIP*.

Incentive systems were a practice that created tremendous amounts of excess WIP. These systems paid individual workers on the amount they produced (piece rate) or gave them a bonus that was proportional to the amount of product produced over a base level. The base level would be the "established standard" or cycle time to complete a task. If you as the operator could complete the task in less time than the base time, you earned a bonus. In theory, this would seem to make the operation more productive. However, its principal effect was to increase excess WIP.

It is important to recognize why excess WIP is a problem. An example will illustrate the problem excess WIP can cause. **Figure 3-1** lists the three tasks used to make a simple product—a painted part with a spot-welded tab.

In this example, all the cycle times are equal so in theory the product can move through the manufacturing sequence immediately. If an order is received for one part and all materials are available, the *theoretical lead time*—the minimum time that will elapse from the moment the order for *one part* is received to the time it is completed and ready to ship—is the sum of the cycle times.

Theoretical lead time = Sum of cycle times
for all tasks

Theoretical lead time = (.00278 hours)
+ (.00278 hours)
+ (.00278 hours)

Theoretical lead time = .00834 hours

If we were building these parts continuously, each workstation would have to be staffed to complete the specific task. When the first task is done, the part would move to the second station. The only delay in the work would be the transportation time, however, in this example we will assume it is zero time. When Task 3 (painting) is completed, the part is ready to ship. The person doing the painting will then immediately begin painting the next part, which will be done in .00278 hours. Therefore, the manufacturing cycle time is .00278 hours. That means every .00278 hours another piece is ready to ship. The production rate is the reciprocal of the cycle time.

Production rate = 1 ÷ cycle time
Production rate = 1 ÷ .00278 = 360 parts
per hour

It would be ideal if all three operations could be operated as previously described. There would not be any excess WIP. In practice however, perfectly balanced manufacturing operations do not exist. Also, organizations that focus on keeping people busy and reducing individual operation cycle times will generate excess WIP. Consider what happens to the system when the cycle time for one task is reduced. Suppose that a time study and work methods analyst discovers a change in the fixture that holds the tab in place during the spot welding operation, Task 2, will reduce the cycle time to .00222 hours (8 seconds). That means that Task 2 will increase production from 360 pieces per hour (1/.00278 hours) to 450 pieces per hour (1/.00222 hours).

What does this improvement in cycle time accomplish? The answer is it will increase excess WIP unless we can find something for the operator to do with this "spare time." The press operation is still producing at .00278 hours; therefore, the spot welder will have to wait for .00056 hours after each tab is welded. "Labor focused" management sees this as a loss of labor or inefficiency. If the "spare time" in the spot welding operation is going to be used, the operation needs to be "decoupled." When a task is *decoupled*, it becomes an independent activity. However, decoupling will generate excess WIP.

To illustrate the implications of decoupling, we need to recap what will happen.

Task 1 (the stamping task) produces 1 part every .00278 hours or 360 pieces per hour. During an eight-hour shift, 2880 pieces will be stamped out (8 hours × 360 pieces per hour).

Task 2 (spot welding) produces 1 part every .00222 hours or 450 pieces per hour. During an eight-hour shift, 3600 pieces will be spot welded (8 hours × 450 pieces per hour).

During an eight-hour shift, spot welding will out-produce the stamping operation by 720 pieces (3600 pieces − 2880 pieces). One approach that a manufacturing manager might take is to have the

Figure 3-1. Tasks and cycle times to make a part.

Task ID	Task	Standard Cycle Time	
		Seconds	Hours
1	Stamp out the part in a press	10	.00278
2	Spot-weld a tab on the part	10	.00278
3	Spray paint the assembly	10	.00278

stamping operation begin work two hours earlier (720 pieces ÷ 360 pieces per hour) to build up the extra parts the tab welder needs to run eight hours. The extra parts would become excess WIP that is needed to keep the tab welder busy for the entire shift. The tab welder could also start later because only 6.4 hours (2880 pieces per shift ÷ 450 pieces per hour = 6.4 hours) is needed to process the 2880 pieces that the stamping operation produces. This approach would create excess WIP as well between operations. It becomes apparent that each operation needs to be decoupled and a buffer of excess WIP will need to be maintained between each operation.

This emphasis on keeping the worker busy can also mean keeping machines busy, which will provide the same result, excess WIP. Manufacturing management during the last 50 years was focused on machine utilization and efficiency. Goldratt and Cox in their novel *The Goal* describe the problems that occur when manufacturers define the effectiveness of a manufacturing system in terms of machine/worker utilization or efficiency.

The amount of WIP that exists day to day is not the problem, but rather the symptom of the problem that the manufacturing management needs to solve. So, if keeping people and machines busy producing is not the way to run a manufacturing plant, then how should it be run?

Product Flow Manufacturing Systems

Ford's production line dealt with a single product. Every automobile produced on the production line was one color, black, and there were no optional "packages." Today, automotive production lines have to accommodate a wide variety of options and even different models of automobiles. Variety in product configurations—from automobiles to computers—is now the rule and not the exception. However, the manufacturing system developed by Henry Ford's engineers and technicians had another very significant attribute—it was product flow oriented.

A manufacturing manager who is product flow oriented is committed to keeping products moving toward being completed and shipped. A symptom of a manufacturing system that is not moving product is excess WIP. Parts, assemblies, and partially completed products are examples of WIP and should ideally be in a workstation being worked on or arriving *just in time* (**JIT**) to be worked on. The

term *just in time* has become synonymous with product flow manufacturing systems. A manufacturing system that is operating just in time will not have any excess WIP.

In the 1980s, manufacturing management became very enthusiastic about JIT but generally saw it as only an inventory-reduction concept. Those companies that tried to implement JIT found that they could no longer operate machines and workstations as decoupled activities. Once the operations and processes are tied together, workers or machines might be idled, waiting for product to flow into the workstation. Managers now had to concentrate on keeping product moving as opposed to keeping people and machines busy. This change was hard for some manufacturing managers to accept, therefore, many organizations were not able to implement the philosophy of product flow manufacturing.

Those companies that were able to institute the concept of product flow manufacturing found that the problems that affect productivity were more apparent—easier to identify with the excess WIP stripped away. With the production operations no longer decoupled, problems that obstructed product flow caused a bottleneck or stopped production since there is no excess WIP to keep downstream workstations busy. This interruption in production is actually good if management is committed to product flow, since the interruption immediately focuses everyone's attention on the machine or workstation causing the obstruction. It is extremely important for the organization to be able to prevent the problem from recurring. If the problem remains unresolved, excess WIP will creep back into the manufacturing system, destroying the product flow approach to manufacturing.

Problems that cause product flow obstructions fall into one or more of the following five categories.

- Quality problem
- Machine breakdown
- Setup or changeover downtime
- Operator problems or unplanned absenteeism
- Materials unavailable

What happens if a problem remains unresolved? In nearly every problem type, there is an increase in excess WIP. As indicated earlier, an increase in excess WIP is a symptom that a problem exists, but the increase is not the problem. The example described earlier will help illustrate how these problems impact WIP and what has to be done to resolve each type of problem.

- **Quality problem**—5 percent of the parts made by the stamping operation are defective.
  - **Symptom**—To overcome the shortfall created by the defects, 105 parts are made for every 100 parts needed and raw material orders are increased by 5 percent. Since the press makes 360 pieces per hour or 2880 pieces in an eight-hour shift, the 5 percent overbuild creates 144 parts of excess WIP. This overbuild has to be maintained since occurrence of a chronic problem is not predictable.
  - **Solution**—Institute a quality improvement process in the organization to eliminate chronic quality problems such as stamping defects running at 5 percent.
- **Machine breakdown**—The press that stamps the part is out of production on an average of .5 hours per eight-hour shift.
  - **Symptom**—To prevent a shortage of parts the press is operated .5 hours on overtime each day to maintain a safety stock of 180 parts (.5 hours × 360 parts per hour).
  - **Solution**—Establish a preventive or predictive maintenance program to eliminate the chronic breakdown problem.
- **Setup or changeover downtime**—The press is down for over an hour each time a different model or style must be produced.
  - **Symptom**—Every time the press is down for a die change, the flow of parts to tab welding is stopped. Overtime is used to create a buffer of excess WIP to keep product flowing to the tab welding and painting operations.
  - **Solution**—Establish a Single Minute Exchange of Dies (SMED) program. This is a team approach to reduce die changeover times.
- **Operator problem**—The operator cannot maintain the established cycle time or consistently produce good parts.
  - **Symptom**—A bottleneck along with parts requiring repair/rework is created by the lower than expected output. Excess WIP has to be generated to cover longer cycle times and the bad parts produced. This WIP is called a *buffer stock*.
  - **Solution**—Provide the operator with the knowledge, support, authority, and responsibility to produce good parts.
- **Delivery of required materials is late**—Suppliers of raw materials and suppliers of components cannot be relied on to deliver materials when promised.
  - **Symptom**—Materials are scheduled to be delivered well in advance of when they are needed, thereby increasing raw material inventory levels. Also, jobs that are not due to be shipped will often be substituted to keep the production facility busy creating excess WIP and an increase in the finished goods inventory.
  - **Solution**—Choose suppliers on their ability to respond on time and provide quality materials. The supplier must be as reliable as any department in the company. Part price is not the only criteria for selecting a supplier.

A company that expects to compete effectively in their market cannot afford to pay for the waste generated by these five problem areas. These problems, if unresolved, will prevent a manufacturing organization from efficiently using the labor, materials, and facilities to create a competitive product. When these problem areas are being successfully addressed, the excess WIP "melts" away creating a *lean manufacturing organization*.

The Basis for Lean Manufacturing

Most manufacturing does not lend itself to Henry Ford's production line factory. The variety of products is greater and the volume will vary substantially from month to month. Consequently, the paced flow and structured order of the production line just does not provide the needed flexibility. Nevertheless, competitive manufacturers are able to operate product flow manufacturing systems without using production lines. However, in order to keep product moving, successful companies need to address five fundamental facets of manufacturing. These facets provide the basis for the solutions to the problems that were introduced in the previous section. As you recall, the problems all had a common symptom, excess WIP. However, this symptom does indicate the magnitude of the cost that these problems can generate. Each of these five problems left uncontrolled can stop the flow of product and cause tremendous amounts of waste. For most companies the most difficult problem area to resolve is product quality.

Quality

Quality programs in lean manufacturing are based on a systematic approach to eliminate and prevent waste (non-value-added activities) in the production of product. Prime examples of non-value-added activities are delays and repair/rework due to defective materials or manufacturing. It surprises some managers when they realize that inspection of parts and products is a non-value-added activity. Inspection does not change or improve the function of the product. In fact, the need for inspection is a symptom of a wasteful manufacturing system that is not effective in producing a quality product. In contrast, any activity that increases the market value (something the customer is willing to pay for) is part of the quality system.

So, if inspection is not involved in the creation of quality, how does a lean manufacturing system establish product quality? To begin with, quality has to be designed into the product. The manufacturing process does not determine quality. Product quality is based on a robust design that ensures the product is fit for its intended use in the hands of the customer. This responsibility rests entirely with the designers. There are four quality characteristics that a designer/engineer must provide in a product to achieve this *fitness for use* criteria. These characteristics are:

- **Appropriate for use**—the product design is suitable for its application and conditions of use.
- **Safe for use**—the product will not harm the persons using it or others by its use.
- **Reliability in use**—the product will perform, as intended throughout its expected life.
- **Value for use**—the product is perceived to be worth its cost.

In the marketplace, quality products that meet these characteristics have a reputation for doing what they are supposed to do safely, reliably, while still being affordable. These products offer good value for the money. Initially the Model T built on Ford's production line met these characteristics. However, as automobile technology improved and customer needs changed, the Model T no longer met these characteristics and, therefore, had to be replaced with another design. Therefore, a manufacturing system that is flawless (waste free) in replicating a design will not be able to make a quality product if the product design does not meet these four quality characteristics.

Then what is manufacturing's responsibility for quality? It has to create a production system that can replicate the product design. A lean manufacturing system replicates the design with minimal waste. Once again, the engineers/designers have a major role in ensuring that a product can be successfully manufactured. This activity is called "designing for manufacture." The equipment and processes selected to make the product must be robust and appropriate for manufacturing the product. In regard to manufacturing, system robustness means that the system can tolerate the natural variability of the manufacturing environment. The manufacturing equipment and processes must also meet the safety, value, and reliability characteristics as well. As you can see, the characteristics for the quality of the manufacturing system are the same as they are for the product. In practice, this means that any properly trained operators using appropriate materials and tools/equipment will be able to complete the tasks necessary to manufacture products that replicate the product design.

The objective is to put together a lean manufacturing system that can make a product that meets the criteria established by the product engineers/designers over and over again. Consistent replication of the product design characteristics is the goal of the manufacturing system. A key method used to ensure that each element (machine, process, or task) of the manufacturing system is able to replicate the intended design characteristics is the application of a technique called *Statistical Process Control (SPC)*. The technique is based on a sequence of steps that, when completed, provides manufacturers with a simple but effective way to confirm that an activity is under control.

There are three basic steps to institute SPC. They are:

1. **Establish control**—Bring the process or activity into control. Make it capable of replicating an acceptable product or task. This means the process or operation is predictable.
2. Establish a means to **monitor the activity**—Create a simple visual method to indicate and track the performance of the activity. Then learn how to recognize when the activity is not performing normally.
3. Provide for **problem solving**—Tsrain the manufacturing personnel to be effective problem solvers so that they can bring the activity back into control if there are indications that the process is not able to replicate satisfactorily.

To help illustrate how SPC works, we will follow a simple product through a manufacturing process. The product is a 2″ square metal plate in which a .125″ diameter hole must be drilled exactly in the center. The manufacturing process requires a drill press with a fixture attached to the table to properly locate the metal plate. **Figure 3-2** shows the plate and the two important dimensions (parameters) that will be controlled. The operator knows that drilling the hole at the specified location (replicating these two dimensions) is essential to the quality imparted by this operation.

The first step in establishing SPC is to determine if the drilling process is capable of meeting the parameters established by the design engineers. To do this, the operator will do a preproduction run of about thirty parts. This will provide the data to determine the capability of the drilling process and will answer two very important questions for the operator.

1. What is the nature of the variation in these dimensions—is the variation special or is it based on common causes?
2. How much variation is there?

To answer the first question, the operator is going to record the data and then construct two "run charts" to determine if there are instances of special (sporadic) variation, **Figure 3-3**. The data plotted is the distance between the center of the hole and a side. The dimensions are labeled (A) and (B). If special variation exists, the operator will investigate to determine why. In our example, a burr was present on the bottom side of the fifth metal plate drilled. This prevented the plate from being located properly in the fixture and was the reason or special

Figure 3-3. Run charts for the drilling operation.

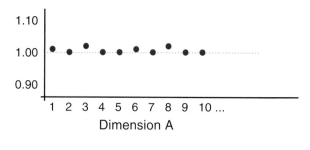

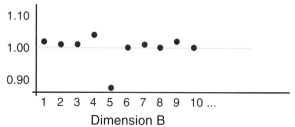

cause (dimension B being extremely small) for the unusually high variation. All the measurements for dimensions A and B are listed in Figure 3-3.

Figure 3-4 has only the first 10 measurements plotted. However, it is apparent that, other than measurement B5, the variability seems consistent. It appears that the average dimension for A and B will be higher than the 1.00″ that is specified. Some adjustment to the fixture will need to be made to shift the average down to meet the target specified.

Figure 3-2. Drawing showing the location of a drilled hole.

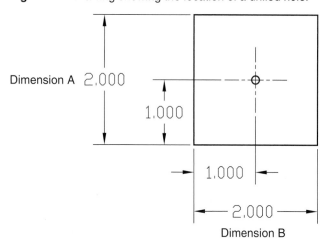

Dimension A   2.000

1.000

1.000

2.000

Dimension B

Figure 3-4. Measurements for dimensions A and B.

Plate Number	A	B	Plate Number	A	B
1	1.01	1.02	16	1.01	1.00
2	1.00	1.01	17	1.00	1.01
3	1.02	1.01	18	1.01	1.00
4	1.00	1.03	19	1.00	1.00
5	1.00	0.89	20	1.02	1.00
6	1.01	1.00	21	1.00	1.01
7	1.00	1.01	22	1.00	1.00
8	1.02	1.00	23	1.01	1.01
9	1.00	1.02	24	1.00	1.00
10	1.00	1.00	25	1.00	1.00
11	1.01	1.00	26	1.03	1.01
12	1.00	1.01	27	1.00	1.00
13	1.02	1.00	28	1.01	1.00
14	1.00	1.00	29	1.00	1.00
15	1.02	1.03	30	1.00	1.01

Once the thirty points are plotted for A and B, the next step will be to compute the average and the standard deviation for each dimension. Item 5B will not be included in the calculations for dimension B since the cause is known and steps can be taken to prevent it from reoccurring. The average and standard deviation will be used to compute the process capability to meet the design specifications based on the present equipment, tooling, and operator skill.

It is important to recognize that designers generally provide some provision for variation. This is the "tolerance" or the acceptable variability from the desired design parameter that the engineer believes can be tolerated without degrading the parameter's fitness for use. The manufacturing process variation, however, must be less than this tolerance if the process is to be considered capable. One of the most commonly used measures of process capability is the ratio obtained by dividing the total design tolerance by the total process variation (typically considered to be 6 standard deviations). A process is considered capable if the ratio is larger than one. Most textbooks on quality and statistical quality control can provide you with the formulas and procedures for calculating the common capability indices. The objective, regardless of the formula used, is to make sure that the manufacturing process has less variation than the design tolerance. Therefore, the operator has to ensure that the drilling process:

- Is subject only to common cause (natural) variation.
- Has total variation less than the total tolerance, thus indicating that the process is capable.

With these two points established, the operator could say that the process can be operated in control and is able to replicate parts according to the design.

The next step makes use of a technique that provides a quick visual method to portray how the process is performing based on the key parameters being tracked. The technique uses charts similar to the run charts in **Figure 3-5**. These charts, however, incorporate guidelines—a centerline and control limits to enable the operator to monitor and interpret if the process is in control. An experienced operator can detect patterns indicating that a problem is developing even before defective parts are actually being made.

SPC charting is based on taking a relatively small sample and then calculating the average value of the parameter being monitored and plotting that average on a chart called an X-bar chart. Next, the operator

Figure 3-5. X-bar and R charts for dimension A.

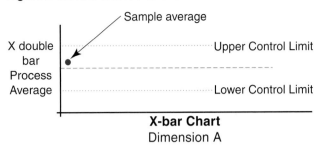

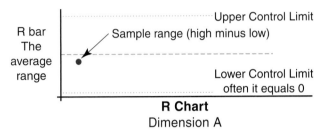

subtracts the lowest value in the sample from the highest value to get a range—an indication of variability. This range is plotted on a second chart—the R chart. Figure 3-5 shows an example of what these two charts might look like for our example. The plotted points will develop a pattern, which must be interpreted by the operator. Patterns that indicate trends other than normal variation can mean that the process is changing, going out of control, which will require action to bring it back into control.

The charting either confirms that the process is in control or it provides a warning that action is needed to make some corrections. It also serves to document how the activity has performed over time. In practice, the charting's major benefit is to provide an early warning to take action to prevent defects from occurring and, therefore, eliminate the need for inspection.

The third step in establishing an SPC program is to create a workforce that is able to quickly and effectively solve problems and take corrective action to bring a process back into control. Some of the commonly used techniques are:

- **Cause & Effect Diagram** (Isakowa Diagram or Fishbone Diagram)—A tool used to analyze all factors (machine, material, operator, design, and environment) that can contribute to a given problem (effect) by breaking down main causes into smaller and smaller subcauses to discover the root cause(s) and not just symptoms of a problem or condition.

- **The 5 Whys**—Used with the cause and effect diagram, a person repeatedly asks the question "Why?" (five is a good rule of thumb) five times in order to get to the root cause(s) of the problem. In this way one can peel away the layers of symptoms, which then leads to the root cause of a problem.

- **Pareto Analysis**—This is typically a bar chart displaying, in descending order, the most frequent to the least frequent problems. This allows the analyst to identify the most important. The concept is based on the work of Vilfredo Pareto, a turn-of-the-century Italian economist who studied the distributions of wealth in countries. He concluded that a minority of approximately 20 percent of the people controlled the majority of 80 percent of a society's wealth. In the *Quality Handbook*, J. M. Juran noted that this 80/20 rule seemed to occur in manufacturing defects—80 percent of the defects are produced by only 20 percent of the problems. The Pareto principle directs one to work on the significant few problems—the 20 percent that cause 80 percent of the defects.

The ability of operators to resolve problems and take corrective action is a major part of the SPC technique. The three techniques listed provide the basics for a systematic approach for identifying the source of the problem. However, this approach requires a knowledgeable and responsible workforce. Each worker on the plant floor needs to be an effective manager even if their management responsibility only covers his or her own actions.

Reliability

Delays and interruptions due to machines malfunctioning and tools breaking are common reasons for lower than expected production output or missing a shipping schedule. To control these problems, manufacturing organizations have maintenance departments ready to respond to these unplanned interruptions of work. This approach is called *breakdown* or *reactive maintenance*. To minimize the length of time these disruptions cause, the maintenance departments stock spare parts and production departments create "safety stocks" (excess WIP) of critical production materials. In addition, production departments frequently use overtime in an effort to "catch up" once the repaired machine is back in production. In general, this approach to maintenance is a "cost adder" to the manufacturing system. Sometimes manufacturing managers, in an effort to reduce maintenance costs, will gamble by reducing the number of maintenance personnel and the amount of spare parts kept on hand for the production equipment. The success of this approach depends on nothing other than the manager's luck.

In 1950, some Japanese engineers popularized what should have been an obvious concept. They instituted a maintenance program based on following the manufacturers' recommendations about the care that should be given to production equipment, known as *preventive maintenance*. This method had an unusual feature because it required operators to take an active role in maintaining their equipment. The maintenance function and the people operating the equipment worked in partnership to prevent equipment breakdowns. As a result, plant managers were encouraged to have their supervisors, mechanics, electricians, and other specialists develop programs for cleaning, lubricating, and servicing equipment. Production workers were trained to avoid practices that would damage equipment and to recognize signs that indicated a potential problem. Although this helped to reduce downtime, it did not eliminate all unforeseen breakdowns. Unfortunately, many of the preventive maintenance activities suggested by the equipment manufacturers were based on "good intentions" that had little to do with minimizing the effect of wear caused by day-to-day production.

During the 1960s, a more effective approach was developed called productive maintenance. This method also involved the maintenance department and the production operators, but it brought in another group—manufacturing engineers and technicians. These individuals were assigned to departments called Plant Engineering or Production Engineering. The tasks that these engineers/technicians performed included quantifying the reliability of the key elements in the equipment and determining the element's mode of failure. *Productive maintenance* focuses on the reliability and design of the equipment as it is used in production. With this information, engineering would establish a preventive maintenance schedule to service equipment based on the probability of failure under actual production stress. This approach went a long way toward reducing unexpected breakdowns and eliminated many unnecessary preventive maintenance activities.

The benefits of productive maintenance have made most modern production equipment more reliable and serviceable even as equipment has become more complicated. The global economy, however, has created a more competitive environment that has raised the standard for acceptable performance of production equipment. These changes require a different approach that has been called *total productive maintenance* or *total participation maintenance*.

Total participation maintenance required a change in the attitude of all the members of the manufacturing organization to accept responsibility for the upkeep of equipment. In a lean manufacturing system, maintaining product flow is the goal. Keeping people or equipment busy is not the objective. Therefore, eliminating obstructions to product flow is a major part of a manager's strategy. When a machine is needed to maintain product flow and a person is not involved in supporting product flow, an opportunity is created for "participation maintenance." In practical terms, this means that a person who has been trained (possesses operator self-control) to operate a machine can and should carry out preventive maintenance and diagnostic tasks on the equipment. One of the most effective diagnostic activities is cleaning equipment. This activity frequently detects symptoms of problems that could cause a breakdown. Oil leaks, loose bolts, and partially clogged filters are examples of what can be found while cleaning a piece of equipment.

Involving people who are not familiar with the equipment and its operation can be beneficial, too. Part suppliers can often provide useful suggestions when they see how their product is used in the manufacturing process. They might be able to suggest a change in packaging that could reduce the time it takes to load a machine, thereby reducing downtime. The machine's builders should see the machine in operation. They may see ways to improve the piece of equipment that may not occur to the people who live with the machine on a daily basis.

The objective of all maintenance programs is to reduce waste caused by interruptions to product flow. Therefore, these programs are not cost-adding activities. However, in the years to come, new approaches to maintenance will be developed and put into practice since, like all aspects of lean manufacturing, the methods being used are continually evolving.

Setup Reduction

Eliminating interruptions due to breakdowns is crucial in maintaining product flow. However, downtime caused by setting up equipment to run a different part or style of part is also an interruption that must be eliminated. Setup downtime is no different than the time lost due to a machinery breakdown. Therefore, in lean manufacturing the objective is to reduce and eventually eliminate setups as a reason for product flow interruptions.

To achieve this objective, manufacturing managers have grouped like parts together to increase the length of a production run. This does not reduce the time required for a setup, it merely reduces the frequency of setups. Unfortunately, this approach results in an increase in WIP and frequently extends the lead-time for a customer's order. This is because the customer's order has to move through the production process with other orders of similar parts.

An example helps to illustrate the detrimental effect that setup times have on WIP and responsiveness. We will look at a fictitious company called Atypical Press, Inc. that has a manufacturing process consisting of just one manufacturing operation. Atypical Press, Inc. stamps out parts for dozens of customers. The details on how the company operates are listed below.

Operation Details
- The company has one eight-hour shift that runs five days a week.
- The only equipment the company has is one press.
- The press can make 1000 parts per hour.
- It takes 2 hours to do a setup (switch dies) in this press.
- Customer demand requires that the press must operate 7 hours a day to complete 7000 parts. To ensure that the 7 hours of operation time is available, the company allows an hour of overtime to get the extra hour required to complete a setup. As a result, the press makes 7000 parts between setups, thus the manufacturing lot size is 7000 parts.
- Every customer order is a unique part requiring a setup. Each of the company's customers usually orders 2000 pieces a month (8 out of 10 orders are for 2000 pieces).

Since each customer requires only 2000 parts per month and the company always makes 7000 parts at a time (the minimum lot size), 5000 excess parts are produced. This amounts to 2.5 months of customer requirements that are stored as finished goods (excess WIP). This inventory is due entirely to the manufacturing lot size dictated by the 2-hour setup time. This manufacturing lot size also means that only one customer is served per shift, thereby limiting production flexibility. Setup time, therefore, is dictating the manufacturing lot size, which generates the amount of WIP and the amount of finished goods inventory that the company must carry to meet customer requirements. The excess WIP and finished goods caused by the setup time are all forms of waste.

How much should the setup time be reduced? Generally, the setup time should be small enough to enable the manufacturing lot size to equal the quantity the customer wants shipped as an order. One "rule of thumb" for establishing the manufacturing lot size is based on the Pareto principle, the 80/20 rule. The application of this rule means that the manufacturing lot size should be equal to or smaller than the quantity required by 80 percent of the customer's orders.

In this example, 8 out of 10 (80 percent) of the orders are for a quantity of 2000 pieces. Therefore, if the setup time could be reduced from 2 hours to twenty minutes, the one-hour of available time during the normal shift could be split up through the day to do three setups. The manufacturing lot size then would drop to 2333 pieces.

$$2333 \text{ parts} = (7 \text{ hours/day} \times 1000 \text{ parts/hour})$$
$$\div 3 \text{ setups per day}$$

Adding an additional 20-minute setup on overtime means that the manufacturing lot size could then be reduced to 1750 pieces, which is well under the average order size in the example. The need to inventory finished parts and generate excess WIP would be eliminated. This means the company will be more responsive. Consider what happens when a customer reduces their ordering frequency. With a 7000 piece manufacturing lot size, the company would have to hold on to the finished goods longer. If the customer needs an order quickly, the smaller lot size (1750 pieces) would allow the company to fit it into a production schedule because it does not have to build excess parts.

How can Atypical Press, Inc. reduce its setup time? One way to do this is to buy an additional press. This means that one press could be running while the other is being set up to run the next order. This

would work and the cost to reduce the manufacturing lot size is the price of a second press. However, there are other ways to reduce setup times that are less expensive than purchasing extra equipment.

Successful companies have organized themselves to specifically address the task of reducing setup time. They approach setup reduction the same way a NASCAR team views the time it takes to do a pit stop. It is a team task, special tools are needed, and each person involved has specific responsibilities. For Atypical Press, Inc. the steps are:

1. Form a setup team, which will include the machine operators, setup people, and plant/ production engineering.
2. Establish baselines.
   - Determine how long it takes to complete a setup now.
   - Calculate the target manufacturing lot size. This is based on customer order quantities.
   - Calculate the number of setups per shift that will be needed to meet the target manufacturing lot size.
   - Determine the target setup time. Divide the available free machine time during the shift by the number of setups per shift to get the target setup time.
3. Video record the current setup process so that it can be analyzed.
4. Separate external/internal setup tasks—Those setup tasks that can be accomplished while the machine is still operating are external tasks and do not add to the downtime caused by the internal tasks. Take advantage of this difference and focus on reducing or eliminating the tasks that keep the machine from producing product.
5. Involve the setup reduction team in developing an innovative solution to achieve the target setup time.

The money invested in reducing setup times is aimed at improving product flow. However, it also has a major effect on making the manufacturing process more responsive. During the 1980s and 90s, managers talked about manufacturing systems that could handle a lot size of one. Setups and changeovers could be done so quickly that model and style changes could be done with ease and would not impede the flow of product. Flow manufacturing no longer requires the ridged inflexibility of the Ford Model T assembly line from the beginning of the twentieth century.

Operator Ability

As we have seen, lean manufacturing requires operators to be involved in a wide range of activities beyond their assigned production work. These activities include monitoring the performance of their equipment to ensure that it is in control, participating in the maintenance of the equipment, and working to reduce setup times. The knowledge and skills needed by workers to maintain continuous product flow is far more extensive than the vision Frederick Taylor had of factory labor one hundred years ago.

Today each person involved in manufacturing has management responsibilities for their own activities. Each man and woman in the manufacturing process is responsible for creating and maintaining defect-free product flow. This form of individual responsibility has had several names over the years. It has been called **Operator Self-Control** or **Jidoka**, which in Japanese is loosely interpreted as autonomous defect control. Regardless of what it is called, ensuring that each person accepts this responsibility is an essential building block in creating a lean manufacturing environment.

Putting this individual management responsibility (operator self-control) into practice is very straightforward. The premise is that each individual is a manager and must be able to effectively manage his or her own tasks. Therefore, to carry out the task(s) assigned, each person must possess the knowledge/skill to:

- Do the task correctly.
- Recognize that the materials being used meet the required standards.
- Determine that the tools, equipment, and other materials required to complete the task are in proper working order.
- Recognize if the task is not being done as it should be and be willing to take corrective action.
- Obtain help to solve the problem when it is apparent that the problem cannot be corrected with the resources at hand.
- Never let a problem/defect move on in the manufacturing process.

If everyone in an organization operates under this concept, the basic management system is in place for lean manufacturing. With quality materials, reliable/flexible equipment, and capable people, a manufacturing facility is almost ready to produce product. What remains to be put into operation is the *system* that organizes the way the facility will operate. The system can be classified as either push or pull.

Manufacturing System

The United States has been a leader in developing production-planning systems. These systems evaluate the resources (material and manufacturing capacity) when an order is placed, then creates a plan detailing when and how the job should be built. These activities are usually carried out by a production planning or production control department. They try to schedule the work to meet the customers requested delivery date(s), check for materials on hand, and ask purchasing to buy the materials needed. When all the materials are available, the job is released to the factory for production. In factories that are mass-producing a product, production-planning departments regulate and monitor the work being done.

With the arrival of computers, software programs were written to carry out most of the detail work needed to track, purchase, and release materials. The work carried out by this software became known as **material requirements planning (MRP)**. MRP is a popular way to schedule and organize production and is widely used by American industry.

The objective of computer-based MRP is to schedule operations in a way that minimizes raw material inventory and excess WIP. As computer capability increased and hardware costs decreased, the MRP based systems became more sophisticated and expanded to control more aspects of the manufacturing operation. The MRP II software was designed to be a manufacturing resources planning system. The programs in this system attempt to coordinate and schedule production departments based on customer demand and the availability of material. Reports from material suppliers on the anticipated delivery of orders along with production statistics are fed into the computer to establish and update production schedules.

However, the production schedules, at best, are an educated estimate of what is available and when a customer's order will be shipped. For critical orders, production control managers, manufacturing department supervisors, and sales managers rely on a group of people (expeditors) to track, confirm, and work out problems. Many plants have teams or groups that meet regularly to review schedules to work out errors and omissions for orders that are not handled properly. This is due in part because MRP systems are not able to accurately model complex production systems. The reasons are many but there are a few common causes.

- The information being reported to the computer system does not represent what is happening in real time. Consequently, the computer-generated reports or the status being reported is not up to date.

- Other than production line regulated manufacturing, few companies have good estimates of how much time it takes to complete the work to be done on a manufacturing lot (order quantity going though the plant as a group). The plans and the shipment dates generated by the computer-based scheduling systems are generally optimistic. The result is that the shipping date originally given to the customer is revised one or more times or includes a cushion of several days or weeks.

- Delays due to breakdowns, absences, and unforeseen setups cannot be forecasted for specific orders by the computer system. Therefore, supervisors view the computer-generated schedules and plans as wish lists or goals.

Companies that have complex or varied product lines can benefit from using sophisticated computer-based planning and control systems to schedule materials and to make shipping date commitments to customers. However, MRP based systems are extremely vulnerable to waste. Companies that did not address these problems found that competition began to take business away from them because their product was not always shipped on time. Competitors that had reduced or eliminated the sources of waste were also able to sell the product for less and with fewer quality problems.

Push and Pull Systems

The terms *push* and *pull* describe how product moves through a factory. When an order is received, a **push system** (an MRP system) begins by ordering materials and then establishing a start date to begin production based on when the materials will be available. Once materials are available, the push system schedules the order through the plant from the starting operation to the final shipping point. Push systems "push" the product through the factory.

Pull systems, however, try to respond immediately by shipping the order when it is received. Shipping the order then creates a demand that moves up the production line to replace what has just been shipped. It is apparent that for a pull system to be responsive, work in process must be available at every operation involved in the production of the product. In practice, a pull system has proven to be very responsive and effective in synchronizing production activity with demand.

The inherent ability of a pull system to synchronize manufacturing activities occurs because production is initiated by responding to *the need to replace product* that has been pulled out of the workstation. The signal or trigger indicating a need for production can be as simple as an empty tote pan. Following a production sequence in a pull system will illustrate how this works. See **Figure 3-6.**

Figure 3-6. The sequence of operations in a pull system.

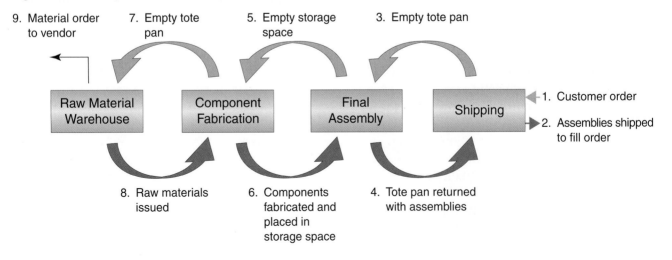

The sequence begins at the finished goods warehouse. The first event is the receipt of the customer's order. In response, the product is pulled to complete the order. This creates an empty tote pan, a signal to final assembly that the shipping department needs a replacement for the finished goods that were just shipped. Final assembly completes the components on hand, places them in the tote pan, and sends it back to shipping. However, the place where the components were stored is now empty, which is the signal to the next operation "upstream" to replace what was just used. That operation, called component fabrication, completes the parts for final assembly and then signals the raw materials warehouse to issue materials. The raw materials warehouse then signals a vendor to make a delivery.

In this example of a pull system, the factory ships on demand and then rebuilds product at each workcenter on its own time. This "short" lead-time is a unique characteristic of a pull system. As a result, the time the customer has to wait for the order to be shipped (the lead-time) is reduced to the time it takes the shipping department to package the product and send it to the customer. By contrast, a push system's lead-time to complete the same order is equal to the total time it takes for the order to pass through all of the production operations. This is why pull systems are considered to be more responsive to customer demand than a push system.

Kanbans

The triggers used to start production in a pull system are called *kanbans*, which literally means "card" in Japanese. The Toyota Automobile Corporation is acknowledged for developing and refining the kanban technique. Supposedly, the kanban approach started when Toyota executives visited an American supermarket and observed that when customers withdrew goods from the stock on supermarket shelves, this triggered replenishment by stock clerks who constantly checked the shelves and replaced only the quantity that was taken. The Toyota executives recognized that an empty space assigned to a specific product could be interpreted as a visual signal for replacement. Based on this experience, Toyota set out to design a visual signal appropriate for a manufacturing facility that would indicate when a specific item and quantity needed to be replaced.

The Kanban Approach

The kanban signal to initiate production consists of a card and a container (tote pan). While there are many variations on the kanban concept, the Toyota system is an excellent example. This system utilizes a specifically sized container for each part. This container cycles back and forth between the *producing department* and the *using department*. The kanban therefore consists of two components:

* Material conveyance—a tote pan or production container.
* Information conveyance—a card containing the part number, the container capacity and other needed data.

When a using department withdraws a container of parts, the card previously attached to it by the producing (upstream) department is detached and placed in a collection box. When the most recently emptied container for the same parts is ready to be returned to its producing department, the card from the collection box is attached to it. When the producing department receives this empty container, a card is removed and placed into a recently manufactured container full of parts that is then sent to the using department. This process repeats itself over and over again.

This kanban approach has three simple rules:

* Producing departments may not make parts unless there is an empty kanban container and a card (kanban) authorizing production.
* A kanban consists of precisely one card and one production container.
* The number of containers is controlled by manufacturing management and is kept to the smallest possible quantity. Furthermore, a container cannot hold more than a fraction of a shift's supply of parts.

Kanbans, which have become synonymous with pull systems, vary depending on the product size and volume. There are kanban systems that use pallets or painted squares on the floor. There are even computer-based kanbans such as those used by General Motors. No matter how the kanban signal varies, the principle is the same—the using department tells the producing department what to do based on demand which starts at the end of the production chain—the customer's order for the product.

The kanban pull system is ideally suited for small lot production in standard lot sizes (the container). For companies with a large variety of products or large fluctuations in demand, the basic pull system will not work or at best, it will be extremely cumbersome requiring a substantial amount of WIP. However, for these situations there is an alternative.

Managers have noted that when production quantities vary significantly, the simple kanban based pull system is overwhelmed and automatically reverts to a push system. Therefore, managers saw that the two systems could be combined, using the computer-based planning and scheduling capabilities found in a push system, to create *virtual kanbans*. The virtual kanban makes it possible to quickly change the number of kanbans in use. In the typical kanban system, the container (production kanban) capacity corresponds to the manufacturing lot size. The demand is represented by the information conveyance kanban (part number, amount, etc.). The number of parts (demand), however, can also be expressed by the time it takes to process a product in the using department. Therefore, we can create virtual kanbans at a computer workstation, which can be easily expanded and contracted to reflect the varying production requirements based on customer demand. When this approach is used, we still follow the three rules established earlier. This hybrid system, however, includes some new features that need to be explained.

In a hybrid pull system, lot sizes and lead-times for the virtual kanbans will reside in computer files. The "system" creates a signal when the kanban(s) for a specific part (a value in the computer database) diminishes to a point that represents the removal of a production kanban (container) from the producing department and the receipt of a virtual information conveyance kanban (card). To generate this signal, the computer system needs to have some information about the production activities before virtual kanbans can be put into use. They are:

- **Cycle time.** The time it takes to complete a task on one unit.
- **Single product lead-time.** The time it takes one product to be manufactured.
- **Kanban cycle time.** The time for a kanban to move between the producing department and the using department and then back to the producing department, assuming no idle time. This is the sum of the transportation time and the time it takes the using department to process the contents in the kanban. Note that the kanban size determines the manufacturing lot size and the manufacturing lot size should be equal to or smaller than the customer order quantity.

- **Active number of kanbans.** The volume of orders and the size of customer orders will determine how many active kanbans there should be between the producing and using department.
- **Forecasts.** Sales forecasting is the critical element in all manufacturing management control systems. In an MRP II system, it is used to set the master schedule that will push the product through manufacturing. In a pull system, sales forecasting is used to predict the number of virtual kanbans that will be needed at specific points in time.
- **Buffer stock.** A buffer is the stock level that creates balance in the system. This is WIP that would be considered as excess WIP but in a kanban it is viewed as *being on standby* and ready to move. At the finished goods level, the buffer stock exists to permit product to be shipped in less time than the final assembly cycle time. For component parts, buffer stocks are designed to permit production of finished goods in less time than the fabrication cycle time. These buffer inventories are based on the forecasted daily production rates and are intended to provide for the least amount of stock on hand to maintain production and shipment reliability.

An example will help to illustrate how some of these terms are used. A manufacturer makes spacers and washers. One product is a piece of aluminum with a hole drilled in it. The manufacturing process consists of three tasks:

Task	Cycle time
Cut the stock to size	45 seconds
Drill hole	35 seconds
Wrap and box	45 seconds

Transportation from the drilling department to shipping is 180 seconds

The total lead-time for one item moving through the manufacturing system is the sum of the task cycle times (45 + 35 + 45 = 125 seconds).

In this production system, the kanban container capacity is 10 pieces. This means that for 10 pieces, the total cycle time for the drilling operation is 350 seconds, or 5.8 minutes.

Kanbans do not have to be the same size between all operations, however, if they are the same size there may be some delay that will have to be factored into the kanban cycle time. In this example, the drilling operation is quicker than the preceding operation by 10 seconds. Therefore, if this manufacturing system is in continuous operation, the drilling operation will have delay time. The total delay time waiting for the replenishment kanban container to arrive will be 100 seconds [10 seconds × 10 parts = 100 seconds]. For the drilling operation, the kanban cycle time will have to include this delay time. Therefore, the kanban cycle time for drilling is [(10 × 35 seconds) + 100 seconds delay = 450] 450 seconds or 7.5 minutes.

The transportation time is 180 seconds or 3 minutes. There is no setup time for the drilling operation so this value is zero. With all this informaton available, the kanban cycle time for a kanban cycling between the drilling (the Producing Department) and shipping (the Using Department) can be calculated.

(operation cycle time × container capacity)
      delay
      transportation time
+ setup time
────────────────
= kanban Cycle Time

     5.8 minutes
     7.5 minutes
     3 minutes
+ 0 minutes
────────────────
= 16.3 minutes

Since the kanban size defines the manufacturing lot size, the manufacturing lot lead-time will be the sum of all the operation kanban cycle times. This lead-time is the realistic time it takes to manufacture the products in a manufacturing facility. Once all this information has been obtained, a computer-based planning system can calculate the number of virtual kanbans required based on customer demand that needs to exist between departments.

Bringing Vendors into the Manufacturing System

Typically, the date that a company said it would ship an order had to include the time it took to procure needed materials from a supplier. Replenishment cycle times were initially based on the lead-time a vendor quoted to a purchasing manager. Customarily, the quoted lead-times were based on the order size and how busy the supplier was at the time the order was received. However, companies adopting lean manufacturing practices started treating vendors as if they were another workcenter. That meant that a vender had to respond to a material order as if they were part of the customer's plant. As a result, vendors had to ship more frequently and the shipments were smaller. Actually, the size of the orders and the number of orders a vendor shipped related to the size and number of the manufacturing lots their customer was processing in a day. The term *JIT supplier* has been used to identify companies that supply manufacturers in this way.

People outside of manufacturing interpreted this as an inventory control and reduction practice. This was not the case. The objective for these companies, whether they were using a push or a pull system, was to reduce the manufacturing lead-time. In a push system, the material replenishment cycle time is added on to all the other operation lead times to obtain the manufacturing lead-time. In a pull system, the replenishment cycle time has to be equal to or less than the kanban cycle time. If the replenishment cycle time is more than the kanban cycle time, the additional amount of time must be added to the kanban cycle time as delay. In an ideal pull system, the manufacturing lead-time is equal the kanban cycle time of the last operation, therefore, added delay time makes the system less responsive.

To keep material replenishment cycle times small, JIT suppliers began keeping materials on hand and ready to be pulled out by their customers. The customer's kanban cycle time plus the transportation time will dictate the amount of material that must be kept ready to ship. In some cases, transportation time can be days or even weeks if the supplier is overseas. Therefore, to minimize delays due to long transportation times, the concept of a JIT warehouse was developed. These warehouses have been built right next to a customer or located geographically so that shipments can be only a day away, and with buffer stocks, there will be no delay time added to the kanban cycle times. In practice, buffer stocks allow lean manufacturing systems to deal effectively with distant vendors, complex product mixes, and fluctuating demand loads that were previously handled only by traditional push systems.

Conclusion

Lean manufacturing is a strategy to enable manufacturers to produce products with a minimum of waste. A manufacturer is not limited to any one method of operation. Any manufacturing operation ranging from a craft producer to a Henry Ford style assembly line can institute a lean manufacturing strategy. The goal of lean manufacturing is to effectively produce what the customer requires with a minimum of waste or delay. A key characteristic of a lean manufacturing facility is *robustness*—the ability to function effectively even under the changing circumstances that occur in the day-to-day operation of the factory. Reliability and quality are also traits of a lean manufacturing facility. Reliability includes people that can be depended on to do what needs to be done as well as suppliers that deliver the right materials on time all the time. Implementing programs that will ensure that equipment does not break down unexpectedly as well as developing and maintaining a quality process to prevent defects are also important parts of a successful lean manufacturing strategy.

The concept of JIT (Just-In-Time) has become associated with lean manufacturing because it minimizes or eliminates excess work in process, which will result in a reduction of inventory, an important objective. However, JIT has a more significant role to perform for lean manufacturers. JIT ensures that any problem impacting product flow will immediately become apparent by causing an interruption in production. This interruption occurs because JIT eliminates the extra stock to replace defective parts, prevents stocking surplus inventory to cover a supplier's missed shipment, or removes the need to maintain an emergency stock to bridge the time it takes to fix a broken down machine. JIT means the manufacturing system does not have any of the "extras" or "surpluses" to hide or work around problems when they occur. Therefore, JIT is an alarm system that prevents problems from being hidden from management. This means that with JIT in place a manufacturer must learn how to quickly solve problems and work to prevent them from reoccurring. JIT is the foundation for a strategy to develop lean manufacturing.

Important Terms

breakdown maintenance
buffer stock
cycle time
decoupled
excess WIP (Work In Process)
fitness for use
Jidoka
just in time (JIT)
JIT supplier
kanban
lean manufacturing organization
material requirements planning (MRP)
operator self-control
preventive maintenance
productive maintenance
pull system
push system
reactive maintenance
robustness
statistical process control (SPC)
theoretical lead time
time-and-motion study
vertical integration
virtual kanbans
work in process (WIP)

Questions for Review and Discussion

1. How does a push system differ from a pull system?
2. What constitutes lean manufacturing?
3. Explain what it means when someone says a process is in control?
4. What does the presence of excess WIP indicate?
5. Explain the purpose of an MRP system?
6. What is the purpose of the various maintenance programs and how do they differ?
7. Explain the reason for implementing lean manufacturing concepts.
8. Why would a company want to adopt JIT?
9. What is the purpose for reducing setup time?
10. Why is quality and SPC so important to a company practicing lean manufacturing?

Identifying Opportunities for Improving Manufacturing Processes

4

Chapter Highlights

- Robustness is a characteristic of a well-designed manufacturing process.
- Measuring quality by the deviation from a specified target value.
- Relying on the quality of the manufacturing process and the product design rather than inspection and rework.
- Using experimentation to improve the quality and robustness of the manufacturing processes.

Introduction

In the early 1920s, Sir Ronald Fisher, a scientist working at a small agricultural research station in England, demonstrated how to set up and conduct useful experiments outside of a laboratory. Fisher's agricultural experiments worked even though he had no control over conditions such as temperature, soil composition, and rainfall. The application of his experimental technique was recognized as an important achievement for agriculture but has proven even more useful for industry.

In recent years, manufacturers have become quite adept at problem solving. However, at the same time, the complexity of manufacturing process and the control of these processes have become more difficult. This is why companies are instituting and training their employees to design and carry out experiments on the factory floor.

Setting up and conducting an experiment is not difficult, and the mathematics needed to interpret the results are fairly simple. There is, however, a definite set of steps that needs to be followed to carry out a successful experiment. Once the experiment has been completed and the significant factors identified, the operators can then adjust these factors to manage the process properly, eliminating waste. Therefore, companies working to be lean manufacturers use planned experimentation to understand how to control new processes to avoid defects/waste and to solve problems as they emerge in existing processes.

Robust Design and Quality

A leader in applying the experimentation concept to manufacturing is Genichi Taguchi, a Japanese quality expert. Taguchi's methods work to improve quality through the *design* of both the product and the process. A significant part of his approach is the use of experimentation as a means to create robustness in the process and product. *Robustness* is the property of a product or process that makes it "insensitive" to the fluctuations in its working environment. The working environment is defined as the conditions in which the process or product operates. Therefore, a robust product will continue to operate satisfactorily under a wide range of circumstances while robust processes perform correctly without continuous adjustment even when materials, operators, and environmental conditions vary.

Taguchi's approach includes quality engineering techniques that embody both statistical process control (SPC) and quality management principles. However, Taguchi's concept of quality is based on two principles:

- Quality should be measured by the deviation from a specified target value, rather than by conformance to preset tolerance limits
- Quality cannot be ensured through inspection and rework, but must be designed into the manufacturing process and the product

The emphasis on design places a large part of the responsibility for quality on the product and manufacturing engineers. However, once the product engineers have created a quality design, the emphasis shifts to the manufacturing area. Obviously the product and manufacturing engineers work together, however, each has specific responsibilities. The manufacturing technicians, managers, and engineers must determine what aspects of the processes and operations will control the specifications and characteristics of the product. There are generally a few critical variables that individually and in conjunction with others must be controlled to limit product quality variability. Experimentation however provides a way to identify these significant variables. An explanation of how this is accomplished is easier to understand by following a real-life application of Taguchi's approach to experimentation.

A Case Study of The Plastics Works Inc. – Planned Experimentation

It is 3:00 in the afternoon at The Plastics Works Inc., and the plastic material supplier has just delivered the pellets for a "rush order" of a new cooling fan blade. The injection-molding machine is being set up to run the new parts. Fifteen minutes later the first dozen blades are laid out to be examined. Every one of the blades is a "bad" part. This order is for the company's largest customer and being late on delivery or producing poor quality parts will cause the company to lose business.

The machine operator, the department supervisor, and the product engineer recheck the machine settings and the key product characteristic (part weight) to determine what is wrong. The operator opens up the "setup guide" that the three of them developed two months ago using experimentation. It takes them only a couple of minutes to figure out the problem. The guide shows them how to set the process variables for this part. The adjustments are made and they allow sufficient time for the machine to reach equilibrium at the new settings. The next batch of parts are fine, the order is completed and shipped on time.

How did these three people create the guide that helped them to solve the problem? They know that injection molding is a common manufacturing process that is used all over the world to form parts from polymers. As a high-volume fabrication process, it is capable of producing identical parts by melting pellets or powdered resin, which is introduced into a mold cavity under high pressure. The process is very similar to metal die casting; however, unlike molten metals, polymer melts have a high viscosity and cannot simply be poured into a mold. Instead, the melted plastic must be forced into the mold cavity. As the plastic cools in the mold, more material must be forced into the cavity as the part solidifies. This "packing" will minimize part shrinkage as well as provide for the proper part dimensions and weight.

It takes a great deal of skill and knowledge to change an injection-molding machine to handle a different part or material. When a change is made, there are some general rules that provide a starting point for setting each process variable. Once the process is underway, these variables generally must be adjusted to eliminate unsatisfactory parts or to reduce cycle times. Therefore, developing a guide to understand how these variables interact to achieve satisfactory results is the objective of experimentation.

Two months ago the machine operator, the department supervisor, and the product engineer got together to develop a guide to help them quickly change their molding machine over and make the necessary adjustments. As a starting point, the group identified exactly what they wanted to control. This meant identifying the output from the process that is critical to the performance or quality of the part, which in this case was part weight. In experimental terms this output is called the ***dependent variable***.

The next step in developing the guide was to detail how the process works and list the factors that they felt influenced the output from the process. These are the factors (process variables) that they felt must be controlled or adjusted during a molding cycle. They listed:

- Melt temperature of the polymer resin
- Injection pressure
- Flow rate
- Cooling rate

Another factor that was suggested was molding *cycle times*. These vary based on material and part geometry. Typically cycle times can range from 10 to 100 seconds. A significant part of the cycle is the cooling time for the thermoplastic or the curing time if it is a thermosetting plastic material.

Once the factors are identified, the next step in the experiment process is to determine if the injection-molding machine is in control. Being in control is another way of saying that the machine, once it is set up properly, is able to mold a part consistently and predictably under a specific set of conditions. The way this is determined is by running the machine under one set of conditions and tracking a key characteristic such as part weight. To do this, the machine needs to run a minimum of thirty cycles. The part(s) from each cycle are measured and plotted on a run chart. The company in this example, like most contemporary manufacturers, routinely uses statistical process control as a way to gauge if their machines are operating in control. Therefore, the machine operator entered the data on the X-bar and R charts and confirmed the machine was stable—in control.

Since the process was in control, the steps to carry out the experiment were recorded. These steps (the ***experimental procedure***) require forms and

Figure 4-1. The first step in organizing an experiment is to list the independent variables as shown.

Independent variables	ID
Cooling rate	X1
Injection (melt) pressure	X2

charts to organize the data to be gathered. The next sections contain these forms and charts and how they are used.

Organizing the Experiment

As indicated, the crucial quality characteristic is part weight. The specification indicates that the part should weigh 10 grams. With this in mind, the operator and the manufacturing engineer reviewed all the factors they had listed and determined that only the injection pressure and cooling rate should influence part weight. These two variables are identified (ID) in **Figure 4-1** as **X1** and **X2**.

Obviously, this example has been simplified with only two variables being examined. However, this simplification will make it easier to follow the experimental procedure. Nevertheless, the procedure works essentially the same way for more complex experiments involving three, four, or even five variables.

All that is left to do is to define the procedure for carrying out the experiment. First we need to establish the upper and lower end of the range. This provides a way to measure the effect that changing the variable has on the output, part weight. The high and low values are shown in **Figure 4-2**.

An important part of the experimental procedure is documenting how the molding machine will be set up and run to collect the necessary data. This frequently is called the experimental design. Specifically, ***experimental design*** has to do with the precise way different settings will be tried and adjusted during the experiment. The following outlines the major steps to be followed.

Figure 4-2. Establish the high and low values for the variables.

Independent variables	ID	Low value	High value
Cooling rate	X1	5°/second	12°/second
Injection (melt) pressure	X2	750 PSI	790 PSI

Review of the Planned Experiment Procedure

An experimental design should contain:

1. A clear statement of the *experimental objective* or problem to be addressed.
2. Description of the *dependent variable(s)*, or output to be measured.
3. List of the factors, *independent variables*, to be varied in the tests to determine their effect on the output.
4. Specification of the *controlled factors* (or settings) during the experiment. Describe the machine being used, part being made, material type, etc., so that someone later on could set up the experiment and run it again to confirm the results obtained.
5. List of the *uncontrollable factors* or variables that might affect the output being recorded or measured for each test. Examples might be ambient temperature, humidity, changes in material batches, etc.
6. An outline listing the *controllable factors* and how they will be handled during the experiment. Although the experiment may span several weeks, the same machine operator is always assigned to do the tests. This is an example of controlling the "operator skill" factor.
7. A set of instructions on how to conduct the experiment. Examples might be:
   - Randomization of the test sequence.
   - Specifying how long the machine should be allowed to normalize after adjusting a variable(s).
   - Data collection. How many parts should be made and how they should be measured.
8. Conduct the experiment and fill in the observed data in the Experiment Test Table.
9. Create a calculation table and calculate the effect each variable has on the output.
10. Create a guide based on the experimental model.

Conducting the Experiment

At this point in our example the first seven steps of the planned experiment procedure are now complete. The next task is to construct the test table mentioned in step 8. The test table (**Figure 4-3**) shows that four tests must be performed in an experiment with two variables set at one of two levels. The table also contains a column (*Y observed*) for recording the

Figure 4-3. The test table is where the observed data is recorded.

Test	Independent variables		Dependent variable
	X1	X2	Y observed
1	Low (5°/Second)	Low (750 PSI)	
2	High (12°/Second)	Low (750 PSI)	
3	Low (5°/Second)	High (790 PSI)	
4	High (12°/Second)	High (790 PSI)	

output (average part weight). The data from the four tests will be recorded in this table.

Obviously, each test will require the molding machine to be adjusted to the new settings and allowed to normalize. Once the molding machine is ready, parts can be molded and their weight recorded. Thirty or more cycles should be run if possible for each test. The weight of every part in a test will be recorded on a tally sheet and the average part weight will be calculated. It is the average part weight that will be entered in the *Dependent variable* column labeled *Y observed* in the test table.

After the average part weight has been collected for each of the four tests, a calculation table needs to be prepared. In this table the *Low* and *High* terms are replaced with a *–1* representing the low level and a *+1* for a high level.

Calculation of the Effects of the Variables

After the experiment is conducted and the data from each test is entered, the test table will look like this, **Figure 4-4**. The Y observed value for each test in this example is the average part weight based on 30 parts (one part per mold cycle). The number of parts molded can be larger but in all cases the number of parts should be the same for each test.

Next, a computational table is created and filled out, **Figure 4-5**. The objective in creating this table is to provide a means to calculate the effect that each variable has on part weight. Once the effects are known, the magnitude that each effect has on the average part weight can be evaluated. This will indicate which variable(s) are important and which are insignificant.

Figure 4-4. Completed test table showing the observed average part weight for each test.

Order in which the tests were carried out (randomized)	Test number	X1	X2	Y observed (average part weight in grams)
3	1	−1	−1	9.5
4	2	+1	−1	9.4
1	3	−1	+1	10.1
2	4	+1	+1	9.9

Figure 4-5. Use the computational table to calculate the effects as shown in the shaded areas.

Test	X1	X2	X1X2	Y Observed	X1	X2	X1X2
Column ID	A	B	C	D	E	F	G
					(X1)(Y)	(X2)(Y)	(X1)(X2)(Y)
1	−1	−1	(−1)(−1)=+1	9.5	−9.5	−9.5	+9.5
2	+1	−1	(+1)(−1)=−1	9.4	+9.4	−9.4	−9.4
3	−1	+1	(−1)(+1)=−1	10.1	−10.1	+10.1	−10.1
4	+1	+1	(+1)(+1)=+1	9.9	+9.9	+9.9	+9.9
			Sum	38.9	−3	1.1	−.1
			Average	9.7			
		Average effect of each variable = Sum/2			−1.5	+.55	−.05

The computational table has four additional columns, which are shaded. The column, *X1X2*, represents the possible interaction created by X1 and X2. *Interaction* is the influence one variable has on another variable as the first variable's level changes. You may have come across the term interaction in reference to medicines where one drug changes how another medication performs. Interaction is not that common but it should be evaluated as part of the experimental analysis.

The other three columns are calculation columns used to determine the magnitude of the effect for the variables X1, X2, and the interaction variable X1X2. The values in these calculation columns for a test is the product of the *indicators* of the high/low levels (−1 or +1) multiplied by the *observed output* for that test. The result is written in the appropriate calculation column. Here is how the values are calculated for the variables of *test 1*. See **Figure 4-6**.

X1, the value of −9.5 in column E is the product of column A times the observed value in column D.

$$(-1)(9.5) = -9.5$$

X2, the value of −9.5 in column F is the product of column B times the observed value in column D.

$$(-1)(9.5) = -9.5$$

X1X2, The value of +9.5 in column G is the product of column A times column B times the observed value in column D.

$$(-1)(-1)(9.5) = +9.5$$

Figure 4-6. Values for the variables from *test 1* are calculated.

Test	X1	X2	X1X2	Y Observed	X1	X2	X1X2
Column ID	A	B	C	D	E	F	G
					(X1)(Y)	(X2)(Y)	(X1)(X2)(Y)
1	−1	−1	(−1)(−1)=+1	9.5	−9.5	−9.5	+9.5

This same procedure is used to calculate the values in columns E, F, and G for tests 2, 3, and 4. Once these values have been computed, the four values in each column (E, F, and G) are added together algebraically to get the sum for the column. All of these values are shown in **Figure 4-5**.

The sum is divided by 2 to get the average effect for that variable. This is the simplest way to calculate the effects but it does not provide any insight into the logic behind the experimental procedure. The next section explains what is happening in the calculation table.

The Logic Behind the Calculation of Effects

To begin with, we will determine the effect that variable X1 has on influencing part weight. The analysis begins when the variable in question (X1) changes from its low level to the high level while the second variable remains constant. As an example, notice in tests 1 and 2 that variable X1 goes from low to high while X2 stays low. In tests 3 and 4, X1 again changes from low to high while X2 stays at the high level. We can calculate the effect that the change of X1 has on part weight by averaging the difference in the change of part weight during these two instances. Therefore the calculation for the effect of X1 is:

$$X1 \text{ effect} = [(9.4 - 9.5) + (9.9 - 10.1)]/2$$
$$X1 \text{ effect} = [(-.1) + (-.2)]/2 = \textbf{-.15 grams}$$

This can be stated as the total effect of X1 when it changes from its low setting to its high setting. The *effect* is a decrease in part weight of .15 grams.

We can do the same type of comparison for X2. In this instance the first comparison is between test 1 and test 3. In these tests X1 stays low while X2 changes from low to high. The second comparison that can be made is test 2 and 4. In this group X1 stays high while X2 changes from low to high. The calculation of the effect for X2 is:

$$X2 \text{ effect} = [(10.1 - 9.5) + (9.9 - 9.4)]$$
$$X2 \text{ effect} = [(.6) + (.5)]/2 = \textbf{.55 grams}$$

This result can be stated as the total effect of X2 when it changes from its low setting to its high setting. The *effect* is an increase in part weight of .55 grams.

Knowing the effect each variable has on part weight is the beginning of the guide we are trying to develop. However, there is one other effect that must be computed. What happens to part weight when the two variables team up? That variable is the combination of the X1 and X2. It is identified as X1X2.

Figure 4-7. A listing of the variable effects helps to uncover which are most influential.

Variable	Effect on part weight as the variable goes from low to high (grams)
X1 (Cooling rate)	−.15
X2 Injection (melt) pressure	+.55
X1X2	−.05

The comparison of X1X2 is done as follows. In this instance the comparison will be between test 1 and test 2. In test 1 both variables are the same (set at their low value) and in test 2 X1 has changed going to its high value while X2 stays high. The second comparison is between test 3 and 4. Again both variables are the same (set at their high value) and in test 3 the variable X2 is high while X1 has gone low. The calculation of the effect for X1X2 is:

$$X1X2 \text{ effect} = [(9.5 - 9.4) + (9.9 - 10.1)]/2$$
$$X1X2 \text{ effect} = [(.1) + (-.2)]/2 = \textbf{-.05 grams}$$

This result can be stated as the total interaction between X1 and X2 represented by X1X2. The effect is a decrease in part weight of .05 grams.

By listing the effects, it is easier to see which has the most influence on the part weight, **Figure 4-7**. It also helps in developing a visual model of the experiment.

The model (**Figure 4-8**) shows the effects and the relationship of the two variables graphically. The X-axis represents the variable X1 and the Y-axis corresponds to X2.

Figure 4-8. A graphical model of the experimental results showing the degree of effect.

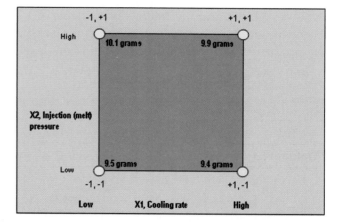

Determining the Average Part Weight

The experiment as described consisted of 4 tests. In each test the same number of parts was molded and the average weight of the part was recorded in the column labeled Y observed. If we sum the average part weight recorded for each test and divide it by 4, we will have the average part weight for the experiment. Based on the recorded observations, the experimental average part weight is 9.73 grams. Obviously this is below the target value of 10 grams. However it appears that the variable X2 has an effect that is large enough to reach our target. The next section describes how to adjust the variables to reach the 10 gram objective.

In **Figure 4-9**, the average weight of the parts when X2 is low is 9.45 grams. When X2 is at its high level the average part weight is 10 grams. Therefore the total effect that X2 has on the part weight is: 10.0 – 9.45 = .55 grams

The creation of the guide begins by entering an average part weight 9.73 grams, which comes from the four tests conducted during the experiment. The question then is how do we factor in the effect of X2? The average part weight as shown in **Figure 4-9** represents the midpoint between the test weights when X2 is low and when X2 is high. Therefore X2's effect on the midpoint average will be only half the total effect or: .55/2 = .275 or .28 grams.

That means to increase part weight we need to set X2 high. The increase then will be the average part weight (the midpoint average of 9.73 grams) plus half the total effect of X2 which is rounded to .28 grams (.55/2 = .275 or .28 grams). This means that the resulting part weight will be 10.01 grams (9.73 + .28), which is essentially the target value. The last step in the experimental procedure is to create a guide using the values that have been calculated. The guide that was developed is shown in the table, **Figure 4-10**.

When using the guide, the operator selects the desired level for a variable and multiplies the level indicator (–1 or +1) times the effect divided by 2. Recall that the effect is divided by 2 because that is the total influence that variable's effect has on increasing or decreasing the midpoint average (Y average observed). The effect's value is the total change from low to high, so going from a midpoint to a high level or from a midpoint to a low level means the value to be added is only 1/2 the total effect for that variable. The operator reviews the guide and selects the

Figure 4-9. Graphical depiction of the influence of X2 on part weight.

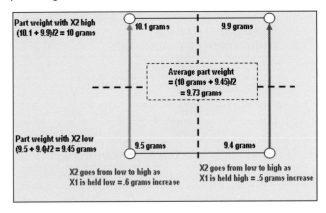

variables and levels that should get the part weight as close as possible to the target. **Figure 4-11** shows how this would appear.

The value for the midpoint average (Y observed) is always included in the calculation when using the guide. To reach the target value, only the major variable(s) need to be included, which in this example was the variable X2. In **Figure 4-11** the injection pressure was set to its high level, which adds .28 grams to the midpoint average, moving us very close to the target value. However, to remove the influence of the cooling rate, the operator had to set its level to a midpoint (8.5 degrees/second). The other variable to be considered is the interaction of pressure and cooling rate (X1X2). Since the operator decided to set the cooling rate at its midpoint value (0) the interaction value also becomes zero.

Determining Significant Effects

Up to this point all the variables in this study, including the interaction variable (X1X2), have been treated as if they influence the part weight of the molded product. That is not an assumption that can always be made, even if the effect has a value. As an example, suppose someone had suggested that the color of the operator's shirt influenced part weight and that variable was used instead of cooling rate. It is possible that the natural variability that occurs in any process might have produced what would appear to be a small effect in part weight due to the operator's shirt color. It is obvious that the color of an operator's shirt will not influence part weight so the variable should not be included in the guide. The question then is how can an experimenter sort the significant variables from the irrelevant ones?

Figure 4-10. Setting up a guide for achieving target part weight.

Description		Level	Effect/2	C
Midpoint average (average of Y observed)				**9.73 grams**
X1, Cooling rate				
Low level	5°/sec	(−1)	−.08 grams	___
Mid level	8.5°/sec	(0)	0 grams	
High level	12°/sec	(+1)	−.08 grams	
X2, Injection (melt) pressure				
Low level (−1)	750 PSI	(−1)	+.28 grams	___
Mid level	770 PSI	(0)	0 grams	
High level (+1)	790 PSI	(+1)	+.28 grams	
X1X2, Cooling rate/injection pressure				
Low level (X1 low & X2 high or X1 high & X2 low)		(−1)	−.03 grams	___
Mid level (either X1 or X2 set at mid level)		(0)	0 grams	
High level (X1 low & X2 low or X1 high & X2 high)		(+1)	−.03 grams	
Total Weight				___

Figure 4-11. Guide showing how variables should be set to close in on the target part weight.

Description		Level	Effect/2	C
Midpoint average (average of Y observed)				**9.73 grams**
X1, Cooling rate				
Low level	5°/sec	(−1)	−.08 grams	**0**
Mid level	8.5°/sec	(0)	0 grams	
High level	12°/sec	(+1)	−.08 grams	
X2, Injection (melt) pressure				
Low level (−1)	750 PSI	(−1)	+.28 grams	**.28**
Mid level	770 PSI	(0)	0 grams	
High level (+1)	790 PSI	(+1)	+.28 grams	
X1X2, Cooling rate/injection pressure				
Low level (X1 low & X2 high or X1 high & X2 low)		(−1)	−.03 grams	**0**
Mid level (either X1 or X2 set at mid level)		(0)	0 grams	
High level (X1 low & X2 low or X1 high & X2 high)		(+1)	−.03 grams	
Total Weight				**10.01**

Statisticians have a number of tests and methods that can identify the significant effects. These tests are not overly complicated, but they will make this discussion more involved than it needs to be. However, one technique is very quick and straightforward. The test classifies the effects into two groups based on their size. This test is based on the order of magnitude.

Here is how the test is done. First, we need to clarify what is meant by an order of magnitude. As an example, the number 10 is an order of magnitude larger than 1 and the number 1.5 is an order of magnitude larger than .15. If a number is multiplied by 10, the result is a number that is an order of magnitude larger. In a similar manner, if a number is divided by 10, then it will be an order of magnitude smaller.

Here is how this test works on the results of the injection-molding example. If we list the effects in our example we can easily pick out the smallest value, **Figure 4-12**. That will be our starting point.

Of the three variables, the interaction term (X1X2) is the smallest. Therefore, this is the term that will be multiplied by 10.

$$(-.05)(10) = -.50$$

The result is −.50, which is now our order of magnitude indicator to be used to identify significant effects. **Figure 4-13** shows each effect listed with the indictor for the comparison next to it. If the effect is

Figure 4-12. A listing of variable effects is used to help select the lowest.

Variable	Effect value
X1 Cooling rate	−.15
X2 Injection (melt) pressure	+.55
X1X2 Interaction	−.05

less than the indicator then the effect is not considered significant. The comment column contains the results of the comparison.

In our example the value of X1 is well under the indicator and would be excluded from the guide. X2, however, is larger than the indicator and, therefore, is identified as being significant. The negative or positive signs on the effects or on the indicator are not a factor when making these comparisons and therefore are ignored. However, remember that this method is very crude, and there will be times when one of the more sophisticated statistical tests will be needed to identify significant effects. This is particularly true when all the significant variables are just barely significant. In this example it is possible that X1 might be significant and should be included in the guide. If X1 were included, the guide would appear as shown in **Figure 4-14**.

Figure 4-13. Table that identifies significant variables by comparing the *effect value*.

Variable	Effect value	Indicator	Comment
X1	−.15	−.50	Probably not significant
X2	+.55	−.50	Significant
X1X2	−.05	−.50	Not significant

Figure 4-14. This guide contains only significant effects.

Description	Level	Effect/2	C
Midpoint average (the average of Y observed)			9.73 grams
X1, Cooling rate			
Low level 5°/second	(−1)	−.08 grams	
Mid level 8.5°/second	(0)	0 grams	0
High level 12°/second	(+1)	−.08 grams	
X2, Injection (melt) pressure			
Low level 750 PSI	(−1)	+.28 grams	
Mid level 770 PSI	(0)	0 grams	.28
High level 790 PSI	(+1)	+.28 grams	
Total weight			10.01 grams

Figure 4-15. The average weight from 13 production lots is shown in this table.

Average part weight based in a sample of 30 pieces from each production lot.													
Lot	1	2	3	4	5	6	7	8	9	10	11	12	13
Wt.	9.95	9.90	10.08	10.0	10.1	9.9	9.9	10.13	10.02	9.9	10.05	10.04	9.95

Comparing the Guide Target to Actual Results

Once the guide is developed, the machine can be set up and run to make the part. During the production run the operator should measure and calculate the average part weight based on a sampling of parts. Recall that the result from each test that was conducted in the experiment was based on thirty pieces. Therefore, to determine the validity of the guide, the actual results of production should be plotted on a chart that is similar to a run chart. You should remember the run chart from the chapter on lean manufacturing. The centerline of the run chart would be the average part weight predicted by the guide, 10.01 grams. An example of this chart is shown in **Figure 4-15**. The data that is used to plot this chart is shown in **Figure 4-16**. The thirteen lots of thirty pieces each were drawn from consecutive production runs that occurred over the past four months.

The plot should be interpreted in the same manner as a statistical process control chart. If the process is stable and in control, the plot of points will be symmetrical around the centerline of the chart, clustered near the centerline, with no discernable pattern to the plot. The plot in Figure 4-16 meets these criteria and therefore the guide based on the experiment appears to be valid.

Residual Plots

A variation on plotting performance on a run chart is the residual plot. The *residual plot* is very similar to the run chart just described but it offers some other advantages. The plot is obtained by subtracting the average weight of the production sample from the average weight predicted by the guide (10.01 grams). The difference is the residual and that is what is plotted on the chart. **Figure 4-17** shows the residuals based on the data in Figure 4-15.

The plot of the residuals is shown in **Figure 4-18**. Note that the centerline of this plot is zero and the shape as compared to the run chart is "flipped." Nevertheless, the plot is similar in that it does not appear to follow any apparent pattern and the number of points above the centerline is about the same as those below the centerline.

Figure 4-17. This table compares predicted part weight with production weight to calculate the residual.

Lot	Predicted weight from the guide	Average part weight from each production lot	Residual
1	10.01	9.95	.06
2	10.01	9.90	.11
3	10.01	10.08	−.07
4	10.01	10.00	.01
5	10.01	10.10	−.09
6	10.01	9.90	.11
7	10.01	9.90	.11
8	10.01	10.13	−.12
9	10.01	10.02	−.01
10	10.01	9.90	.11
11	10.01	10.05	−.04
12	10.01	10.04	−.03
13	10.01	9.95	.06

Figure 4-16. A run chart that shows the actual part weight plotted against the predicted weight.

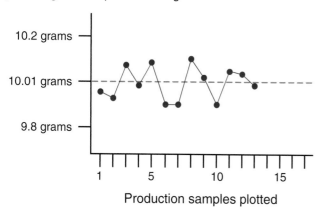

Production samples plotted

Figure 4-18. The values of the residuals are plotted in a run chart.

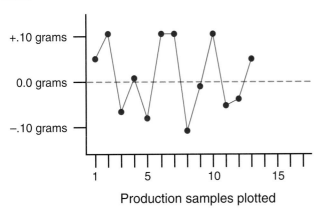

Production samples plotted

The magnitude of the residuals ranges from a positive .11 grams to a negative .12 grams. Recall that the size of the effect for the interaction term that was considered insignificant was –.05. The random variation of the residuals greatly exceeds this value, which provides some support for our assumption that there is not any interaction and, therefore, this variable (X1X2) does not influence part weight. It would be interesting to include X2 (cooling rate) to see if this variable can reduce the magnitude of the residuals. If it does reduce the variation in the residual plot and symmetry and randomness (lack of a discernable pattern) are not affected then it needs to be included in our guide.

The ultimate goal is to consistently achieve the target value, which would result in a residual plot of a straight line right on the centerline of zero. Usually, the plot of residuals leads the operators and technicians to launch another experiment.

As everyone becomes more comfortable with the methods of experimentation, the scope of the experiments can become more extensive. It is not unusual for three, four, or even five variables to be investigated at one time. This provides a more comprehensive approach but the number of tests needed doubles for every variable added. Fortunately, there are some techniques and software that can make the experimental process very manageable for everyday use on the factory floor. Details can be found in the many good references devoted to practical planned experimentation for industry.

Important Terms

controllable factors
controlled factors
dependent variable
experimental design
experimental objective
experimental procedure
independent variables
interaction
residual plot
robustness
uncontrollable factors

Questions for Review and Discussion

1. Discuss the value of planned experimentation in industry.

2. Why would a manufacturer want to create robust manufacturing processes?

3. A company has a heat-treating process to harden a gear that is used in an appliance. After heat treatment, a large number of the gears are found to be below the minimum level of hardness. Technicians have identified two factors that they feel influence the heat treating process. One factor is the steel supplier. Currently they purchase from two companies—Acme Steel and Universal Steel. The second factor is the type of oven used. Oven A is gas fired and Oven B is heated electrically. Using the DOE approach, outline how the technicians should conduct the experiment.

4. Technicians completed an experiment on a painting operation. The operation is supposed to provide a coat of paint that is at least 2 mils thick. They conducted an experiment examining two variables. The first variable (X1) was part temperature. The low value was the part at room temperature and the high value was heating the part to 50°C. The second variable (X2) was paint temperature. The paint was applied at 25°C (low value) and at 45°C. Construct a computation table for the experiment.

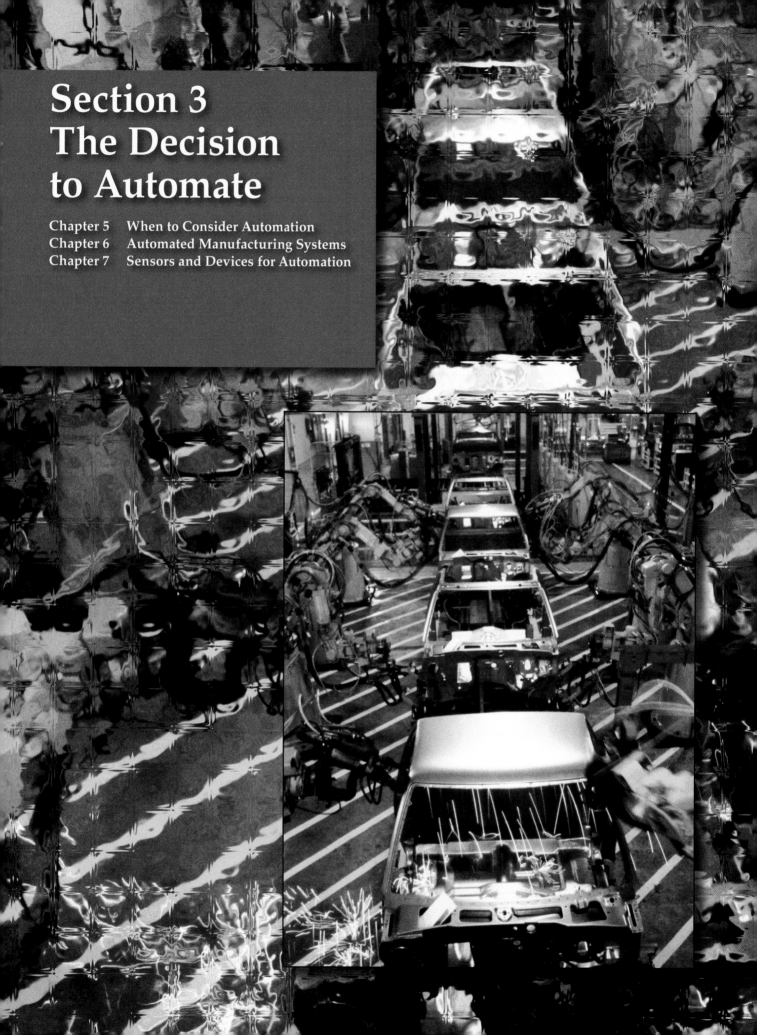

Section 3
The Decision to Automate

5 When to Consider Automation

Chapter Highlights

- An automated manufacturing process is designed to consistently replicate a product.
- Productivity and quality improvement programs must be completed before automation is attempted.
- Products must be designed for manufacturability (manually or automated).
- Know the rules to design for manufacturability.
- If cost reduction is the major goal then automation *may* not be necessary after all the pre-automation steps are completed.

Introduction

Several years ago, the Minneapolis School of Art and Design completed constructing a large high bay building to be used as a studio for creating sculpture. The facility looked like it should be part of a manufacturing facility or a small metal fabricating plant. Whatever it might be called, it was truly state-of-the-art. There was a modern foundry, plasma-arc metal cutting equipment, CNC machines, and other manufacturing equipment, all serviced by a high bay gantry crane and fork trucks. I commented to our host that their equipment rivaled the best manufacturing facilities I have visited. She replied that the technology they use is the same as used by manufacturers but with one major difference. "The difference," I asked "would be what?" She replied, "In manufacturing, it's all about being able to repeat what you do over and over again while in art one doesn't make any provision for replication". She was right; engineering departments expend a considerable amount of time documenting a product's characteristics and limiting variability (specification tolerances) to ensure that the parts or products are identical.

Although the technology may be the same, art strives to be unique while manufacturing makes every effort to make things the same. A successful manufacturer (replicator) is able to:

- Make parts and products that are identical or interchangeable.
- Create and maintain a production system that is predictable.

In short, manufacturing is the practice of producing a product with consistency and predictability. The effort to minimize variability and to replicate products according to the engineering design is referred to as quality control. During the last two decades of the twentieth century, the control of quality became a company wide effort called *total quality management* (TQM). More recently, people in manufacturing have recognized the link between quality and productivity. These two efforts (quality and productivity) share a common objective—the elimination of waste. Minimizing waste, however, is not dependent on a single practice or concept. What seems to work for one company may need to be substantially modified to work in another organization. Nevertheless, some basic techniques and practices are common to all successful manufacturers. These activities apply to labor-intensive production as well as to process-oriented manufacturing. However, one of the most commonly used techniques has been automation. In the last century, automating the labor operations needed to make a product became an unchallenged precept for manufacturing management.

When Automation Is Needed

In the United States, automation is considered to be a cost reduction practice. Automation equipment is typically justified by the amount of direct labor that can be eliminated. This reduction in labor is assumed to require less operating costs. Surprisingly, many companies that were able to justify automation did not always succeed in reducing costs or improving productivity. One of the most well-known instances occurred during the 1980s. The largest automobile manufacturer in the world, General Motors, was investing heavily in robotics and automation as a way to reduce the cost of making an automobile. One of the measures they used to determine if their cost reduction program was working was productivity. Productivity is the total labor hours needed to produce an automobile. Surprisingly, their investment in robots and automation did not make them more productive than their competitors that were using seemingly more labor-intensive manufacturing processes. Companies like General Motors discovered that automation does not necessarily improve productivity because it does not reduce waste. Observers remarked that automation oftentimes just serves to increase the amount of waste. So when should a company consider automation?

Automation makes sense if:

- The manufacturer needs to increase the rate of production substantially
- A human being needs to be removed from an unsafe/unhealthy production process

If either of those two conditions does not exist, then it is unlikely that automation is needed. What is always needed, however, is the elimination of waste. In fact, even if automation is needed, the first task in any automation project is the elimination of waste prior to automating. So, what exactly is waste? It shows up in all kinds of ways. **Figure 5-1** lists the categories and some common examples.

Figure 5-1. Categories and examples of waste.

Category	Example
Non-value adding activities	Inspections, repair/rework, taking parts in or out of storage.
Unneeded materials	Parts that can be eliminated or have no function in the product.
Unneeded operations	Making, using, or assembling unneeded parts.
Delays and downtime	Materials arriving late, machines that breakdown, unplanned absences of workers.
Loss	Spoiled, scrap, or unusable/defective materials that are thrown away.
Duplication	Making extra product in anticipation of defects or scrap. Making/using two or more styles of product/components that have/serve the identical function.

Pre-Automation Process

A company that has productivity and quality improvement as objectives should organize itself to eliminate the five forms of waste listed in **Figure 5-1**. The steps to eliminate this waste constitute a *pre-automation program*. If automation were not the end objective, this would be a *productivity/quality improvement program*. For either name, the same steps apply. The first step is to identify who should be involved in the program.

A review of the list of waste categories indicates that most of the major departments in a company have some direct responsibility for the source of the waste. To eliminate parts in an assembly will require engineering's input. Identifying and removing non-value added activities would involve management in the departments responsible for these activities. Downtime and material delays are going to bring in the maintenance department and the people involved in material procurement. At this point, it should be apparent that this has to be a company wide effort. However, in practice, the number of people involved is not as unwieldy as it might appear. With top management's commitment and support, several small task groups can be formed to address specific waste targets. These groups can be organized around the waste reduction techniques. One of the most important is designing the product for assembly or *manufacturability*.

Design for Assembly

A current manufacturing dictum is *never design and build a part that can be purchased*. This approach is based on the concept that a company buys the best and if the best is not available, that part must be made. The definition of *best* includes price and function. As a result, the dominant activity in most manufacturing plants is assembly, specifically assembling purchased parts. Since the number of fabricated parts is kept to a minimum, companies can focus their resources on a limited number of technologies and processes to build just a few parts.

The potential for waste in assembly is considerable. Specific areas that almost certainly will need to be addressed are the elimination of scrap, rework, unneeded parts, and unneeded operations. Fortunately, there are several techniques to guide manufacturers in improving their assembly operations. The most significant technique is called *design for assembly* (**DFA**). A product that meets the DFA criteria often times can be assembled manually so inexpensively that automation cannot be justified. The set of principles that make up the technique were originally developed by Boothroyd and Dewhurst and published in a book titled *Design for Assembly*.

They established the following guidelines:

- Design the parts so that they are symmetrical. If possible, each part should be symmetrical so that it does not have to be oriented for assembly. In manual assembly, symmetrical parts cannot be installed improperly. In automatic assembly, symmetrical parts do not require special sensors or tooling to orient them correctly. See **Figure 5-2A**.
- If the part cannot be made symmetric, then make it *very* asymmetric. An operator is not likely to install the part incorrectly because the right way will be more obvious. It may be easier for automation machinery to be able to orient the part. See **Figure 5-2B**.
- If a part tangles or jams together at any point in the manufacturing process, change its design. See **Figure 5-2C**.
- Eliminate parts that are slippery, delicate, or awkward to handle. See **Figure 5-2D**.
- Make insertion and fastening of parts simple and fast.
  - Design a single nest (fixture) to hold the chassis component or serve as the chassis component. This chassis holds or orients all the other assembled parts. The nest supports this component and serves as the carrier as the assembly moves between assembly operations. See **Figure 5-3A**.
  - Provide chamfers, insertion guides, and ample clearances for the part being inserted. See **Figure 5-3B**.
  - Create a nested progressive sequence of assembly along one axis. Use gravity to help, not hinder, the placement and support of the parts. This approach eliminates the need to manipulate the assembly when placing other parts. See **Figure 5-3C**.
  - Design the parts to be properly oriented before they are released. See **Figure 5-3D**.
  - Standardize on commonly available fasteners and fastening techniques.

Figure 5-2. Examples of DFA guidelines for part design.

Different designs of a thrust washer.
The washer fits in a recess to
prevent it from turning.

A. An asymmetrical design requires orientation

A symmetrical design

B. A design that is obviously asymmetric

C. A part that will jam together or wedge together when stored making it difficult to pick up

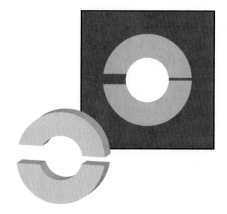

D. A two-piece washer that would be difficult to handle and orient during assembly into the recess shown

Width

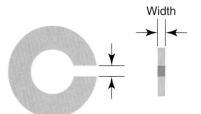

E. Opening is larger than the width

Figure 5-3. Examples of DFA assembly guidelines.

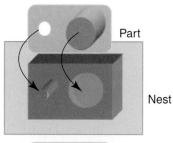

Part

Nest

A. A nest or fixture is used to hold the component for a manufacturing operation(s).

Chamfer edges, if possible, on the part

Part

Definitely chamfer edges and round posts on the nest

Nest

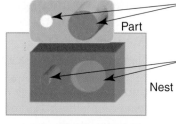

B. Provide chamfers and guides to make it easy to place the part properly.

C. Create a nested progressive assembly along one axis.

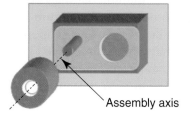

The second part to be assembled is symmetrical and is located by the post on the nest

Assembly axis

D. Design the parts so that they are properly oriented before they are released.

Reducing the Number of Parts

One of the most beneficial waste reduction activities that will improve the assembly process is eliminating parts and reducing the variety of parts in a product. This approach requires creativity, innovation, and technical skill. Determining the minimum number of parts required can be accomplished by asking three questions of each part. The questions are:

- Does the part have to move relative to other parts in the assembly when the product is in operation?
- Does the part have to be made from a material that is different than the other parts in the assembly? As an example, does the part have to be an insulator or serve as an electrical conductor?
- Does the part have a unique function even though the answer to the first two criteria was no? A unique function might be that the part serves as a drain plug or an inspection plate.

If the answer to the three questions is "no", the part can be eliminated or combined with another part.

The success of this approach can be illustrated by the experience a bathroom scale manufacturer had in meeting a competitive challenge. This Midwestern company's most popular mechanical scale used a mechanism that consisted of 17 separate stamped metal parts held together with six spot welds and seven fasteners. At first, the design was very profitable. However, over time the assembly became too costly compared to the off shore competition. The company initially investigated automating the assembly of the mechanism without any change in the design. Unfortunately, they found that even if all the direct labor used to manufacture the scale were eliminated, their selling price would still be higher than the competition.

The company's engineering department started a DFA program to redesign the scale mechanism. The result was a new design that had only six parts, of which two were stamped and assembled in the press. The number of fasteners was reduced to five from seven and all six spot welds were eliminated. The new mechanism was competitive and the cost of retooling (replacing all the old die sets) was just a fraction of the estimated cost to automate the assembly of the old design.

The DFA program was a team project that involved representatives from:

- Product engineering—they had the overall design responsibility.
- Production engineering—they were involved in the design of the prototype tooling and assembly operations.
- Purchasing—a buyer developed the material costing information from suppliers.
- Sales—spoke for the customer and their needs.
- Tool and die supplier—an engineer from the tool and die supplier worked with the company's design engineering department to design the die to create the two-part component.
- Cost accounting—they worked with purchasing and plant personnel to estimate the product cost for the new design.
- Plant personnel—the assembly department supervisor made the pilot runs and did the transition planning.

From the list above, it is apparent that the major design change untaken by this manufacturer involved every major function in the company's organization. In a program as extensive as this, management support and involvement is essential to obtain success.

A major problem that happens with some company's efforts to reduce costs occurs because the cost reduction techniques just "cheapen" the product. As an example, a company determines it can save money by replacing the current material with a weaker and less expensive material that they believe will only slightly degrade the products performance. Although the manufacturing cost has been reduced, so has the product value. Unfortunately, competitive market forces will cause the cheapened product to be sold at a lower price, forcing the company to initiate another round of cost reduction.

However, eliminating waste does not cheapen a product. The DFA technique is aimed directly at eliminating waste by eliminating unneeded parts and simplifying product design. Therefore, it is important to keep in mind that cost cutting and waste reduction are two entirely different programs. An organization that implements a DFA program has to make sure that everyone involved understands this distinction. In addition, the DFA technique should be adopted as an ongoing activity as should be all efforts in waste reduction.

Mechanization of Assembly

The process of assembly consists of two tasks. The first task is acquiring a component and the second is inserting, attaching, joining, or fastening that component to an assembly. This assembly sequence is repeated until the product is complete. Even this simple sequence has conditions that must be met to be effective. There needs to be an established hierarchy or sequence for assembly that meets the DFA guidelines for making assembly simple and fast. The starting point for building this hierarchy is to identify the part that will serve as the chassis for the product. The *chassis* serves to locate and orient the product to make assembly easy and logical. If that part does not exist, then a nest or an assembly jig (tooling) needs to be provided. This form of tooling is the most rudimentary form of mechanization. Mechanization, however, is not the same as automation. *Mechanization* means making or assembling things with machines and tooling to increase the rate at which work is being done. Mechanization also serves to reduce variability.

If the parts are properly designed for assembly, the next concern is how they are stored and dispensed for assembly. As an example, consider the parts that are commonly used to hold two sheets of paper together. One is a paper clip and the other is a staple. Ignoring that they serve slightly different functions, just review how each is usually stored. Paper clips are typically kept jumbled in a box or tray. Joining the two sheets of paper together requires the operator to separate a clip from the jumble of clips and then slip it over the two sheets of aligned paper. The staple, however, is stored in what is termed a *magazine*, a chamber inside the stapler that keeps its contents secure and ready to be dispensed. To assemble the two sheets of paper the operator aligns the paper, inserts them into the throat of the stapler, and sharply presses the stapler to complete the stapling operation. The acquisition, feeding, and placement of the staple to join the two sheets of paper have been mechanized.

Mechanizing fastening operations is very common. Nail guns, staplers, riveting machines, and screw guns are some of the examples that can be found in use today. Part feeding, especially fasteners, is one of the two common areas of mechanization. The other popular area for mechanization is positioning materials for assembly. Conveyors, index tables, transfer lines, and a variety of specialty tooling are all devoted to moving and positioning parts for assembly. An example of an indexing table used for assembly is shown in **Figure 5-4**.

Mistake Proofing

Repair and rework are the most blatant forms of waste. These sources of waste must be eliminated before automation can be considered. Organizations now recognize the need to formally and continually work to eliminate the errors and omissions that cause repair and rework. The Japanese call the process *Poka-yoke* or *mistake proofing*. The DFA techniques certainly support the concept of mistake proofing however, mistake proofing goes beyond the parts and components that go into the product. Poka-yoke begins with the product's design and extends to the arrangement of the work area. The mistake-proofing process involves six activities. Some of these activities are used when a new product is being designed while others apply to existing production. The first two listed are based directly on the principles of DFA:

● **Eliminate the opportunity for error** by redesigning the product or process so that the operation or part is no longer necessary.

Figure 5-4. An index table with assembly nests to hold components being assembled manually.

An assembler places the bezel and reflector assembly for a flashlight in a nest. This is the first of four assembly operations carried out on a rotary index table. The table will not index until the operator's hands are clear of the assembly area.

Notice that the nests are mounted on insert plates. These plates with the nests can be quickly changed to handle other styles of flashlights.

- **Minimize the opportunity for error** by combining two or more parts or activities into one.
- **Prevent the possibility of error** by designing the product or process so that it is impossible to make a mistake. An example is the plugs and connectors used with desktop computers. Their distinct configuration prevents them from being plugged in incorrectly.
- **Reduce the chance for error** by making the work easy. This can be accomplished by providing the proper tools, fixtures, and jigs so the operators and assemblers can complete their tasks easily and consistently. Examples are color-coding parts and providing fixtures that lift and tilt assemblies so that the operator has easy access to work.
- **Detect and correct the error** before further processing occurs. Each assembly step should include a test that confirms the prior step was completed properly. The operator should not consider the task complete without this confirmation. A confirmation might be as simple as the snap sound when two parts are properly joined together or as complex as a built-in self-test.
- **Create an atmosphere for eliminating errors** in the organization. Production should not flow around a problem and a defect should not be pushed aside to be dealt with later. Management must support and reward operators that detect and solve problems.

After reading this list, it is evident that the steps for mistake proofing are very similar to the design for assembly precepts. Consequently, mistake proofing can be accomplished more effectively during the design and development of a new product. This is why production engineering must have a major role during the design phase to create mistake proof processes and production lines. Once the product is designed and the manufacturing process is selected, the opportunities for mistake proofing are more limited. Regardless, the organization must develop the will and the skill to eliminate opportunities for errors that cause repair and rework. This is critical because even with a dedicated effort to mistake proof a new design, it is nearly impossible to foresee all the opportunities for error. Therefore, it is essential that the manufacturing organization develop an ongoing approach for identifying and eliminating the source of mistakes.

Identifying and Eliminating Variability

Another important part of the pre-automation process is identifying and controlling what causes the variability in the time it takes to complete a task. Tasks that can be done in a consistent and predictable amount of time are less prone to errors or problems. Making the time it takes to complete an operation (operation cycle time) predictable increases manufacturing productivity and product quality.

Operation Cycle-Time Variability

Manufacturers too often focus on reducing operation cycle times instead of first reducing the variability of the cycle time. An example will help to illustrate the importance of reducing variability. Suppose an assembly operation takes one minute to complete, making its cycle time one minute. Therefore, in five minutes, the operator would expect to have completed five units. However, at the end of five minutes of work, only four units are completed. The operator explains that a part jammed and it required a minute to free it up before it could be assembled. That means that during those five minutes, three pieces had a cycle time of one minute and one piece had a cycle time of two minutes. The cycle time in this example varies from one minute to two minutes per piece. Another way to explain this increase in cycle time is that the organization wasted a minute of the operator's time.

If this happens repeatedly, then there is a chronic problem. If this is an unusual occurrence, the problem is sporadic. Regardless the variability in cycle time points to a problem that impacts productivity.

The steps to reduce cycle time variability begin by observing the operation and listing the reasons for the delays and downtime. Once the sources and reasons for the waste are known, the individual or team can apply problem-solving skills by attacking the most frequently occurring causes first.

Removing the causes for cycle time variation is a critically important pre-automation activity. People are very effective in minimizing the magnitude of disruption caused by material variation. Most operators handle these problems so well that often the only symptom is some variability in the operation cycle times. Unfortunately, the tooling and intelligence in an automated machine is much more limited and less able to handle material inconsistency and defective parts. That is why the causes for cycle time variability have to be identified and solved before an operation is automated.

Cycle times are often determined using *time study*, which records how long it takes to complete an operation. In a time study, usually only ten to twenty cycles are timed unless the time study analyst finds the process is not consistent. Time studies, unfortunately, may exclude a reading that is unusually long due to a chronic or sporadic problem. Therefore, the calculated time study cycle time may not convey the variability that would be found in day-to-day operations. Therefore, to effectively determine an operation's cycle time variability requires more than a time study. The analysis should also include tracking the operation for a week or more to determine the average daily cycle time and having the operator logging the delays and interruptions that occur. The average daily cycle time is computed by dividing the hours worked during a shift by the number of parts produced by that operation. As an example, a soldering operation that worked for 6.0 hours producing 1200 soldered parts has an average operation cycle time of .0050 hours (6.0÷1200). A well-designed operation should have consistent average daily cycle times.

Product Characteristic Variability

Another form of variability that must be measured is the variation in key product characteristics that an operation imparts to the component. Well-maintained process control charts (Xbar and R chart) can be used to determine if the process is stable and in control. A process that is not in control cannot be automated successfully. Many companies have discovered that much of the modifying and debugging of new equipment can be directly attributed to *chronic cause* variation that existed prior to automating the process. When automation cost overruns are analyzed, it is apparent that the elimination of the chronic cause would have been relatively inexpensive if it had been done prior to automating. Bringing a process into control after the automated equipment is in place is expensive. A change in the material or the product/process design means modifying the nests, escapements, and associated tooling that were constructed for the automation project. If a company is not successful in bringing and holding manual operations in control, then it will probably have difficulty in carrying out an automation project.

Is Automation Needed?

The ways companies evaluate an automation project vary tremendously. Most large companies have a formal proposal process that describes the project, its costs, and the benefits. At many companies, the engineers and technicians that prepare these proposals are asked to include a second approach or competitive proposal. Companies that are being pressed in the marketplace by competition or slowing demand may not have the financial and sometimes even the technical resources to carry out a major automation project. Therefore, a less costly proposal that is able to reduce waste can be as effective as a capital-intensive automation project. If the alternative proposal is to implement a pre-automation process, the company is taking a positive step in reducing waste. This makes the company more competitive and at the same time is laying the foundation for future automation. In general, automation only makes sense when a company needs a substantial increase in output or to remove people from unsafe production processes.

Important Terms

chassis
design for assembly (DFA)
magazine
manufacturability
mechanization
mistake proofing
poka-yoke
pre-automation program
productivity/quality improvement program
time study

Questions for Review and Discussion

1. When should automation be considered?

2. List the steps that a company should take prior to automating an operation.

3. Under what circumstances would a company decide that automation is not needed?

4. Should a company apply DFA techniques on all their products or just those parts that are going to be automated?

5. How does mistake-proofing relate to automation?

6 Automated Manufacturing Systems

Chapter Highlights

- Automation is a repeatable operation or process that relies on several components working together.
- Both hard and soft automation have a place in manufacturing.
- Automation has advanced quickly with the availability of inexpensive computed-based control systems.
- Programming control systems for automated machines and processes has been facilitated with the integration of product design software and manufacturing software systems.

Introduction

The word *automation* was coined in the first half of the twentieth century by the automotive industry to describe the process of applying automatic control devices to production equipment. In those early days, automation was generally limited to controlling a single machine or process.

When we compare the earlier methods used to automate machines and operations in the automotive industry with the sophisticated automated systems used today, there seems to be little similarity. However, all forms of automation—regardless of their degree of sophistication—have three basic ingredients. These can be classified as the *building blocks of automation*. These building blocks are:

- A repeatable manufacturing operation or process.
- A control system.
- A material placement system.

If a particular manufacturing operation or process utilizes all of these building blocks and requires no human intervention, it can be referred to as an automated operation or process. Linking several of these processes or operations together produces an *automated manufacturing system*. See **Figure 6-1**.

One of the most difficult obstacles to overcome when creating an automated system is *variability* (an inconsistency in a process) when manufacturing a particular product. Variability can exist in either the manufacturing operation or process, or it can be present in the materials being used. If automation is to be successfully implemented, all sources of variability must be controlled. Sometimes this is difficult to accomplish.

The previous chapter discussed the appropriate first steps for planning and preparing for automation. A key step is minimizing the variability of the operations being considered for automation. This can be accomplished by implementing the concepts of statistical process control (SPC). In most cases, the process of automation takes place in stages—an evolution, rather than a revolution. Once the sources of variability are controlled, other manufacturing operations and processes can be automated and then linked together to create a more comprehensive and integrated automated system. In later chapters of this book, hundreds of manufacturing processes and operations will be described in detail. Virtually all of these processes or operations can be automated once they are under control.

After all the pre-automation steps have been completed and the automation project is still viable, the design phase of the control and material placement systems can begin. This chapter will emphasize methods used to create these systems. It will

also touch on design considerations that affect the selection of hardware and control systems used in automation.

The Nature of Automation

To develop a better understanding of the nature of automation, it will be helpful to review what is and is not required in manufacturing to achieve automation. Automation does not require large batch sizes, or a high volume of mass-produced parts. However, high-volume production is based on the concept of *interchangeability* of components and materials, a concept that is necessary for automation. This means that modifications or adjustments to individual parts are unnecessary because each part in a particular job lot is identical to every other part in that lot. Before interchangeability can be accomplished, components and materials must be standardized. Once this is done, any part, assembly, or material in a manufactured lot can be substituted for any other one in the lot.

To achieve interchangeability, you must control variability. Controlling variability, in turn, is necessary to implement automation. Thus, reducing variability in an operation can provide an improvement in quality and, therefore, a reduction in costs, even without automation. It is important to remember that neither high volume nor mass production is required for automation.

Automation is not the same as mechanization. Often, mechanization increases the rate at which work is being done by reducing variability in the characteristics and cycle times of that operation. Theoretically, mechanization also enhances the machine operator's abilities. This is why it is often viewed as one of the fastest ways to improve worker productivity. Automation relies heavily on mechanization, but it does not require an operator who functions as an integral part of the production cycle.

Automation does share the primary goal of mass production and mechanization—increasing productivity and reducing cycle times to improve throughput in a manufacturing system. *Throughput* is the amount of product that moves through a production operation, from start to finish, in a given period of time. Automation may also achieve a reduction in the cost for direct labor, since it uses the concepts of mechanization and control without human intervention.

Figure 6-1. An automated manufacturing operation.

The primary goal of automation is to improve productivity. Improvements in productivity will be realized if the manufacturing system is able to directly translate the automated system's output into throughput. This translation is dependent on manufacturing management's direction and support, as well as the reliability, availability, and maintainability of the hardware and tooling in the automated system.

Now that some of the fundamental requirements for automation have been identified, it will be useful to examine four criteria that shape the nature of automation. Two of these criteria, replication and volume, are dependent entirely on the product being manufactured. The other two, control system sophistication and manufacturing flexibility, pertain to the manufacturing process.

Figure 6-2 is a diagram that illustrates the effects that the four criteria have on the manufacturing process. All types of manufacturing systems, ranging from those that are manual to those that are totally automated, can be portrayed in such a diagram.

Cases that would fall in the upper-right corner of the diagram would include processes with the most sophisticated control system and highest flexibility. This type of system would most closely approximate the artistic capability of a human being. It can be argued, of course, that machines that can mimic the capacity of the human brain cannot be created at this time. It might also be argued that a highly skilled human being will not be able to achieve the high

volume and uniformity of output that can be realized from an automated system with dedicated tooling. See the lower-left portion of the diagram in Figure 6-2.

These two extremes illustrate points on a continuum that are defined by the four criteria of automation. This provides a framework for identifying appropriate control and material placement methods for automation in manufacturing. With the general framework for defining automation established, it is now possible to focus on the other factors that shape the control and placement systems used in automation.

Automation is important to both primary and secondary manufacturing firms. ***Primary manufacturers*** are concerned with the production of industrial stock. These manufacturers produce petroleum products, pharmaceuticals, chemicals, and food products. Production at these facilities is usually highly automated. In these industries there is a well-established technology for material control and movement with piping, valves, pumps, and fluid processing equipment standardized and readily available.

Secondary manufacturing firms are concerned with the production of discrete products. These companies manufacture products that have specific characteristics and functions, such as appliances, circuit boards, or power tools. The control and material placement systems used to manufacture discrete products are generally tailored more to the specific application than the systems in process, or primary, manufacturing. In primary manufacturing, automation can often be purchased as a ***turnkey system*** (a package that is standardized and ready to operate once purchased). Secondary manufacturers often must create systems that are tailored to their unique needs.

In general, the three building blocks of automation can be found in automated systems employed by both primary and secondary manufacturing firms. The design emphasis by primary firms is generally on the process and control system. Secondary manufacturers concentrate on the control and material placement systems. However, in both cases, the design and development of an automated process must address each of these basic building blocks: repeatability, control, and material placement.

The two major types of automated systems, hard and soft, were identified in Figure 6-2. In ***hard automation***, the machines and tooling are designed to produce a specific part or a family of similar parts or assemblies. Consequently, the control system must

Figure 6-2. Diagram showing the relationship between volume and flexibility.

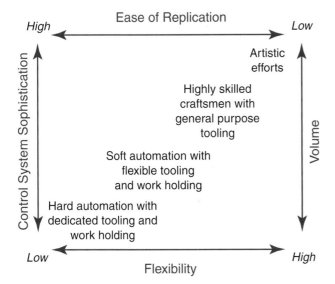

deal with only a finite sequence of logic and decision points defined by the structure of the hardware and tooling. This type of automation is dedicated to production that requires only infrequent changes in product or component configuration. Examples would be food canning, bottling operations in the beverage industry, and nail-making. **Figure 6-3** shows part of a hard automation system where closures for metal beverage containers are being distributed from one operation to another.

Hard automation focuses on the mechanical components of the automated system. Because of this, the relative complexity and cost of automation is concentrated in the hardware and tooling components, not in the control system.

Soft automation refers to a system that can handle a wide variety of different shapes and material characteristics. Typically, the control system is very sophisticated. The hardware, which is designed to accommodate a wide range of components and materials, can appear to be deceptively simple. This form of automation makes extensive use of modern computers and software that can accept a variety of inputs from sensors. A *sensor* is usually an electrical device that receives different types of information. This will be discussed in greater detail later in the next chapter.

The software that defines the logic of the control system can call from memory the subroutines that are appropriate for the circumstances. Generally, the control system software can be easily changed to accommodate different material or assembly requirements.

A more sophisticated adaptation of soft automation includes an element of *artificial intelligence (AI)*. As a control system concept, AI provides a great degree of flexibility. The inputs from sensors, coupled with predefined rules and relationships stored in the computer's memory, allow the control system to provide expert control in well-defined situations. Some of the AI software in use actually enables the control system to "learn," giving the control systems of this type a limited but human-like reasoning ability. It can provide simple perception systems that are able to process information in various forms, including vision, speech, and movement. Robots are a good example of the type of hardware often used with this form of intelligent control.

The preceding discussion on the nature of automation made it clear that hardware and control systems are closely tied to each other. This means that one system cannot be designed without giving consideration to the other. It should also be evident that there are several types of control systems and they incorporate many forms of technology. It is important now to look more closely at the unifying concepts that form the basis for all control systems. These concepts will be useful when you must design or select control systems appropriate for specific automation situations.

Control Systems

Control systems contain the *logic*, or principles of reasoning, that is designed into the automated system. A control system must also be able to make appropriate decisions to operate a particular process, operation, or production system. As an example, if material runs out, the system should shut down or sound an alarm rather than continuing to create bad parts or just waste time.

Manual control is the simplest type of control system. In a manual control system, an operator is required to start, stop, or adjust the process by pushing buttons, turning knobs, or engaging levers on the machine. *Automated control systems* for manufacturing also must be able to start, stop, and sequence production (advance parts). However, automated control also must simultaneously monitor the quality of the product and the functioning of the system.

Once the sequence of operations has been initiated, the control system must be able to carry out the predetermined functions, regardless of the number or complexity of these tasks. Consequently, safe operation must be an integral part of the control system; it must not be left to human intervention.

Figure 6-3. An example of hard automation.

The basic problem in designing a control system is determining what standards or references must be adhered to in order to satisfy the operation and complete the cycle. Consider the following example. Does a control system determine if a bottle is full on the basis of its time under the spigot, or the height of the material in the bottle? Your response to this question will define the type of control system that will be needed. There are two basic types of control systems—open-loop and closed-loop.

Open-Loop Systems

An **open-loop control system** attempts to meet a preset standard without monitoring the output or taking corrective action. With the example just presented, an open-loop system would be appropriate if you chose *time under the spigot* as the standard for control. It is easy to visualize that the actual output, material dispensed to the bottle, may deviate from the desired output, since there is no feedback in this type of system. An open-loop system is composed of an input, controller, actuator, and an output. See **Figure 6-4**.

The **input** part of the system represents the value for a characteristic such as time, temperature, or pressure. The **controller** in such a system provides the logic and governs the action of the actuator. The **actuator** (output device) is a piece of equipment or mechanism that responds to the signal from the controller. Typical output devices are electric motors, solenoid valves, and relays.

Examples of open-loop control systems can be found in such household appliances as dishwashers and clothes dryers. Satisfactory action depends entirely on the appropriateness of the characteristic selected by the operator. The actual results may vary from the expected to the extent that other variables can influence the operation or process under control.

Closed-Loop Systems

Closed-loop control systems include the same functional components as an open-loop system, but have one significant additional feature: feedback. A closed-loop control system consists of an input, error detector, controller, actuator, output, and a feedback circuit. See **Figure 6-5**.

With this type of control system, the actual output is measured and then compared to a preset standard or reference. This is called *feedback*. For

Figure 6-4. An open-loop control system.

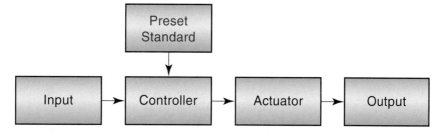

Figure 6-5. A closed-loop control system.

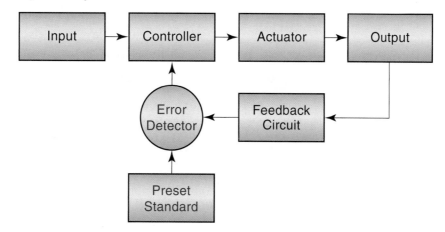

example, in the bottle-filling operation, the height of the liquid in the bottle is detected and compared to the standard. Any measured deviation from the standard (detection of error) results in a compensating action from the control system. With this compensation, the bottle will be filled to the desired height.

There are four criteria used to evaluate the performance of a closed-loop control system. They are transient response, steady-state error, stability, and sensitivity. These are shown in **Figure 6-6**.

In general, *response criteria* provide a means to illustrate the system's ability to achieve a desired output. A *transient response* describes how the output device under control responds to an input before steady-state conditions are achieved. Ideally, the output should immediately match the input requirement. However, as shown in Figure 6-6, there are several possible response conditions.

Two different applications will help illustrate why a control system must respond differently. In an application such as a drying oven, it is crucial to avoid overheating items near the heat inlets; an *overdamped response*, or slow response, therefore is desirable because the temperature of the oven will rise slowly. Temperature adjustment does not have to be made rapidly to match input and output.

In a different type of application, such as maintaining tension on a wire feeder in a coil-winding machine, the system may require a *critically damped response*. In this case, tension adjustments must closely follow input requirements to avoid snarls.

Figure 6-6. Different types of transient response by a closed-loop control system are suited to different situations. A critically damped response, for example, is needed where little deviation from the input requirement is permitted: response must be fast and accurate. Underdamped and overdamped reponses are allowed under less stringent conditions.

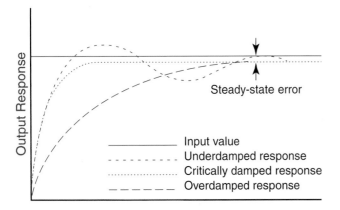

Steady-state error is another characteristic that is present in some degree in every control system. *Steady-state error* is the difference between the actual output and the desired output required by the control system. Often, this error is minimized by recalibration of the input signal.

Stability is a term that refers to how well the control system is able to maintain steady-state conditions. If the control system lacks stability, it will oscillate around the desired output level.

Sensitivity is the ratio of change in output to the change in input. This characteristic can be used to gauge the ability of the system to correct for small changes in input. In *adaptive control systems*, systems that provide control in response to changing conditions, sensitivity is a particularly important design characteristic. This will be discussed in greater detail later.

The four criteria used to evaluate the response of control systems have a direct relationship to the type of actuators that are chosen for use in the system. Actuation can be expressed in one of two ways: on/off or proportional.

On/Off Control

Although other factors can influence the criteria for designing control systems, the most important consideration is the mode of actuation that is chosen by the designer. In many cases, actuators are on/off devices such as ac motors, solenoid valves, and relays. Many control systems are based on the on/off mode of operation.

Relay logic, a system based on opening and closing electromechanical relays, served for many years as the basis for the majority of on/off control systems. Now, *programmable logic controllers (PLCs)* have almost completely replaced relays in automation control systems. This occurred for a number of reasons. As the basis for designing relay control systems, relay logic, also called ladder logic, worked so well that it also became the basis for programming PLCs. Technicians familiar with relay logic found it easy to understand PLCs and adapt them to existing automated systems.

Another reason for using PLCs is their reduced cost. The price of a PLC is very competitive with relays for small control systems. Often, PLCs are less expensive than relay-based systems for large, complex automated operations. However, the major reason for using PLCs does not relate to cost or to understanding the technology. When compared to relay-based systems, a PLC control system has a significantly higher level of reliability. For this reason, PLCs can be more cost-effective than conventional relay-based systems for on/off modes of control. Consider the following example:

If a machine makes 50 parts per minute, and you allow for 50 production minutes per hour, the machine will produce 5,200,000 parts per year from a one-shift, five-day per week operation. The life expectancy of a relay ranges from 200,000 to 2,000,000 operations, depending on the electrical load and quality of the relay. Thus, you might anticipate that each relay will fail and need to be replaced two to three times a year. In a small eight-relay system, assuming an average of 2.5 replacements per relay, this amounts to 20 relay failures per year.

The situation is complicated further by the fact that relays do not just quit. They fail intermittently, and then fail again. This means that there is a lot of time lost by maintenance personnel in troubleshooting and correcting problems. For this reason, lost production time to handle each relay failure might average four hours.

This boils down to 10,000 parts that are not manufactured because of each failure. With 20 failures per year, this means that 200,000 parts will be lost each year due to system downtime.

Under the same conditions, a machine under PLC control would typically have a failure rate of less than two per year. Thanks to the built-in error-checking capability of the PLC and its light-emitting diode input/output registers, downtime might be reduced to less than two hours per failure. Lost production would be only 10,000 parts annually.

An added advantage for using PLCs is that they are easier to modify or expand than corresponding relay-based systems. Consequently, whenever an on/off mode of operation for an application is appropriate, a PLC should be considered.

Proportional Control

The proportional mode of control also has an established place in automation control systems. Often, these systems have a dedicated electronic control system with an appropriate analog interface to actuate the device. **Proportional control** is heavily used by primary manufacturing firms to control automated processes. In most cases, if the operation requires variation of speed, tension, or temperature, the proportional control mode should be considered.

Control System Design

Regardless of the *mode* of control selected, the design of the control system generally begins with the identification of the outputs to be controlled and their states of operation. The **states of operation** describe the action or condition of an output device. Examples are an ac motor with two states (on and off) or a single solenoid valve for an air cylinder, also with two states (energized and deenergized).

Let us take a moment now to apply some of the concepts that have been presented. Before you can design an automation control system, there are some preliminary steps to be taken.

- First, you must identify the output devices that are to be activated or controlled.
- Next, you must determine the states of the output devices.
- Finally, you must identify an input that would cause the control system to activate the output device and result in the desired change in state.

After all of this has been done, a control system for the automated operation or process can be developed.

Numerical Control

One of the most important discoveries in terms of hardware related to automated systems is numerical control. The notion of numerical control of machines is often traced to the invention of punched cards used to control the weaving of complex textile patterns on a loom. The control system was developed in 1801 by Joseph M. Jacquard, a French textile weaver. It was almost 150 years later, however, that a practical numerical control system was applied in the metalworking industry.

The practical application of numerical control really began in 1949, when John Parsons, owner of a Michigan machining company, developed the concept of using numerical control to produce contoured parts using information stored on punched tape. At that time, his company was using tabulating equipment to generate the incremental elements for tool paths to machine patterns for helicopter blade airfoils. The process that Parsons established became the basis for **numerical control (NC)**. Numerical control, in and of itself, is not a manufacturing process, but a method for automating manufacturing processes.

Parsons recognized that a tool path can be described using a series of vectors. This means that the tool moves along a path comprised of the resultants of all the axes being controlled on the machine tool. Here is how Parsons tested out his approach, which will help you to visualize this process.

Parsons decided to test his theory of control by using a three-axis milling machine. To do this he needed the help of three machinists. One machinist ran the X-axis handwheel. This controlled movement of the table from left to right. A second machinist controlled the Y-axis. This moved the table from front to back. A third machinist controlled the Z-axis, the up-and-down movement of the spindle. Next, Parsons read the data, line by line, defining the tool path and called it out to the operator controlling each axis of the machine. After a line of data was read, the machinists would turn their handwheels according to the instruction. This was the beginning of numerical control.

The first actual machine tool to be operated by numerical control was developed in 1952 by the Massachusetts Institute of Technology (MIT). The machine was able to adapt a computer to read the encoded data and actuate motors that moved the machine the desired amount along each axis. The first commercially available NC machining center was produced by Kearney and Trecker in 1958. This machining center was able to change tools automatically so that milling, drilling, tapping, and boring could be done on one machine.

How NC Controls a Machine Tool

Next, we will look more carefully at how the principle of numerical control can be used to control a machine tool. Think of the lead screw that can be found on virtually all milling machines. Generally, each revolution of this screw advances the table 1/10". The dial on the lead screw wheel is divided in 100 equal markings so that as each mark goes by on the dial, the lead screw advances 1/100 of 1/10", or 0.001". To move the table 1/4", the operator rotates the lead screw wheel two full turns and then one additional half turn. This makes it very easy to move the table of a milling machine to a specific position.

A *numerical control program* contains precise instructions for each move along each machine axis. Each instruction is sent by the machine control unit (MCU) to activate the specified output, such as an electric motor attached to the lead screw. When the motor is energized, a sensor on the lead screw records a pulse that occurs as each division on the dial passes a reference point. Once the number of pulses representing the distance to be traveled is reached, the MCU will no longer actuate the motor turning this lead screw. The MCU is now free to act on the next instruction.

NC Programming

Numerical control is a type of programmable machine tool automation. The machine is controlled by a program containing a set of instructions that can be stored on a magnetic tape, diskette, compact disc, hard disk, or in computer memory. The program defines the tool-path as well as auxiliary operations.

In the transition from manual to NC machine control, the skill and knowledge needed by the machinist to machine a part had to be included in the programmer's set of instructions. The effective operation of the NC machine tool thus became the responsibility of the programmer. The programmer determined the moves, sequences, tools, and overall motions that the machine tool used in machining the part. Under NC, the primary responsibility of the machine operator is monitoring of the machine tool's operation, rather than controlling it.

As NC became more popular, programs were written to simplify the control of the machine tools. Programming languages provide a means for describing the machining operations so that it is understandable to technicians and machinists and able to be translated by a computer into NC code. Some of the programs available to do this are APT (Automated Programmed Tool), ADAPT, IFAPT, MINIAPT, AUTOSPOT, COMPACT II, ACTION, SPLIT, NUFORM, and UNIAPT. Of all the languages available, APT is the most commonly used, and was the first developed to use English-like statements. Many of these languages are still in use because companies have not replaced the machine tools and their controllers that utilize these languages. However, software is available that is more sophisticated and will generate the instructions that control the machine tool directly from the CAD drawing.

Examples of these sophisticated programs are TekSoft's PROCAD/CAM, MasterCam, and Pro/ENGINEER Complete Machining software from PTC. These software products integrate computer aided design (CAD) software with powerful programs that develop machining parameters appropriate for the part geometry and the capability of the machine tool that will make the part. If there is a change in the part geometry or machining parameters, the software can easily and efficiently regenerate the tool-path. Thus, how the part is made has become an integral function of the design of the part. How this is done is fairly simple.

When the part design is completed the designer then switches the program to the machining mode (CAM) and begins to define how it should be made. Once all the surfaces and operations are defined, the designer can view a simulation of the part being made. Since virtually all major CAD/CAM software is now three dimensional, the simulation provides an excellent representation of the finished part. With this solid modeling feature, many of the production problems that could only be found during an actual pilot run can be discovered and solved on the computer. The simulation also allows the programmer/designer to define convenient ways to locate the part, establish entry/exit points for tools, and specify rapid traverse paths. Once these parameters are established, the file is *post processed*. This is the process of generating machine code without any further involvement from the designer.

CAD/CAM software has been a tremendous cost and time saver for manufacturers. Currently the CAD/CAM software market is very fragmented. There are several dozen significant producers of CAD/CAM software packages. However, most of the software programs have provisions for importing and exporting part geometry to other programs, limiting problems of incompatibility.

NC Machine Tool Control Systems

There are some fundamental differences in NC machine tool control systems that can be useful in helping to classify them. There are two types of NC control systems: *p-t-p sys* and *c-p sys*.

The first type of NC control is called a positioning or *point-to-point system*. This type of system moves the working spindle to specific locations on the workpiece, and then performs the programmed sequence of events. These events are typically drilling, reaming, threading, or spot-facing. The location for the operation is specified in a two-axis coordinate system. Movement is typically confined to one axis at a time. When the table or the tool moves to a new location, it generally does so without regard to a specific path.

In point-to-point control, the software defines only the centerline of the tool. Consequently, there is no concern for specifying the tool diameter or tool offset.

The second type of NC control system is called contouring (surfacing) or *continuous-path control*. This system incorporates two requirements that are not present in point-to-point control:

- The tool feed rate is controlled at all times.
- Cutting tool offsets or compensations are specified for every point on the tool path.

The continuous path generated by the NC part program defines the tool path and the appropriate feed rate for the entire machining operation. The constraints on feed rate input by the programmer for a specific machine tool are predicated on the size of the cut, rigidity of the machine, characteristics of the tool, and the material being machined. In application, continuous-path control runs the gamut from very simple machine control to extremely sophisticated systems that are able to generate complex curved paths.

There are two variations to the numerical control process that have been made possible through refinements in technology. The first of these was used in the 1960s. Initially called direct numerical control, the term was later changed to *distributed numerical control (DNC)*.

In DNC, individual NC machine tools are connected through communication lines to a central command computer. Most contemporary applications of DNC utilize a small personal computer (PC) located near the machine tool and connected to a larger command computer. Such a configuration permits enhanced memory and computational capability. The advantage of DNC over conventional NC is distributed control, resulting in improved efficiency.

Today, a more popular adaptation of the basic process of NC is called Computer Numerical Control, or CNC. With CNC, a microprocessor is built into the control panel of the machine tool. This enables the CNC operator to modify programs when necessary or to prepare programs for unique parts.

With CNC systems, the master program is driven by a large computer at a remote site. CNC systems provide greater flexibility, accuracy, and versatility than NC or DNC systems.

Adaptive Control

Adaptive control enhances the capability of manufacturing processes that utilize numerical control and computer numerical control. Adaptive control systems can respond to the immediate conditions being encountered by a particular machining operation and make necessary changes in feed rates or other factors. Adaptive control became possible with the advent of miniaturized sensors and transducers that could measure the forces and temperatures created by the cutting tool. These input devices, used with low-cost industrial computers, have made it possible to design and implement a control system that responds to a time-varying operating environment while providing feedback in a closed-loop system.

Adaptive control systems are able to change *input values*, arbitrary standards or references that reflect circumstances at the time they were established. This reduces the degree of error between the system's output and the real-time conditions that are impacting the system. In practice, adaptive control would assess a current condition, such as the heat generated at the cutting tool, and then make adjustments to the machine if necessary. This might result in slowing down the speed that the material is revolving, decreasing the depth of cut, or even adding more coolant.

Computer-Integrated Manufacturing

When numerical control is coupled with automated part handling and tool changing systems, tremendous potential for use in many manufacturing applications is provided. Numerical control can be viewed as a major means for integrating the total manufacturing operation, since it enables the processing of data from such dissimilar functions as design engineering, manufacturing engineering, and machining. Today, numerical control is a major influence in the design of automated systems and integrated manufacturing operations.

The concept of numerical control is frequently and appropriately used as a model of integrated manufacturing. Because of this, it is often seen as an example of *computer-integrated manufacturing (CIM)*. However, such a perception is inaccurate. NC, in and of itself, is not CIM. Computer-integrated manufacturing systems can be developed with or without numerical control.

The major difference between NC and CIM is magnitude. In addition to the design and manufacturing operations that are served by numerical control, CIM systems generally manage data planning, sales, accounting, and management functions.

Computer-integrated manufacturing involves using computers to link together the various control systems that are found in manufacturing. When properly implemented, CIM ties every aspect of a factory into a structure that makes all of its component parts and control systems visible to any user at any point in the system. This is necessary before firms can accurately understand what happens when a particular product is manufactured.

Computer-integrated manufacturing is now viewed by many manufacturing firms as an essential strategy for remaining competitive in a global market. Effectively designed CIM systems can result in significant reductions in engineering design cost and overall lead time. The goal that many manufacturers have for implementing CIM is to make the company a responsive and lean manufacturer, increasing its manufacturing flexibility. Large production lots therefore are not necessary; in fact, it is often said that with CIM, the ideal lot, or batch, size is one piece.

Control System Design Requirements

Control systems can be designed around a variety of different types of actuation devices. The most popular types are mechanical, hydraulic, fluidic, or electronic. A system may also combine two or more of these types of devices.

Electronic systems may have the logic and input/output functions hardwired, or they may be programmed through software. Software-based systems can involve the use of PLCs, dedicated industrial computers, or a computer system integrated with other operations.

Regardless of the hardware and type of actuation devices used, a control system for automation must be capable of carrying out some basic operations or functions.

- It must be able to initiate, coordinate, cycle, and stop the various motions of the machine and its individual stations.
- It must have the means to ensure that the machine does not begin or continue to operate in an unsafe condition.
- It must be able to monitor and identify incomplete or defective products.

The first two functions, addressing machine actuation and operational safety, are standard requirements for machine control systems. Standards are covered by OSHA (Occupational Safety and Health Administration) regulations and industry standards such as ANSI (American National Standards Institute) specifications.

The third function emphasizes monitoring to assess production output. Monitoring should include techniques for incorporating the concepts of statistical process control (SPC) into the system. This approach provides a basis for control that is predicated on preventing unacceptable production. It also helps the control system designer focus on inputs and methods for detecting errors. Both of these elements are necessary in order to meet the final operational requirement of the control system.

Material Handling Systems

There are many different types of devices and systems that are used to provide material to automated processes and assembly machines. In general, material-handling methods can be grouped into three major types or classifications: bulk systems, feeding/orienting methods, and magazine feeders. See **Figure 6-7**.

Bulk systems are designed to dispense liquids, gases, and granular solids. These types of handling systems need only to contain and convey the material to the dispensing mechanism. The material is such that it does not need to be oriented or sorted. The dispensing mechanism only needs to distribute the desired amount of stock to the machine.

A *feeder system* contains and conveys discrete components while simultaneously orienting and sorting them. Typically, parts to be handled have a specific geometry that must be oriented and sorted from other components that do not conform to the specified configuration. This **orientation**, or positioning, allows the dispensing system to capture and hold the part until it is ready to be placed and assembled or processed. A vibratory feeder with internal tooling is probably the most popular type of feeder in this class.

Magazine systems are material handling systems that contain and convey pre-oriented parts. This method is used when bulk or vibratory feeding methods could cause the parts to become tangled or damaged on the way to the dispensing system.

A variation of magazine feeding is called *reel feeding* or *strip feeding*, **Figure 6-8**. With this type of feeding, the parts or materials are linked together to provide easier handling and control. The reel or strip is fed into a processing or assembly operation. Electrical terminals are frequently handled in this way, since they have the tendency to jam if fed individually.

Another common example of a magazine feeding system application is supplying coiled strip steel to a die stamping press. In this case, the coil is an economical and convenient method for packaging and handling the material.

Material-Feeding Techniques

Material feeding consists of orienting and providing a method for propelling components. Vibratory bowl feeders can move components and, with the help of bowl tooling, can separate and orient them. However, the act of orienting does not in itself control or place a part for acquisition by the

Figure 6-7. The three types of material handling systems.

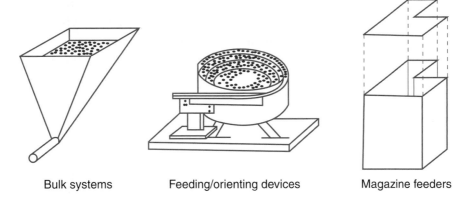

Bulk systems   Feeding/orienting devices   Magazine feeders

Figure 6-8. Two types of electrical terminals are provided to this connector assembly machine, using the strip-feeding method. Note the two reels at the top of the machine, each feeding a strip of connectors. The operator is checking some connectors after assembly. (AMP Inc.)

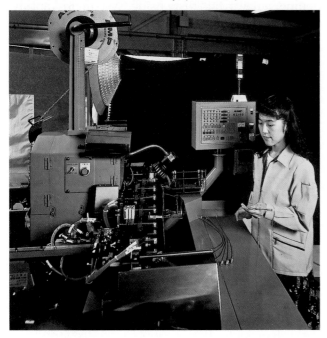

tooling. This is the role of the dispensing system or *escapement*. Specifically, an escapement is a device for orienting and dispensing individual parts or components to a workholding fixture.

Typically, parts coming from a feeder are fed into a track. Once in the track, the parts must come to a stopping point where an individual component can be picked and placed or stripped onto a workholding fixture or nest. After a part is removed, the escapement immediately replaces it with the next one in line.

The component feeding system is always designed to provide more parts to the escapement than are actually required by the operation. Consequently, the escapement must be able to accept parts only when needed. It must also be able to withstand pressure from the feeding system without jeopardizing the orientation and control of the part that is being presented for selection.

Stopping parts at the pick-up point and withstanding the feeding pressure are major challenges in designing an escapement. Sometimes parts are fragile or tend to tangle. In cases such as these, the escapement must be able to isolate the controlled part from the part feeding system.

Once a part has moved into position at the pick-up point, there are several ways to eliminate feeding pressure. One method senses when the escapement is filled, and activates a gate or chute that causes the feeder to spill off the parts, relieving pressure on the escapement. A second method eliminates feeding pressure by using a switch to shut down the feeder.

There are many types of escapements in general use. The simplest is the *dead-end escapement*. This type of escapement tracks components to a fixed point, the dead end. This point is defined by a barrier, which contains the part until it is picked.

The part contained by the barrier also acts as a barrier by blocking the next part from moving into the pick-up point. After the part is removed at the pick-up point, the remaining parts in the track move ahead until the next part encounters the barrier.

A more positive approach uses an escapement similar to the one shown in **Figure 6-9**. This escapement has two fingers that are able to capture and place individual parts. In this instance, the parts are machine screws that are being supplied by a vibratory feeder aligned to a track leading to the escapement. It takes the escapement two cycles to advance a part into the pick-up position. In the first cycle, it captures the part; in the second cycle, the screw is placed into position. The pick-up position, in this case, is the jaws of an automated screwdriver, which assembles the screw into the product, a wiring block.

Figure 6-10 shows a similar assembly operation involving the placement of a machine screw in a fuse block. This photograph clearly illustrates the use of

Figure 6-9. A view of an escapement.

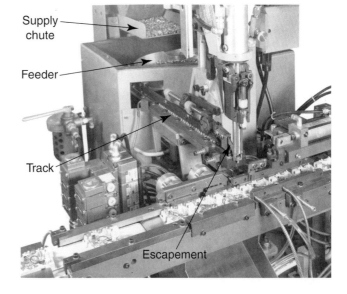

Figure 6-10. Two escapements in use for an assembly operation.

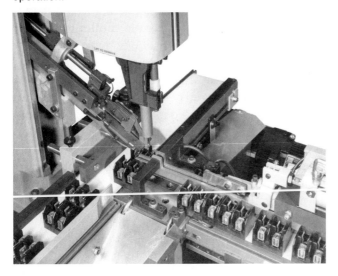

two escapements—one for the machine screws, the other for the fuse blocks. These escapements are able to locate and position both parts, eliminating the need for nests or other fixturing to hold the assembly. Escapements used in this manner can simplify and reduce overall costs, if the components lend themselves to this form of placement and control.

A third variation in escapement design provides for the tracking and control of parts in one orientation, then placing them in a different orientation at the pick-up point. This change in orientation may be a shift in the plane in which the part is located or a change in the angle of the part. A familiar example of this type of escapement is the cassette transport mechanism in front-loading VHS tape decks. This escapement moves the oriented cassette into the tape deck and then drops. This change in plane positions the cassette in alignment with the tape heads.

Despite the fact that escapements can become very complicated mechanically, their function is really quite simple. In general, escapements are mechanical devices working in tandem with feeders. Their purpose is to present oriented components at a fixed point and at a rate that exceeds the processing rate of the automated process or assembly operation.

After a part or component has been fed, oriented, and then positioned by an escapement, it is ready to be placed for assembly or processing. The means used to move and place components are termed *part-transfer mechanisms*. These mechanisms are often considered to be extensions of escapements. However, because of the variety of methods in use today, they merit consideration as a specific technique.

Part-transfer mechanisms take the part from the escapement pick-up point and place it in a work-holding fixture or nest. The methods used to make these transfers can be as simple as dropping the part into the nest, or they can be very complicated. The more complicated the transfer action, the more costly the device.

One of the simplest and most widely used transfer methods is a wiper mechanism. The part to be transferred is placed in position by an escapement, and then wiped or swept away to a nest or fixture by the transfer mechanism. A good example of this type of transfer can be found in a magazine-fed bolt-action rifle. The wiper is the bolt, which moves the cartridge from the escapement into the barrel of the rifle, the nest.

A variation on this method uses a mechanism in the escapement to hold back the part until the nest or fixture is in place to accept it. When the part is released by the escapement, it drops into the nest. In some cases, the part drops into a shuttle, which carries it from the escapement to the nest. The shuttle is similar in concept to a coin slide on a gumball machine.

Shuttles

A more complicated type of shuttle is shown in **Figure 6-11**. In this instance, the shuttle arm is able

Figure 6-11. A shuttle can be used to move and position a part.

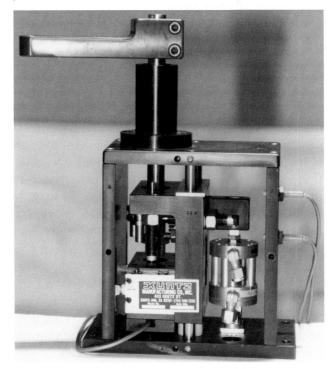

to move up and down and also rotate. Although this is a standard shuttle, it can be easily customized to accept tooling for specific applications. The use of standardized shuttle mechanisms is popular with machine designers.

Shuttles are widely used to link escapements to a process. Although they add complexity and cost to the material placement system, they are cost-effective in many applications. The use of a shuttle between the escapement and the operation or process being served should be considered if any of the following conditions is present:

- The escapement's operation is jeopardized by heat, overspray, or other dangers from the processing equipment being served.
- The part must be reoriented before it can be placed.
- Other parts are also fed into position at this station.
- The part being fed is an option, and thus, is not always required.
- The escapement or operation is subject to jams or frequent maintenance. In this case, the shuttle allows the feeder and escapement to be more accessible.
- Manual feeding of components must be done at one station of an otherwise fully automated system. The shuttle provides an interface between a manual operation and an otherwise automated process.
- The type of material or the configuration of the part being fed at this station is frequently changed.

In the first six conditions outlined above, the design, operation, or safety of a specific station may be improved by incorporating a shuttle in the process. Shuttles may also simplify the service and maintenance requirements of automated systems.

In situations where the configuration of the part is frequently changed, the shuttle provides a transition point that can significantly improve the flexibility of the system. Feeders and escapements are generally dedicated to a specific part or component configuration. Therefore, if a part change is needed, the feeder and escapement mechanism can be moved out and replaced with a new unit. The shuttle tooling serves as the transition to the remaining portions of the automated system. Specifically, the shuttle provides an easily accessible alignment point. It also facilitates testing the operation of the escapement without the need to operate the entire system.

Robots

When robots are used in an automated system to transfer material, they play a role similar to shuttles. In general, *pick-and-place robots*, those that simply pick up a part and move it to another location, can carry out two important functions. They are effective in transferring material between operations. They are also able to load and unload completed components.

In order to accomplish these functions, the robot is equipped with a gripper-type end effector. In most cases, the gripper is custom designed to handle the specific part or material being moved. See **Figure 6-12**. In nearly all applications where robotic systems are involved, the parts must be presented to the robot in a known position or sequence, and in a known orientation. An escapement can be used to provide the needed positioning and orientation of the part for the robot.

To successfully unload the part, the robot can place the part in a nest or fixture or to a defined position and orientation. Using a robot to transfer parts can provide even greater flexibility than a shuttle. The robot provides the advantages of soft automation (the ability to change easily), whereas the shuttle is limited by the fact that it is a hard automation device (designed for a specific part or operation). In job shop operations, the flexibility of soft automation is extremely important.

Moving Parts Between Stations

There are various types of mechanisms used to move a part or assembly between machining operations or material placement stations. All of these mechanisms use a fixture or a nest to control the part's orientation and location, so that it is properly

Figure 6-12. A pick-and-place robot picks up punched parts and stacks based on configuration. (Phoenix Fabrication, Inc.)

positioned at the next operation. This type of movement enables various operations to be performed on the part without having to re-orient and recapture that part. Two types of mechanisms are used to achieve these objectives: rotary and in-line transfer systems.

Rotary Indexing Tables

Indexing tables are the most commonly used methods for rotary transfer. An indexing table provides intermittent motion that accurately moves parts to successive operations located at the periphery of the table. The number and physical size of the operations performed is a function of the circumference of the table and the number of *dwell points*, or stops, in one table revolution.

A rotary indexing table with its tooling, escapements, and vibratory feeders is shown in **Figure 6-13**. The view is taken from directly overhead. This particular machine assembles a health-care product. As with most automated assembly machines, functional testing is incorporated as part of the sequence of operations.

The selection process for a rotary index table is not overly complicated. First, the designer must determine the sequence and types of operations to be accomplished. Once this information is available, the remaining tasks to be completed are identifying the number of dwell points required, calculating the weight of the parts and tooling, and determining the type of mechanism needed to provide the desired rotary motion.

There are three basic driving mechanisms in general use to provide intermittent rotary motion. As shown in **Figure 6-14A**, the *ratchet and pawl* is the simplest and least costly of these systems. It is also the least accurate, however.

Figure 6-13. An automated assembly machine using a rotary index table.

The simplicity of the ratchet and pawl driving mechanism begins with the power source used to create the intermittent motion of the table. The driver for this mechanism is frequently a single-acting air cylinder. The ratchet and pawl indexing apparatus has the advantage of rapid changeover from one index ratio to another. This allows the designer to use the same drive mechanism with several dial plates, each having a different number of points of dwell. Since the ratchet and pawl are not mechanically interlocked, the table or dial can be moved freely in one direction. This can be an advantage in clearing jams or changing tooling on the dial. However, this also creates its major disadvantage. This type of drive can overrun its dwell position. This decreases its precision and limits its usefulness for many forms of automation.

A second popular method for achieving rotary motion is called the *Geneva mechanism*, **Figure 6-14B**.

Figure 6-14. Shows three popular types of index mechanisms.

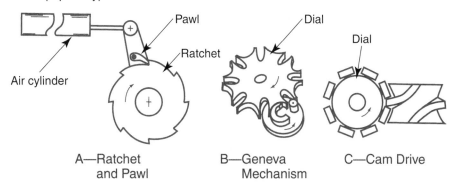

A—Ratchet and Pawl   B—Geneva Mechanism   C—Cam Drive

This method uses a continuously rotating driver to index the dial. The driven member has a radial slot machined in it for each point of dwell. Because of this slot, the mechanism has an inherent locking feature when in dwell. This locking feature ensures good positioning and accuracy for most assembly operations.

The primary drawback for this type of mechanism is the severe acceleration and deceleration caused by the basic geometry of the Geneva drive. This motion can shake or even damage components being carried in the fixtures on the dial. These high acceleration rates also cause a great amount of stress on the mechanism. Consequently, Geneva drives must be very robust to achieve the precision and long life expected of an automated system.

The *cam drive*, **Figure 6-14C**, is a third type of mechanism that provides an accurate and reliable means for achieving rotary motion. Like the Geneva mechanism, this design locks the dial when in dwell. The cam drive mechanism has relatively simple construction and exhibits excellent acceleration/deceleration characteristics.

Cam-type mechanisms used on rotary machines are also used to drive in-line transfer systems, such as continuous conveyors. Although cam drives are expensive, they have become a very popular drive mechanism for automated assembly.

In-line Transfer Mechanisms

In-line transfer systems are generally chain or belt conveyors that convey a *pallet*, a support base, frame, or tray, containing a parts-holding nest or fixture between operations. In comparison to rotary index tables, in-line mechanisms significantly expand the size and complexity that can be achieved in automated systems. **Figure 6-15** provides a view of a transfer line that uses a pallet with a standard footprint to convey products between workstations. This type of conveyance is very popular for developing in-line systems and for linking together in-line automated systems.

There are also in-line transfer mechanisms that move intermittently, bringing pallets into an approximate location at a workstation. Once the pallet is sensed as being in position, a set of locating pins accurately aligns and fixes the pallet at the station. Without locating pins, positioning accuracy is dependent on the rigidity and precision of the conveyor and the drive mechanism.

Figure 6-15. A transfer line.

The intermittent motion that causes a transfer line to increment between stations is frequently achieved with a drive mechanism similar to the cam drive used on rotary index tables. This means that the conveyor chain and tooling acceleration and deceleration is accurately controlled as is the dwell position. Once in position, the transfer line is locked during the period of dwell by virtue of the drive mechanism geometry.

There are other, less-expensive drive mechanisms that can be used. One type is a rectangular-motion device, or *walking-beam mechanism*. These mechanisms are commonly used with in-line machines where flexibility is important. In operation, the walking-beam transfer mechanism uses a transfer bar to lift the work in process (WIP) out of a workstation location and move it one position ahead to the next station. The transfer bar lowers the part into a nest, which positions it for the next operation.

An alternate approach with this method is to transfer a pallet containing the work in process. The *workholding pallet* is relocated at each station by guide pins. With this method, there is no need to place the part in a nest. Since the pallets are not attached to the transfer line, they can be changed quickly, increasing the mechanism's versatility. See **Figure 6-16**.

Overhead or track conveyors are also used to carry parts or assemblies to a nest or fixture. In these instances, the work is generally oriented and has locating points that can be easily found by an escapement that captures the component at each station.

Selecting the most appropriate method for moving materials and tooling between stations or

Figure 6-16. A transfer line using pallets.

Load-unload station

AGV

operations depends on the general characteristics of the operations and on the components being automated. The decision on which approach to take is influenced by several factors. Answers to the following questions will influence design of the conveying system:

- How many stations or operations need to be linked by the transfer mechanism?
- What is the total weight of the components and tooling to be transferred by the mechanism?
- Is it necessary for the transfer mechanism to provide location and position accuracy?
- What is the cycle time or production rate for the system?
- Do the assembly or processing stations have to operate sequentially, or will the order vary?
- Are there operations that are redundant or that can be bypassed in the assembly sequence?
- Are there any operations in the manufacturing sequence that will be done manually?

In most cases, rotary methods are appropriate for relatively small, lightweight parts that require ten or fewer operations. In-line systems are able to accommodate larger components and a greater number of stations. Although variation in sequence and the inclusion of redundant stations or pass-by operations tends to favor the use of in-line transfer mechanisms, they will not be the primary determinants for any one method of transfer. The use of a shuttle can provide for flexibility or redundancy in either mechanism. In practice, shuttles can also be used to link different transfer systems together.

Conclusion

Automation today encompasses a wide variety of technologies to create hardware and control systems that have the capability of operating manufacturing machinery without human effort or intervention. There are few limitations to the application of automation. Any operation or process can be automated if the result justifies the cost of adding the necessary control and material placement systems.

Linking several automated systems together is also feasible providing that a coherent control system can be established and variability can be controlled. When this stage of automation is reached, the information necessary for carrying on automatic operations becomes part of the management control system for the manufacturing facility. At this level of sophistication, automation has truly become computer-integrated manufacturing.

Important Terms

actuator
adaptive control systems
artificial intelligence (AI)
automated control systems
automated manufacturing system
automation
building blocks of automation
bulk systems
cam drive
closed-loop control systems
computer-integrated manufacturing (CIM)
continuous-path control
controller
critically damped response
dead-end escapement
distributed numerical control (DNC)
dwell points
escapement
feedback
feeder system
Geneva mechanism
hard automation
in-line transfer systems
input
input values
interchangeability
logic
magazine systems
manual control
material feeding
numerical control (NC)
numerical control program
open-loop control system
orientation
overdamped response
pallet
part-transfer mechanisms
pick-and-place robots
point-to-point system
post processed
primary manufacturers
programmable logic controllers (PLCs)
proportional control
ratchet and pawl
reel feeding
relay logic
response criteria
secondary manufacturing
sensitivity
sensor
soft automation
stability
states of operation
steady-state error
strip feeding
throughput
transient response
turnkey system
variability
walking-beam mechanism
workholding pallet

Questions for Review and Discussion

1. Can variability be reduced by automating a manufacturing operation?

2. Describe the difference between mechanization and automation.

3. What are the distinguishing elements of automation?

4. Discuss the differences between hard and soft automation. Provide examples.

5. List several devices you are familiar with that could serve as examples of either open-loop or closed-loop control systems.

7 Sensors and Devices for Automation

Chapter Highlights

- Sensors are devices that gather information for control systems.
- Transducers gather and convert information.
- Plug-and-play sensors make it easier to create and modify computer-based control systems.
- Wireless sensors simplify the design and modification of control systems.

Introduction

The control system for an automated process needs information to enable the machines to carry out its functions. This flow of information tells the control system where the part is, what the state of the machine is, and when a task has been completed. The ability of a machine in an automated process to sense and do its work can be almost humanlike. And, like humans, these automated machines must be able to sense the condition of its work environment to carry out a task. The components that gather this information for an automated system are its sensors.

Sensing Devices

Sensors attempt to perform the same functions that our own senses of smell, sight, hearing, touch, and even taste do for us. Automated machines without sensors would be limited to open-loop control systems (no feedback). There would not be any feedback that sensors provide in a closed-loop control system.

The powerful, yet inexpensive control systems available today have a considerable amount of intelligence and are able to handle a huge amount of information. The variety and sophistication of sensors has grown to support the computing capability of these control systems. Therefore, the number of sensors found on automated equipment has increased tremendously over the past decade.

Sensors

A *sensor* is a device that can detect a characteristic or the condition of a variable. An example of a variable is the presence of a part in an escapement. The variable has two levels, the part is present or the part is not present. The sensor detects if a part is present (the state of the variable) and sends a signal to the control system when the part is present. If the control system does not receive a signal, it interprets that to mean a part is not present.

Years ago, electrical switches were the most common form of sensors found on automated assembly equipment. These switches were called limit switches and microswitches. These switches are designed to be momentary. A *momentary switch* is one that is only active while pressed. When released, the switch returns to its *normal state*, the position during no activity. These switches can be wired so that their normal state is either *normally closed (NC)* or *normally open (NO)*. This means that the switch can be operated to close and complete an electrical circuit or operated to open and break the circuit.

Switches were certainly appropriate as sensors when used with control systems based on relay logic. They were still the most commonly used sensors when relay-based control systems were replaced with programmable logic controllers (PLCs) in automated assembly machines. During the 1980s, PLCs gained popularity because their function as well as their programming language was based on relay logic (ladder logic). Therefore, a PLC could replace a relay panel on an automated machine without having to rewire or retool all the sensors (switches) on the machine. However, by the end of the last century the desktop computer had been adapted to serve as the basis for many, if not most, of the control systems. This enabled the control systems to process a wider variety of information, communicate with other systems, and utilize programming languages that were far more sophisticated than relay logic. These new control systems, along with the availability of inexpensive large-scale integrated circuits, created a variety of sophisticated sensors.

Although computer-based control systems are the most commonly used, there are other types of systems. These control systems are based on pneumatics, mechanics, or hydraulics and do not rely on any electrical or electronic components. As an example, there are sophisticated pneumatic control systems designed to work in explosive atmospheres where a comparable electrically based control system would be a hazard. The components used in these systems are physically very different from the electrical computer-based control systems, but the logic of their operation is similar.

Transducers

Sensors react to a physical stimulus and turn it into a signal that can be measured or recorded. A *transducer* converts one form of energy to another form of energy. As an example, a microphone converts sound energy into electrical energy. The transferred energy may be in the same form or in a different form. All transducers contain a sensor. Most devices called sensors are actually transducers.

It is often difficult to distinguish how sensors differ from transducers. Specifically, a sensor measures a physical attribute or event. It responds to a stimulus, however, its function is not to convert energy from one form to another. As an example, visualize a pushbutton switch mounted to sense when a carton comes to the end of a conveyor. When the box arrives, it pushes the switch closed and completes an electrical circuit. It might be argued that the switch is a transducer because it is converting the force that the box exerts into an electrical current, but it actually is just responding to the presence of the box. Unlike a transducer, the switch is not able to vary the current flow as pressure changes, as would happen when a heavier carton comes to rest against the switch. A pressure or force-sensing transducer is able to vary the current flow in proportion to the weight of the carton.

Classifying Sensors and Transducers

Sensors and transducers are more often classified as being either analog or discrete. In the example on the previous page, the momentary switch sensing the presence or absence of a carton is a discrete device. A *discrete device* has specific defined values. The switch is either closed and completing the circuit or open and providing information on a discrete variable. Because it has only two states or values, it is considered to be a *binary device*.

An *analog device* produces a signal that is proportional to the characteristic being measured. A *thermocouple*, a type of transducer that generates a current proportional to its temperature, is an analog device. In the example, if the switch is replaced with a force-sensing transducer, the output from the transducer could have any number of values based on the weight of the box. See **Figure 7-1**. The variable would be a *continuous variable* since it could have any value from zero (no box present) to the maximum force the transducer is able to measure.

Computer-based control systems require binary, or digital, signals. However, analog signals can be used if they go through an analog-to-digital converter. In practice, nearly any type of signal from a sensor can be accepted and interpreted by commercial control systems.

Figure 7-1. A force-sensing transducer.

Types of Sensors

An automation engineer once remarked that any physical, biological, or chemical phenomena could be used to create a sensor. Sensors provide an interface between the computer-based control system and the physical world. They enable the control system to see, hear, smell, taste, and touch the physical world by converting objective physical or chemical input into electrical signals. In almost every situation, an automation engineer or technician has a choice of several different types of devices that can serve the purpose. The choice will depend on many factors such as availability, cost, power consumption, environmental conditions, and control system type.

The first step in choosing a sensor is identifying the signal type or stimulus being sensed. Although the variety of sensors available is enormous, there are only six major categories. The categories are based on signal type.

Sensor Categories and Stimuli	
Category	**Stimulus**
Thermal	Temperature
Mechanical	Position (linear or angular), acceleration, force, stress, pressure, torque
Chemical	Gases (oxygen, hydrocarbons, carbon monoxide, etc.) compound type, ions, etc.
Magnetic	Magnetic field strength
Electrical	Voltage, current, resistance
Optical	Light (visible, infrared, and ultraviolet)

Sensor performance is based on the physical property that is altered by an external stimulus. In many sensors, the altered property is an electrical characteristic. As an example, a thermocouple generates a small electrical current that is proportional to its temperature. This electrical current produced by the thermocouple is not sufficient to be used as an input to a computer-based control system. Therefore, the current must be amplified, which is called *conditioning*. If the sensor does not produce an electric signal, the output needs to be transformed and conditioned for the control system.

Sensor Performance

The performance of a sensor is dependent on its sensitivity to the measured stimulus, its accuracy, and its range of operation. Each of these measures of performance has multiple key indicators. The *sensitivity* of the sensor, a measure of the smallest change that can be detected accurately, must be accurate enough to detect the smallest unacceptable change. Any change below this point will not be detected.

The ratio of the change in output of the sensor per unit change in the parameter being measured is *absolute sensitivity*. The factor may be constant over the range of the sensor (linear), or it may vary (nonlinear). A cylindrical beaker with a constant diameter throughout its height is an example of a linear measurement. Every inch of water contains the same volume. An example of a nonlinear measurement is a gas tank that is shaped to conform to the automobile's frame. The gas gauge may indicate that the tank is half full, based on the height of the fuel, but there may be far less if the tank is narrower at the bottom.

The smallest change that can be detected by a sensor in a change of the output signal is a measure of the sensor's *resolution*. Resolution is a measure of fineness or detail. One sensor in a gas tank may be able to detect if a tank is empty, half full, or full. A sensor with better resolution could measure the amount of fuel at empty, a quarter, half, three-quarters, or full.

The ability of a sensor to detect change is not the only important feature. Since the sensor is such a vital part of the manufacturing process, the sensor must be able to detect changes accurately and consistently.

The performance of sensors is described by several factors. The designer has to choose the right values for each of these parameters if the sensor is going to meet the requirements of the control system. Some of the key factors are:

- **Accuracy.** Generally defined as the largest expected error between an actual and ideal (expected) output signal, accuracy is often expressed as the percentage of the full range output. A typical expression of accuracy is "Accuracy is ± 5 percent of the full range output."

- **Error.** This is the difference between the true value of the quantity being measured and the value indicated by the sensor. Units of the error quantity are those of the quantity being measured. As an example, a sensor whose resistance changes as temperature changes might have an error value of $+2°$ C. Typically, the error stated is the maximum for the sensors range of operation.

- **Sensitivity.** The sensitivity of a device is the relationship indicating the amount of output per unit of input. Generally, it is the ratio of a small change in the output value compared to the input (physical signal) value. As an example, a sensor whose resistance changes as temperature changes would have sensitivity expressed as .5 ohms per degree Celsius. A larger change in resistance to each degree change would indicate a more sensitive device.

- **Zero drift.** This occurs when the signal level varies from its *set zero value* even when there are no inputs present to cause a change. This drift introduces an error into the measurement equal to the amount of drift or variation from its zero point. A drift from zero may be caused by changes in temperature or aging of the device.

The sensor's ability to provide the same output signal whether the stimulus is increasing or decreasing is the sensor's *hysteresis*. The sensor should be able to provide the same signal regardless of the direction of operation.

The distance between the maximum and minimum measurable values of a sensor is the *dynamic range*. The *span (input span)* is the portion of the dynamic range that can be converted by a sensor without causing an unacceptable amount of error.

Designers of automated equipment have always been able to choose from a large variety of sensors. However, as recently as twenty years ago the vast majority of sensors used in automation were switches. They were easy to incorporate into a control circuit. As control systems and sensors became more sophisticated, a considerable amount of work has been done recently to produce sensors and their associated interface circuitry that are easier to integrate into modern computer-based control systems. The term *plug-and-play sensor* is being used to describe these new devices.

The *Institute of Electrical and Electronics Engineers (IEEE)*, best known for developing standards for the computer and electronics industry, has developed a series of *Smart Transducer Interface Standards* known as IEEE 1451. These standards were established to create a comprehensive set of hardware and software protocols. It was hoped that the standards would pave the way for a seamless connection between smart sensors, the control systems, and the software used by these systems. One of the standards is *IEEE 1451.4*, which describes the *sensor and software protocols* that provide for the uniformity and interoperability between control system components even when these components are made by different manufacturers. Essentially, *P1451.4* outlines a relatively simple approach for building plug-and-play technology into traditional analog sensors. This standard specifies the configuration and the identifying parameters of the sensor in the form of a *transducer electronic data sheet* **(TEDS)** that is contained in an EEPROM located on the sensor. EEPROM is the acronym for *electrically erasable programmable read-only memory* and is pronounced *double-e-prom* or *e-e-prom*. Like other types of PROM devices, EEPROM retains its contents even when the power is turned off.

Now imagine being able to quickly replace a transducer with one made by another manufacturer, or use that same transducer on a different piece of equipment without any configuration changes to the control system. This is what *IEEE 1451.2* is designed to accomplish. This means manufacturers of sensors will design sensors and transducers to a specific standard interface to work with different types of field buses. The *bus* is the pathway that connects all the devices in a control system.

A standard transducer interface module (STIM) described by the standard covers the sensor interface, signal conditioning and conversion, calibration, linearization, and basic communication. Some of the benefits for using sensors made to this new standard are:

- Connecting transducers, sensors, or actuators to control systems or networks is significantly simplified.
- The "plug-and-play" characteristic of P1451-compatible sensors and actuators make it easy to service or modify control systems.
- The standard allows sensor manufacturers and users to support and use multiple control networks such as remote monitoring, remote actuating, collaborative control, and distributed control.

The *P1451.4* standard makes provisions for a variety of sensor types, including IEPE (Integrated Electronics for PiezoElectric) accelerometers and microphones, IEPE pressure sensors, Wheatstone bridge sensors, strain gauges, load and force transducers, thermocouples, resistance temperature detectors, thermistors, linear variable differential transformers/rotary variable differential transformers, resistive sensors, frequency output sensors, and amplified sensors.

Wireless Sensors

For control systems and sensors, the laboratory provides an ideal environment—the atmosphere is free from contaminants, the temperature is relatively stable, the area is free from vibration, and the chance of physical damage is minimal. However, on the floor of a manufacturing plant these ideal conditions are seldom found. Therefore, sensors must be physically protected to withstand the stress of the manufacturing environment. This is especially true for the electronics and electrical connections needed for communications. The signal produced by the sensor must be capable of being transmitted to the control system, which may be a considerable distance away. The wiring between the sensor and the control system will often need to be shielded from electrical noise. The entire system must also be protected from transients (voltage spikes). Although it is common for most sensors in industrial applications to be hardwired, there are many instances where a wireless connection between a sensor and the control system would be an advantage. A wireless system makes designing simpler for an engineer, cleaning and loading easier for an operator, and servicing and upgrading faster for maintenance personnel.

The *P1451* series of standards being developed will include a protocol for wireless communication between a sensor and a control system. Work is going on to develop wireless communication interfaces and protocols for sensors in industrial applications. This will be an open wireless transducer communication standard that can accommodate the various existing wireless communication technologies. Once this standard is developed, it should enhance the acceptance of wireless technology for transducers. The major hurdle for wireless communication between a sensor and the control system will be eliminating the problems caused by interference and noise that is prevalent in a factory.

Important Terms

absolute sensitivity
accuracy
analog device
binary device
bus
continuous variable
discrete device

dynamic range
electrically erasable programmable read-only memory (EEPROM)
error
hysteresis
Institute of Electrical and Electronics Engineers (IEEE)
momentary switch
normal state
normally closed (NC)
normally open (NO)
plug-and-play sensor
resolution
sensitivity
sensor
span (input span)
thermocouple
transducer
transducer electronic data sheet (TEDS)
zero drift

Questions for Review

1. A float in a tank closes a valve when the tank is full. Is this a sensor? Explain your answer.

2. Gas tanks have a fuel level sensor to indicate how much fuel is in the tank. Is this sensor a transducer? Is the fuel level sensor an analog or discrete device? Explain your answers.

3. Why were switches so popular as sensors?

4. The IEEE has prepared a standard, *P1451.4*. Explain why this is important to manufacturers building or updating automated equipment.

5. Determine where wireless sensors are being used in industry.

A wireless sensor system is used to transfer information from critical areas of the production floor to server. These systems can be purchased in kits that contain sensor-equipped pods, a gateway to act as a junction between the pods and a data-collection server, and mounting hardware. (Accsense, Inc.)

Section 4
Selecting Appropriate Materials

8 Behavior and Characteristics of Manufacturing Materials

Chapter Highlights

- Materials engineering as a career field.
- How to identify engineered materials by their unique properties.
- Environmental legislation regulating use of industrial materials.
- The importance of Materials Safety Data Sheets (MSDS) for identifying and classifying materials.
- Overview of methods for regulating generation of hazardous materials.
- Laws regulating disposal of hazardous waste.

Introduction

It is difficult to imagine living a day without any exposure to manufactured products. Infants live in an environment surrounded by cribs, playpens, high chairs, car seats, toys, medications and other things designed to suit their unique needs. Many adolescents enjoy engineered products such as bicycles, DVDs, electronic games, medications, and products specifically marketed to adolescents. Adults use products such as cars and trucks, laptop computers, machinery at work, and health care products that become important to their daily lives. Thankfully, most of these products are well-designed and pose no safety risk to individuals using them. However, in the race to get products to market faster than the competition, some products are poorly designed, are made of inappropriate materials, or are poorly constructed. Problems with poorly designed and unsafe consumer products have become so prevalent that the federal government created the Consumer Product Safety Commission (CPSC) to provide increased protection from dangerous products. A recall of Rival Slow Cookers, because of a poorly designed handle that can break causing burns to the user, was issued in 2005. This recall affected approximately 2.6 million units.

Another example of products recalled by the CPSC are window blinds with pull cords that form a loop. In 1999, the CPSC began a new investigation of window blind deaths. They found that children could become entangled in both inner and outer cords used to raise the slats of blinds. **Figure 8-1** shows a copy of a recall notice from the CPSC on blinds with pull cords. Look at the design of the cords to see how young children could become entangled. **Figure 8-2** shows how the CPSC designed repair kits for consumers to use in correcting these design defects.

Aggressive investigation by the CPSC has resulted in the industry continuing to redesign window blinds. Window blinds sold since November, 2000 have attachments on the pull cords so that the inner cords cannot form a loop. Repair kits are also provided for older products.

Product design usually begins with a detailed study of the quality and performance of a product. Knowing what the product needs to do is not enough. It is also necessary to know what percentage of the time the product has to work during its *life cycle* (the time it will be in service up until it wears out and is discarded or recycled). If someone's life is at stake, as in the case of NASA astronauts traveling into space or race car drivers traveling at 200 miles per hour, the product has to be designed to work 100 percent of the time. Other products such as cars or trucks are designed to work most of the time. If they had

Figure 8-1. This bulletin describes the product recall for window blinds with pull cords.
(Consumer Protection Safety Commission)

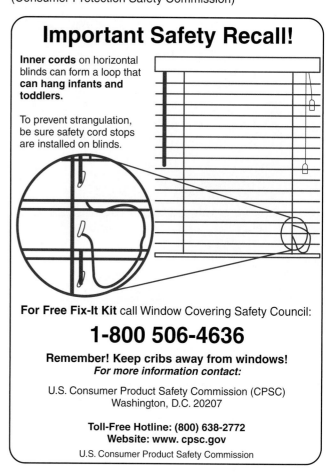

Important Safety Recall!

Inner cords on horizontal blinds can form a loop that **can hang infants and toddlers.**

To prevent strangulation, be sure safety cord stops are installed on blinds.

For Free Fix-It Kit call Window Covering Safety Council:

1-800 506-4636

Remember! Keep cribs away from windows!
For more information contact:

U.S. Consumer Product Safety Commission (CPSC)
Washington, D.C. 20207

Toll-Free Hotline: (800) 638-2772
Website: www. cpsc.gov

U.S. Consumer Product Safety Commission

Figure 8-2. Repair kit offered free by the CPSC. (Consumer Protection Safety Commission)

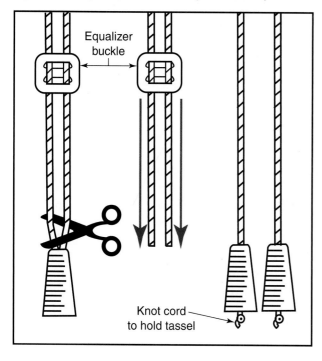

Equalizer buckle

Knot cord
to hold tassel

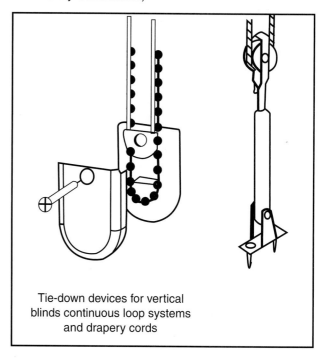

Tie-down devices for vertical
blinds continuous loop systems
and drapery cords

to work 100 percent of the time, the manufacturer's warranty would be for the life of the vehicle, not for 100,000 or fewer miles. Other products, such as hand drills, shoes, and fishing reels are designed to work for a much shorter time before wearing out. The length of time that something has to work influences the selection of the materials that go into the product. In many cases, cheaper materials will not last as long as more expensive engineered (human created) materials. However, just picking a material because it is the most expensive is not a good move either. The material has to be selected based on its ability to do the job intended.

Making appropriate choices about material selection involves more than just picking *off-the-shelf* materials that are readily available and then changing these materials using manufacturing processes. At first glance, picking the right materials might appear to be a very simple thing to do. Why would a company not start by looking at other products on the store shelves to identify materials that might be appropriate? Well, this is done, but what currently exists might not represent the best solution. More thought is needed to develop products that will sell better than those offered by the competition. For example, one type of sunblock cover for dog pens that is purchased by many consumers is made of lightweight blue plastic. These covers are less costly than engineered synthetic or canvas covers. The problem is that the low-cost covers often will not withstand continued exposure to severe weather conditions, and may deteriorate faster than covers made of other materials. What is popular may not be best.

Selecting the best material to be used in a product is an exciting activity for engineers and technologists. Often, they will need to look at different types of products entirely to get new ideas for materials to be used in the product they will be making. Selecting the best materials for the job is one of the most important things a firm can do in the quest to make great products and generate profits. Material selection is one thing that should not be hurried to get a product out the door.

There is more to the story than just using materials and processes to make great products. There are requirements that must be addressed in dealing with materials. Materials have to be effectively stored before use. Waste materials produced during manufacturing must be stored or disposed of. Treatment of these materials may have special requirements in terms of handling. Many of these requirements are controlled by laws addressing safety in the workplace. All of these things are important to people working in the field of materials engineering.

The Field of Materials Engineering

The Bureau of Labor Statistics describes many different types of engineers who are involved with the extraction, development, processing, and testing of materials used in manufacturing. Engineers dealing with the behavior and composition of metals are called metallurgical engineers. Some engineers are concerned with specific materials such as ceramics, polymerics (plastics, woods, and elastomers), and alloys. Other types of engineering emphasize performance and applications. Examples would be chemical engineers, process engineers, and product engineers.

Most materials engineers work in various types of manufacturing. These types are primarily metal production and processing, electronic and electrical equipment, transportation equipment, and industrial machinery and equipment. Materials engineers also work in service industries such as research and testing, and for federal and state governments.

Of all the materials in use today, the most common is metal. Its importance is evident when one realizes that not a moment in our day goes by that we are not using something made of metal. Metal is also the most recycled material in the world. The American Iron and Steel Institute points out that steel is the most recycled material in the United States, with the overall recycling rate typically over 65 percent. This percentage is higher than any other recycled material and total tons recycled are greater than all other materials combined.

Why Some Materials Are Used Instead of Others

There are thousands of different materials used today in the production of hard good manufactured products, also known as durable goods. Twenty years ago it would have been easy to make a fairly accurate statement about the percentage of materials that could be attributed to the final cost of a product. In the time period from 1980 to 2000, about

50 percent of the final cost of most products was for materials. Today, many specialized markets such as aerospace, defense, and transportation require high performance composites and exotic metals, raising the material costs well above the 50 percent level.

One of the major factors that influence material selection is tradition. If a company has been making a product like a cell phone, of plastic, then it is difficult for them to shift to another material. This is because the company has invested in equipment, knowledge, and training dedicated to processing and forming plastic materials. It is much more likely that if changes are anticipated, the company will gradually transition to the use of new materials.

Market research is also a major driver influencing selection of materials for a particular product. Surveys are conducted to determine what type of product has the most appeal, and what types of materials are preferred by the end users. Seldom do manufacturers engage in a bold new venture, investing in products made of unique materials, without extensive market research. Usually small production runs of a new product are conducted to test the market before significant investment is made. When the time appears right for a shift to new materials, a careful analysis must be made in terms of the cost, availability, performance, and environmental impact.

There is no need to consider materials for a product if it will fail under use. It must pass the test of reliability, and perform consistently throughout the product's life cycle without failure. Therefore, it is important to evaluate the performance of the material being considered under the same conditions as would be expected in the user's environment.

There are generally three performance criteria that are important to the selection of materials: mechanical properties, chemical properties, and physical properties.

- *Mechanical properties* include hardness, tensile strength, wearability, and toughness. An interesting aspect of mechanical properties is that it is the only one of the three performance criteria that can be altered by a manufacturer. As an example, when a screwdriver is being made, the blade is formed when the metal is soft. The blade is then heat-treated to make it hard so that it does not distort or bend during use. The mechanical property of hardness is controlled by the toolmaker.

- *Chemical properties* address the density and classification information such as corrosion resistance and flammability. Data such as atomic number, atomic symbol, atomic weight, electron configuration, and oxidation states are found in the *Periodic Table of Elements,* also known as the *Periodic Table of Chemical Elements.*

- *Physical properties* include the material's melting point, boiling point, and its behavior when exposed to heat, light, or electrical energy.

When it is determined that the material will perform reliably as desired, then it is time to move to the next level of analysis—cost. If the material is expensive, naturally the product costs will escalate, possibly to a point where the product may not be sold. Cost of the material is a factor that relates directly to availability of the material. Some firms purchase materials at a lower cost from offshore providers. Surprisingly, they may find the cost of transportation of the material to the manufacturing plant makes the final cost of the material greater than it would be if the material were purchased locally.

Other costs should also be considered when selecting materials. One of these is storage. Any extra materials must be stored before use. This is why materials should be purchased just before they are needed, and preferably from a reliable source. This process, called just-in-time, means that less money is tied up storing and protecting materials, replacing damaged items, and disposing of out-of-date materials.

Another hidden cost in materials is waste. **Waste materials** are those that are lost or consumed in the manufacturing process, but do not end up as material present in the completed product. Examples of waste are materials on a casting that are trimmed from the product. Material waste also results from overproduction and product defects. Overproduction actually reduces product quality and productivity. Making more than can immediately be sold creates work-in-process inventory (WIP) that must be stored. It also forces firms to require excessive lead times for activities to occur. This reduces the manufacturing plant's ability to reduce defects and respond to changing market conditions.

Materials that are purchased for manufacture but not used are referred to as **surplus** or excess materials, not waste. Waste is something that can drive the cost of a product sky high. If the material is difficult to form, fabricate, separate, condition, or finish, there will likely be more material waste. Materials that are difficult to process should be avoided unless the material offers unique performance characteristics.

Environmental and Safety Aspects of Materials

The federal *Hazard Communication Rule (CFR 1910.1200)*, also known as the **Hazcom Standard** or **Right-to-Know Law**, was first introduced by the Occupational Safety and Health Administration (OSHA) in 1983. Industry has been obligated to comply with this standard since 1988. The OSHA law requires employers to identify hazardous chemicals in the workplace and effectively communicate information and train employees how to recognize, use, and handle hazardous chemicals or products. This includes commonly used products such as diesel fuel, lubricants, pesticides, and a wide range of products used and produced within the manufacturing plants.

In order to comply with the law, companies must prepare a written hazard communication plan, keep an up-to-date inventory of hazardous chemicals or products used in the facility, label containers stored in inventory, prepare and post **Material Safety Data Sheets (MSDS)**, and provide training to employees on the handling of hazardous chemical and products. This includes proper training on factors such as melting point, boiling point, flash point, toxicity, health effects, first aid treatment, storage and disposal, protective equipment, and spill-proof leakproof containers. The Occupational Safety and Health Act defines an MSDS as "A compilation of information on the identity of hazardous chemicals, health, and physical hazards, exposure limits, and precautions."

Material Safety Data Sheets do not have to be prepared for all industrial materials. The following materials do not require MSDSs:

- Materials not chemically or physically hazardous, such as water, ice, saline solution, water emulsion-based white glue (most things that are formed using pressure, heated, or eaten are exempt).
- Drugs in solid, tablet, or capsule form.
- Solid products such as tools, electronic components, fasteners, or fabricated parts that do not require processes that create toxic dust or fumes. Some metals do not require an MSDS. However, galvanized steel has a zinc coating that gives off toxic fumes during welding. Galvanized steel requires an MSDS to provide information to the welder.

- Consumer products housed in the same packing container are protected by the Consumer Product Safety Commission, not OSHA. Therefore, an MSDS would not be required. Protection would be covered on the product label.

A Materials Safety Data Sheet contains detailed information on the hazards of materials or products. MSDSs must be available to all employees at the worksite during each shift. Information required by OSHA on each MSDS is as follows:

- Section 1: Chemical Identify (Using both chemical and common names)
- Section 2: Hazardous ingredients
- Section 3: Physical and Chemical Characteristics
- Section 4: Fire and Explosion Data
- Section 5: Reactivity Data (How it reacts to other chemicals)
- Section 6: Health Hazards
- Section 7: Precautions for Safe Handling and Use
- Section 8: Control Measures

The National Institute for Occupational Safety and Health (NIOSH) is currently working on a project to develop *International Chemical Safety Cards*. The project is being developed by the International Programme on Chemical Safety (IPCS) in collaboration with the Commission of the European Communities. The IPCS is a joint venture of three International Organizations: the United Nations Environment Programme (UNEP), the International Labour Office (ILO), and the World Health Organization (WHO). The major goal of the IPCS is to evaluate and provide information on hazards posed by chemicals to human health and the environment. The IPCS intends to generate about 2000 cards in the next six years. (NIOSH)

International Chemical Safety Cards (ICSC's) provide important health and safety information on the chemicals for use by workers and employers. These cards provide information on hazards posed by the chemical and the risks for producing and using the chemical substance. The identification of the chemicals on the cards is based on the *UN numbers*, the Chemical Abstracts Service (CAS) number and the Registry of Toxic Effects of Chemical Substances (RTECS/NIOSH) numbers. UN numbers are four-digit numbers that identify hazardous substances and products that have commercial importance. This numbering classification is used in international commerce to label contents in shipping containers. Similarities and differences of Material Safety Data Sheets and International Chemical Safety Cards are shown in **Figure 8-3**.

Figure 8-3. Notice the similarities and differences between the information found on an MSDS and an ICSC. (NIOSH)

International Council of Chemical Associations (ICCA) Headings of Material Safety Data Sheets	International Programme on Chemical Safety (IPCS) Headings of International Chemical Safety Cards
1. Chemical product identification, and company identification	1. Chemical identification
2. Composition/Information on ingredients	2. Composition/formula
3. Hazards identification	3. Hazard identification from fire and explosion, and from exposure by inhalation, skin, eyes, and ingestion, and prevention measures (with personal protective equipment)
4. First-aid measures	First-aid measures
5. Fire-fighting measures	Fire-fighting measures
6. Accidental release measures	4. Spillage, disposal
7. Handling and storage	5. Storage
	6. Packaging, labeling & transport
8. Exposure controls/Personal measures	See 3. above
	7. Important data:
See 15. below	Occupational exposure limits
9. Physical & chemical properties	See 8. below
10. Stability & reactivity	Physical & chemical dangers
11. Toxicological information	Routes of exposure Effects of short- and long-term exposure
See 9. above	8. Physical properties
12. Ecological information	9. Environmental data
13. Disposal considerations	See 4. above
14. Transport information	See 6. above
15. Regulatory information	See 7. above
	10. Notes
16. Other information	11. Additional information

Title 1 of the *Superfund Amendments and Reauthorization Act* mandated OSHA to development training requirements for employees handling hazardous wastes. OSHA developed the *1910.120 CFR* standard, commonly known as *Hazwoper (Hazardous Waste Operations and Emergency Response Standard)*. The standard requires employees to complete a minimum 40-hour off-site training course with at least three days of on-the-job training before they are permitted to handle hazardous wastes. Supervisors of these workers must also receive 40 hours of training.

Employers are also required to prepare and follow a written hazardous communication plan. This must provide an inventory of the hazardous chemicals or products in each department or work area, and show how the employer informs employees of the hazards associated with the materials. The employer must also be able to provide, on written request, a written plan to employees and OSHA representatives. Training must be provided whenever a hazardous material is introduced to the work area.

Types of Hazardous Materials

There are two common methods for determining whether a material can be classified as a hazardous waste. The simplest method is to consult a listing referred to as the *EPA Notification of Hazardous Waste Activity*, provided by the Environmental Protection Agency to establish the code in a particular state or locality. If the material is noted on the EPA list, then it can be classified as a *listed waste*.

Another approach for determining whether the material is hazardous is to consider the material in terms of its characteristics when humans and animals are exposed to the material. These materials are considered to be *characteristic waste*, and are seldom included on EPA lists. Examples would be materials that are readily ignitable, corrosive (has a pH level of 2 or less or 12.5 or higher), reactive (normally unstable and readily undergoes violent change), or toxic.

Hazardous Waste Generation by Manufacturers

The **Resource Conservation and Recovery Act (RCRA)** of 1976 controls the generation, storage, transportation, management, and disposal of hazardous wastes. The Environmental Protection Agency defines *hazardous waste* as "a waste with properties that make it dangerous or capable of having a harmful effect on human health and the environment." The Environmental Protection Agency also states that there are two types of hazardous waste: listed waste, or nonlisted waste. Listed waste is viewed as hazardous if it appears on one of four lists: P, U, K and F-lists published in the Code of Federal Regulations (40 CFR, Part 261). There are now more than 400 wastes listed. Examples of listed hazardous wastes are shown below:

- **F-List Wastes.** Hazardous waste from non-specific sources such as solvent mixtures.
- **K-List Wastes.** Hazardous waste from specific sources such as certain manufacturing operations.
- **U- and P-List Wastes.** Hazardous waste from discarded commercial chemical products. Off-spec species, container residues, and spill residues.
- **P-List Wastes.** Hazardous wastes, even though they may be properly managed, are so dangerous that they are called acutely hazardous wastes.

A waste that is not on the hazardous waste lists might still be considered hazardous if it demonstrates one or more of the following characteristics:

- **Corrosive.** It corrodes metals or has a very high or low pH. Chemicals in this class are: alkaline cleaning fluids, battery acid, and rust removers.
- **Ignitable.** It catches fire under certain conditions. Chemicals in this class are: degreasers, paints, and solvents.
- **Reactive.** It is unstable and explodes or produces toxic fumes, gases, and vapors when mixed with water or under other conditions such as heat or pressure. Chemicals in this class are: cyanides or sulfide-bearing wastes.
- **Toxic.** It is harmful or fatal when ingested or absorbed, or leaches toxic chemicals into the soil or groundwater when disposed of on land. Chemicals in this class are: wastes that contain high concentrations of pesticides, cadmium, lead, or mercury. (Florida Department of Environmental Protection)

The Resource Conservation and Recovery Act established what is referred to as *cradle to grave* responsibility for the material by manufacturers generating hazardous materials. This means that manufacturing firms are liable for their hazardous materials once they take purchase of these materials, and that they must know how and where their waste is being transported, and how it is being disposed of. The RCRA is now regulated by the Office of Solid Wastes (OSW).

The RCRA classifies producers of hazardous waste, referred to as *waste generators*, into three categories:

- **Conditionally Exempt.** These are generators that produce less than 220 pounds of hazardous materials per month.
- **Small-quantity Generators.** Firms producing between 220 and 2200 lb/month of hazardous waste, no more than 2.2 lb/month of acutely hazardous waste on the "P list," and no more than 220 lb/month of waste from spills and cleanup are in this category.
- **Large-quantity Generators.** These generators produce over 2200 lb/month of hazardous waste, and over 2.2 lb/month of acutely hazardous waste.

The RCRA specifies how manufacturers must store, treat, dispose of, and transport hazardous waste materials. Small- and large-quantity generators must first obtain an EPA identification number, *EPA form 8700-12, Notification of Hazardous Waste Activity* to initiate the process. After the firm obtains its identification number, they are only required to notify EPA if there is a change in the materials generated.

Storage of materials by small- and large-quantity generators is also regulated by the RCRA. Small-quantity generators are permitted to store up to 13,200 pounds of waste for 180 days. Large-quantity generators can store hazardous waste for only 90 days.

Transportation of hazardous materials by small- and large-quantity generators requires a manifest, of the materials being transported. A *manifest* is a "one-page form used by haulers transporting waste that lists EPA identification numbers, type and quantity of waste, the generator it originated from, the transporter that shipped it, and the storage or disposal facility to which it is being shipped." (Environmental Protection Agency) A manifest is completed using EPA form 8700-22, requiring generators to list the generator's name, transporter's name, and list of all materials being transported. The manifest must also provide a signed statement by the generator that they are attempting to reduce hazardous waste generation, and a copy of the firm's annual report providing evidence of hazardous waste reduction.

Transportation of hazardous materials is also regulated by the *Hazardous Materials Transportation Act (HMTA)*. The Department of Transportation requires labels to be placed on all containers before shipment. The label must provide the following information: proper shipping name, the UN (Universal) Hazard Code number for the material, name of shipper, and the name of the receiver.

Disposal of Hazardous Wastes

There are other laws regulating the disposal of hazardous wastes in groundwater systems. The *National Pollution and Discharge Elimination System (NPDES)* permit program addresses the discharge of hazardous and toxic wastes into the nation's water system. Materials to be discharged must be pretreated and rendered safe before disposal in public sewage treatment systems. The *Federal Water Pollution Control Act (FWPCA)* provides information on pretreatment requirements for wastewater to be discharged into rivers, streams, lakes, and other water sources.

Burning or incineration of solid hazardous wastes is regulated by the *Clean Air Act.* The concept of burning, as it is to be interpreted in this act, is for purposes of energy recovery, destruction, material processing, or as an ingredient. The Clean Air Act provides specific levels that must not be exceeded in terms of gallons to be burned per month based on the TESH Formula (Terrain-adjusted Effective Stack Height). The TESH Formula is described in Section 726.206 Standards to Control Emissions from burning metals. Feed rate screening limits for metals are specified as a function of terrain-adjusted effective stack height and terrain and land use in the vicinity of the processing facility.

There are many laws, acts, and requirements that manufacturing establishments must understand to comply with legislation and be good corporate citizens in today's world. Programs must be developed and implemented to deal with issues ranging from initial identification and storage, through off-site transportation and disposal. Training programs must be in place to address these and many other issues related to hazardous and toxic materials.

Now that you are more informed about safety related to materials, we need to shift our emphasis to the study of the behavior of materials used in manufacturing. You will recall from chapter 2 that there are five major manufacturing materials: metals, plastics, woods, ceramics, and composites. Each of these materials is unique and behaves differently when used with different types of manufacturing processes. Let us start with metals, and then gradually work through each of the other materials. This should provide a much better foundation for your study of the processes that can be used with specific materials.

Important Terms

characteristic waste
chemical properties
Federal Water Pollution Control Act (FWPCA)
Hazardous Materials Transportation Act (HMTA)
hazardous waste
Hazcom Standard
Hazwoper (Hazardous Waste Operations and
  Emergency Response Standard)
listed waste
manifest
Material Safety Data Sheet (MSDS)
mechanical properties
National Pollution and Discharge Elimination
  System (NPDES)
physical properties
Resource Conservation and Recovery Act (RCRA)
Right-to-Know Law
Superfund Amendments and Reauthorization Act
surplus
waste generators
waste materials

Questions for Review and Discussion

1. If a company can be classified as conditionally-exempt in terms of hazardous materials, what does this mean?

2. If a company is classified as a small-quantity generator, what are the management's responsibilities in terms of handling and disposal of hazardous materials?

3. What is an MSDS? Does your firm have any responsibility in terms of preparation and display of these?

4. Can waste generated by manufacturers be burned? What are the restrictions pertaining to this?

9 Characteristics of Metallic Materials

Chapter Highlights

- Properties that are used to classify metals for use with specific manufacturing processes and product applications.
- There are four major types or classifications of metals—ferrous, nonferrous, high-temperature superalloy, and refractories.
- Discussion of the physical and mechanical properties of metals.
- Iron and steel, the two major types of ferrous metals.
- Methods for classifying superalloys according to base metals.
- Techniques used to identify types of steel.

Background

There are thousands of traditional and exotic industrial materials that have been used over the ages, and many new engineered and exotic industrial materials available today. In spite of this huge selection of materials, manufacturers of industrial products choose a metal more frequently than any other material. Six of the top employment sectors listed in NAICS (North American Industry Classification System) depend on a metal as their primary manufacturing material. These six sectors account for about half of all employment in manufacturing.

- Primary metals
- Fabricated metal products
- Machinery
- Computers and electronics
- Transportation equipment manufacturing
- Electrical equipment, appliances, and component manufacturing

Not only are metals used in the products made by these and other industries, they are also used to make the machines that produce their products. Of the known chemical elements, more than 50 (almost half) are classified as metals.

Structure of Metallic Materials

Metals are particularly useful in manufacturing because of their molecular composition, which distinguishes them from other engineering materials. Unlike most plastics, ceramics, and woods, metals can survive drastic changes in the environment in which they are used. At one extreme, being subjected to heat, they remain strong and rigid enough to support heavy loads. At the opposite extreme, in frigid environments, metals retain flexibility and can still be easily formed. Few other materials retain their essential properties when subjected to the same range of hot and cold temperatures.

The atomic structure of *pure metals*, metals found in nature in their pure form, are easy to describe because the atoms that form these metals appear as identical perfect spheres. The atoms of a pure metal are packed closely together in an orderly (lattice-like) arrangement and are held together by electrostatic forces. A listing of pure metals is shown in **Figure 9-1**. Few pure metals are used in their natural form, because they are either too hard or soft, or too expensive because of their scarcity. Alloys, a blend of metals and other elements are used more frequently than pure metals to make products. The structure of alloys is more complex than that of pure metals.

The scientific community has identified and officially named 116 known chemical elements, and 91 occur naturally. It is interesting to note that all man-made elements are radioactive with short half-lives. The *half-life* of an element refers to the gradual process of exponential decay where the element exhibits only half of its initial value. Quantities subject to exponential decay are represented in periodic tables with the symbol, *N*. The first element created by humankind was *Technetium* in 1937.

Chemical elements are the fundamental materials of which all matter is composed. From the scientific point of view, a substance that cannot be broken down or reduced further is, by definition, an element. These elements exist as atoms. Elements consist of protons, neutrons, and electrons. Elements are identified by a two-letter symbol and a number. Iron, one of the more common elements on earth, has an atomic number of 26 (the atomic number corresponds to the number of protons in the nucleus of the element's atom) and its symbol is Fe. Copper's symbol is Cu and its atomic number is 29, indicating that it has a higher density than iron. The larger the atomic number, the heavier the atom.

The major metals are aluminum, copper, iron, lead, tin, magnesium, nickel, titanium, and zinc. Other metals such as steel, brass, and bronze are alloys of these metals. In manufacturing, alloys are used more frequently than pure metals. An *alloy* consists of a blend of two or more elements, with at least one of the materials being a metal. The primary metal is the base material of the alloy. Brass, for example, is a simple alloy of copper and zinc, two metals. Steel, however, is a mixture of iron, carbon, magnesium, vanadium, nickel, and chromium. Carbon, which is not a metal, is a key ingredient in steel.

Why is alloying so important? An alloy can have the best attributes of the base metal and be significantly enhanced by the addition of another element(s). For example, adding tin to the base metal copper makes the alloy bronze stronger than either tin or copper alone.

Because of alloying, metals can be created that are able to withstand exposure to just about any environment. There are more than 25,000 different types of steel, and more than 200 standard copper alloys, not including the many specialty brasses and bronzes.

Figure 9-1. Pure metals are those that are found in nature.

aluminum	antimony	beryllium	bismuth	cadmium
cerium	chromium	cobalt	copper	germanium
gold	hafnium	indium	iridium	iron
lead (metal)	magnesium	manganese	mercury(metal)	molybdenum
neodymium	nickel	niobium(columbium)	osmium	palladium
platinum	rhodium	ruthenium	silver	tantalum
tin	titanium	tungsten	uranium	vanadium
zinc	zirconium			

Physical Properties

All materials have their own unique physical properties that distinguish one material from another. There are basically four physical properties used to classify metals: weight, color, conductivity (electrical and thermal), and reaction of the material when exposed to heat.

A heavy metal such as lead is denser and has a tighter molecular structure than a lightweight material such as aluminum. Lightweight, less dense materials have a more open or loose molecular structure.

Other metals can easily be distinguished by their color. The brilliant color and reflection of platinum, gold, and silver are quite different from the color and shine from stainless steel, or the dull gray color and shine of iron.

Metals have different properties when they are subjected to temperature extremes. Each type of metal has a different thermal conductivity and its own coefficient of thermal expansion, depending on its molecular structure. Copper and aluminum expand much more than cast iron or steel. Designers need to overcome these characteristics when making products that combine these different materials. The selection of the metal is an important factor when designing any product.

Often, the physical property of a metal is more important than the actual mechanical property. A property such as the specific heat of the material is important to know when a particular metal is to be cast. The *specific heat* is the amount of energy necessary to raise one gram of material 1°C. The specific heat is important since this enables a machine operator or technician to calculate how much heat energy is needed to raise an object to a desired temperature.

Thermal conductivity is the ability of a substance to conduct heat. This value is determined by measuring the rate of heat flowing through a substance. This is a measure of the effectiveness of a material as a thermal insulator. Metals with good electrical conductivity will also have good thermal conductivity. As an example, copper exhibits high thermal conductivity and is an excellent conductor of electricity.

The property of *thermal expansion*, the change in volume of a product at different temperatures, is particularly important to consider when metals are being cast and cooled. Most molten metals shrink during solidification and cooling. Manufacturers must take this amount of shrinkage into consideration during the tooling stage of production, or products that are created will be undersized.

At the other end of the spectrum, some metals and *intermetallic* (ceramic/metallic) compounds become superconductors when they are exposed to *ultra cold* (near absolute zero) temperatures. Oakridge National Laboratories describes intermetallics as "a unique class of materials, consisting of ordered alloy phases formed between two or more metallic elements where the different atomic species occupy specific sites in the crystal lattice." (Nuclear Regulatory Commission, 1984; Anton et al., 1989). Intermetallics have characteristics of both metals and ceramics, and their mechanical properties are intermediate between metals (which are generally softer and more ductile) and ceramics (which are generally harder and more brittle). Intermetallics have simple chemical names like $TiAl$, Ti_3Al, $NiAl$, Ni_3Al, $CuZn$, Cu_3Au, and Nb_5Si_3. (Oakridge National Laboratories)

A *superconductor* is a material that is able to conduct electricity without exhibiting any resistance to the flow of current. At absolute zero (0 Kelvin or –273.15°C), which is defined as the absence of all heat, a metal becomes a superconductor. It is also true that as a metal heats up, its electrical resistance also increases. Superconductivity was discovered by the Dutch physicist, Kammerlingh Onnes, in 1922. Onnes cooled mercury to below 4.1 K (Kelvin) and found that it lost all of its electrical resistance. Unfortunately, a metal will decay rapidly if resistance is removed. However, a superconductor could exhibit a decay constant of many years.

Let us look more closely at what superconductors really accomplish. In 1986, a major discovery was made by two IBM scientists in Zurich, Switzerland: Georg Bednorz and Alex Müller. Bednorz and Müller studied hundreds of oxide compounds and narrowed in on a particular class of metal oxide called *perovskites*. Initially, they found that ceramics made of lanthanum, barium, copper, and oxygen exhibited superconductivity at 35 K. This was 12 K above the old record. One year later, a perovskite ceramic material was found to superconduct at 90 K. This was a very significant discovery, because it made it possible to use liquid nitrogen as a coolant. Since perovskites superconduct at significantly higher temperatures, they are called *high temperature superconductors*.

In the past twenty years, scientists throughout the world have experimented with different forms of perovskites and have developed materials that function as superconductors in excess of 130 K. It has become a much simpler task to cool superconductors. This is greatly influencing the development of new materials and processes for joining materials. It is also leading to new discoveries on the behavior of superconductors at relatively high temperatures. This is starting to lead to new opportunities for using high temperature superconductors for electrical power applications. The future of superconductors is only just now beginning.

Mechanical Properties

Physical properties are often critical to the selection of a metallic material for a particular end use, but there are many times when the mechanical properties are even more important. *Tensile strength* (the ability of a material to resist being pulled apart) and hardness (the resistance to penetration) are examples of mechanical properties. When a manufacturer buys 1000 pounds of one-inch diameter steel rods, the physical properties are set. However, the manufacturer can change the mechanical properties. The rods can be cut, machined, and heat-treated to increase their tensile strength and hardness. Alloys generally give the manufacturer more control over mechanical properties, but the mechanical properties of pure metals can also be controlled. Pure copper can be work-hardened to improve its strength and hardness. The manufacturer can also reverse the process. Once the steel is hardened through heat-treating, it can be softened using another heating process called annealing. For the manufacturer there is real magic in mechanical properties.

It is the mechanical properties that determine how the part or product will withstand continued use or abuse in the user's work environment. Mechanical properties can be measured using standardized testing procedures. Typical tests used to assess hardness-related mechanical properties are the Rockwell test, Brinell test, or Charpy impact test. The test used to determine tensile strength is called the tensile test. Both ductility and toughness are measured with a tensile test. Flexibility or elasticity is measured using Young's Modulus of Elasticity. Fatigue is measured using the fatigue failure test.

Behavioral characteristics must be taken into consideration when selecting metals for use in a manufactured product. Stresses arise when a load is applied to a metal structure, part, or component. Stress occurs when the material attempts to respond to deformation created by an applied force. The addition of heat also increases the effect of stress on metals.

Another characteristic of metals that must be considered in the design of a product is called fatigue. *Fatigue* results in the breaking of a piece of metal after the stress is continually removed and reapplied. An example is the process of folding a thin sheet of metal together, then straighten it out again, then fold it back together. Doing this a number of times would eventually lead to fatigue and breakage. With metal parts, most breakage is caused by fatigue.

Another term that is related to stress is *creep*. This is something that needs to be addressed when making durable products from metal or plastic. Creep is the elongation of material that occurs when the material is exposed to elevated temperatures while under stress. If a material is exposed to heat for an extended length of time, and this exposure continues to produce elongation, then the material will eventually separate, or rupture. This is called *creep failure*. The creep rate of steel can be reduced by adding elements such as nickel and chromium to the alloy.

Let's think about a more practical application of what has just been discussed. If a heavy enough weight is placed on the middle of a length of bar stock that is suspended by its ends, the weight will cause deformation of the bar (creep). If this deformation occurs over an extended length of time, it will result in some lengthening of the bar stock. The change in dimension of the bar is over and above the bar's initial elastic deformation. The amount of stress resulting from a particular load depends on the characteristics of the metal. What is particularly important to know is the plasticity of the material. *Plasticity* refers to the ability of the material to change shape or size as a result of force being applied. Information related to plasticity is useful to know when shaping and forming metal. These are critical issues in the design of metal buildings and load-bearing structures.

A similar concept, called ductility, must be considered. *Ductility* is the ability of the material to be formed plastically, without breaking. This is an important characteristic when selecting a manufacturing process. If the material is not ductile, and fractures easily, then it will be more difficult to form. Imagine trying to squirt hard and brittle toothpaste out of its tube. This is the same type of thing that would occur when trying to extrude metal that is not ductile out of a small opening.

Classifications of Metals

There are four major classifications, or types, of metals. These are ferrous metals, nonferrous metals, high-temperature superalloys, and refractory metals.

Most of the metal products manufactured today are made from either ferrous or nonferrous metals. However, many advancements are being made in the fields of high-temperature superalloys, exotic metals, and composite materials. Some of these materials are designed to survive exposure to conditions such as extreme heat, cold, and pressure. The materials will need to perform reliably beyond the limits placed on products used in the earth's atmosphere. They may have to travel into deep space, into the ocean depths, and beyond.

To put this into perspective, let us take a moment to think about how strong products have to be to survive in the ocean. Imagine that your test equipment is inside an exploration vessel in the Marianas Trench, located in the Pacific Ocean, just east of the Philippines. This is an area that was actually surveyed in 1951 by the British navy vessel, *Challenger II*. Your vessel is 36,000 feet below sea level, in the deepest known region of the world's oceans. What do you think the pressure would be like at this depth? Well, pressure is normally discussed in terms of atmospheres. One atmosphere at sea level is equivalent to a weight of 14.6 pounds per square inch. Pressure increases about one atmosphere for every 10 meters of water depth (30 feet). So at the bottom of the trench your vessel would be at 36,000 feet ÷ 30, or 1,200 atmospheres. Multiplying 1,200 atmospheres × 14.6 pounds per square inch, your craft would have to withstand a force of 17,520 pounds per square inch. That is a lot of pressure!

Superalloys, or high-performance alloys, are often used in load-bearing situations for structural applications. Product applications range from hot sections of turbine engines and rocket engines to coatings on saw blades, and components used in chemical and petroleum plants. Superalloys are designed to exhibit superior mechanical strength, good surface stability, corrosion resistance, and the ability to withstand high temperatures without oxidizing or losing mechanical properties. Superalloys are typically based on nickel, cobalt, or iron. Other elements can also be present. Chromium, molybdenum, tungsten, aluminum, and zirconium are examples.

Ferrous Metals

Ferrous metals, those that contain iron, are the most commonly used metals in the world. The word ferrous comes from the Latin word *ferrum*, which means iron. Pure iron is rarely used. Steel (an alloy of iron) and cast iron are two types of ferrous metals that are often used today. **Figure 9-2**.

There are two basic types of iron: wrought iron and cast iron. *Wrought iron* is tough and ductile, because it contains very little carbon. Wrought iron is easy to bend, even without heating, which makes it a prime candidate for use in ornamental ironwork. Wrought iron also includes *slag,* a mixture of impurities. Surprisingly, slag is used to improve the corrosion resistance of wrought iron.

Cast iron is a very useful material in manufacturing and can be readily cast to make heavy equipment bases and machine stands. Cast iron is made by pouring molten iron, which has been mixed with between 1.7 percent and 4 percent carbon, into a mold. Cast iron may be the material of choice when compression strength and wear resistance are important.

Cast iron is hard. This makes it brittle and easy to crack. Cast iron cannot be bent, stretched, or formed by forging. The high carbon content gives cast iron

Figure 9-2. There are two common types of ferrous metals: iron and steel. Iron can further be divided into the classifications of wrought iron and cast iron; steel into hot-rolled and cold-rolled.

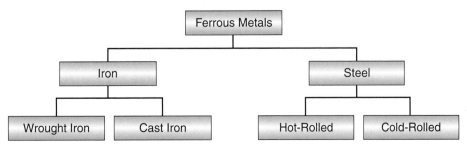

some natural lubricity. It also has good sound deadening properties and corrosion resistance. Typical uses of cast iron include engine blocks, machine frames, machine parts, and gear cases.

There are several different types of cast iron. The most popular type is called *gray iron*. It is easily cast and is less expensive than the other types. In making cast iron, *pig iron* (refined wrought iron) is mixed with scrap iron or steel to help control the carbon content. Most cast iron production is accomplished in a *cupola*, or an induction-heating furnace.

Another type is called *white cast iron*. White cast iron is very hard and is used for making parts that must combat fatigue from extreme wear and abrasion conditions.

A third type of cast iron, called *malleable iron*, is made by heating white cast iron to a specific temperature, and then cooling it slowly. This is a softening process called annealing. Annealing is an important process in the metal manufacturing industries and will be discussed in greater detail later in this section. A fractured piece of malleable cast iron will have a white rim with a dark center.

Another major type is called *ductile cast iron*. It is also known as nodular, or spheroidal grade (S.G.), cast iron. Ductile cast iron is heat-treatable and is used for making parts such as crankshafts, camshafts, and connecting rods for both gasoline and diesel engines. This is one of the newest forms of cast iron and has replaced much of the malleable and white cast iron production in recent years. Ductile iron has the ductility of malleable iron, the corrosion resistance of alloy cast iron, and tensile strength exceeding that of gray cast iron. Because ductile iron is a nodular iron, it can be arc welded.

Steel, an alloy of iron and carbon, was one of the most significant developments in the nineteenth century. However, carbon is not always the primary alloying additive. Elements such as tungsten, molybdenum, or vanadium are often added to make the steel harder and tougher. Alloying iron with nickel and chromium creates a highly corrosion resistant material called *stainless steel*. See **Figure 9-3**.

When steel includes carbon as an alloying element, it is called *carbon steel*. There are three types of carbon steel: low-carbon, medium-carbon, and high-carbon. The *Unified Numbering System (UNS)* classifies steel according to the quantity of carbon in the steel, stated in hundredths of a percent.

Figure 9-3. These are the basic ingredients of steel.

Types of Steel	Ingredients
Carbon steel	Iron + carbon
Harder and tougher steel than carbon steel	Iron + tungsten, molybdenum, or vanadium
Stainless steel	Iron + chromium

- Low-carbon steel, or mild steel, usually has from between 0.05 and 0.30 percent carbon. Since it is very soft, it can be easily formed and machined. The major disadvantage of mild steel is that it doesn't respond well to heat-treating.
- Medium-carbon steel has between 0.30 percent and 0.60 percent carbon. It is more difficult to bend and shape, but it can be hardened through heat treatment. Hardening causes the material to become more brittle.
- High-carbon steel is often referred to as *tool steel*. It has from 0.60 percent to 1.50 percent carbon. Tool steel is hard and difficult to bend. It can be made even harder through heat-treating. Tool steel is used to make tools, such as forging dies, screwdriver shafts, chisels, and milling cutters.

There are also other methods used to classify steel. The method used by the American Iron and Steel Institute (AISI) and Society of Automotive Engineers (SAE) classifies steel according to its properties.

Steel is made from purified iron (free of oxides and impurities) and with a carbon content of less than 1.7 percent or less. The carbon content enables the steel to be heat-treated by the manufacturer to change the mechanical properties. The amount of change, however, is dependent on the carbon content of the steel.

Steels are also classified according to how they are shaped. Steel furnaces produce ingots or billets. Ingots began as a mass of metal that is heated beyond the melting point, then recast in the form of a bar or block. Billets are the feedstock for long products with small cross sections. Billets are cast by a continuous caster or rolled in billet mills.

Ingots and billets are formed into useable shapes such as angle and bar stock, I-beams, and plates. There are two processes that are used to create these shapes—hot-rolling or cold-rolling. *Hot-rolled steel* is squeezed between rollers while it is hot, and can be

Figure 9-4. Nonferrous metals are those that do not contain iron. In effect, all metals other than iron and steel are nonferrous.

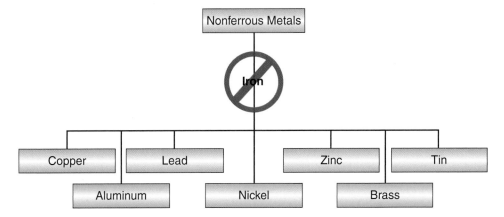

Nonferrous Metals

Metals with no iron content—aluminum, copper, lead, magnesium, nickel, and zinc—are called *nonferrous metals*. See **Figure 9-4**. Nonferrous metals are softer and easier to form. In spite of the fact that they are softer than iron and steel, nonferrous metals are usually more durable since they are resistant to corrosion.

Copper and copper alloys are classified in a Unified Numbering System specified by the Copper Development Association. The numbers C10100 through C15500 are reserved for copper alloys with 99.5 percent or greater purity. Higher numbers represent alloys with a smaller percentage of copper. Copper and its alloys are useful in product applications that require superior electrical and thermal conductivity. These materials are soft, easy to form, and resistant to corrosion.

Aluminum is another nonferrous metal that is soft and easy to form. But its major advantage over most other metals is light weight. Like copper, it is resistant to corrosion and is a good conductor. Aluminum and aluminum alloy stock are classified by their own numbering system. Aluminum is classified in a four-digit index system to designate the nature of the alloy. See **Figure 9-5**.

Classifying aluminum is a relative complex undertaking. Just identifying aluminum according to the four-digit classification number above is only part of the story. Aluminum is also graded in terms of a hardness classification number or temper designation. The first four digits are followed by the hardness classification number. The first four digits refer to the purity of the metal and the nature of the alloying. An aluminum alloy with the number 1000-H12 is interpreted as follows: "1000" means that the principal ingredient is aluminum of 99 percent or greater purity. The "H12" means that the aluminum was strain hardened only, and that it was 1/4 or 25 percent hardened. Let us take a closer look at the hardness classification system.

The hardness or temper designations are different between non-heat-treatable alloys and heat-treatable alloys. With non-heat-treatable alloys the letter, "H" is always followed by 2 or 3 digits. The first digit indicates the method used to obtain the temper. For example, H1 means strain hardened only. H2 means strain hardened, then partially annealed. H3 means strain hardened. The second digit indicates the extent of the hardness or temper. Here is how it is represented: 2 means 1/4 hard, 4 means 1/2 hard; 6 means 3/4 hard, 8 is fully hardened, and 9 is extra hard.

With heat-treatable alloys the first digit is either an "F," "O," or "T." F means "as fabricated." O means that it will be annealed, and T means that it will be heat treated. The letter "T" is always followed by one or more digits. These indicate the manufacturing process used to produce the stable temper. See **Figure 9-6**.

High-Temperature Superalloys

The term "superalloys" was first used following World War II to refer to a group of alloys that were used to make turbosuperchargers and aircraft turbine engines that operated at high temperatures.

Figure 9-5. Classification groups for numbering aluminum alloys.

Number Designation	Major Alloying Element	Characteristics
1xxx	Pure unalloyed (greater than 99% Aluminum)	Provides excellent corrosion resistance, high thermal and electrical conductivities, low mechanical properties, and excellent workability.
2xxx	Copper; other elements such as magnesium are sometimes used	Requires heat treatment maximize properties. After heat treatment, mechanical properties are similar to those of low-carbon steel. 2xxx alloys do not have as good a corrosion resistance as most other aluminum alloys.
3xxx	Manganese	Not heat treatable, but has more strength than number 1xxx series alloy.
4xxx	Silicon	Contains silicon, which is used to lower the melting temperature. Used in welding wire and for brazing wire used to fabricate aluminum.
5xxx	Magnesium	Medium to high-strength work-hardenable alloy. Easy to weld and exhibit good resistance to corrosion in marine environments.
6xxx	Magnesium and Silicon	Heat treatable. The least expensive of the heat treatable alloys. Although not as strong as most 2xxx and 7xxx alloys, 6xxx aluminum has good formability, weldability, machinability, and relatively good corrosion resistance, but is not as strong as 2xxx or 7xxx.
7xxx	Zinc and a smaller percentage of magnesium	Heat treatable and high strength.
8xxx	Others elements, including Tin and Lithium combinations	Designed to perform in specialized environments.

Figure 9-6. The hardness number of an aluminum alloy indicates the heat-treating process used to produce the temper. (Aircraft Spruce and Specialty Company)

Hardness Number	Manufacturing Process to Temper
T3	Solution heat treated and then cold worked.
T351	Solution heat treated, stress-relieved stretched, then cold worked.
T36	Solution heat treated and then cold worked (controlled).
T4	Solution heat treated and then naturally aged.
T451	Solution heat treated and then stress-relieved stretched.
T5	Artificially aged only.
T6	Solution heat treated and then artificially aged.
T61	Solution heat treated (boiling water quench), then artificially aged.
T651	Solution heat treated, stress-relieved stretched and then artificially aged (precipitation heat treatment).
T652	Solution heat treated, stress relieved by compression and then artificially aged.
T7	Solution heat treated and then stabilized.
T8	Solution heat treated, cold worked, and then artificially aged.
T81	Solution heat treated, cold worked (controlled), and then artificially aged.
T851	Solution heat treated, cold worked, stress-relieved stretched and then artificially aged.
T9	Solution heat treated, artificially aged, and then cold worked.
T10	Artificially aged and then cold worked.

Today superalloys are used by the defense and space-related industries in applications requiring materials that can withstand abuse from extreme forces and severely oxidizing, high-temperature environments. Superalloys are used in the manufacture of rocket engines, turbines, jet engines, and space vehicles. They are also being used in the chemical processing and nuclear fuels industries.

Superalloys are classified according to the base metal in the alloy. In most cases, this is iron, nickel, or cobalt. Thus, superalloys can be iron-based, nickel-based, or cobalt-based. Other alloys such as chromium, titanium, aluminum, or tungsten are then added to the base metal.

What all of these *superalloys* have in common is their ability to survive, without degradation in temperatures as great as 2200°F (1200°C) for reasonable periods of time when used in a nonloadbearing structure. Under load, most superalloys can be used at 1800°F (1000°C). See **Figure 9-7**.

While superalloys are able to retain their strength in high temperatures, they do have weaknesses to environmental elements. They are made with alloying elements that are influenced by oxidation, hot corrosion, and thermal fatigue. In some product applications such as turbine vanes and blades, the parts are often coated with other materials to improve their environmental resistance.

Refractory Metals

Refractory metals, the fourth major category, are high-temperature metals. They can withstand heat and maintain their strength at temperatures ranging from 4474°F (2468°C), for niobium, to 6170°F (3410°C), for tungsten.

Refractory metals include niobium (Nb), tungsten (W), and molybdenum (Mo). All of these metals have very strong interatomic bonding, which is the reason why they can withstand such high melting temperatures. Tungsten has the highest known melting point of any metal.

Refractory metals are used in many ways. When molybdenum is alloyed with stainless steel, the material has greatly improved resistance to corrosion. Often tantalum is also added, creating an alloy that is impervious to chemical attack in environments below 302°F (150°C).

These metals are used in products such as incandescent light filaments and welding electrodes. They are also used for tools, dies, rocket engines, gas turbines, and containers for holding and dispensing molten metal.

Nature of Industrial Stock

As noted earlier, the primary emphasis of this book is on the secondary manufacturing industries—those that start with industrial stock and transform it into a useful consumer product. Industrial stock is the product of primary manufacturing firms. Hard goods for sale to the consumer are the products of secondary manufacturing firms. In the case of metals, primary industries are concerned with mining, refining, and production of metals.

Steel is a very important form of industrial stock. The steelmaking industry shapes molten steel into solid ingots. Ingots are produced in standard lengths, and are normally sold by the pound. Ingots are later rolled or extruded into various shapes such as bars, hot-rolled strip (band iron), plates, round or hexagonal rod, tubing, wire, or angle stock. These shapes are used in manufacturing.

There are other types of stock that are widely used. Metal stock in the form of powder or billets (for example, aluminum bricks for casting) is also available. Stock in powder or billet form is normally sold by the pound. Nonferrous metals such as aluminum and copper are usually sold according to the weight per square foot, the thickness in decimal equivalents, or the gage number.

Stock is available in sheet form, as well. Steel plate or sheet stock is sized in thickness by a *gage number*. The smaller the gage number, the thicker the stock. However, care is needed when purchasing different types of sheet stock, since not all plate has the same numbering system.

Figure 9-7. Superalloys are used for many aircraft and aerospace parts. This turbine component for a jet engine is being checked for machining accuracy by a coordinate measuring machine.
(Garret Engine Division, Allied Signal Aerospace)

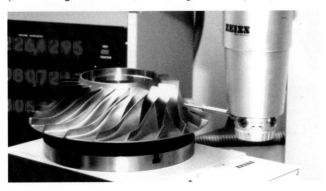

Tin plate (mild steel coated with tin) is sometimes referred to as *electrolytic tin plate* (ETP). Electrolytic Tin Plate is *black plate* that is electrolytically coated with tin. Black plate is light-gage cold rolled steel without any metallic coating. Tin plate has a durable, bright surface finish. By far, the largest use for tin plate is in the foods industry for can making. It is perfect for this application because it is nontoxic, light in weight, strong, and corrosion resistant. It is also easily formed, soldered, and welded.

There are several different methods used to classify tin plate. Sometimes it is sold by the pound in a *tin plate base box*. This is a container measured in terms of pounds of sheets (112 sheets 14″ × 20″), a unit unique to the tin industry. This corresponds to its area of sheet totaling to 31.360 square inches of any gage (thickness) and is applied to tin plate weighing from 55 to 275 pounds per base box.

ASTM International (originally known as the American Society for Testing and Materials) developed another approach for classifying tin plate. This is detailed in A623-03 Standard Specification for Tin Mill Products. Here, plate is identified according to its temper and method of annealing. A typical number might be (T1 BA). The basic methods of annealing are: Box Annealed and Continuous Annealed (BA or CA). Tin plate is recognized by its temper as follows:

- **T1.** Soft for drawing
- **T2.** Moderate drawing, when some stiffness is required
- **T3.** General purpose, when some stiffness is required
- **T4.** General purpose with increased stiffness
- **T5.** General purpose with increased stiffness
- **T6.** *Rephosphorized steel* (high strength steels with addition of phosphorus), providing great stiffness

T1 BA is soft and good for drawing and was box annealed. In addition to these two features, the Rockwell 30-T hardness number is probably provided. The tempering scheme to identifying hardness for Tin Mill products is directly opposite than that used for steel. For example, #1, or (T-1) tin indicates a dead soft condition, whereas #1 temper in cold-rolled strip and sheet steel indicates a full hard-rolled condition.

Determining the Type of Steel

Most of the time it is impossible to look at stock and determine what type of steel it is. However, there are three basic techniques that you can use to determine the type of steel.

The American Iron and Steel Institute (AISI), in cooperation with the Society of Automotive Engineers (SAE), has developed a simplified system that classifies steel according to its intended purpose or unique properties. See **Figure 9-8**.

The second method is more accurate. This system was briefly mentioned earlier and is called the Unified Numbering System (UNS). The UNS was developed by the American Iron and Steel Institute and the Society of Automotive Engineers. This system uses a series of four or five digits to classify steels according to the primary alloying element and the percentage of carbon in hundredths of one percent. Here is how this system works:

- The first number stands for the type of metal.
- The second number indicates the percentage of alloy.
- The last two digits tell how much carbon is in the metal in hundredths of a percent.

For example, the number *1020* for steel would be interpreted as follows: *1* means that it is carbon steel, *0* that it contains no alloy, and *20* that it has 0.20 percent carbon.

A simple way to remember this is to move the decimal point on the percentage of carbon two spaces to the right. For example, tool steel with 1 percent

Figure 9-8. Steel is classified according to its intended purpose or unique properties and assigned an appropriate letter designation.

Designation	Properties
P	Mild and low carbon steels
F	Carbon/Tungsten, and special purpose steels
L	Low alloy/special purpose
M	Molybdenum alloy and high speed steel
T	Tungsten alloy and high speed steel
H	Hot working , chromium, tungsten, and/or molybdenum
D	Die steel, air hardened, and high chromium steel
A	Air hardened steel
O	Oil hardened steel
S	Shock resistant steel
W	Water hardened steel

Figure 9-9. The four and five-digit Unified Numbering system devised by AISI is widely used in industry to classify steels.

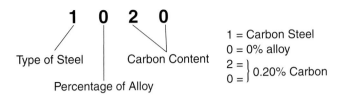

Figure 9-10. Types of steel listed in the Unified Numbering System.

UNS Number	Type of Steel
1	Plain carbon steel (no alloy)
2	Nickel steel
3	Chromium and nickel steel
4	Molybdenum steel
5	Chromium steel
6	Chromium and vanadium steel
7	Tungsten steel
8	Nickel, chromium and molybdenum steel
9	Silicon and manganese steel

carbon content is classified as *100 carbon steel.* Tool steel with 1.5 percent carbon is classified as *150 carbon steel.* Medium-carbon steel with 0.60 percent carbon is *60 carbon steel.* See **Figure 9-9.**

The different types of steel used by the Unified Numbering System, and shown as the first digit in the UNS steel number, are listed in **Figure 9-10.**

A third method for determining the type of steel is by a color code. Most manufacturers paint the end of the stock with one or more colors to indicate the particular type of steel. The end of a 1020 carbon steel rod is brown. Other types of carbon steel are painted white, red, blue, green, orange, bronze, or aluminum. Alloy steel normally is painted with two colors. Nickel steel, for example, is painted red or another color, depending upon its chemical composition.

However, in spite of all these attempts to identify the type of steel, sometimes no information is available. If the identification tags or marking is not on the steel, the material will have to be identified based on its properties. This can be done effectively with training and experience. An old-time method used by some craftsmen requires grinding off a small section of stock and studying the pattern and color of the sparks. Mild (low-carbon) steel creates a long and consistently distributed pattern of sparks. High-carbon steel produces a wider and more-tightly clustered pattern. Wrought iron produces a smaller cluster of sparks. High-speed steel creates a more-scattered long and thin pattern of chrome yellow sparks. See **Figure 9-11.**

Important Terms

alloy
black plate
carbon steel
cast iron
cold-rolled steel
creep
creep failure
cupola
ductile cast iron
ductility
electrolytic tin plate I(ETP)
fatigue
ferrous metals
gage number
gray iron
half-life
high temperature superconductors
hot-rolled steel
intermetallic
malleable iron
nonferrous metals
pig iron
plasticity
perovskites
pure metals
refractory metals
rephosphorized steel
specific heat
stainless steel
steel
superalloys
superconductor
tensile strength
thermal conductivity
thermal expansion
tin plate
tin plate base box
tool steel
ultra cold
Unified Numbering System (UNS)
white cast iron
wrought iron

Figure 9-11. The spark color and pattern that results from grinding is one of several methods used to identify different types of steels.

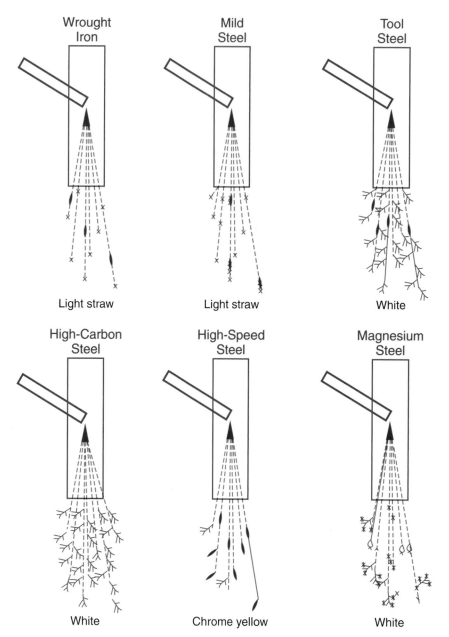

Questions for Review and Discussion

1. Alloys are used more often than pure metals in manufacturing. Alloys are also more costly than pure metals. Why are alloys so popular?

2. Discuss the concept of plasticity in terms of a material's ability to withstand changes in its shape when it is being formed under pressure.

3. How do the material qualities of *plasticity* and *ductility* differ?

4. List some products made with refractory metals. How are the metals made?

5. Describe the types of information you would need to know in order to purchase tin plate.

6. Explain how the AISI numbering system is used to classify types of steel.

7. How can grinding be used to identify the type of steel?

10 Characteristics of Plastic Materials

Chapter Highlights

- Plastics as an engineered or human-made material.
- Cost advantages often gained by substituting plastics for other manufacturing materials.
- The difference between resins and plastics.
- When commodity resins are used instead of engineered resins.
- Thermoplastics formed through addition polymerization, enabling reheating and reforming after cooling.
- Thermosetting plastics formed by condensation polymerization, involving chemical change through cross-linking of polymer chains.

Introduction

Plastics are synthetic (engineered, or human created) materials that are capable of being formed and molded to produce finished products. In fact, the term *plastics* is derived from the Greek word *plastikos*, meaning "to form." Although plastics can be derived from many types of organic and inorganic materials, they are most often made from petroleum base stocks or natural gas. For example, the ingredients used to make the plastic known as *polystyrene* are benzene and ethylene. Both of these chemicals are derived from petroleum.

Innovative companies are searching for new ways to make plastics that do not depend on petroleum-based products. Cargill Dow, LLC invented the first commercially viable plastic made from naturally occurring sugar found in resources such as corn and sugar beets. See **Figure 10-1**.

Figure 10-1. This manufacturing plant in Blair, Nebraska produces compostable drink cups from NatureWorks™ PLA, a polymer made from corn. (NatureWorks, LLC)

To create this polymer, a renewable resource such as corn is milled and the starch is separated from the raw material. Unrefined dextrose (sugar) is then produced from the starch. Next, the dextrose is converted into lactic acid using a fermentation process much like that used to make beer and wine. The lactic acid is purified through vacuum distillation. Polymerization (hardening) of the lactide polymer is then accomplished when the product is formed using a solvent-free melt process such as thermoforming, sheet and film extrusion, blown film processing, fiber spinning, or injection molding.

Unique Characteristics of Plastics

Plastics cover a vast array of applications ranging from devices and components in the International Space Station, communication systems, and memory polymers, to prosthetic limbs and medical implants, high-performance racing boats, and everyday products such as beverage containers, shrink wrap, and automobile fenders. Plastics exhibit many characteristics that provide design advantages over other industrial materials.

Most plastics weigh about one-eighth as much as steel. This makes plastics an ideal candidate for product applications where weight affects performance or cost. Design improvements have resulted in continued weight reductions in plastic products. In 1977, the 2-liter plastic soft drink bottle weighed 68 grams. Today it weighs only 51 grams, a 25 percent reduction per bottle. This results in a savings of more than 206 million pounds of packaging each year. According to the American Plastics Council, the one-gallon plastic milk jug weighs 30 percent less than what it did 20 years ago. Lightweight materials have also opened up many new wearable products to the consumer such as mobile handsets, ultra portable computers, and digital music devices.

Saving weight through the use of plastics is an issue critical to the automotive industry. Reduced vehicle weight translates into cost savings at the gas pump. The content of plastics used in today's cars and trucks is now between 250 and 300 pounds per vehicle. If other materials were used, the vehicles would probably weigh hundreds of pounds more than they do now.

In August of 2001, the American Plastics Council provided an exhibit during the Management Briefing to the automotive industry in Traverse City, Michigan. Two product applications from their presentation are listed in the table in **Figure 10-2**. The

Figure 10-2. Compare these two weight-saving applications of plastics in automotives.

AUTOMOTIVE APPLICATION	PRODUCT	COST SAVINGS
Fan/shroud reservoir	2000 Dodge Dakota and Durango	$1.50 cost savings per vehicle, and reduction in part count from five to one; reduction in weight of 1.1 pounds per vehicle
Composite Truck Box	2002 Chevy Silverado	Savings of 50 pounds per vehicle; the box requires no bed liner, therefore reducing total weight by another 25 pounds

two applications shown may, at first glance, appear to be of little significance in terms of weight savings. But, let us take a closer look at how this reduction in weight influences gas costs to fleet owners operating these vehicles.

In the case of the Dodge Dakota and Durango, the reduction in weight by 1.1 pounds per vehicle resulted in a total fleet lifetime fuel savings of 210,000 gallons. The truck box on the Chevy Silverado saved 50 pounds, and possible total fleet lifetime fuel savings as great as 9.9 million gallons. Calculation of this data was conducted by the American Plastics Council, based on research from an Oak Ridge National Laboratory (ORNL) study.

Our earth's environment is precious and space is limited for us to live and prosper as members of a world community. It is imperative for all of us to think carefully about the impact of manufacturing and use of materials on our environment. An interesting environmental comparison can be made when considering the difference between plastic and paper bags. Not only do plastic bags require less total energy to produce than paper bags, they conserve fuel in transporting them to the end user. It takes five trucks to transport the same number of paper bags as could be carried in one truck of the same number of plastic bags. (The American Plastics Council)

Plastics are highly resistant to corrosion, and for all practical purposes, do not deteriorate when exposed to the environment. This makes them ideal materials for use in a wide range of products, from dent-resistant body panels to boats used in salty environments.

Plastics are also good thermal and electrical insulators, making them ideal for cooking-utensil handles or housings for portable electric tools. In the area of packaging, plastics are essential to the lives of most consumers. Products cover a wide array of applications, from plastic wrap used to keep meat fresh in the market and heat-sealed plastic pouches used for storing food, to heat shrink tubes used to keep water out of electrical joints, and thousands of variations of shipping containers and packing materials.

Another design advantage of plastics is that they can be created with smooth surfaces for minimal friction. This makes plastics an ideal choice in applications where the product has moving parts, many of which may never need to be lubricated.

In many applications, energy consumption can also be reduced with plastic materials. The U.S.

Department of Energy indicates that using plastic foam insulation in homes and buildings each year will ultimately save a million barrels of oil over other kinds of insulation. The same benefit applies to consumer appliances such as refrigerators and air conditioners.

The Development of Plastics

The first synthetic plastic, called *Parkesine*, was introduced in 1862 by the English chemist, Alexander Parkes at the Great International Exhibition in London, England. What was exciting about this new discovery was that once the material was heated, it could be molded, and once cooled, it retained its shape. Parkes claimed that his material could serve as a good substitute for rubber at a lower price. However, Parkes lost much of his financial backing when investors discovered the high cost of the raw materials needed in the production of Parkesine.

The basic material of Parkes's plastic was nitrocellulose, softened with camphor and a vegetable oil to make it possible to form into the desired shape. But plastic, as a commercially successful material, had its real birth in the United States almost a decade later. In 1866, when the game of billiards was gaining great popularity, a New York manufacturer offered a $10,000 prize for a usable substitute for the ivory used to make billiard balls. A young printer, John Hyatt, had experimented with many different chemical combinations. Finally, the search for a substitute material ended when he independently discovered the combination of camphor and nitrocellulose used by Parkes. Hyatt produced a plastic mass that he could form into any shape and allow to harden in a matter of minutes. On April 6, 1869, Hyatt patented his discovery, calling it *celluloid*. The material was later used for such diverse products as dental plates, vehicle windshields, and motion picture film. Celluloid's greatest drawback was flammability. Celluloid has been replaced by plastics that do not burn so easily.

The really significant industrial application of plastics began with the development of Bakelite in 1907. An American chemist, Dr. Leo Baekeland, developed the first commercially successful thermoset phenolic resin, a combination of the chemicals phenol and formaldehyde, while seeking a synthetic substitute for shellac.

This new material would not burn, boil, melt, or dissolve in any known acid or solvent. This was an important discovery. Baekeland's Bakelite was a thermoset material, which means it could not be remelted. Once the product became hard, it was permanently cast. Parkes's and Hyatt's material, however, was a thermoplastic that could be remelted and used again.

Bakelite was widely used for telephone housings, cooking-pan handles, and similar applications. It was used as a coating on other materials, such as softwood, which made the wood more durable and effective. Bakelite was used by the U.S. military extensively in the Second World War. It was also used for domestic applications such as electrical insulators. It is chemically stable, heat-resistant, shatterproof and resistant to cracks, fading, or discoloration from exposure to sunlight or damp environments.

By 1940, many industrial and consumer products that had previously been made from metals, wood, glass, leather, paper, and vulcanized natural rubber were made of plastics. The use of plastics has permeated all aspects of daily life since.

From that point onward, there were many innovations in the field of plastics. Major developments include rayon, cellophane, nylon, PVC (polyvinyl chloride), Saran™, Teflon®, polyethylene, Velcro®, and others.

Back in 1981, Malden Mills and Patagonia marketed a synthetic sheepskin called Polar Fleece. Today the finely knit polyester called *Microfiber* is used in many high-tech fashions. Plastics are also constantly being made more environmentally friendly. Modern-day biodegradable polymers, derived from starches and proteins, are helping to improve plastic's environmental image.

There is a great deal of improvement needed in terms of programs to promote plastic recycling. In 1999, more than 750,000 tons of plastic bottles were recycled. The Association of Postconsumer Plastic Recyclers, APR points out that the quantity of collected plastic bottles has been between 750 and 800 million pounds for the past several years. However, the percentage of **PET (polyethelene terephthalate)** bottles that are recycled against those that are produced is decreasing. The recycling rate for all postconsumer PET bottles fell from near 40 percent of products sold in 1995 to about 20 percent in 2002. In 2002, the quantity of high density polyethelene (HDPE) bottles available for recycling did not keep up with total demand for bottles. This led to a shutdown of about 17 percent of reclaiming capacity between 2000 and 2002. In 2002, the quantity of net exports of

PET bottles was more than 34 percent of bottles that were collected. In 1998, this figure was 12 percent. It was also anticipated that the U.S. demand for bottles for recycling could be more than two billion pounds in the next two years. (Association of Postconsumer Plastic Recyclers, APR). Today, there are close to 1,800 businesses handling and reclaiming post-consumer plastics.

The two most common types of plastics that are recycled are PET (polyethylene terephthalate) and HDPE (high density polyethylene). PET is used to make clear, blow-molded containers such as soft drink bottles and some plastic sheeting products. Other major markets for PET are carpet yarns, fiber-fill (pillow stuffing), and manufacturing geotextiles (fabrics used in road construction). HDPE plastics are used to make more durable bottles for juice, milk, water, laundry, and cleaning products.

New end use applications for recycled plastic materials are rapidly developing. Great promise is being realized for the use of recycled PET as a coating on corrugated paper and other waterproofing applications such as shipping containers. For HDPE, applications now include garden products such as edging and lawn chairs.

One of the major limitations to using recyclable PET and HDPE is the supply of these materials. There is more demand today for plastic products than materials available through recycling programs. According to the American Plastics Council, more than 20,000 American localities (63 percent of our population) have access to a plastics recycling program. Nearly all of these programs include PET and HDPE plastic bottles in their collection programs. Approximately 95 percent of plastic bottles are manufactured from PET and HDPE resins.

Many structural applications markets are being targeted by the plastics industry. Some of these major product areas are dimensional lumber, railroad ties, and marine pilings. Today, plastics are an integral part of our society—it would be hard to imagine life without them.

Making Industrial Stock

The terms *resin* and *plastic* are often confused with each other. The *Plastics Engineering Handbook*, produced by The Society of the Plastics Industry, Inc., defines **resin** as: "Any of a class of solid or semi-solid organic products of natural or synthetic origin, generally of high molecular weight with no definite melting point." Plastic, on the other hand, is defined as one of many natural or synthetic, high-polymeric products (excluding rubber), that are capable of flowing under heat and pressure, into a desired shape. (Adapted from *The Plastics Engineering Handbook*)

From these definitions it may be difficult to understand where resin ends and plastic begins. In the plastics industry, it is common practice to view all processed material up to the point where industrial stock is created as *resin*. The transition from resin to plastic occurs with the manufacturer of the industrial stock. The manufacturer starts with resin in liquid, pellet, granule, or powdered form, and uses it to form or create the plastic material that will be used in the process to create the final product. See **Figure 10-3**.

Commodity resins

Today, most of the thermoplastic and thermoset resins produced are referred to as *commodity resins*. Commodity resins are used for molding or fabricating such everyday items as household accessories and containers, refuse bags, toys, decorative items, and automotive accessories. Commodity resins include standard-grade resins such as low-density polyethylene (LDPE), polypropylene homopolymer (PP), crystal polystyrene (PS), and rigid polyvinyl chloride (PVC).

Engineering resins

Another category of resins is referred to as the *engineering resins*. Many firms process high-performance resins in various grades such as advanced, intermediate, and commodity. Examples of these resins are nylon, polycarbonate (PC), and polyphenylene sulphide (PPS). The advanced grades of resins are most resistant to chemical attack, extreme heat, and impact. Examples of these applications might be football helmets, scientific laboratory equipment, reheatable food containers, and industrial conveyor rollers. See **Figure 10-4**.

Resin as Industrial Stock

Commodity and engineering resin can be purchased in several different forms. Stock is selected in the form that is most appropriate for the manufacturing process and application. See **Figure 10-5**. Resin is manufactured in pellet, granule, powder, and liquid form. Most makers of thermoplastic resins convert fine particles into BB-sized pellets. The pellets are then shipped in bags, drums, special containers, trucks, and train cars to the manufacturing plants that produce the plastic product—monofilament line, coatings on paper, sheets, and rods. One innovative company, Insta-Bulk of Houston, Texas has developed several different types of custom bulk

Figure 10-3. Plastic is the material that results when a resin is changed into a product or part by forming or molding. These returnable, reusable milk jugs are molded from a strong polycarbonate resin. They can be sterilized and refilled up to 60 times, then ground up and recycled. (GE Plastics).

Figure 10-4. The design requirements for these individual pallet flow rack wheels included impact strength and the ability to stand up to loads in both hot and cold temperature extremes. An engineering plastic was selected to precisely fit these requirements. The design shown illustrates the use of steel bearings, but for use in corrosive environments, the wheels could contain bearings made from engineered plastic.

Figure 10-5. Major types of resins and their product applications.

Type of Resin	Properties	Product Examples
Polyethylene Terephthalate (PET)	Tough, excellent gas and moisture barrier properties	drink bottles
High Density Polyethylene (HDPE)	Excellent protective qualities and resistance to chemicals	milk and detergent bottles
Polyvinyl Chloride (PVC)	Excellent clarity, puncture resistance and cling	wrap for protecting meat
Low Density Polyethylene (LDPE)	Good flexibility	grocery and trash bags
Polypropylene (PP)	High tensile strength	threaded lids for hot products, ketchup and syrup
Polystyrene (PS)	Clear, hard, or foamed	protective cases on appliances
Expanded Polystyrene (EPS)	Lightweight foam	egg cartons, coffee cups, and packaging materials
Ultramid® glass reinforced nylon 66	Resistant to attack from engine temperatures, hot oil, gasoline and diesel fuel	99 BMW Turbo Diesel, first all-plastic air-intake module, consisting of air-intake manifold, engine cover, and air-filter housing

liner systems for ocean containers and over-the-road trailers used by the plastics and petrochemical manufacturing industries. These systems are used to convert train cars and other shipping containers for waterproof shipping worldwide.

Most plastic industrial stock goes on to manufacturers who use it directly to make consumer products. Resins in liquid form may be combined with reinforcing material such as fiberglass mat, chopped fiberglass, cloth, roving (strips), and expandable foam.

Structure of Plastics

The *Modern Plastics Encyclopedia* lists nearly 500 different types of compounds and resins that can be used in the manufacture of plastics. Each manufacturer of resins is willing to modify or tailor their product to meet specific applications. Consequently, combinations of different resins mean that there are thousands of different types of plastics available.

Plastics Are Polymers

Despite the many variations in type, all plastics are based on a high-molecular-weight molecule called a *polymer*. The polymer is a long chain made up of thousands of smaller molecules linked together. These smaller, simple molecules are called *monomers*. See **Figure 10-6**.

Different polymers result from varying the combinations of monomers. For example, *polyethylene*, a polymer used for many types of containers, consists entirely of linked *ethylene* monomers. The polymer *acrylonitrile-butadiene-styrene* (usually referred to as ABS) is the plastic widely used for plumbing pipe. It is made up of three different monomers: acrylonitrile, butadiene, and styrene. See **Figure 10-7**.

Figure 10-6. A monomer is a simple molecule of a substance. This ethylene monomer consists of two carbon atoms and four hydrogen atoms. If a number of ethylene monomers are joined together, end-to-end, the result is a long chain molecule (polymer) known as polyethylene.

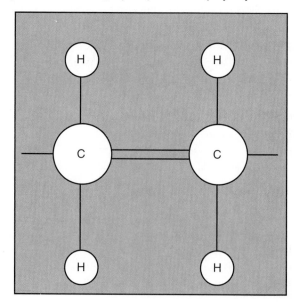

Figure 10-7. A copolymer is a molecule made up of two or more kinds of monomers. If monomers of acrylonitrile, butadiene, and styrene join together, the result is the polymer ABS. The actual polymer chain, or course, would consist of literally thousands of repetitions of the A+B+S sequence shown here, all strung together in a chain.

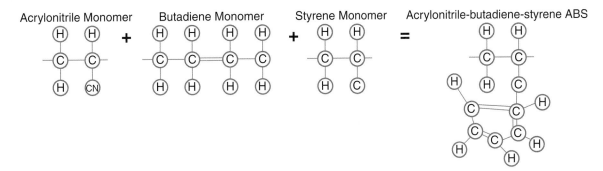

The term polymer is a combination of *poly*, meaning many repeating, and *mer*, meaning segment. Therefore, a polymer is literally many repeating segments. When two or more kinds of monomers are combined to form a long-chain molecule, such as ABS, the resulting molecule is normally called a *copolymer*.

Natural and synthetic polymers

Polymers are found in both natural and *synthetic* (human-made) forms. Wood, rubber, and cotton, for example, are natural polymers. The group of commercial plastics called *cellulosics* is made from a natural polymer, cotton cellulose. However, most polymers used in industry are synthetic and are based on chemicals derived from oil or natural gas.

Another class of plastics is referred to as *elastomers*. These highly resilient materials are more like rubber than plastic. Under normal room temperature conditions, an elastomer can be elongated to at least twice its original length and, on release, will return to its original dimensions. Elastomers can be processed with faster cycle times than other plastics in conventional injection molding machines. Neoprene, used for faucet washers, o-rings, and other types of sealing applications, is an example of an elastomer. Manufacturing related to the production of elastomers is considered to be part of the rubber industry. See **Figure 10-8**.

The Polymerization Process

The chemical reaction that creates polymers is called *polymerization*. In order to create this chemical reaction, monomers must be subjected to heat, pressure, or a *catalyst* (agent that causes a chemical reaction). When polymerization takes place, the monomers blend chemically, linking their molecular chains. The polymerization process is critical; without it, there can be no plastic.

The degree of polymerization in a particular plastic influences the molecular weight of that material. In most cases, the greater the degree of polymerization, the greater the molecular weight. The molecular weight of a particular sample of plastic is an important factor to consider, since it directly affects *viscosity*, the thickness or ability of liquid material to flow. This is especially important when products are being produced by casting or molding processes.

Figure 10-8. Elastomers are resilient materials that are often used for seals and gaskets. Different polymer formulations are used to meet specific application requirements. (Apple Rubber Products, Inc.)

Types of polymerization

Essentially, there are two different types of polymerization: addition and condensation. In *addition polymerization*, the monomer molecules join together to form long chains. The chains may be linear, like beads in a necklace, or branched. See **Figure 10-9**. The chains in addition polymerization link with one another. However, the molecules do not change chemically and their molecules do not lose or gain atoms.

Condensation polymerization is different in that there is chemical change involved. When this type of polymerization takes place, chains become *cross-linked* to each other. See **Figure 10-10**. In the process of cross-linking chains, some atoms are lost by the monomer molecules. The lost atoms combine to form other compounds, such as water molecules, as a byproduct of the process. These new molecules do *not* become part of the polymer.

The linear and branched polymers resulting from addition polymerization are typically thermoplastics, such as polyvinyl chloride and acrylic. Cross-linked polymers formed by condensation polymerization are usually thermosets, such as polyurethane and urea formaldehyde.

Major Classes of Plastics

The two major classes of plastics are thermoplastics and thermosetting plastics (thermosets). Thermoplastics can be heated to make them fluid enough to be formed. Thermosets can be heated to speed up the rate of curing and polymerization. The thermoplastics can be reheated and reformed repeatedly; thermosets cannot.

Thermoplastics

Thermoplastics consist of long, discrete chains of molecules that melt to a viscous liquid at a specific processing temperature, as low as 240°F (116°C) to a high of 700°F (371°C). The plastic behavior of these polymers is influenced by the arrangement of their molecules (morphology). Polymer molecular arrangements are either amorphous or crystalline.

- *Amorphous* molecules are arranged randomly and are intertwined.
- *Crystalline* molecules are arranged closely and in a distinct order.

Most thermosets are amorphous. Thermosetting resins undergo a chemical reaction that creates an infusible and insoluble network. Essentially a product made from a thermoset resin becomes a

Figure 10-9. The polymers resulting from addition polymerization are long chains of either the linear or branched form. Polymers of the linear and branched forms are typically found in thermoplastics. These materials can be melted and recycled into new products.

Figure 10-10. Cross linking of polymers occurs during condensation polymerization. A chemical change in the monomer molecules takes place during the process. The thermosetting plastics that result, unlike thermoplastics, cannot be melted down and reformed.

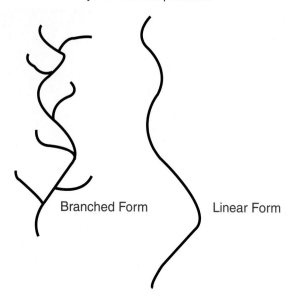

Branched Form   Linear Form

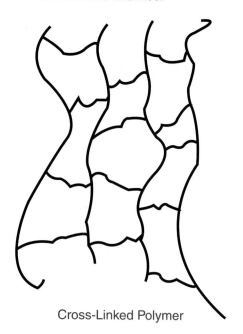

Cross-Linked Polymer

large molecule. Thermosets, which do not melt upon reheating, can be dimensionally stable up to the temperature at which chemical degradation begins.

Thermoplastics however, may be both amorphous and semicrystalline. *Semicrystalline* materials display crystalline regions, called *crystallites*, within an amorphous matrix. Thermoplastic materials retain their molded shapes up to a point called the *glass transition temperature* (this temperature differs for each particular polymer). Below the glass transition temperature, the molecules are frozen in what is known as the *glassy state.* In this state, there is little or no movement of molecules past one another, and the material is stiff or even brittle. Above the glass transition temperature, the amorphous parts of the polymer enter the rubbery state in which the molecules display increased mobility. It is at this stage that the material becomes plastic and even elastic.

If you have ever chewed gum, you probably have had a first hand experience with the glass transition temperature. At body temperature chewing gum is soft and pliable, which is characteristic of an amorphous solid in the rubbery state above the glass transition temperature. However, when you drink a cold beverage or suck on an ice cube while chewing gum, the gum becomes hard and rigid. The gum has dropped below the glass transition temperature.

The glass transition temperature is not the same thing as the melting point. The melting point occurs when the polymer chains fall out of their crystal structures, turning the plastic to a liquid. For partly crystalline polymers, such as low-density polyethylene and polyethylene terephthalate, the liquid state is not reached until the melting temperature is passed. Beyond the melting point, the crystalline regions are no longer stable and the rubbery or liquid polymers can be molded or extruded.

The crystallizing process requires a tremendous amount of cooling capacity. During crystallization the temperature should remain stable for a long time, which means a longer cooling cycle must be provided. The cooling process of an amorphous polymer, however, is significantly faster because there is no crystal formation. In plastic molding, the degree of crystallization in a thermoplastic significantly affects the material's physical and mechanical properties. That is why the cooling cycle, the last stage of the molding process, is so critical. Materials with a high degree of crystallization have molecular chains that are packed more tightly together, resulting in plastics that are typically rigid, strong, and translucent or opaque.

Polymers with an amorphous structure cannot pack as tightly together, so the resulting plastics tend to be less rigid, have less strength, and are more transparent than crystalline plastics.

Thermosets

Thermosets, as described in the preceding paragraphs, behave very differently from thermoplastics. Thermosets are chemically reactive in their fluid state and harden through a further reaction called *curing,* in which cross-linking takes place. Cross-linking results in growing polymer molecules into a spatial network, with strong bonding between the molecular chains. These three-dimensional bonds provide high dimensional stability, high temperature resistance, and excellent resistance to solvents. In injection molding, heat is applied to liquefy the thermoset plastic pellets to complete the cure of the partially cured polymer. After a thermoset plastic has cross-linked, it cannot be returned to a fluid state by heating, although heat may cause it to soften slightly. The addition of more heat will typically char and decompose a thermoset plastic.

One of the first commercially important plastics, Bakelite™, was a phenol-formaldehyde thermosetting resin created through a reaction of phenol and formaldehyde. Other thermosets include the unsaturated polyesters, melamines, epoxies, ureas, and silicones. Products which must be able to function in high heat or ultra cold environments are normally made of thermosets, such as the centrifuge shown in **Figure 10-11**.

Thermoplastics are more resistant to cracking and impact damage than thermosets. However, thermoplastics are usually inferior to thermosets in terms of high-temperature strength and chemical stability. An exception is one of the newer thermoplastics, polyether ether ketone (PEEK), which has a semicrystalline microstructure with excellent high-temperature strength and chemical resistivity.

What should be remembered about all thermoplastics is that the softening and hardening cycles are reversible. The polymer can be formed into a desired shape—pellet, tube, bar, sheet, film, or rod—cooled, and shipped as stock to the secondary manufacturing firm. There, the stock is used to make the final product by transforming it once again to the fluid state using heat, then forming and cooling.

There are many different types of plastics, materials for making these plastics, and types of

Figure 10-11. This centrifuge, designed for use in small medical laboratories, uses three different formulations of Lexan™ polycarbonate to meet different requirements. The basic housing used a foamable resin that provided structural strength, impact resistance, and sound-deadening qualities. A retainer used to hold motor wiring safely away from spinning parts was injection molded from a resin formulated to provide flame-retardant capabilities. The resin from which the transparent cover was injection-molded was selected for high impact resistance. It was designed to protect workers from glass fragments and contents from test tubes if they were to shatter while being spun at high speed. (GE Plastics)

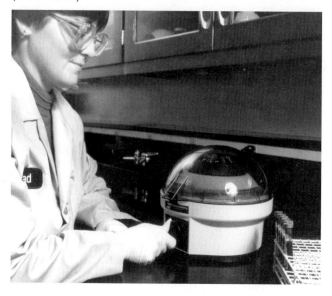

processing. Knowing which plastic and process to use for a specific application is a skill that will be developed with education and experience. For now, you must understand how the overall process occurs in the plastic industry. See **Figure 10-12**.

Plastic Memory

Imagine that you caught a favorite pair of sunglasses in your car door as it closed, severely deforming them. By taking advantage of the unique characteristic of thermoplastic materials called "memory," the glasses could be returned to their original shape by heating. *Memory* is the capability of a material to return to its original shape after it is bent or formed. Polyvinyl chloride (PVC) materials, for example, have very high plastic memory. If the material is distorted at slightly elevated temperatures and then cooled, it will maintain this shape until it is again heated to nearly the heat distortion temperature. Then it "remembers" what its shape was originally and returns to that shape.

Shape memory materials are available in both plastics and metals. A spring can be formed from thermoplastic rod or straight Nitinol (nickel-titanium alloy) wire. If that spring were exposed to sufficient heat, it would uncoil and return to its original straight shape.

Shape memory materials can be used in applications where there is a need to cover joints, protect wires, or even apply tension to a window covering. When heat is applied to the film it attempts to shrink to its original size.

Shape memory fabrics are now being developed. Mitsubishi International Corporation, has developed an intelligent fabric called *DiAPLEX*. According to the manufacturer, this fabric is made of shape memory polymer that adjusts to changes in the user's environment. Applications range from flexi-grip spoons that can conform to the shape of the arthritic fingers using only the heat of the user's hand to ski parkas that tailor themselves to the user.

Some manufacturers who are producing large, colorful, three-dimensional signs for gasoline stations, offices, and other businesses have also discovered memory plastics. Here is how they take advantage of its unique characteristics.

First, a flat sheet of memory plastic is vacuum formed in the shape of the final sign. This will become the template for the signs that later will be produced in high volume. Then the graphics, including colorful designs and text, are screen printed on this 3-D sign. This is very close to what the final product will look like—but we are not there yet. Next, the sign is heated. Since it is made of memory plastic, it returns to its original flat shape. The problem is, now the graphic image has shrunk and is all distorted. Nevertheless, this is the design that will be used for the production of the new signs. This 2-D image is photographed to create a stencil. Then this stencil is used to print images on multiple flat, unprinted work pieces that have previously been vacuum formed and reheated (to make them flat). The final operation is to vacuum form the printed sheet, and the end result will be a sign with perfect graphics. Now, this may seem like a complicated way to produce a sign. It would be, if only one sign was involved. However, the advantage of memory plastics for this application is that it permits printing of a flat sheet, rather than having to print on a 3-D object. It also enables complicated 3-D images to be mass-produced economically. Without memory, the original template would not have returned to its flat shape.

Figure 10-12. This is a systems model, or taxonomy, showing the major types of manufacturing activities in the plastics industry.

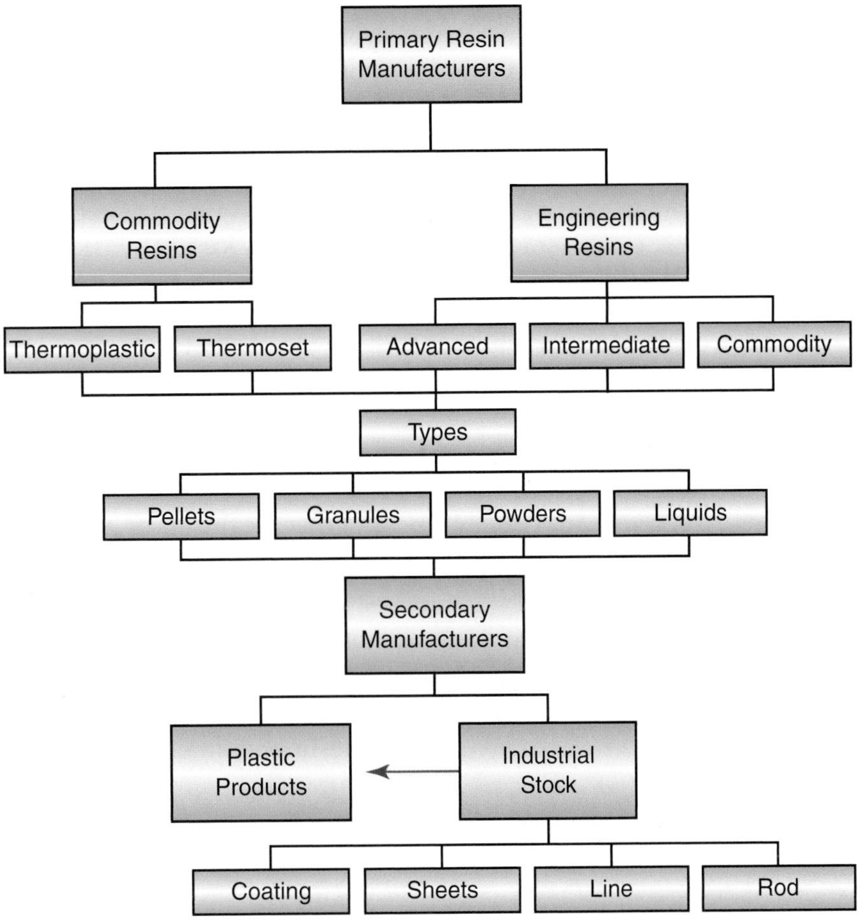

Creep

A major negative characteristic of thermoplastics is that they, like metals, are subject to creep. *Creep* is the slow and continuous increase in length at the point of deformation, over a period of time, when a material is placed under a steady load and constant temperature. When creep takes place at room temperature, it is called *cold flow*.

Let us consider how creep takes place. Imagine that you placed a plastic bar that is two-inches square by six feet long on two cement blocks. The ends for the bar are fastened securely to the blocks. A one-pound weight is then placed in the center of the bar. The plastic will naturally sag to some degree. This is called *elastic deformation*. If you remove the weight, the bar returns to its original shape. However, if the weight is allowed to remain in place for a period of time, the plastic will continue to sag at a slow rate. The additional sagging is a result of the gradual displacement of material. If you remove the weight and

examine the bar, you will find that it remains in its deformed shape. This is an example of *creep failure*, which occurs when the dimensions of the material change. With creep failure, the material will not return to its original size and shape once the load is removed. It is an important factor to consider when designing plastic parts that must withstand constant loads and stresses.

Creep failure cannot be accurately predicted by a short-term tensile or compression test. The creep stress limit that can be tolerated in a plastic material is often as little as 40 percent of the failure stress that might be shown in a short-term tensile test. This percentage would be even lower if the temperature of the plastic were increased above room temperature. Creep increases more rapidly at higher temperatures. Plastic materials can withstand high stresses for a short length of time. However, they must be used conservatively in applications where continued high loadings and temperatures are necessary.

Important Terms

acrylonitrile-butadiene-styrene (ABS)
addition polymerization
amorphous
catalyst
cellulosics
cold flow
commodity resins
condensation polymerization
copolymer
creep
creep failure
crystalline
crystallites
curing
elastic deformation
elastomers
engineering resins
glass transition temperature
glassy state
memory
monomer
polyethelene terephthalate (PET)
plastic

polyethylene
polymer
polymerization
polystyrene
resin
semicrystalline
shape memory
synthetic
thermoplastics
thermosets
viscosity

Questions for Review and Discussion

1. What effect would the application of heat from a molding process have on polymerization?

2. What are the major differences between thermoplastics and thermosets? When would you select one over the other?

3. List several common thermoplastics and several common thermosetting plastics.

4. What is the difference between a commodity resin and an engineering resin? List some uses for each.

11 Characteristics of Wood Materials

Chapter Highlights

- Hardwoods are not necessarily hard, nor are softwoods necessarily soft.
- Cellulose and lignin are the principal components of wood, with cellulose molecules constituting about 70 percent of the volume.
- Moisture content affects the strength and size of wood, and also causes swelling and shrinkage.
- Wood is classified into various groups and quality grades, depending on its end use.
- There are seven major sub-industries making up the broad category of the woods industry.

Background

The phrase *polymer production* brings to mind big chemical plants with miles of piping and huge storage tanks. However, there are many polymers that are used in manufacturing that come from farms, forests, and oceans. Examples of these polymers are starch from potatoes and wheat; cellulose and natural rubber from trees; and the shells of shrimp and other crustaceans. For thousands of years, the most important natural polymer in terms of manufacturing has been wood.

Unique Characteristics of Wood

Wood is one of the few natural materials that humans have used throughout history without finding it necessary to drastically change its properties. Wood requires little modification to make it useful for most industrial applications, and its warmth and beauty are unmatched by other materials when used for residential and furniture applications. Thanks to careful harvesting and reforestation programs, wood has become a bountiful renewable resource. See **Figure 11-1**.

The Forest Inventory and Analysis (FIA) program of the U. S. Department of Agriculture (USDA) Forest Service has been in continuous operation since 1933. The program's goal is to maintain an inventory on the conditions and requirements of renewable resources in the 155 forests and rangelands in the United States. Land covered in these areas is found in 44 states and constitutes about 8.5 percent of the total land area in the United States.

The FIA is the nation's census taker of forest resources. The FIA census reports the status and trends of each forest. This data includes information on species, size, and health of trees, and information on ozone indicator plants, vegetative diversity of the forest, and coarse woody debris. The program is managed by the Research and Development organization within the USDA Forest Service, working in conjunction with state and private forestry and national forest systems. The Forest Service is the largest forestry research organization in the world, and provides technical and financial assistance to state and private forestry agencies.

There are approximately 747 million acres of forests in the United States. Private individuals own about half of this area. Timber producers such as Weyerhaeuser and Georgia Pacific own only about 10 percent of the total forests.

Even with aggressive conservation and reforestation efforts, it is increasingly difficult to keep pace with the manufacturing and construction industry's demand for woods. To make things more challenging, the requirements for wood extends far beyond the traditional uses in manufacturing and construction. Wood from managed forests is not just used for lumber, engineered wood products, and building materials. It is also used for paper, packaging, and pulp products such as hand towels, napkins, and diapers. Wood fibers and wood derivatives are also used to make hundreds of different products ranging from ice cream to clothing.

The following statistics will illustrate the importance of wood. More than 90 percent of all products in the U.S. are shipped in corrugated paper boxes. The average American consumer uses about 750 pounds of paper products and another 18 cubic feet of wood products each year. That is the equivalent of two to four trees grown in managed forests. (www.weyerhaeuser.com)

Structure of the Woods Industry

There are nearly 2 million people employed in the woods industry. The woods industry can be subdivided into seven major industries: forestry, lumbering, millwork, furniture making, construction, wood processing, and distribution of wood products. Construction is classified as a separate industry and is not included in employment and production data from NAICS on manufacturing. The NAICS classification system identifies four major sectors in the woods industry. See **Figure 11-2**.

Figure 11-1. When carefully managed, wood is a renewable resource. Reforestation programs are an important activity for most logging companies, helping to ensure lumber supplies for the future. (Weyerhaeuser Company)

Figure 11-2. Major sectors in the wood industry listed by NAICS.

NAICS Code	Name of Sector
321	Wood Product Manufacturing
322	Paper Manufacturing
337	Furniture and Related Products
3211	Sawmill and Wood Machinery Manufacturing

The woods industry includes firms that manufacture veneer and plywood, wood composition board, paper, wooden containers, engineered wood, trusses, reconstituted wood products, and many others. The paper industry is almost totally dependent on wood. About 70 percent of our paper is made from new growth trees and 30 percent from recycled material.

The Forest Service manages our public forests to ensure that our supply of timber will continue. The Forest Service is the governmental agency concerned with the overall management of our timber reserves. Foresters and forester assistants are responsible for planting and conservation programs. Typical occupations in this industry include lumberjacks, fellers, limbers, and buckers.

The lumber industry is responsible for selecting, cutting, and transporting trees to saw mills. When the timber has been transported to the sawmill, it is cut into boards and graded. Then it is air- or kiln-dried. The sizes of the cuttings that are made depend on the purpose of the mill. Some mills produce *veneer* (a thin sheet of wood for laminated surfaces) and ship it to furniture plants for final processing. Other mills specialize in stock wood or plywood.

There are more than 5000 industrial woodworking firms in the United States and Canada. Some of these are furniture-making companies. Others make panels, doors, windows, cabinets, sporting goods equipment, and other consumer products. More than 100 companies manufacture machinery for use by the woods industry.

Construction firms use woods to build structures on site for commercial, residential, and industrial purposes. About two-thirds of all people in the construction industry are carpenters. This is the largest of all the skilled construction occupations.

Wood is also processed for use by other industries. The paper industry is one of the largest users, consuming more than 15 million cords (a cord equals 128 ft^3) of pulpwood each year to make paper and boxes.

Trees are also used as a source of material to make finishes and other chemical products. Gum and resin harvested from living trees are used to make varnish, paint, printing ink, insecticides, pharmaceuticals, chewing gum, and wax. Chemical processes are used to change wood cellulose to lacquer, synthetic fibers, photographic film, floor tile, and solid fuel for rockets.

The wood distribution industry is concerned with the distribution and sale of lumber and wood products. There are more than 30,000 retail building supply centers that deal specifically with the distribution of wood materials. See **Figure 11-3**.

Wood as a Manufacturing Material

Wood can easily be formed, shaped, and smoothed with many different types of manufacturing processes. This makes it a particularly useful material in manufacturing. At first glance, one might think that wood has utility as a manufacturing material only for the furniture, cabinetmaking, and the construction industries. While it is true that these industries depend on wood materials, nearly all manufacturing firms capitalize on the unique processing characteristics of woods to produce design prototypes, molds, jigs and fixtures, and other needed items.

During the past few years, wood has even been used in product applications that previously required plastic or metal. The cellular structure of wood makes it practical to impregnate the open pores of the material with synthetic polymers, thus improving stiffness, water repellency, strength, and stability. Many new types of materials have been created from a combination of wood and plastic. These *engineered woods* are even more durable than hardwood and more attractive than plastic.

There are hundreds of different species of wood used commercially throughout the world. They exhibit a wide variety of grain patterns, tones, and surface textures. Some types of wood are strong and durable, others have little structural strength.

Figure 11-3. Large building supply centers stock many kinds of dimension and structural lumber. Often, they prefabricate building components, such as these roof trusses, to meet customer demand. (Hoge Lumber Company)

Classification of Wood

Wood is classified, according to its cellular structure and type of tree, as either a *hardwood* or *softwood*. However, this classification is sometimes misleading and confusing. Just because a wood is classified as a hardwood does not mean that it is hard in physical terms. As an example, balsa, one of the world's softest and lightest woods, is classified as a hardwood. It is the structure, variety, and arrangement of wood cells that makes one type of wood different from another. Depending on their cellular structure, woods differ in terms of hardness, porosity, density, moisture content, and strength.

Wood fibers are different from synthetic polymers (plastics). Wood fibers are hollow, while synthetic polymers are solid. The cells in wood are arranged in a bundle of hollow tubes, much like soda straws bound together with glue. Cellular structure affects the appearance of the wood, forming its *grain*. The grain is the appearance of the annual rings and fibers viewed longitudinally.

Deciduous trees, the broad-leafed species that typically shed their leaves each fall, are hardwoods. Oak, walnut, maple, birch, and ash are well-known hardwoods. Most of the hardwoods that are harvested come from the region east of the Mississippi river.

Conifers, the cone-bearing trees that have needles and remain green all year long, are softwoods. Pine, fir, hemlock, cypress, redwood, and red cedar are common softwoods. Most of the softwoods are harvested in the southern and western regions of the United States.

Composition of Wood

There are two major ingredients in wood: cellulose and lignin. About 70 percent of the volume of the wood is *cellulose*. The long-chain cellulose molecules are arranged in a nearly parallel orientation into units called *crystallites*. Crystallites are linked together in bundles called *microfibrils*. About 100 microfibrils join in the cell wall to form *fibrils*, or lamella. The strength of wood is influenced by the angle of the fibrils against the long direction of the cell; the smaller the angle, the stronger the cell. However, mechanical properties of wood are directly related to the physical properties. A better estimate of strength can be made by determining the wood's moisture content, density, and specific gravity. Additional strength estimates can be made using data from the United States Forest Products

Laboratory, U.S. Department of Agriculture for individual species.

Adjacent layers of cells are bonded together with a hard adhesive that is called *lignin*. Lignin constitutes about 25 percent of the total volume of the wood. Normally hardwoods have less lignin and more cellulose than softwoods. This is what makes hardwoods more dense than softwoods.

When a tree trunk or limb is viewed in cross section, the layers of cells appear as concentric circles. Each year's growth results in a new *annular ring*. Cells that are formed in the spring and early summer have a thinner wall and are lighter in color. Growth is slower in late summer and fall, so the cells formed then are thicker and darker in color. The darkness of these cells and the pattern of the annular rings creates the attractive grain in wood.

In addition to cellulose and lignin, minerals and extractives in the tree also influence the composition of the wood. *Extractives* are materials that, when removed from the tree and processed, result in products such as: starches, oils, tannins, coloring agents, fats, and waxes. Cellulose represents about 50 percent of the total composition of the wood. *Hemicellulose*, a low molecular weight polymer formed from glucose represents about 25 percent. Lignin constitutes approximately 20 percent of the wood. Extractive materials normally account for about 5 percent of the total composition.

Cellular structure of hardwoods

Wood is made of long thin cells with tapered ends. **Figure 11-4** illustrates what a block of yellow poplar about 1/32″ thick would look like under a microscope. Yellow poplar is a hardwood. The cross-sectional face on the drawing corresponds to a tiny area of the top surface of a stump or end of a log known as *end grain*.

The radial face, at lower left, corresponds to a cut made close to the surface and parallel to the radius. The tangential face, at lower right, shows what the cells look like when they are exposed tangentially within the log. The annual ring, the amount that the tree grows each year, is the area from the earlywood to the latewood.

Hardwoods have cellular structures called *vessels* for carrying sap vertically. The vessels (labeled SC on the drawing) are made up of large cells with open ends positioned one above the other and continuing as open passages or tubes for relatively long distances. The area labeled K is where the

Figure 11-4. The cell structure of hardwood. Note the varied sizes of cells, especially the large vessels (#9) that permit vertical movement of sap. In open-grained woods like oak, these are visible as large pores.
(Forest Products Laboratory, USDA Forest Service)

Figure 11-5. Cell structure of softwood. Note that the cells are much more uniform in size. Sap movement is through heavy-walled cells called tracheids (#10) that also provide strength to the wood. (Forest Products Laboratory, USDA Forest Service)

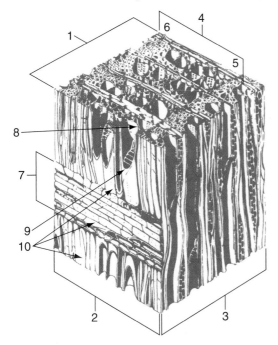

Hardwood Cell Structure

1. Cross-sectional face
2. Radial face
3. Tangential face
4. Annual ring
5. Earlywood
6. Latewood
7. Wood ray
8. Vessel
9. Sieve plate
10. Parenchyma cells

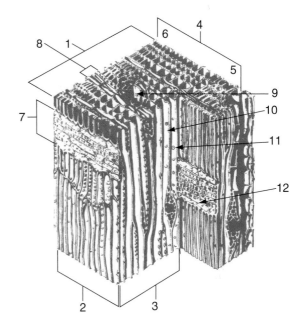

Softwood Cell Structure

1. Cross-sectional face
2. Radial face
3. Tangential face
4. Annual ring
5. Earlywood
6. Latewood
7. Wood ray
8. Fusiform ray
9. Vertical resin duct
10. Tracheid
11. Bordered pit
12. Simple pit

sap drips from one vessel to another. When exposed on the end of a log, these vessels appear as holes or *pores*. This is why hardwoods are often referred to as *porous* woods. The pores vary considerably in size. They are readily visible in some species of wood, but cannot be seen without a magnifying glass in others.

Hardwoods with large pores (ash, mahogany, oak) are classified as ***open-grain woods***. Small-pore hardwoods (birch, cherry, maple) are classified as ***closed-grain woods***. Normally, open-grain wood pores must be closed with ***wood filler*** to achieve a smooth surface before finishing materials are applied.

Cellular structure of softwoods

The structure of softwood is quite different from that of hardwood. The illustration in **Figure 11-5** shows a block of white pine approximately 1/32"

in thickness. The cross-sectional face shows the area parallel to the end surface of a log. In softwoods, sap is transferred vertically through long tubular cells called ***tracheids***, identified in the center of the drawing. In addition to transporting sap, the tracheids provide strength to the wood.

The walls of the tracheids are thicker than the walls of cells in hardwoods. In effect, they form the bulk of the wood substance in softwoods. Cells are arranged in more orderly rows in softwoods than they are in hardwoods. The thin ***intercellular layer*** (adhesive material between the cell units) in softwoods can easily be dissolved using certain chemicals. This is a basic process making paper. See **Figure 11-6**.

Keep in mind that it is not the type or physical hardness of the wood, but the cellular structure that determines whether a given wood can be classified as a hardwood or softwood.

Figure 11-6. In papermaking, the fibers of softwoods are separated by dissolving the glue-like substance that holds them together. They can then be rearranged and formed into a thin sheet on huge papermaking machines like the one shown. (Hammermill Paper Group)

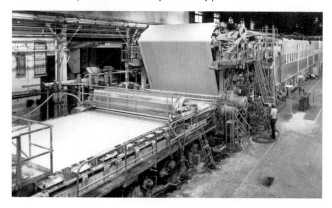

Effects of moisture

We know that hardwoods and softwoods differ in terms of their porosity. Remember that hardwoods have pores and softwoods do not. The moisture content of wood is something that is important to firms manufacturing wood products and contractors who may be using these products in a variety of different environments. Moisture content is directly related to cellular structure. The more open grain that is available, the more moisture that can be absorbed.

Because of its porosity, hardwood is particularly susceptible to the absorption of water, in both vapor and liquid form. Softwood does not have pores, but is also influenced by water. Any wood that has not been filled, treated, or finished will absorb and lose water with changes in temperature and humidity.

Let us take a moment to consider what happens in terms of moisture absorption and retention with wood siding and windows. Studies performed at the Forest Products Laboratory (FPL) in the 1930s and early 1950s found that rain can wet the back of horizontal wood lap siding. In these studies, rainwater reached the space between the siding and sheathing. Their recommendation to the industry was to use a water-resistant barrier, such as asphalt, behind the wood siding. In the 1980s, foam sheathing became popular as a barrier because of its superior insulative capabilities. Research from the FPL indicates that water wets the back of wood siding installed over foam sheathing and stays inside the siding longer than if the siding were placed on top of furring strips such as plastering lath. (Advanced Housing Research Center, USDA Forest Products Laboratory, Madison, Wisconsin)

One approach introduced in the 1950s, and still recommended by some architects, is to prime the backside of solid wood siding with paint or water repellent. This is called *back priming*. Research conducted by the FPL found that this was an effective technique for improving overall performance of horizontal lap siding. However, the benefits of back priming have not been experimentally verified. Opponents of this approach state that the water will be held inside of the siding by the primer, and will not be able to evaporate.

More research is needed to find a perfect solution. What should be evident is that the porosity of wood is a factor that must be addressed, particularly when wood is used in a moist and humid environment.

Many consumers do not worry much about the moisture content of the wood they purchase at a lumberyard or retail distribution center. However, knowing the moisture content of a particular piece may be important. Wood, as is also the case with ceramics, shrinks as it loses moisture. Wood, unlike ceramics and many other materials, also expands or swells with the absorption of moisture. To understand the importance of calculating moisture content, consider this scenario. You have decided to build a deck on the back of your house. You buy some salt-treated lumber that looks great for the project. In reality though, it is heavy and full of moisture. You nail the boards tightly together side by side. With the job done, you decide to go on a two-week vacation. When you return, to your chagrin you find that the boards are spaced about 1/8″ apart. The sun's rays did a job on the project, drying out moisture in the boards. The boards shrunk across their width but not their length. The maximum amount of shrinkage occurs in a tangential direction to the annual rings. You could have reduced the amount of shrinkage by coating the wood with a sealer or exterior finish, or by laying the boards out in the sun without fastening them to the deck supports for two weeks to acclimate (adjust to the local temperature and humidity).

It is often necessary to know the moisture content of the stock being used. An unscientific approach is to test the weight of each piece of lumber. It doesn't take much skill to tell which piece of 2 × 4 is heavier than others. However, approaches that are more accurate are needed in manufacturing.

There are two primary methods used by consumers and contractors to calculate the moisture content (MC) of wood. The easiest approach is to use a moisture meter.

One popular type of handheld moisture content meter measures electrical resistance of the wood. *Resistance type moisture meters* require the user to drive a pin-type probe into the wood. These meters measure resistance between the pins inserted into the wood. Normally readings are within 1–2 percent accuracy. Variables such as wood temperature, knots or resin buildup, salts from the preservative, or the presence of seawater at the test area can influence the accuracy of the test.

Another type of handheld moisture meter that does not use pins, and therefore does not create damage to the sample, is the *dielectric*, or *pinless moisture meter*. It relies on surface contact with a flat electrode that does not penetrate the wood. The depth of field from these meters ranges from 0.5″ to 1.0″, and will be influenced by moisture deeper in the test sample. See **Figure 11-7**.

In the commercial environment, the moisture content of wood is normally calculated using the oven dried method. The advantage of this method is that it accurately measures the entire range of moisture content in the wood. Resistance meters can only measure from 7 to 30 percent of the material being sampled.

With the oven dried method, small samples of the wood are initially weighed and then dried in an oven at 215°F (101.7°C). This continues until the piece is completely dry (when no further weight loss occurs) and then the sample is reweighed to find its oven dry weight (ODW). The loss in weight during drying indicates how much water was originally present in the sample. The process is illustrated in **Figure 11-8**.

For example, if the initial weight (IW) of a piece of oak was 34.1 g, and its oven dry weight (DW) was

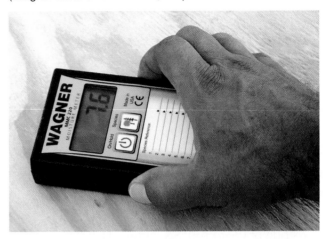

Figure 11-7. Wagner pocket sized pinless-type moisture meters for determining water content in wood uses electromagnetic waves to determine moisture content. (Wagner Electronic Products, Inc.)

22.2 g, then the weight of the water (WW) would be 11.9 g. Moisture content would be calculated as follows:

Initial weight minus oven dry weight equals the weight of the water.

$$IW–DW = WW$$
$$34.1–22.2 = 11.9$$

The weight of the water divided by the initial weight of the oak sample…

$$11.9 / 34.1 = 0.349$$

multiplied by one hundred to display as a percentage

$$0.349 \times 100 = 34.9 \text{ percent}$$

Most shrinkage in wood occurs *tangentially*, or around the tree, in the direction of the annual rings. There is little shrinkage *radially* (across the rings) or *longitudinally* (lengthwise, or in the

Figure 11-8. Procedure for calculating moisture content of wood using the oven dried method.

STEP	DESCRIPTION
1	Samples are weighed to obtain initial weight (IW)
2	Samples are dried in oven until the point where there is no more loss of moisture. This oven dry weight (ODW) is recorded
3	The ODW is subtracted from the IW to determine the weight of the water (WW) that was initially present
4	Data is substituted in the formula to determine moisture content (MC) of the wood being sampled
5	MC (Moisture Content) = WW (Weight of Water) / ODW (Oven Dry Weight) X 100

Figure 11-9. It is important to consider the different directions that wood shrinks. The greatest shrinkage is tangential (around the tree, in the direction of the rings). Radial (across the rings) or longitudinal (lengthwise) shrinkage is much less extensive. (Forest Products Laboratory, USDA Forest Service)

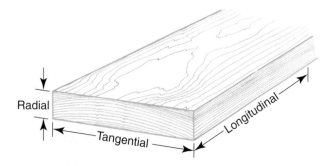

Figure 11-10. Tangential shrinkage of wood is more or less severe, depending on the way the annual rings are related to the board's dimensions. Note the severe cupping effect of the flat-sawn board at the top, and the even shrinkage and minimal distortion of the flat-sawn board at center left. (Forest Products Laboratory, USDA Forest Service)

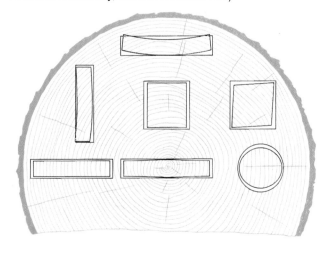

direction of the grain). See **Figure 11-9**. The illustration in **Figure 11-10** shows how tangential shrinkage, or warpage, and distortion of flat, square, and round stock occurs in relation to the annual rings.

The strength of wood normally increases as it dries. Stock that has just been cut from a log is called *green wood*, and contains a great deal of moisture, or sap. It has only about half the strength (in endwise compression) of wood that has been dried to 10 percent moisture. However, reducing moisture content does not improve all characteristics related to strength. Flexibility, or the bending strength, of wood decreases significantly as it dries. Compression strength is particularly important in the construction industry and other applications where load-bearing structures are involved. Normally, dry wood is more brittle and not able to withstand as great a pressure under compression as wet wood.

Drying methods

As previously noted, when wood is first cut into lumber it is classified as green wood. A green tree often retains as much as 20 percent moisture in the cavities of its cells. We know that if the wood is used at this stage, it will shrink and undergo drastic shifts and distortions as it gradually dries through exposure to air. For this reason, wood must be dried or seasoned before use. Seasoning often results in a reduction of moisture by as much as 20 percent. Sometimes, however, the drying process is not enough to keep the stock from further drying and warping. Stock should always be checked for warpage prior to purchase and use. Warpage results in twisting and bending of the wood in all directions, and occurs when green wood loses and gains

moisture. There are four common forms of warpage. See **Figure 11-11**.

- *Bow warp* occurs with the length of the board.
- *Crook warp* causes the board to twist in an arc or partial circle.
- *Twist warp* involves twisting of the board from one end to the other.
- *Cup warp* occurs across the width of the board.

Figure 11-11. These are the type of warp that occur in wood. (Forest Products Laboratory, USDA Forest Service)

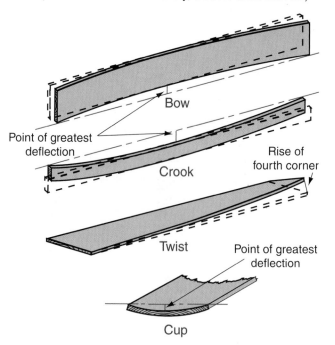

Wood is usually air-dried (AD) or kiln-dried (KD). *Air drying* reduces the moisture content to around 15 percent. Normally, air drying is done for wood that is to be used outdoors. See **Figure 11-12**.

Kiln drying is accomplished in temperature- and humidity-controlled ovens. Kiln-dried wood has only about 7 percent moisture content. In addition to removing a larger amount of moisture from the wood, kiln drying relieves much more stress in the material than would be possible with air drying. Kiln-dried wood is intended for indoor use.

One of the newer processes gaining popularity for drying hardwood is called *radio frequency (RF) dielectric drying*. The process is called dielectric drying because the wood to be heated is placed in a high-frequency field between two parallel plates or electrodes. Here, the wood becomes the dielectric of a capacitor. Let us take a moment to think about this.

Capacitors are used in radios and other electrical devices to store energy. Capacitors consist of two plates that serve as conductors of electricity. These plates are separated by an insulator, called a *dielectric*. Capacitors store electrical energy between the plates. This capability to store energy is referred to as *capacitance*. The basic unit of capacitance is the *farad* (*F*). The size of the plates, the distance between them and a characteristic of the dielectric (called the *dielectric constant*) determine a capacitor's capacitance. The dielectric constant refers to the extent to which a substance concentrates electrostatic lines of flux. A material with a high dielectric constant would result in greater capacitance.

Figure 11-12. Wood that is to be used outdoors, such as the siding on this contemporary house, is normally air-dried. Often, products like siding will be treated with preservatives during manufacture to provide longer service life when exposed to weather. (Benjamin Moore)

Now, let us apply what has been learned to the application of RF radio waves to dry wood. Here is what happens: In a radio frequency heating system, the RF generator creates an alternating electric field between two plates. The wood to be heated is placed between the plates. The energy field causes the molecules in the material to continuously reposition themselves, much like the way bar magnets behave in an alternating magnetic field. Friction caused by this molecular movement causes the material to rapidly heat throughout its entire mass. The amount of heat generated in the wood is determined by the frequency, the square of the applied voltage, the size of the product, and the dielectric loss factor of the material. When heat is applied to the wood, this is immediately transferred to the core or interior of the material. This high-frequency energy creates frictional heat inside the wood. This heat drives out moisture and results in uniform shrinkage. Most dielectric heating is conducted at radio frequencies of 10 to 100 MHz. RF drying improves the shear strength, impact resistance, checking resistance, and surface finish of the material. It also reduces the amount of warpage and is much faster than kiln drying. A typical batch of wood might take 4 or 5 days to be kiln-dried while RF drying can be completed in one day.

Wood Grades

For softwoods, there are two subclassifications within the basic classifications (dimension, factory and shop, and structural). These subclassifications are select and common.

Select lumber is graded from A to D, with A presenting the best quality surface appearance. Structural *common lumber* is graded by number, ranging from 1 to 4. Number 1 common is the best grade, with no knots or knotholes. Each lower grade (2, 3, 4) exhibits an increase in the number of defects. Defects include knots, stains, split areas, holes, and warping.

The National Hardwood Association establishes hardwood standards. The best grade of hardwood stock is *FAS*, meaning firsts and seconds. Thus, the highest quality grade is called *firsts,* and the next best grade, *seconds.*

In order to meet the exacting standards for FAS grading, a board must be at least 8′ long and 6″ wide. It must also have at least 83.33 percent clear cuttings. No knots or defects are acceptable in FAS-grade lumber.

The next hardwood grade is select. *Select hardwood* boards must be at least 6′ long and 4″ wide. One side, called the face, must be FAS quality; the other can have some defects and blemishes.

The next lower grade of hardwood is called Number 1 Common, which must have 66.66 percent clear cuttings. Number 2 Common must have at least 50 percent clear cuttings; Number 3 Common, 33.33 percent clear cuttings. The term, *clear cuttings* refers to the felling of all trees in a chosen area in one operation. It is still practiced in many areas, but new techniques for forest management are rapidly being introduced. Change is taking place, largely due to pressure from environmentalists and others who are concerned about the impact of this practice on our ecosystem. Major negative impacts of clear cutting include the elimination of the native forest ecosystem, increased erosion, impairment of recreational values of forest lands, and susceptibility of the forest to insect damage, diseases, acid rain, and uprooted trees.

As an alternative to clear cutting, many foresters are now practicing techniques of *selection management*. With this approach, individual trees are marked and cut in rows, creating small clearings or openings that allow for growth of new trees through natural reseeding from remaining trees. Such an approach has already been proven to yield great benefits to wildlife and to the productivity of forestland.

When wood is purchased from the mill or in quantity from a supplier, it is sized according to the number of board feet in the stack. A *board foot* is the basic unit of measurement in the lumber industry, and is 1″ thick by 12″ wide by 1′ long. To determine the number of board feet in a piece of lumber, multiply the thickness in inches by the width in inches by the length in feet, and then divide by 12. When stock is less than 1″ thick, it is figured as 1″. The formula for computing board feet is:

Thickness × Width × Length / 12 = board feet

As an example, suppose that we have a 1″ × 6″ piece of wood that is 5′ long. We want to know how many board feet are in this piece of wood:

(1″ × 6″ × 5′) / 12 = 2.5 board feet

Nature of Industrial Stock

Wood *stock* is normally purchased by a manufacturing firm as yard lumber. Yard lumber consists mainly of softwoods, particularly white pine. There are three major classifications of yard lumber: dimension lumber, factory and shop lumber, and structural lumber. There are several quality grades in each of these classifications.

Dimension lumber comes in standard-length boards up to 1″ thick and 12″ wide. Dimension lumber is ready to use, without any additional sanding or surface preparation.

Factory and shop lumber is used primarily for remanufacturing purposes in mills that produce fabricated doors, windows, cabinets, moldings, and trim items. The poorest grade of softwood, referred to as Number 4, is used for this purpose.

Structural lumber includes light framing material, such as 2 × 4s, and widths up to 6″ and wider for joists, rafters, and framing uses. See **Figure 11-13**.

Structural lumber is primarily purchased by the construction industry for use as beams, stringers, posts, and timbers for heavy structural applications, and for factory and shop applications.

Products made from fine quality hardwood and softwood are purchased in rough-sawn form directly from a lumber mill. *Rough-sawn lumber* needs to be planed or surfaced to smooth the board prior to use. The dimension across the board or face is smoothed by running it through a planer. Normally, the rough-sawn edges are smoothed by another process, called *jointing*. These processes will be discussed in a later chapter.

Wood stock is also produced in chip form for making particle board, wafer board, and oriented strand board (OSB). Wood fibers are used to make

Figure 11-13. Structural lumber, such as the 2 × 4s used for wall framing in this photo, is the basic stock of the construction industry. (Milwaukee Electric Tool Corporation)

hardboard for paneling and floor coverings. Panels made from laminated wood to produce plywood and beams are used extensively in construction. Plywood covered with a veneer of fine wood is widely used in manufacturing furniture and as interior paneling in homes and offices. Wood is also used as a raw material to manufacture a variety of products, including paper and cardboard, fuel, synthetic fibers, turpentine, and waxes.

Important Terms

air drying
annular ring
board foot
bow warp
cellulose
clear cuttings
closed-grain woods
common lumber
conifers
crook warp
crystallites
cup warp
deciduous
dialectric
dimension lumber
end grain
extractives
factory and shop lumber
FAS
fibrils
grain
green wood
hardwood
hemicellulose
intercellular layer
kiln drying
lignin
longitudinally
microfibrils
open-grain woods
pinless moisture meter
pores
radially
radio frequency (RF) dielectric drying
resistance type moisture meters
rough-sawn lumber
select hardwood
selection management
select lumber
softwood
stock
structural lumber
tangentially
tracheids
twist warp
veneer
vessels
wood filler

Questions for Review and Discussion

1. What is the major difference between a hardwood and a softwood?
2. What type of surface preparation must be done before finishing hardwoods? What will happen if this is not done?
3. What advantages does RF drying offer when compared to kiln drying?
4. If you went to the local lumber mart and purchased salt-treated boards for use in making a deck on the back of your house, you would notice that there is a lot of difference in the weight of the individual boards. Would it be better for you to select a heavy or light board? Do not make your choice based on the labor necessary to carry the boards. Explain.

Wood materials are used extensively in the furniture industry, as shown in the construction of these sofa frames.

12 Characteristics of Ceramic Materials

Chapter Highlights

- There are two major divisions in the ceramic industry, clay and glass.
- Many types of ceramic stock are used by manufacturers including synthetic materials such as high temperature carbide and alumina titania, and natural materials such as soapstone and clay feldspar.
- Bioceramic devices have been developed that bond to bone and are gradually assimilated into the bloodstream of humans and animals.
- Humans throughout history have used traditional ceramics.
- Industrial ceramics evolved in the 1960s and 1970s with the advent of engineered and composite materials.

Background

Ceramic products are made of clay; an inorganic, nonmetallic solid material that is derived from naturally decomposed granite. Clay contains a white mineral, *kaolinite*, that consists of aluminum oxide, silica dioxide (quartz), and water. The fusion of silica dioxide produces glass. Clay and glass are major divisions of the ceramics industry. Both natural and synthetic minerals are used in the industry.

Historically, ceramics have been viewed as materials that behave very differently from metals or organic materials. The variety of ceramic materials used in industry today is so wide-ranging that it is difficult to describe ceramic products in terms of their composition. Ceramic products, such as brick, pottery, and artware have been made from sand and clay since the Iron Age and Bronze Age. In fact, many of the same manufacturing processes that were used in antiquity are still used today. However, in the field of industrial ceramics, manufacturing has undergone many dramatic changes. The clay used for making dishes, pottery, and artware bears little resemblance to the stock that is used today by the advanced ceramics industry. Today's advanced ceramics companies are manufacturing high temperature cutting tools, body armor, electrochemical devices, artificial joints, heat transfer parts and much more.

In February of 2000, at the National Press Club in Washington, DC, astronaut and engineer Neil Armstrong read a list identified by the National Academy of Engineering as the twenty engineering achievements that had the greatest impact on the quality of life in the 20th Century. These achievements are changing the way we live and work today. The table in **Figure 12-1** lists these achievements along with an explanation of how ceramics were involved in the achievements.

Sometimes we forget just how important ceramics are to our way of life. The clay and glass industries manufacture quite a diverse array of products—from glassware and tiles to concrete, to high-temperature refractories and oxide ceramics, to electronic glass and laser crystals produced from a melt. Without ceramics, the miniaturization permitted by oxide ceramic components in wireless devices and systems; such as garage door openers, digital smart phones, gaming devices, WiFi access

Figure 12-1. Notice the importance of ceramics in this list titled *Top 20 Engineering Achievements of the Twentieth Century.* (National Academy of Engineering.)

Rank	Achievement	How Ceramics Contribute
1	Electrification	Insulators for power lines and industrial/household applications
2	Automobile	Engine sensors, catalytic converter, spark plugs, and windows
3	Airplane	Antifogging/freezing glass windows, jet engine components
4	Safe Water	Water treatment filters
5	Electronics	Substrates and IC packages, capacitors, piezoelectrics, insulators, magnets, superconductors
6	Radio and TV	Glass tubes (CRT's), glass faceplate, phosphor coatings, electrical components
7	Agriculture Mechanization	Refractories used to melt and form ferrous and nonferrous metals
8	Computers	Electrical components, magnetic storage, glass for computer monitors
9	Telephone	Electrical components, glass optical fibers
10	Air Conditioning and Refrigeration	Glass fiber insulation, ceramic magnets
11	Interstate Highways	Cement for roads and bridges, glass microspheres used to produce reflective paints for signs and road lines
12	Space Exploration	Space shuttle tile, high-temperature resistant components, ceramic ablation materials, electromagnetic and transparent windows, electrical components, telescope lenses
13	Internet	Electrical components, magnetic storage, glass for computer monitor
14	Imaging: X-rays to film	Piezoceramic transducers for ultrasound diagnostics, sonar detection, ocean floor mapping, ceramic scintillator for X-ray computerized tomography, phosphor coatings for radar and sonar screens
15	Household Appliances	Porcelain enamel coatings for appliances, glass fiber insulation for stoves and refrigerators, electrical ceramics, glass-ceramic stove tops
16	Health Technologies	Replacement joints, heart valves, bone substitutes, hearing aids, pacemakers, dental ceramics, transducers
17	Petroleum and Natural Gas Technologies	Ceramic catalysts, refractories and packing media for petroleum and gas refinement, cement for well drilling, drill bits for well drilling
18	Laser and Fiber Optics	Glass optional fibers, fiber amplifiers, laser materials
19	Nuclear Technologies	Fuel pellets, control rods, high-reliability seats and valves, containerization components, spent nuclear waste containment
20	High-Performance Materials	Ceramic materials cited for advanced properties such as wear, corrosion, and high temperature resistance, high stiffness, light weight

points, and electronic pet containment systems, might still be a figment of someone's imagination.

High-frequency capacitors and magnetic components, ceramic substrates and microwave chip modules made from advanced ceramics are used in these wireless devices.

In the telecommunication industry, ceramic fiber optics (glass fiber cables) are replacing copper wires because of their light weight and ability to carry more information at high speeds. As an example, one fiber optic pair of fibers, about the thickness of a human hair, can carry more than 50,000 telephone conversations with no interference (*noise*) and little resistance. Conventional copper wires are able to carry only twenty-four simultaneous telephone conversations.

Unique Characteristics of Ceramics

Ceramics have many advantages over other materials. These include high melting temperatures, high hardness, high modulus of elasticity, high compressive strength, and low electrical and thermal conductivity. When compared to metallic parts, ceramic parts exhibit better wear resistance, greater strength at high temperatures, and higher resistance to harsh chemicals. Ceramic materials can be made with a wide range of densities—low enough to float on water or high enough to rival lead. Their strong chemical bonding makes them hard and able to withstand high temperatures without melting, and able to survive attacks by corrosive gases, molten metals, and acids. Ceramics are well suited for use in equipment that is exposed to corrosive substances for extended periods without deteriorating or requiring frequent maintenance.

However, ceramics also have some distinct disadvantages. See **Figure 12-2**. Ceramics are typically less tough than metals and many composites. Low impact strength and a high degree of *brittleness* is also a weakness of ceramics. This may cause them to break without warning, after stress causes cracks that start as microscopic flaws in the material.

It might appear that the characteristic of brittleness would severely hamper the usefulness of most ceramic products. However, this tendency can be engineered out of the material. New technology, such as the development of *ceramic-matrix composites*, offers methods for toughening the material. These composites incorporate reinforcing fibers or whiskers

Figure 12-2. Advantages and disadvantages of ceramic materials.

Advantages of Ceramics	Disadvantages of Ceramics
Hard	Brittle
Chemically stable	Susceptible to thermal shock
Electrical insulator	
Heat resistant (refractory material)	
Nonmagnetic	
Oxidation resistant	
Thermal insulator	
Wear resistant	

(high-strength single crystals with a length at least ten times their diameter). Another popular technique to provide strength is to apply ceramic coatings to metal substrates.

In other cases, *fatigue* (the tendency of a material to crack and fail due to repeated stress) can be reduced by using processes that compact powder under pressure to improve the density of a part. Most ceramic forming processes start with these steps before firing the parts at high temperatures in a furnace called a *kiln*.

Not all ceramics require heat to create a molecular bond, however. Some ceramic materials develop their strength at room temperature, usually through chemical reactions occurring with the addition of water. Cement paste and concrete materials behave in this manner.

Product specifications that are appropriate for advanced ceramics materials and processes include high-temperature stability, corrosion resistance, and increased toughness. Uses include diesel engine exhaust valves and heat exchanger tubes, hot gas filters, and gas turbine combustion chamber liners.

Ceramics in Space

The first thing that most people think of when considering the use of ceramics in space is the heat-resistant tiles that cover the underside of the space shuttle. These tiles are part of a thermal protection system (TPS) that insulates each orbiter's ship and crew from temperatures that approach 3,000° F as the orbiter returns to Earth. The TPS also includes insulating blankets, white ceramic tiles that withstand lower temperatures, and a reinforced matrix composite material along the craft's nose and wing edges.

One example of how the space program plays a role in terms of advanced ceramics research was NASA's USMP-4 mission, conducted in 1997. A furnace NASA called the *Advanced Automated Directional Solidification Furnace* (AADSF) was used to process two types of semiconductor materials believed to hold significant promise for the future: mercury cadmium telluride and lead tin telluride. The furnace was successfully used to process two samples. The conditions of the samples may result in the design and manufacture of better infrared detectors.

Space is an important laboratory for conducting materials research. When semiconductors are made on Earth, the process involves melting and then resolidifying the material into the desired form. Any impurities or irregularities in the material interfere with performance. In most cases, problems are caused by the gravity-driven flow of materials when the semiconductor is formed. The quality of the material is greatly improved when produced in the weightless environment of Space.

Another interesting application area is for ceramic textiles. The NASA Marshall Space Flight Center in Huntsville, AL, and Johnson Space Center in Houston, TX have developed a heat and debris shield called a *whipple shield,* using 3M™ Nextel™ ceramic fabric. This has proven to be a lightweight and effective alternative to conventional shielding used to protect vehicles from collisions with space debris during reentry.

Other Applications of Ceramics

Today, ceramics play a major role in the advancement of a wide range of fields of importance to society. The military has been one of the first areas to apply and transfer ceramic materials and processing technology to practice. The NATO C-130 transport aircraft is one example. C-130s are equipped with aramid fiber and glass ceramic cockpit armor.

Ceramics are used for gun liners, ceramic armor, missile and ammunition parts, and many parts on aircraft and attack helicopters. Ceramic armament (armor) is used to protect vehicles from missiles, bombs, and bullets. Ceramic *anti-armor* is also incorporated into explosives and other devices used against combatants. One type of anti-armor that was used in Southern Iraq is called the *BAT (Brilliant Anti-armor Technology* submunition). These cluster bombs or artillery rockets are designed to break apart over a target, scattering hundreds of bomblets over acres of ground.

In the medical field, there are ceramics designed specifically for use in the human body. These materials are called **bioceramics**. In the future, bioceramics may account for a significant percentage of the world ceramic market.

One of the biggest areas of application for bioceramics is in medicine. There are three major types of bioceramic applications in medicine. These are referred to as nearly inert, surface-active, and resorbable. ***Nearly inert*** ceramic devices can be implanted in the body without causing toxic reactions. These materials include silicon nitride–based ceramics, zirconia, and alumina. ***Surface-active*** ceramics are those that form a chemical bond with the surrounding tissue and encourage growth. They allow the implant to be held in place and help prevent rejection due to dislocation. Surface-active glass-ceramics and surface-active glass are able to bond directly to bone. ***Resorbable*** ceramics materials have the ability to dissolve and be assimilated into the blood stream. In these applications, ceramic materials act much like bone tissue in the body.

Doctors are experiencing success with ceramics in hip and knee joint replacements. Ceramics are being used for ceramic heart valves, dental implants, and coatings on surgical instruments. A particularly interesting discovery is that ceramics implanted in the human body can stimulate bone and tissue growth.

Ceramics are also being used to contain environmental spills and to encapsulate hazardous wastes. Containment booms using ceramic materials are being used to surround oil spills. These booms are then towed away to be siphoned off or burned. Hazardous materials can be recycled, mixed with clay and other materials, and formed into harmless consumer products such as floor tile and bricks.

In the electronics field, ceramic components can be created that have lower electrical and thermal conductivity than parts made from most other materials. Capacitors for storing electricity can be made of ceramics. Ceramics are also important to manufacturers of consumer electronic products. Ceramic gas igniters are being used in nearly all appliances and grills. Ceramic superconductors have been designed that enable the flow of electricity with little resistance.

Other important applications include the manufacture of ceramic sensors, actuators, semiconductors, and electro-optic devices.

One of the areas where ceramics is used extensively is automotive manufacturing. The ignition systems for automobile engines designed in the 1920s used ceramic spark plug insulators that are very similar to the spark plugs of today. Other automotive applications of ceramics include catalytic converter, oxygen sensors (emission control), computer control systems (substrates and components), magnets, brake rotors, valve seats, and ceramic fuel cells.

Residential and commercial construction has become one of the largest markets for ceramics today. Ceramic materials are used in products ranging from floor tile, brick, and roofing to cement, gypsum board, sewer pipe, and glass. Approximately three billion square feet of glass is used each year on windows alone. To put this into better perspective, consider the fact that this volume of glass would be sufficient to construct a highway of glass 200 feet wide, stretching from Los Angeles to New York City. (American Ceramic Society)

Materials for Ceramics

One of the most attractive features of ceramics is that they are made from readily available materials. Some of the most plentiful minerals are the silicates and aluminum silicates. These minerals, together with oxygen, constitute the majority of naturally occurring ceramic raw materials. The other materials used in ceramics, such as clay, kaolin, sand, feldspar, carbonate, and soapstone, are also widely available.

Most ceramic products are made from natural silicon (sand), but synthetic materials are also used to produce special purpose glass or porcelain ceramics. Many industrial ceramic products also use new raw materials such as carbide, fused cordierite, and pure alumina titania. Some of the other materials also important to industrial or advanced ceramics include alumina, zirconia, tungsten carbide, boron nitride, and boron carbide.

Structure of the Ceramics Industry

According to the American Ceramic Society, the U.S. market for ceramics is estimated to be more than $35 billion. There are many manufacturing firms producing different types of ceramic products. In most cases, these firms can be classified in terms of the type of product they manufacture. The best way to describe this industry is by analyzing it in terms of two major sectors, traditional ceramics and advanced ceramics.

Traditional Ceramics

Manufacturing firms in the ***traditional ceramics*** group are generally concerned with the production of clay and glass products. See **Figure 12-3**. While this includes potteries and glassmaking establishments, it also extends into other product lines, such as concrete, and sandpaper. There are six major industry segments for traditional ceramics:
- Structural clay products
- Whiteware
- Refractories
- Glass
- Abrasives
- Cement

Advanced Ceramics

The ***advanced ceramics*** sector is sometimes referred to as *industrial ceramics*, or *structural ceramics*. These titles are not very accurate descriptors because the products these companies make are not used exclusively by industry. The one trait that advanced ceramic firms have in common is that they manufacture products from engineered (human-made) materials. The ceramic stock they use is carefully selected to meet the needs of unique and highly specialized environments. There are four major industry segments in advanced ceramics:
- Structural
- Electrical
- Coatings
- Chemical and environmental ceramics

Advanced ceramic products are used in different fields, from sports medicine and racing catamarans, to communication satellites and energy power generators. Advanced ceramic materials are used to make fiber optic cable, capacitors, electrical insulators, integrated circuits, and even dentures. See **Figure 12-4**. Several of the fastest-growing product applications for ceramics are ceramic engines and engine parts; medical devices; impellers for the pump industry; ceramic scintillators employing rare-earth oxides used in spiral CT (computed tomography) x-ray machines; drug delivery systems and biosensors; and in ceramic matrix composites.

Figure 12-3. The structure of the traditional ceramics sector of the ceramics industry. Raw materials used in this industry segment, such as clay and sand, are naturally occurring and plentiful.

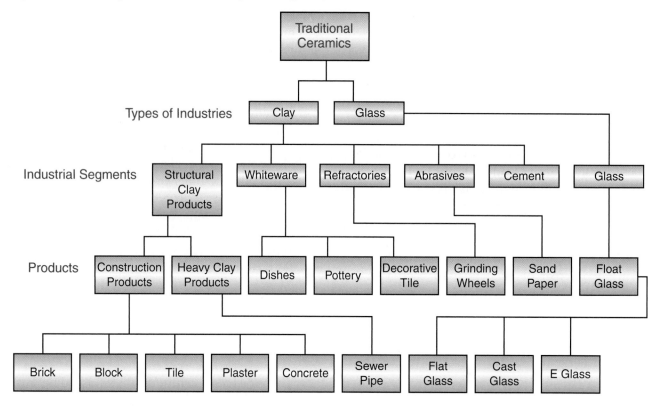

The use of alumina, silicon nitride, and silicon carbide for *wear parts* (moving parts exposed to extensive friction) is gaining popularity in the automotive and aerospace industries. Typical products are reentry nose cones, turbine components, heat exchangers, piston rings, valves, and emission control devices.

On the horizon is the development of the *ceramic engine*. Research on this innovation indicates that such a development is probable, and that it will be able to perform at higher operating temperatures with better efficiency than conventional engines. One reason why this has taken so long to develop and commercialize is that researchers are still trying to resolve problems related to the lubrication of the engine. Conventional lubrication processes are often ineffective with ceramics.

Figure 12-4. Advanced ceramics markets and products.

Markets	Products
Aerospace and Automotives	Reentry nose cones, turbine components, heat exchangers, emission control devices, high temperature tile, automotive sensors
Bioceramics	Dental implants, orthopedic implants, prosthetic devices
Technical and Scientific	Scientific glassware, technical glassware
High Temperature	Advanced refractories, kiln furniture, crucibles
Nuclear	Fuel rods, fuels, control devices
Electronic	Substrates, components, electrical porcelain and insulators
Miscellaneous	Optical fibers, cutting tool inserts, coatings, wear parts

Today, all industrialized countries are manufacturing engine components that are either entirely ceramic or ceramic-coated metal. Countries that appear to have the lead on this technology are the United States, Japan, Germany, and Russia. Cooperation between scientists and engineers in these and other countries has helped to advance developments in this field. There are already many different types of ceramic engine parts being manufactured and used today:

- **Gas turbine engine parts.** Burners, heat exchange devices, rotors, and starters
- **Turbocharger parts.** Housings and rotors
- **Diesel engine combustion parts.** Cylinder heads and piston rings
- **Engine chamber parts.** Cylinder liners and swirl chamber pot liner
- **Diesel engine moving parts.** Bushings, rocker arms, valves and valve guides
- **Other engine parts.** Bearings, mechanical seals, rotary engine seals

New materials are being rapidly developed for use in manufacturing products such as cutting tools, coatings, and bearings. Companies are now able to combine metals and ceramics to make *cermet* (ceramic/metal composite) cutting tool inserts. The insert is made from either aluminum oxide or titanium carbide to create a very hard shock resistant cutting surface. The advantage of inserts is that they can be quickly removed from the toolholder and replaced when they are worn or chipped. One interesting application for cermet inserts is on a rotating index head used in lathes and machining centers. See **Figure 12-5**. Cutting tools with cermet inserts enable high-speed material removal.

The Nature of Industrial Stock

To understand the nature of stock used to make ceramic products from clay, it is important to briefly trace the historical uses of ceramics in manufacturing. Long before the building of the Great Pyramids in Egypt, craftsmen made ceramic containers or "pottery" to carry and store food and water. They combined local clay and water, shaped the plastic material by hand, and then permanently removed the water from the clay using direct fire or heated furnaces. The field of ceramics consisted primarily of clay materials and processes. The use of glass to produce artifacts is thought to have first occurred around 3000 BC, during the Bronze Age. Egyptian glass beads were discovered that date back to about 2500 BC.

For centuries, there were few changes in the materials or manufacturing processes used to make ceramic products. The major change was from shaping the product by building it up with coils of clay to "throwing" and shaping the clay mass on a rotating potter's wheel.

Figure 12-5. Cermet cutting tool inserts are fastened to special holders for a variety of uses.

Stock about to be machined

Cermet insert mounted in an index head

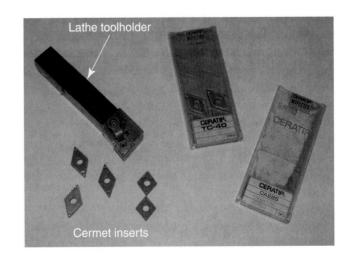

Lathe toolholder

Cermet inserts

Major changes in both materials and processes were introduced in the 1960s and 1970s, when new discoveries in science and technology led to the birth of the advanced or industrial ceramics industry. Manufacturing processes were developed to produce the great diversity of products demanded by expanding markets. New processes, such as thin film tape casting on belts of Mylar® (used to make electrical capacitors) and isostatic hot pressing of fine powders under a vacuum, were introduced. Other processes, such as blow molding and injection molding, were refined and modified.

As the demands for new product applications increased, new types of ceramic raw materials had to be made available for use with the more sophisticated manufacturing processes. In many cases, the wet clay that lent itself so beautifully to the processes used in the ceramic industry of 1940 or 1950 did not work well with the automated processes and improved technology of the new era.

Traditional ceramic products with high water content were very elastic and easily formed, but they were difficult to move without destroying their shape. Once the pieces were placed in an oven (kiln) for firing, the most difficult task was still unfinished: the water had to be gradually removed from the workpiece. If the product was too wet when fired, it would crack or explode in the kiln. Shrinkage occurring during firing was excessive, and achieving a perfect product was often a combination of skill and luck.

Today, stock for making clay products is usually purchased by the manufacturer in a dry powder, or semidry, plastic state. The type of stock that is selected depends on the type of manufacturing process that will be used, the end use for the product, and the manufacturer's preference for the basic raw material. Advanced ceramics are made of high purity oxides such as silicon carbide, silicon nitride, tungsten carbide, alumina, boron carbide, and boron nitride. Typical applications are shown in **Figure 12-6**.

Ceramic powders can be compressed to produce dense and durable parts. In this process, the very finely granulated powdered metal is pressed to create a part called a *green compact* or a *green body*. The green compact is a fragile part that has not been fired in an oven.

What is important to understand is that "greenware," no matter whether the stock is metal or clay, or some other combination of other materials, describes the part or product in a pre-sintered condition, up until the point when it is fired. It is called

"green" because it is fragile. In terms of machining, the terminology is also used. *Green machining* refers to machining a fragile part that has not been sintered (fired). Machining in the green state helps control costs because it can be done faster than machining finished materials and is easier to do on parts with complex geometries. Currently, machining (hard grinding) fired parts to achieve the net desired shape can account for 50 to 90 percent of the total component cost.

Dry pressing is best suited when a large number of mass produced smaller parts are needed. The parts need to be relatively flat, with a height to wall thickness ratio less than 4 to 1, and be relatively simple with no blind holes.

The major reason for pressing ceramic powders to form a dense workpiece, and then sintering the green compact at high temperature, is to increase its strength. The finer the size of the individual grains in the powdered mix, the stronger the finished product.

Not all manufacturers use powdered stock with their manufacturing process. Many firms use liquid or semiliquid clay. When the stock is more liquid, or in semiliquid state, other types of forming processes may be required. For example, it would not be practical to use injection molding to squirt dry powder into a mold prior to compression. If the stock contains a small amount of water, squeezing the material through an opening with a process such as extrusion would be fine.

Mixing the basic clay power starts by adding the binding agents, other additives, and a small amount of water to the powder. Additional water is slowly added to make the clay mixture thinner. Too much water will make the material weak and difficult to control.

One of the major sectors of the ceramics industry is glass. Glass manufacturers do not start with dry,

Figure 12-6. Advanced ceramic materials and applications.

Ceramic Oxide	Products
Silicon carbide	Grinding wheels and abrasives
Tungsten carbide	Cutting tools, cermets, blasting nozzles
Alumina	Electrical insulators, railings, polish
Boron carbide	Abrasives, cutting inserts
Boron nitride	High temperature lubricant, resin bonded grinding wheels

semiliquid, or liquid ceramic stock. Glass is a solid that can be turned into a liquid when it is heated. Glass as a raw material is created by fusion. It has been heated and then cooled too rapidly to permit crystallization. The melting point of glass depends on its composition. It can have a melting point of from 1400° to 1600° C, but some glasses melt at 500° C. The low-melting-point glasses have more limited applications in industry. Glass behaves much like plastics. Glass manufacturing companies usually purchase stock in the form of beads or sheets as their basic raw material.

The basic raw material in glass is silica, but the properties of a particular product vary greatly depending on the application. There are many thousands of different compositions of glass. Four of the most popular types of commercial glasses are soda-lime, lead, borosilicate, and high silicate. Soda-lime glass is used extensively for lightbulbs, bottles, and windows.

Important Terms

advanced ceramics
bioceramics
brittleness
ceramic
ceramic engine
ceramic-matrix composites
cermet
fatigue
green body
green compact
green machining
kaolinite
kiln
nearly inert
resorbable
surface-active
traditional ceramics
wear parts

Questions for Review and Discussion

1. What is the major limitation for using ceramics as a design material? How can this be engineered out of the material?

2. What is the most common form of stock used in industrial ceramics? Why is this used more frequently than other forms of stock?

3. Why must products made with highly liquid clay have the water removed before firing?

4. Glass becomes increasingly conductive as its temperature increases. Is the same true for ceramics? Why or why not? What clues does this provide in terms of using these materials for consumer products?

Products exposed to high temperature, thermal shock, high wear, and electrical environments are often made with ceramic materials. Kiln furniture, bus bar insulators, pouring cups for molten metal, and heating element supports are some of the products shown. (Metsch Refractories, Inc.)

13 Characteristics of Composite Materials

Chapter Highlights

- Composites have unique characteristics that make it possible to build products that otherwise would be unavailable or cost prohibitive.
- A variety of fibers can be made from glass to create fiber reinforced plastic.
- The fiber used in composites is available in a variety of forms—most often it is in the form of a cloth, mat, or roving.
- Resins form the composite matrix.
- Additives, fillers, and colorants are added to the matrix to enhance the properties of a composite.

Introduction

Several thousand years ago, our ancestors created a composite by mixing straw and clay together to make bricks. The straw was the fiber reinforcement and the clay was the binding medium. Many centuries later, the construction industry developed more composites. Composites covered interior walls of buildings. In this case, the reinforcement was closely spaced rows of wooden slats that were locked together with plaster (binding medium or matrix) creating a smooth hard surface that served to deaden sound and provide fire resistance. Bridge decks and supporting structures for buildings consisting of concrete reinforced with steel bars produced another composite that has been a basic part of construction for over one hundred years. However, it was not until the middle of the twentieth century that composites truly became popular in industries outside of construction. The development of synthetic polymers as a binding medium and the expansion of glass fiber options created the high performance composites that are in common use today.

Unique Characteristics of Composites

So, what exactly is the definition of a composite? A *composite* is a substance consisting of a minimum of two materials. At least one of the materials serves as a solid reinforcement while a second serves as a binding material (matrix). The binding material begins as either a liquid or slurry to saturate the reinforcement material(s) before it solidifies. Once the binding material solidifies it holds the solid reinforcement in a desired orientation. The position of the reinforcement and its orientation in the composite determine the mechanical properties of the composite. The mechanical properties are also influenced by the product's physical shape and the manufacturing process used.

A composite is a new material with unique properties. The properties of a composite result from the different roles assigned to the component materials. The reinforcement in a composite carries the loads that are imposed while the matrix material serves to distribute the loads across the structure and protect the reinforcement. This composition enables a designer to create a composite that can stand up to a wide variety of service and performance conditions. Composites depend on the right selection of reinforcement materials properly positioned and held in place with a suitable matrix material. Today's designers are using composites to create strong and lightweight components and products that range from bathtubs to aircraft wings.

Structure of Composite Materials

The most widely recognized composite is *fiber reinforced plastic* (*FRP*) which is replacing the older term of *glass reinforced plastic* (*GRP*). FRP consists of glass fibers and a polymer resin. An important property of FRP is its high tensile strength. This is provided by the thin glass fibers. However, these fibers are very susceptible to damage. Although the polymer resins have low tensile strength, they are extremely tough and can provide excellent protection to the glass fiber.

The FRP composite is usually made up of layers of glass fibers. This provides the basis for the term *lay-up*, which is used to describe the FRP manufacturing process. In open molding, a common low-cost manufacturing process, the FRP composite is fabricated by spraying or rolling a resin (polymer) on a mold, and then adding a layer of glass fiber that is then "wetted" with resin before another sheet of glass is laid down. See **Figure 13-1**. This sequence continues until the desired buildup is achieved. The *lay-up schedule*, or *buildup* as it is often called, can be likened to a recipe. The lay-up schedule describes the type and weight of the reinforcement that is needed to impart the unique properties and performance characteristics required by the designer. The variety of fibers available enable designers to meet a range of conditions that make FRP competitive with woods, metals, and other plastics. Although the cost of composite materials is often greater than the cost of conventional materials, the final product may be less expensive. The savings, in part, are due to the wide variety of low-cost manufacturing processes that are available to the composites industry.

Reinforcements

As we now know, there are two primary components in a modern composite—the resin matrix and the reinforcement material. Although the variety of reinforcements that has been used is tremendous, all reinforcement materials fall into one of only three categories:

- Fillers
- Fibers
- Solids

Fillers

Fillers are added to resins to change its flow characteristics and increase the volume. Fillers range from dust-like particles to short fibers, which create an easily handled material that will produce a composite with modest mechanical properties. The addition of fillers to a resin usually produces a paste, which can bridge gaps and fill in voids. In general, a resin with fillers is used to improve the surface finish of a composite. Sometimes, the filler's only purpose is to reduce cost by extending the volume of the resin with less costly filler. When used in this way, the filler is referred to as an *extender*.

Almost anything that can be made into a particulate or granule has probably been used as a filler. There are currently three classes of fillers in use—mineral, organic, and specialty (manufactured) fillers. Each of these classes contains various materials. See **Figure 13-2**.

Figure 13-1. A composite boat being built in an open mold.

A stream of chopped fiberglass and resin is being applied to the bottom of a boat hull to serve as a base for a framework reinforcement that is about to be put into place.

After the framework is placed, it is then covered with chopped fiberglass and resin. The HVLP gun being used is designed to reduce overspray and styrene emissions.

A layer of fiberglass mat is next placed over the chop and then rolled out. The rolling out process saturates the mat with resin from the chop and squeezes out air pockets, creating a composite that conforms to the framework and hull surface.

Figure 13-2. There are three classes of fillers, each containing various materials.

Classes of Fillers		
Mineral fillers	**Organic fillers**	**Specialty fillers**
Calcium carbonate	Wood flour	Microspheres
Calcium sulfate	Calcium carbonate	Solid glass spheres
Talc	Walnut shells (ground)	Hollow glass spheres
Mica	Corncobs (ground)	Ceramic spheres
		Thermoplastic spheres
		Phenolic spheres

Fibers

Fibers, particularly glass fibers, are the mainstay for the reinforcement of composites. These fibers are economical and provide good mechanical properties for FRP. The strength and performance of FRP is dependent on the type of glass fiber and the way it is oriented.

Silica sand is the primary ingredient in making glass fiber. Metal oxides and other ingredients are often added to the silica to improve the physical properties of the fiber. An individual fiber or filament will range in diameter from 3.5 to 24 microns.

The most economical glass fiber for composites is called *E-glass*. It offers sufficient strength in most applications and is low in cost. *E-glass* fibers contain at least 50 percent silica oxide. The remainder is composed of oxides of aluminum, boron, calcium and other compounds, including limestone, fluorspar, boric acid, and clay. Because of its good performance and economical price, E-glass accounts for nearly 90 percent of the glass-fiber reinforcements being made today. The "E" in E-glass refers to the outstanding electrical insulating characteristic of this glass fiber. E-glass is also transparent to radio-signals. Electrical applications include aircraft radomes, antennae coverings, and computer circuit boards.

In the 1960s, some military applications required higher strength and lighter weight composites than could be provided by E-glass. This stronger and lighter replacement is known as S-glass in the United States, R-glass in Europe, and T-glass in Japan. *S-glass* has appreciably higher silica oxide, aluminum oxide, and magnesium oxide content than E-glass has. S-glass is 40 to 70 percent stronger than E-glass.

However, both E-glass and S-glass lose up to half of their tensile strength as temperature increases from room temperature to 1000°F (538°C). Other limitations necessitated the development of additional fiber compositions.

Chemical-resistant glass

Most glass fiber has good chemical resistance; however, exposure to hot water will cause some erosion of exposed glass fibers. To slow erosion, a silane-based sizing compound is applied to the fiber at the time of its manufacture. When this enhanced fiber is coupled with a corrosion-resistant resin, the resulting composite has good chemical and corrosion resistance. However, it is important to select the appropriate fiber based on the specific chemical exposure.

Specialty fibers

There are also some specialty fibers that exhibit higher tensile strength and stiffness compared to standard glass fibers of a similar size. These fibers are made from carbon, boron, and aramid materials. The physical properties of these high-strength fibers are superior to typical glass fiber, but their cost is considerably higher. The carbon, boron, and aramid-based fibers are found only in applications demanding exceptional performance for which the customer is willing to pay a premium. A hybrid of one of these high-strength fibers and glass fiber can be used to increase the performance of the composite at a more affordable price.

A combination of glass fibers and thermoplastic (polypropylene or polyamide) filaments is called ***commingled fiber roving***. This blend of fibers can be processed quickly by heating the mixture of fibers

Glass Fiber Selection	
Glass Fiber	**Properties**
C-glass or E-CR glass (corrosion-resistant glass)	Loses less of its weight when exposed to an acid solution than E-glass.
C-glass and	Good corrosion-resistance to sulfuric acid.
S-2	Good corrosion-resistance to sulfuric acid. Much more resistant to sodium carbonate solution (a base) than is C-glass.
E-glass	Much more resistant to sodium carbonate solution (a base) than is C-glass.
Boron-free glass fiber	Comparable in price to E-glass, but demonstrates greater corrosion resistance in acidic environments. Boron-free glass has somewhat better strength and high-temperature performance than E-glass.

and filaments. The filaments melt and flow around the glass fiber to create the matrix. The benefits of this process include not having to use a liquid resin and reduced emissions during curing.

Manufacturers are also substituting natural fibers, made from flax, hemp, jute, sisal and other plants, for glass fiber. These come from renewable resources and are easily recycled. Natural fibers are also attractive because of their low-cost. In general these fibers have a low specific weight and are adequate for many applications where high strength and impact resistance are not required. These fibers, however, have many drawbacks such as poor fire resistance, high moisture absorption, and processing limitations.

Reinforcement Enhancements

Single strands of fiber are difficult to handle and keep in place until the binding matrix (resin) solidifies. By themselves, the fine hair-like glass filaments are nearly impossible to work with in a lay-up. The fibers are fragile, difficult to handle, and hard to control when the resin is applied. The result is a composite with highly variable mechanical properties that is very time consuming to manufacture. Therefore, to be useble, glass fiber must go through further processing to make it a practical and consistent reinforcement material.

The processing begins by gathering the glass filaments into bundles or strands. A **strand** is a collection of more than one continuous glass filaments. A common form of glass reinforcement is a **roving**, which is a bundle of untwisted glass strands packaged like thread on a large spool. A **yarn**, however, is like a roving except that the strands are twisted. Rovings, which are the most common form of glass fiber, can be chopped or woven to create mats, woven fabrics, braids, knitted fabrics, and hybrid fabrics. Rovings are supplied by weight and classified by the filament diameter. The term **yield** refers to the number of yards of glass fiber roving per pound.

Fabrics are a popular form of fiberglass reinforcement. One of the simplest forms is a **mat**, which is a nonwoven fabric made from fibers held together by a chemical binder. The mat can be made from either chopped or continuous strand. A **chopped mat** contains randomly distributed fibers cut to lengths typically ranging from 1.5″ to 2.5″ while **continuous-strand mat** is formed from swirls of unbroken fiber strands. Because their fibers are randomly oriented, mats are **isotropic**, which means the mat possesses equal strength in every direction. Chopped-strand mats are a low-cost reinforcement that is used in hand lay-up, continuous laminating, and some closed-molding applications. Continuous-strand mat is inherently stronger than chopped-strand mat and is used primarily in compression molding, resin transfer molding, and pultrusion applications.

A more sophisticated form of reinforcement is woven fabric. These are made on looms in the same way as cloth. The fiberglass fabric can be made in a wide variety of weights, weaves, and widths. The **plain weave** is the simplest fabric. In this fabric the **fill yarn** (the yarn is oriented at right angles to the fabric length) alternately crosses over and under each **warp yarn** (the warp yarn runs the length of the fabric, X axis). The warp is also called the *machine direction*, and the fill yarn is sometimes called the **woof**, **weft**, or **cross direction**. See **Figure 13-3**. In the plain weave fabric, the yarns are interlaced in an alternating fashion over and under each other, providing maximum fabric stability and equal strength in both the warp and fill directions. This means that the plain weave has bidirectional strength that creates good tensile strength along the X- and Y-axis. The resulting tensile strength of a woven fabric, however, is reduced somewhat because the fibers are crimped as they pass over and under each other during weaving. When the composite is placed under load in tension, these fibers try to straighten, causing stress within the finished part.

There are other weaves such as twill, crowfoot, and satin that also have the yarn or roving crossing over and under multiple warp fibers. See **Figure 13-4**. These weaves, satin in particular, tend to be more pliable and conform more easily to curved surfaces than plain weaves. Because of their conforming properties, these fabrics are used when superior appearance

Figure 13-3. Plain weave with fill yarn, warp, and the machine direction.

The **warp** is the set of threads that represent the straight grain of the fabric and runs along machine direction. The **weft** or fill is the set of threads that runs perpendicular to the warp. (D. Hillis)

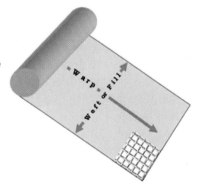

Figure 13-4. Diagram showing twill, crowfoot, and satin weaves.

The **twill weave** is characterized by a diagonal rib or twill line on the face of the fabric. Each warp yarn floats over at least two consecutive weft yarns. The twill weave yields a higher density fabric while maintaining stability with excellent strength.

Satin weaves offer a very pliable fabric which readily conforms to intricate contours. The five harness satin weave is produced by one warp yarn traveling over four and under one weft yarn.

The **crowfoot weave** is achieved when one warp yarn weaves over three and under one weft yarn. Crowfoot weaves offer more pliability and can conform to a complex mold surface better than plain weave fabrics.

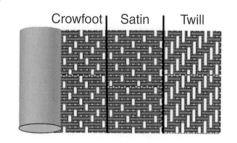

and strength are required. Very fine glass fabrics are used in electronic applications such as circuit board and protective covers. Coarse fabrics made from relatively thick woven roving are used when heavy reinforcement is required. Due to its comparatively coarse weave, woven roving wets quickly and provides an economical composite.

Hybrid fabrics can also be constructed with varying glass types and strand composition. For example, high-strength strands of S-type glass or small-diameter filaments may be used in the warp direction, while less-costly strands compose the fill. A hybrid can be created by stitching woven fabric and mat together to improve strength.

Multiaxial fabrics allow tailored directionality, because the fibers can be laid atop one another in any orientation. Multiaxial fabrics are formed by laying fibers in the desired arrangement and then knitting or stitching them together. Laying fibers on top of each other in this way, rather than weaving them together, avoids crimping and the resulting internal stresses that can develop. In general this approach, which is somewhat more expensive, does make better use of the fibers' inherent strength. Heavyweight nonwoven fabrics can reduce the number of plies required for a lay-up. This speeds up the molding process and may compensate for the extra expense of the material.

An alternative to woven fabrics is a braided fabric. These fabrics are generally more expensive, due to a more intricate manufacturing process, but they are generally stronger by weight than woven fabrics. The strength results from the intertwining of three or more yarns without twisting any two yarns around each other. Braids are continuously woven on the bias and have at least one axial yarn that is not crimped

in the weaving process. This arrangement of yarns allows for highly efficient load distribution. Braids are available in both flat and tubular configurations.

Protecting the glass fiber during processing

Bare glass is not abrasion resistant. Abrasion that occurs during the processing of the glass strand into yarn and fabric and during the lay-up process can cause an unacceptable amount of breakage of the long glass filaments. To reduce this damage, the hair-like glass filaments are coated with an aqueous chemical mixture called *sizing* before they are made into a fabric. Sizing acts as an adhesive to hold the filaments together and improve the integrity of the finished cloth. Sizing also improves the adhesion of resin to the glass fiber.

Moreover, sizing improves the ability of a resin to penetrate and coat glass cloth. This characteristic is termed ***wet out***. Recent developments in sizing formulations have created fiberglass cloth that wets out more efficiently, which allows composite manufacturers to use polyester resins with reduced styrene. Lowering the amount of styrene needed reduces styrene emissions and improves working conditions. Although sizing accounts for only a very small percent of total fiber weight, it plays an important role in creating a glass fiber fabric that has consistent mechanical properties.

Solid Reinforcement

Solid reinforcement materials are used extensively throughout the composites industry to create stiff yet lightweight composite products. The structure of a composite as discussed earlier in this chapter (the matrix material and the reinforcement) is frequently referred to as a ***laminate***. In a laminate, this solid form of reinforcement is generally referred to as a *core material*. A laminate with a core material is described as sandwich construction. The sandwich consists of a face skin laminate that will generally contain fiber reinforcement, the core material, and the reinforced back skin laminate. This type of construction is intended to create a thicker laminate and since stiffness is a function of thickness, the result is a stiffer laminate. Increased thickness is usually not a problem in most designs; however, weight is almost always a concern. Therefore, most core materials are lightweight.

Many suitable core materials that meet the requirements of being lightweight and economical are available. In addition, core materials must be able to be bonded by the resin being used. As an example, polystyrene foam is not a suitable core material when used with polyester resin. The resin dissolves the polystyrene foam.

The size, thickness, and flexibility of the core material must also be considered. Core materials are usually available in large sheets that can be cut and formed to fit a mold. However, since core materials are inherently stiff, something must be done to make them flexible so that they can be used in laminates with complex curves. This can be done by cutting the rigid core materials into small squares and joining the squares together with a fabric scrim. A *scrim* is an open weave fabric that is glued to the back of the core material. See **Figure 13-5**.

Balsa wood is a popular core material used in marine hulls and decks to create a lightweight but rigid composite. Balsa has a closed-cell structure and weighs between 6 and 16 pounds per cubic foot. Balsa is classified as a lightweight core material. Balsa has been a popular core material because of its excellent stiffness and bond strength with popular matrix resins. The balsa block is set in the laminate so that the top and bottom laminate bonds to the end grain of the balsa. This improves the crush strength of the balsa core.

End-grain balsa wood is available in sheet form for flat panel construction or in a scrim-backed small block arrangement that is able to conform to complex curves.

Polyvinyl chloride (PVC) foam cores are made by combining a polyvinyl copolymer with stabilizers, plasticizers, cross-linking compounds, and blowing agents. The resulting PVC foam offers a good combination of strength and low weight. PVC foam weighs between 4 and 30 pounds per cubic foot.

Linear PVC foam is produced mainly for the marine industry. Its unique mechanical properties are a result of a non-cross-linked molecular structure, which allows significant deflection before failure. In comparison to the cross-linked (nonlinear) PVC, linear PVC is not as rigid, but has better impact absorption capability.

Polyurethane foam is available in either sheet stock or liquid that can be foamed in place. This material is often used in the cavities of boat hulls to add stiffness and provide buoyancy. See **Figure 13-6**. Because of its relatively low sheer strength, this foam is generally not used in critical structural applications. Polyurethane foam is, however, used frequently as a core material in the wall panels of refrigeration units. Besides stiffening the panel, it also provides thermal insulation. Polyurethane foam can be blown in a wide range of densities, from 2 pounds per cubic foot to over 20 pounds per cubic foot.

Various types of ***honeycomb*** cores are used extensively in the aerospace and transportation industry. The honeycomb can be made from a wide range of materials, which includes paper, aluminum, and glass reinforced, phenolic. Honeycombs are very lightweight, ranging from 1 to 6 pounds per cubic

Figure 13-6. A cross section of a box section used to stiffen a boat hull.

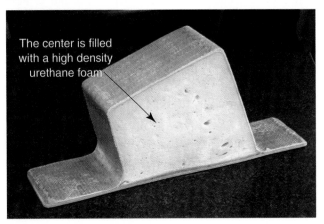

The center is filled with a high density urethane foam

Figure 13-5. Balsa core material held together by a scrim.

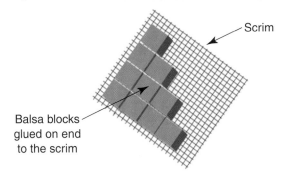

Scrim

Balsa blocks glued on end to the scrim

foot. Their physical properties depend on the specific material and the cell size. Honeycombs create very stiff lightweight laminates that can have very high crush strengths.

There are other materials that are used to increase the thickness of a laminate. Core fabrics and *laminate bulkers* are the primary examples. The purpose of these products is to build laminate thickness quickly or to create a barrier to prevent a problem called print-out. *Print-out* occurs when the *gel coat* (a resin designed to protect the matrix resin and provide exterior gloss and color) and matrix resin shrink during curing causing the pattern of the fiberglass reinforcement to show. Generally, print-out is apparent only in very high gloss finishes such as those required by automobile and yacht builders. Contrary to its name, core fabrics are not generally considered a core material since they are placed on top of the highest layer of reinforcement. In most cases, core fabrics are nonwoven materials using polyester filaments that are bonded into a mat-like blotter configuration. These products wet-out easily with resin and lay-up similar to fiberglass reinforcement.

One form of laminate bulker is *syntactic foams*. This material is made by mixing hollow microspheres with resin. The lightweight microspheres reduce the density of the resin and create a thick mixture that can be applied by hand or sprayed to increase the thickness between layers of reinforcement. Syntactic foams have also been used as a barrier coat between the gel coat and the structural laminate to minimize print-out.

Plywood should also be mentioned as a structural core material, although fiberglass is generally viewed as merely a protective sheathing when used in conjunction with plywood. Exceptions to this application include local reinforcements used for mounting hardware or where plywood replaces a lower density core to improve compression properties of the laminate. Unless the laminate is able to hermetically seal the plywood, there is a possibility that the plywod can aborb water and rot. Consequently, plywood should not be considered as a core material for composites exposed to weather. Also, depending on the type of wood, plywood does not always bond well to the laminate.

Creating the matrix in the composite

In most composites, the matrix is generally a resin. The resin's function is to transfer the load to the reinforcement fiber and protect the fiber from environmental effects. The most popular matrix materials are plastics, which fall into one of two classifications: thermosets and thermoplastics. Common thermoplastics are nylon, polystyrene, polyethylene, and acrylic, while familiar thermoset plastics are polyester, vinyl ester, and epoxy resins.

Since thermoplastic resins are not cross-linked, thermoplastics can be melted and formed and then remelted and reformed. The process is reversible, whereas thermoset resins cannot be re-formed. As an illustration, a thermoset resin can be likened to concrete and a thermoplastic resin can be compared to ice. Both the concrete and ice are solids and both started out as a liquid. The ice, however, can return to its liquid form with the addition of heat and become a sheet of ice again when chilled. The concrete, however, once it has solidified will remain a solid.

Composites manufacturing primarily uses thermoset resins. Thermoset resin is a *polymer* made up of chain-like molecules. The suffix *mer* in *polymer* means segment, therefore the word polymer literally means *many segments*, referring to the repeating chain of molecular units. Many polymers with short chains are liquid at room temperature. However, when these short chains are linked together the polymer becomes a solid. The process of joining polymers into long chains is called *cross-linking*. The application of heat or adding an initiator starts the chemical reaction of cross-linking a thermoset resin. Once a thermoset resin has cross-linked the process cannot be reversed.

Polyester resins are the most commonly used resin systems in FRP composites. This resin is economically priced, has good physical properties, and is fairly easy to handle. Five of the most common resins used in composites are listed in **Figure 13-7**.

Gel coat

Gel coat is a resin formulated to provide a high quality finish to the outer surface of a composite part. It also serves to improve the durability of products that are exposed to weather. Gel coat is not paint, although it serves many of the same functions of paint. It provides color, gloss, and environmental protection for the matrix resin. Gel coats are made from polyester or vinyl ester resins depending on the specific end-use of the composite. Additives to the gel coat include fillers, pigments, and thixotropic agents. A *thixotropic agent* inhibits dripping and sagging. This is an especially important feature for curved and vertical surfaces, since gel coats are usually applied very thick.

Figure 13-7. Popular resins and a description of their properties.

Popular Resins	
Resin Type	**Comment**
Polyester resin Orthophthalic acid based resin	This resin is also called a general purpose or GP resin. It is a low cost polyester resin that is used in applications requiring moderate physical properties, temperature stability, and corrosion resistance. It was at one time the dominant resin in the marine industry. Orthophthalic resins have styrene contents ranging for 35 to 45 percent by weight.
Polyester resin Isophthalic acid based resin	This resin has better physical and corrosion resistant properties than orthophthalic resins, but is more costly. Isophthalic resins are more viscous and therefore require more styrene to make the resin workable during lay-up. This resin's styrene content generally ranges from 42 to 50 percent by weight.
Polyester resin Dicyclopentadiene based resin	This resin is referred to as DCPD. Because DCPD resin shrinks very little during curing, it does not pull away from mold surfaces or show the pattern of the fiber reinforcement. These properties lend this resin to applications where cosmetic appearance is important. The physical properties of DCPD resins are similar to orthophthalic resins. DCPD, however, tends to be very rigid and lacks the toughness of other resins. The cost of DCPD resin is more than orthophthalic but less than vinylester resin. Styrene content is usually 35 to 38 percent.
Vinylester resin	This polymer has characteristics resembling both polyester and epoxy. Vinylester resins are processed similar to polyester resins, which means they are easier to handle than epoxy resins. This resin, however, has superior corrosion resistance as well as toughness. Even though vinylester resin is more expensive, it has become popular in the marine industry because of its resistance to water absorption and environmental degradation.
Epoxy resin	Liquid epoxy resins were first synthesized in Europe in the mid-1930s. Epoxies consist of two components that react with each other, forming a hard, inert material. Part A consists of an epoxy resin (a pre-polymer) and Part B is an epoxy curing agent, sometimes called a hardener. When the two components are mixed, the hardener cross-links with the resin creating a hard, strong, glass-like, chemically resistant polymer.

Curing Polyester and Vinylester—Catalysts/Initiators/Promoters

Resin cures by a process called free radical polymerization. The curing process begins when a catalyst is mixed with the resin. The term *catalyst*, however, is not the proper name for the chemical that initiates the resin polymerization process. The correct term is *initiator*, although the composites industry almost universally refers to an initiator as a catalyst. Technically, a **catalyst** causes a chemical reaction but does become part of the reaction while an **initiator** *initiates* or speeds up a reaction but becomes consumed in the process. In the case of polymerizing polyester resins, the initiator is consumed in the cross-linking process.

At the beginning of the polymerization process, the initiator decomposes into free radical molecules. This starts the process of cross-linking the polyester molecules using the styrene molecules in the resin. Increasing the amount of initiator in the resin will increase the rate of polymerization. The process is exothermic, meaning that it generates heat. As heat builds the process speeds up. In thick lay-ups, the heat can cause distortion and will cause the composite to pull away from the mold surface. To reduce distortion, the amount of heat generated during the curing process must be monitored and controlled.

At room temperature, many of the initiators commonly in use will require the addition of another chemical called a promoter. Without the proper amount of promoter, the resin would not cure within a reasonable amount of time. Most resin suppliers furnish their resin with the promoter already added. The amount of promoter needed is determined by the ambient temperature in the molding area. Generally, the amount needed matches the season of the year. Consequently, there are "summer" and "winter" blends of resin.

A composites manufacturer has the ability to manage the curing time by balancing the amount of initiator, the laminate thickness, and controlling the working temperature of the resin, mold, and production area. The curing time needs to be sufficiently long to work the resin into the reinforcement before the resin starts to gel, but the cure must occur quickly enough so that the mold can be cycled at a rate to meet production demands.

Fortunately, there are several types of initiators available that allow manufacturers to fine-tune the cure time of the polyester and vinyl ester resins. Some of the more common initiators available are ketone peroxide, acetylacetone peroxide, benzoyl peroxide, and cumine hydroperoxide. Occasionally, different initiators can be mixed together to provide a blend that optimizes the cure time. However, changing the type of initiator or blending initiators should only be done with the guidance of the resin manufacturer and the initiator supplier.

Styrene

In plants manufacturing fiberglass products, it is styrene that produces that unmistakable odor so closely associated with fiberglass. The human nose can detect styrene at levels as low as a few parts per million. Styrene is referred to as a **monomer** meaning it is a single (*mono*) segment (*mer*) molecule. It performs several roles in the creation of a composite. First, as a solvent (*diluent*) for ester resins, styrene reduces the viscosity of the polymer to make it a workable liquid. Second, the monomer becomes the link in the chemical reaction that cross-links the liquid resin changing it from a liquid to a solid. This creates the matrix for the composite. Other monomers can be used; however, styrene is currently the most common monomer in use with polyester and vinyl ester resin systems.

Most of the styrene monomer is consumed in the cross-linking or trapped in the cured composite. However, some of the styrene evaporates when the resin is being applied and during the curing process. The amount of styrene that is emitted during the lay-up of the laminate is a concern for manufacturers. Lowering styrene emissions to reduce environmental pollution and worker exposure is a major objective of the composites industry. Efforts to reduce styrene emissions are based on reducing the styrene content needed by the resin and favoring fabrication processes using closed systems that prevent the styrene from escaping to the atmosphere.

Resin Additives

Additives are used solely to improve the chemical and physical properties of the matrix resin for a particular application. There is a huge variety of additives available, which explains why composites work so effectively in applications ranging from snow shovels to circuit boards. A basic polyester resin system includes a polymer resin, a monomer (for cross-linking), and a promoter to speed up curing. However, once the resin manufacturer knows what the application of the final product will be, the appropriate additives can be blended into the basic resin to modify and enhance the material's properties. These additives include thixotropes, fire retardants, suppressants, UV inhibitors, and conductive additives.

In open molding, the tendency for resin to drain from vertical areas and form pools of excess resin in horizontal areas is a serious problem. The property that causes a resin to adhere to a vertical surface is *thixotropy*. Therefore, an additive that modifies the viscosity and flow characteristics of a resin is called a **thixotrope**. The most common thixotrope used in polyester resin is fumed silica.

Pigment dispersions and color pastes can be added to resin for cosmetic purposes or to enhance the resin's ability to withstand continued exposure to the weather. Adding pigments to the matrix resin to the extent that the resin's gel time is forced to increase can harm other properties of the material. Therefore, a separate resin system that is not required to serve as the matrix for the reinforcement was created to provide weather protection and color for the composite. This gel coat is applied to the outer surface of the product.

Most thermoset resins are combustible and create a noxious smoke when burned. In critical applications such as aircraft interiors or mining equipment, adding a fire retardant to reduce combustibility is critical. Additives such as alumina trihydrate and antimony trioxide can reduce the flame spread and the amount of smoke generated from a burning composite.

Many polyester resins remain tacky on the surface after curing. This is due to a characteristic called air inhibition, which prevents a very thin surface layer of the resin from properly curing. Adding a small amount of paraffin wax to the resin enables a wax film to form on the surface during curing allowing the resin to completely harden. This surface film has some effect in reducing styrene emissions. However, if other parts are to be bonded to the cured laminate, the wax film will have to be removed in order to achieve a satisfactory attachment.

Polymer resins will degrade and break down over time when continually exposed to the sunlight's ultraviolet (UV) rays. To combat this damage a UV inhibitor is added to slow the effects of UV exposure.

Composite laminates (except carbon fiber) are inherently nonconductive, which is usually desirable. However, in some cases it is necessary to make a laminate conductive to reduce static charge or to enable electrostatic painting. Carbon black, carbon fibers, metallic/metallized glass fibers are conductive additives that can create an electrically conductive laminate.

The Nature of Composite Stock

Many types of stock are available for manufacturing products using composite materials. Often, the manufacture of composite stock is accomplished during the forming or fabrication process of the composite. At other times, the stock is purchased from an outside vendor to create a composite product. An example of a vendor supplied composite stock is a prepreg. *Prepregs* are sheets of fiber saturated with resin. The resin in the prepreg has an initiator mixed with it so that the manufacturer only has to position the material in a mold and heat it to complete the cure. Prepregs are made to order so that they are the right size for the product. This stock is very heat sensitive, so it must be shipped and stored at a temperature that will not initiate curing.

Prepregs may also be purchased in tape form, with the unidirectional fibers suspended in an epoxy matrix. The tape has a removable backing that keeps the material from sticking together when it is rolled. Again, this is a heat-sensitive stock.

Composite stocks can also be made on-site. Large manufacturers of boats, truck bodies, and tanks may elect to purchase the resin and additives separately. These materials are mixed to suit the current environmental conditions and the type of product being manufactured. The reinforcement going into the matrix may also be made by the composite fabricator. This is a common practice in the aircraft industry.

Engineered composites have assumed a growing importance since the 1960s, when organic-matrix composites (polymers) and glass fiber were considered to be "state-of-the-art." Carbon fiber technology coupled with other materials to serve as a matrix make composite applications almost limitless. The next several decades will see these basic structures change and be applied in new and innovative ways. All of the industrialized countries in the world are making their own unique contribution in composite technology.

Important Terms

additives	lay-up schedule
balsa wood	linear PVC foam
buildup	mat
catalyst	monomer
chopped mat	plain weave
commingled fiber	polymer
roving	polyurethane foam
composite	polyvinyl chloride
continuous-strand mat	(PVC) foam
core material	prepregs
cross direction	print-out
cross-linking	roving
E-glass	scrim
extender	S-glass
fiber reinforced plastic	strand
(FRP)	syntactic foams
fillers	thixotrope
fill yarn	thixotropic agent
gel coat	weft
honeycomb	wet out
initiator	woof
isotropic	warp yarn
laminate	yarn
laminate bulkers	yield
lay-up	

Questions for Review and Discussion

1. If a sheet of metal is given a coat of paint, does this create a composite? Explain.

2. What is a matrix in a composite? Explain what a matrix accomplishes in a composite.

3. What is the difference between a filler and an additive?

4. What is the difference between a yarn and a roving?

5. When an initiator is added to a resin explain what happens.

6. Discuss the major advantages of composites over conventional materials. If they are more expensive, why would they still be desirable in many applications?

Section 5
Manufacturing
Process Database

14 Processes Used to Form Metallic Materials

Chapter Highlights

- Some forming processes take advantage of a metal's property to flow like plastic to conform to a new shape.
- Some forming processes shape metal by cutting away material that is not needed.
- Metal can be melted and cast in the desired shape.
- Each process used to form metals imparts some characteristics that are desirable and some characteristics that must be changed before the part can be used.

Introduction

There are literally hundreds of processes that can shape or form metallic materials to create a part or product. Some of these methods such as swaging, extrusion, roll-forming, and drawing stretch or squeeze the metal as if it were a plastic material. In fact, when metal is forced to move into a new shape it is called *plastic flow*. There are other forming processes that mold metal. These processes heat the metal to a liquid state so that it can be poured into a mold or die. There is, however, an interesting variation of metal molding. This process involves compressing powdered metal in a die to form a part. Finally, there are press-working operations such as stamping, blanking, and shearing that can create amazing metal shapes and parts.

Many of these processes are thousands of years old, but a few are so new that they are just now coming into use. However, choosing a forming process to use depends on the type of metal, the number of pieces that need to be made, and the cost. Another factor that designers use in picking a forming process is whether or not a particular process will impart the desired mechanical properties to the metal. All of these considerations go into the selection of a metal forming process.

Forging

Hand forging is an ancient metal forming process dating back to about 4000 BC. In those times, forging was initially used to shape simple metal tools. Forging was later refined for use in making fine jewelry and coins. Forging is considered a foundational technical development. That is, it played a major role in shaping the development of our technological culture and led to the development of many other forming processes.

School children often learn about hand forging by watching a blacksmith forming metal. The blacksmith shapes metal, such as a horseshoe, by heating it until it is cherry red and hammering it against an anvil. Today, the most extensive use of hand forging is by crafters working with ornamental or wrought iron. Forging is usually a hot forming process, but it can also be used to cold-form metal as well.

Forging (hot working) is applied to several processes in which a piece of metal, such as bar, billet, bloom, or ingot, is shaped to the desired form by plastic deformation. The process uses heavy presses and drop-forging hammers that force heated metal into the desired shape. All forging processes involve successive hammering, resulting in shaping hot or cold metal by compressive force. In forging, little material is lost in the forming process. The shape of the raw material is changed by repositioning metal, rather than removing it. Forgings can range in size from a few grams to components that weigh hundreds of tons. One of the advantages of forging is that the flow of the metal and structure of the grain in the workpiece can be controlled. This results in a stronger and tougher part than could be produced with most other forming processes. Today, forging is often used to produce bolts, rivets, connecting rods, gears, and structural members for equipment.

Open-Die Forging

Open-die forging, also called Smith forging, is the simplest of all of the forging processes. In this process the workpiece is formed between flat dies that compact, but do not completely enclose, the heated metal part. Open-die forging could be used in the case where a round part is needed, and flat bar stock is available to make the parts. Open-die forging is a hot forming process that uses flat, V-shaped, concave, or convex dies in presses. The process can be used to form a range of component sizes from a few pounds to over 300 tons. The workpiece is heated to improve its plastic flow characteristics and to reduce the force required to work the metal. During the process, the part is systematically deformed by a series of strokes from the upper die while being supported on the lower die. See **Figure 14-1**. The position is changed between strokes by a tool called a manipulator.

Open-die forging processes allow the workpiece freedom to move in one or two directions. The workpiece is typically compressed in the axial direction (direction of movement of the upper die) with no lateral constraint. The open-die method is also used to produce rectangular shapes from round stock. In this type of application, heated round bar stock is placed between the flat dies. The dies close, creating a rod with two flat sides. The rod is then turned over in the die by hand and the flat dies press the part again. This forms two more flat sides, resulting in a square produced from round rod. This process may be used to compact metal along the entire length of the rod. Open-die forging is time-consuming and the quality is dependent on the skill of the smith performing the work. Because of this, it has limited application in high-volume production. It is sometimes used to produce simple shapes by replacing the flat open dies with special-shaped dies.

Figure 14-1. In open-die forging, the workpiece is compacted and formed between two die halves that do not completely enclose it.

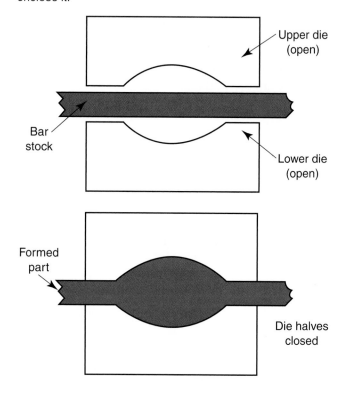

Closed-Die Forging

Open-die forging creates flashing and generally cannot shape the part completely. Therefore, the excess material has to be removed in subsequent machining operations. This excess material, which is actually waste, can be as high as 70 percent for forged gear blanks. However, *closed-die forging* can produce a near-net or net shape forging, which reduces the amount of material needed to produce the part. In this type of forging, a preheated metal billet, or preform, is shaped by being placed between dies that completely encompass the billet and restrict its flow. Closed-die forging is usually done with horizontally opposed precision die sets. These opposing dies simultaneously hammer against the billet until the forming process is complete. See **Figure 14-2**.

This process accounts for the vast majority of all commercial forging production. Impression die forging is sometimes called "closed-die forging." Closed-die forging provides three-dimensional control of the workpiece and furnishes closer dimensional control than open-die forging. This dimensional control is achieved by a pair of matched dies with specially fabricated impressions.

The tremendous abuse from hammering and contact with heated stock requires the two-part dies be made from hardened steel. Note, as shown in Figure 14-2, that the male and female die halves fit together perfectly.

In some closed-die forging applications, more stock is used than is needed to ensure a completely filled and compacted part. This helps to produce a strong product with relatively uniform density. If this is the case, the dies are made with a groove (called a gutter) cut into both mold halves. When the top and bottom die halves compress, the *flash*, or excess material, is forced out of the cavity and into the gutter. Some forging dies do not have gutters. Instead, a thin wing or seam of flash is produced on the outside of the part where the dies close. This seam of flash forms where the dies come together, which is called the *parting line*. The extra material becomes flash when it "squirts" out along the parting line.

The flash that is produced around the edge of the part must be removed by grinding or some other material removal process. The time lost in removing flash adds to production cost. For this reason flashless forging is more popular. In theory, there will not be any flash or voids in the forged part if the preform has exactly the same volume as the die cavity. Therefore, precision preforms are essential. However, careful control of the preheating of the preform is vital along with other factors such as impact and force which causes the plastic flow of the material to fill out the die cavity.

Some metals are easier to forge than others. The softer alloys, such as aluminum, magnesium, and copper can be easily forged at relatively low temperatures 750–1650°F (400–900°C). Metals such as titanium and tungsten require temperatures in excess of 1740°F (950°C) for forging. Other characteristics such as ductility, tensile strength, and frictional behavior of a particular metal also influence its forgeability.

The major weakness of forged products is that they may have surface cracks caused by the forging process. Overlapping of material and complex radii in a particular part can cause stress risers or internal and external cracking.

Figure 14-2. Closed-die forging is a process that completely encloses the workpiece between the two die halves.

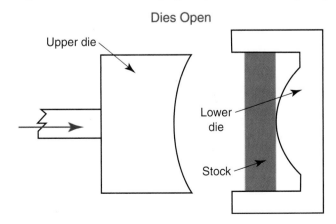

Dies Open

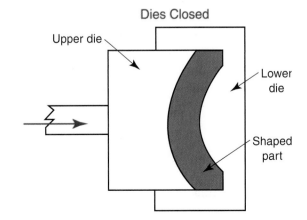

Dies Closed

Coining

Coining is the process of squeezing metal while it is confined in a closed set of dies. It is a procedure that requires high pressure and precision closed-dies to forge such items as coins and medallions. Coining causes metal to move, as the die halves are closing, from thinner to thicker areas. For this reason, coining is often the process chosen when a part requires built-up corners or edges.

The process begins when a blank is placed in a confined die. A movable punch is located within the die. The movement of this punch cold works the material and can form intricate features with very fine detail and finish. Pressures as high as 200,000 psi is required to generate very fine features. To achieve the required detail, coining may require more than one operation. No lubricant is used because it would be trapped in the die and would prevent the reproduction of fine detail.

Figure 14-3 illustrates vertical coining. This method of coining is similar to closed-die forging,

but requires much more pressure to ensure the production of fine detail. Coining can also be done using round dies turning against round stock, *rotary coining*, or with flat dies contacting round stock, *roll-on marking*. Examples of parts produced by rotary coining or roll-on marking are letters found on sheet metal shrouds and guards.

Coining is also used with forged parts to improve surface finish. When coining is conducted in this manner, it is called *sizing*. Sizing forged parts is also done when more accurately rendered detail is needed on particular surfaces, or if improved dimensional accuracy is desired. Sizing is not used to move masses of material, as is the case in vertical or rotary coining.

Rotary Forming Processes

There are many forming processes that use rollers to squeeze and shape metals. Some of these involve progressive roller forming of long lengths of stock while others involve rotary bending.

Roller Forming

Roller forming is also called *contour roll forming* and *cold-roll forming*. Any ductile metal can be roller formed without heating. Roller forming is used to form straight lengths into nearly any imaginable shape. No dies are needed—rollers progressively squeeze the continuous strips of metal into the desired shape. Some of the products made are truck frame members, metal building components, and metal studs.

A roll forming machine consists of several sets of rolls that are used to grip and then form the strip as it passes through the rolling machine. Stock is usually purchased in large coils, which permits continuous feeding through the machine. Roller forming machines consume enormous amounts of floor space, because the forming process stretches out to a long line to accommodate the many sets of rollers needed to form the shape.

The thickness of the metal that is roller formed remains fairly constant throughout the process; only its *shape* is changed. Each set of rollers produces a small change (a little more bend) in the stock. When the part reaches the end of the rolls, it is in the desired form.

Roller forming is a fast process that is suited to continuous production. Sometimes, it is combined

Figure 14-3. Coining is a form of closed-die forging, but is done under great pressure to reproduce fine detail.

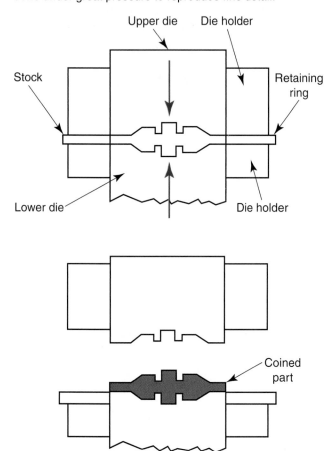

Upper die Die holder

Stock

Retaining ring

Lower die

Die holder

Coined part

with other processes to produce a final product. An example would be in the production of welded tubing. Many firms turn flat stock into tubes using roller forming. The edges are then automatically joined in a continuous welding operation. Another common process that is often combined with roll forming is a sawing operation to cut the formed item to a required length.

Roll Bending

Roll bending is a process that is used to bend circular, curved, and cylindrical shapes from bar, rod, tube, angle, and channel stock. Roll bending is a variation of roller forming.

There are various bending roll arrangements used on a machine called a *roll bender*. The pyramid type has one roll on top of a pair of side-by-side rolls. The position of the top roll is adjustable; the other two rolls are fixed. Stock is fed between the top and two bottom rolls. As the top roll turns, the material is moved through the rolls and bent.

A widely used roll bending configuration is the *two-roll machine*. The configuration is unique because it can produce bends of any diameter using only two rolls and a slip-on tube. The two rolls are aligned with a slight gap between them. A tube is inserted over the top roll, as shown in **Figure 14-4**. The upper roll applies pressure, pushing the stock down against the flexible lower urethane or rubber-covered roll. The tube controls the bending diameter of the workpiece: the larger the tube, the bigger the radius that can be bent. Stock is fed between the rolls as the top roll turns, pushing the material through and bending it.

Another roll arrangement, called the *three-roll double-pinch machine*, is the most popular. The top roll of the three-roll machine turns freely, and the bottom two rolls are powered and adjustable. Rolls on the three-roll machine can be arranged in pyramid fashion for symmetrical bending of products such as barrel hoops, as shown in **Figure 14-5**. Irregular bends can be formed by moving the bending role in or out as stock is being pulled through the rolls. Architectural shapes such as curved panels and oval columns are made this way.

Another process used to shape sheet stock is *stretch forming* or as it is sometimes called, *bullnose forming*. The process begins by inserting a flat metal sheet into the gripper jaws of the machine. The sheet is held over a table that supports a forming die or mandrel. Once in place, the sheet is then stretched to its yield point. With the sheet stretched, the table rises, wrapping the sheet around the forming die, or mandrel. During the stretching process the gripper jaws automatically move in as the die moves up to maintain the desired stretch-force. The sequence of operations on these machines is controlled by either a PLC or CNC system.

Stretch-formed parts can be found in appliances, aircraft, and automobiles. A large stretch forming

Figure 14-4. The diameter of the slip-on tube used in two-roll bending controls the diameter of the bend produced in the stock.

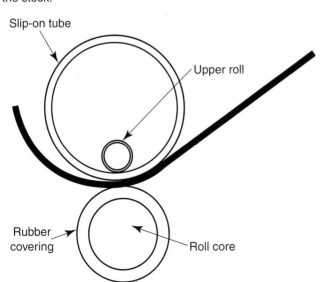

Slip-on tube

Upper roll

Rubber covering

Roll core

Figure 14-5. With a three-roll double-pinch machine, stock can easily be bent symmetrically to form a hoop.

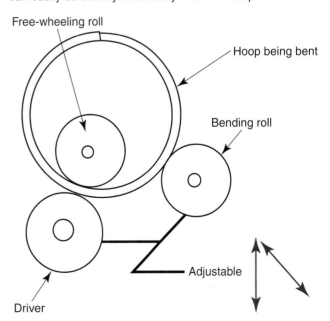

Free-wheeling roll

Hoop being bent

Bending roll

Adjustable

Driver

Figure 14-6. In bullnose or stretch forming, the clamped material is shaped by a forming mandrel (center). The die table raises the mandrel or form to stretch the metal to the desired shape. (The Cyril Bath Co.)

machine is shown in **Figure 14-6**. The mandrel in this photo is in the upright position.

Thread Rolling

Creating threads to make fasteners is one of the most common metal forming processes. External threading on the outside of a rod creates a bolt or screw. An internal thread forms a nut. One of the processes often used to accomplish this is called thread rolling.

Thread rolling is a chipless cold-forming process that can be used to produce either straight or tapered threads. Thread rolling uses rotating hardened steel dies that are pressed against the surface of cylindrical blank. As the blank rolls against the die faces, the material is displaced to form the roots of the thread. This displaced material is then pushed outward to form the thread's crest. Unlike other threading processes, no material is removed so no chips are produced. Therefore, less raw material is needed to create a rolled thread. As an example, it takes 27 percent less material to roll a 3/8"-16 UNC-2A thread than to cut a thread. In addition to the savings in material, cold forming creates a thread that has a higher fatigue strength. A properly formed rolled thread has no sharp corners in the root or tool marks on the flanks to cause stress risers.

To roll a thread, the stock, called a *blank*, is placed between the dies. See **Figure 14-7**. One of the dies moves perpendicular to the blank's axis while the other remains stationary. Flat dies are used in thread-rolling machines (bolt making machines). However,

Figure 14-7. These flat, grooved dies are used for rolling threads. In the thread-rolling machine, one die is stationary and the other moves, rotating the round stock as the threads are formed.

there are many different types of dies that can be used on thread-rolling machines. See **Figure 14-8**.

Threads can also be rolled using three dies positioned in a pyramid arrangement. The equipment used to hold the dies varies, depending on the size and type of the part. Most cylindrical thread rolling is accomplished using thread-rolling machines, automatic screw machines, or automatic lathes. Automatic thread-rolling machines, similar to the one in **Figure 14-9**, are highly cost-effective for producing large quantities of production parts. Some firms are producing in excess of 90 parts per second using rotary dies.

Figure 14-8. These are some of the many different types of dies and attachments available for use in thread-rolling machines. (Reed Rolled Thread Co.)

Figure 14-9. This automatic thread-rolling machine is highly productive. Parts to be threaded are stored in the hopper at the upper right, and fed into a supply chute by vibratory action. After passing through the threading dies, they are ejected into a discharge chute. A container for quantities of finished threaded items would be placed below the chute, at left. (Reed Rolled Thread Co.)

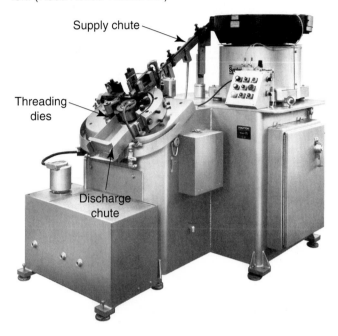

Supply chute

Threading dies

Discharge chute

When many parts must be produced, thread rolling is the fastest way to form threads. Thread-rolling machines can also be used to knurl cylindrical surfaces. The dies used for *knurling* create a rough-textured pattern usually used as a gripping surface on tool handles. See **Figure 14-10**.

Figure 14-10. Knurling dies produce a pattern of raised lines or diamonds on a surface, usually to allow better gripping of a tool or similar device. (Reed Rolled Thread Co.)

Swaging

Rotary swaging, also referred to as *radial forging*, is a process that takes a solid rod, wire, or tube and progressively reduces its cross-sectional shape through repeated impacts from two or four opposing dies. An alternative swaging process can be used to change the internal diameter by placing a mandrel inside the tube and applying radial compressive forces on the outer diameter. Thus, the inner diameter can be a different shape, for example a hexagon, while the outside of the tube is still circular.

Swaging is a noisy process that gradually squeezes away the desired amount of material. Operators must wear ear protection to prevent damage to their hearing. **Figure 14-11** illustrates the process of rotary swaging. Dies rotate around the workpiece, opening and closing rapidly to generate

Figure 14-11. In rotary swaging, the dies rotate around the cylindrical or tubular workpiece, reducing its diameter with repeated blows. Stock is fed continuously through the swaging machine.

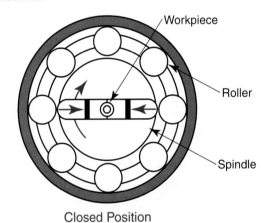

Workpiece

Roller

Spindle

Closed Position

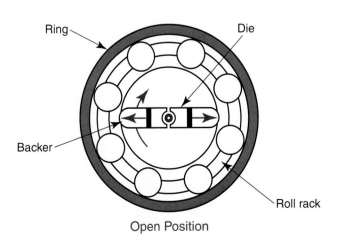

Ring

Die

Backer

Roll rack

Open Position

a hammering action against the rod or bar stock. The blank is fed continuously into the die opening in the swaging machine. There, a revolving spindle creates centrifugal force, throwing the opposing dies outward. The dies hit against rollers carried in a circle around the spindle. This impact causes the dies to rebound against the workpiece. Some swaging operations create as many as 5000 impacts per minute.

Swaging is normally used to produce round parts up to 2″ in diameter. Special machines have been constructed that permit swaging of tubes to 6.5″ and bars to 4″. The swaging machine shown in **Figure 14-12** is equipped with an automatic feeder, although some machines are fed by an operator.

The rifling in gun barrels is done by swaging. This is accomplished by deforming a metal tube placed over a mandrel with spiral grooves. Swaging is used for a great variety of products ranging from screwdriver tips and fasteners to ballpoint pen caps to metal chair legs. Swaging can also be used as a means to attach parts in an assembly. Imagine a cable or a shaft is inside the workpiece before the swaging process begins. This method is used to attach a bushing to a shaft or a fitting to a table.

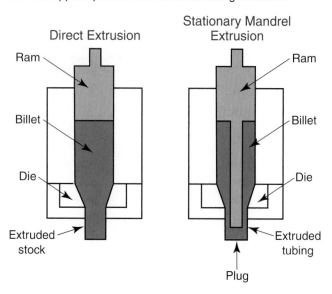

Figure 14-13. Two popular types of extrusion. In direct extrusion, the ram applies pressure to force the hot metal out through the die opening. In the stationary mandrel method, the mandrel keeps the inside of the tube open as the ram applies pressure to the metal being extruded.

Extrusion

Extrusion is a manufacturing process that compresses metal beyond its elastic limit before it is forced through a die opening. Openings in extrusion dies can be a simple round shape or a custom and complex shape. Metals such as aluminum, copper, magnesium, and stainless steel are easily extruded.

Extrusion is a continuous-pressure forming process. In concept, it works much like squeezing toothpaste out of its tube. **Figure 14-13** illustrates two of the most popular forms of extrusion: direct extrusion and the stationary mandrel method. There are also several variations of each of these methods.

In *direct extrusion*, the ram or punch moves down on a heated metal billet. The pressure from the ram forces the stock to extrude (flow) out of the die opening.

The *stationary mandrel method* is used to make hollow tubing. With this method, the ram is attached to a stationary mandrel. As the ram travels down into the metal billet, the mandrel contacts the stock first and guides it through the die opening. The mandrel also keeps the inside of the tube open.

Impact extrusion is commonly used to make collapsible tubes (used to dispense glues and grease) and cans. The process uses soft materials such as aluminum, lead, and tin. Usually a lubricated slug is placed in a die cavity that is then struck by a punch, forcing the metal to flow back around the punch.

Figure 14-12. Rod or other material to be processed is fed into the round opening of this rotary swaging machine. (Abbey Etna Machine Co.)

Hydrostatic extrusion is a form of impact extrusion that uses a fluid instead of a mechanical ram to extrude the metal. This method is useful for making parts out of materials such as molybdenum and tungsten, which are relatively hard to extrude using normal extrusion methods.

Hot extrusion is used to apply a plastic insulation coating to copper electrical wire. This application is explained in more detail in the chapter covering plastic processes. In the wire-coating process, both the wire and plastic material are fed through the die orifice.

When hard metals such as stainless steel are being extruded, glass is often used as a lubricant. Glass is applied to the opening of the extruder by placing a circular glass pad at the die opening. The glass is kept in a molten state, so it serves as both lubricant and as a coating for the metal that is forced through the die opening. Powdered glass is also applied to the inner liner of the chamber of the extruder before the metal billet is introduced.

Metal products that are normally extruded include ladders, window frames, lipstick cases, soft drink cans, and fire extinguishers. Long lengths of aluminum and stainless steel channel are also produced by hot extrusion. Extrusion is even gaining some popularity with larger products such as refrigerators, washing machines, and air conditioners.

Upsetting

The process called *upsetting* (also known as *cold forming* or *cold heading*), is a type of forging that thickens or bulges the workpiece while also shortening it by compression. Upsetting is actually a combination of forging and extrusion. It is one of the fastest processes used by industry for high-volume production of nails, bolts, and rivets.

Cold heading forms a head on the end of an unheated metal rod by compressing its length in a die cavity. The machines that used to perform this action are called *headers*. There are many types of machines and dies that can be used for cold heading. In some operations, the stock is sheared to length and then compressed inside a closed die. Other operations use an open die and a forming punch that hammers the stock to conform to the die shape.

A typical closed-die arrangement is shown in **Figure 14-14**. A closed die consists of a hardened cylinder with a hole running through its center. This

Figure 14-14. Upsetting, or cold heading, is the process of forming a head on a bolt, nail, or other fastener through a combination of forging and extruding. As shown, the punch forces the stock downward into the die, so that the metal bulges out to form the head. When the punch retracts, the ejector pin pushes the workpiece out of the die.

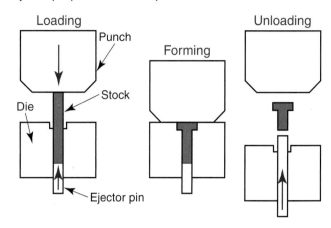

is called the *die block*. The second die half, called a *punch*, automatically pushes wire stock in precut lengths through the hole. With the desired amount of blank projecting out of the hole, a punch hammers down on the stock. One or two blows are used to form the head. Then the punch opens and an ejector pin pushes the part out of the die.

Production bolt-makers combine several operations—forming, head trimming, and thread rolling—into one continuous process. A finished part is generated with each stroke of the machine. A forming center with a robot-assisted tool changer shown in **Figure 14-15** is an example of state-of-the-art cold-forming machine. This equipment can take preforms, coiled wire, or bar stock and produce a

Figure 14-15. This automated forming center, equipped with a gantry-type robotic tool changer, is an example of a state-of-the-art cold-forming machine. (National Machinery Co.)

Figure 14-16. These are some of the varied types of parts that can be formed at high production rates by automated cold-heading machines. Such parts seldom need surface finishing. (National Machinery Co.)

multitude of complex parts requiring no surface finishing. See **Figure 14-16**.

Cold forming is a process that automatically feeds coiled wire into the machine, transfers cutoff blanks through a series of progressive forming dies, and creates a finished part. Cold forming has been widely adopted for use in the high-volume production of small- to medium-sized parts from low alloy steels. One of the best examples of cold forming is the production of spark plug shells, **Figure 14-17**.

Traditionally, these shells were produced using screw machines. Now, in addition to the spark plug shell, many manufacturers are cold forming the electrodes and center posts as well.

Cold Forming Advantages

There are several advantages cold forming has over traditional machining and forging methods. These include higher production rates, little or no scrap, production to size with close tolerances, good surface finish, and an increased part strength. One manufacturer sells a six-die cold former that has a production rate of 120 pieces per minute. The builder claims that cold forming results in a reduction of waste and increases part strength.

Most basic cold-forming machinery can be modified for warm- and hot-forming processes. Parts that are frequently produced by warm forming include automatic transmission gears, planetary pinion gears, lifter rollers, and outer bearing races. The automotive industry is a primary user for parts that are automatically hot formed. Typical hot-formed products are connecting rods, track links for off-the-road vehicles, and large gear blanks. The press shown in **Figure 14-18** is used primarily for warm and hot forming.

Figure 14-18. This large transfer press is used for warm and hot forming. The automotive industry is a major user of such equipment, especially for hot forming. (Schuler Incorporated)

Figure 14-17. Metal spark plug shells can be produced at the rate of two per second by automated cold-heading equipment. (National Machinery Co.)

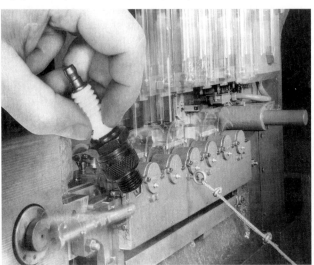

Stamping

Stamping is almost synonymous with mass production. A product that is stamped will bring to mind an item that is inexpensive and readily attainable. Stamped parts, however, are found in very expensive products such as automobiles. A car uses hundreds of stamped parts. Some of these parts are exquisite in their shape while others are crude but functional. During the last century, stamping became a major manufacturing process.

Stamping is a cold-forming process that uses a set of matched molds in a stamping press to compact stock under pressure. This is a chipless process that produces a sheet metal part with one downward stroke of the ram in a stamping press (usually referred to as a *punch press*). Parts are typically stamped cold. One of the largest markets for stamped products is the automotive industry, which uses many thousands of medium-sized and large body panels each year. See **Figure 14-19**.

In addition to performing stamping operations, these presses are used to punch, notch, pierce, and trim sheet metal parts. This somewhat complicates the meaning of the term *stamping*. In actual industrial usage, stamping generally refers to all of the press-work operations related to sheet metal processing. A stamped part may be cut to size at the same time that slots, holes, and notches are generated. All of these operations can be done with a single downward stroke of the ram.

Stamping presses are large, often several stories in height. The large automated transfer press shown in **Figure 14-20** is used to produce body panels for the automotive industry from sheet stock. A press can be designed to hold as many die stations as desired.

Figure 14-19. Automotive body panels, such as these door components, are typical products of large stamping presses. (Schuler Incorporated)

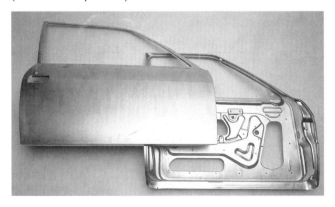

Figure 14-20. Highly automated equipment, such as this transfer press for stamping and forming large body panels, is widely used in the automotive industry. (Schuler Incorporated)

In most stamping presses, the ram is at the top and travels up and down while the material is fed into the die automatically. There are instances where the material is hand fed and the die is manually activated using an operator safety control system. The punches and die block assembly comprise the *die set*. A typical die block assembly is shown in **Figure 14-21**.

Stamping Press Classifications

Stamping presses are classified in different ways. Some are classified by the types of parts that are to be stamped. Presses also may be classified according to the type of power, the structure of the press, or the type of ram that they employ.

Figure 14-21. The die block assembly consists of a ram that forces punches down into a die block mounted on the press bed. Note that stock moves through this press as a continuous ribbon of sheet metal, advancing each time the ram rises after stamping. (The Minster Machine Co.)

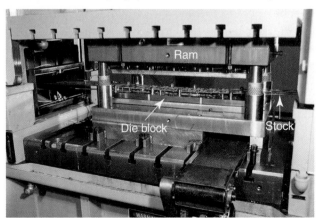

Figure 14-22. A numbering system describes each press. The identifying number of this large stamping press (E2-400-96-42) is shown next to the manufacturer's name on the top front, above the ram.

The Joint Industry Conference (JIC) system has been established to aid in classifying stamping presses. With this system, presses are given an identifying letter and number code. A typical number might be D-700-90-72. This number designates a double action, 700-ton capacity press with a bed width (left to right) of 90″ and a depth (front to back) of 72″. The heavy stamping press shown in **Figure 14-22** is numbered E2-400-96-42. This is a two-point, eccentric-shaft 400-ton press that is 96″ wide and 42″ deep.

Stamping became a high-volume production process capable of producing thousands of parts per hour at a time when high volume mass production was the rule in industry. Today, manufacturers are working to *decrease* lot sizes. New strategies and technologies are being used to change stamping operations from *mass* production to *demand* production. One of the major areas being worked on is reducing the time it takes to change die sets. Conventional die changes used to take several hours. Some companies have come to view a die change the same way auto racing teams treat pit stops during a race. An acronym that has been applied to this approach is SMED (Single Minute Exchange of Dies).

Embossing

Embossing is a process that produces raised areas on a flat sheet of metal. A common product application would be placing names or numbers on such items as military dog tags or machine identification plates. One side of the tag would have raised

printing, while the other would be depressed. This is where the similarity between stamping and embossing ends. The two processes use different types of dies and produce differing levels of quality in the impression that results.

In simplest form, stamping of data on metal could be done by striking the workpiece with a punch. Embossing requires a two-part matched die. The male die half has a raised impression surface, while the female die half has sunken (depressed) detail that conforms to the raised areas of the male punch. While stamping punches deform or stretch the metal, embossing compacts the metal in the area of contact. This produces finer detail and improved strength in the part. Embossing can be performed in a punch press or stamping press. It can also be done using rotary dies to continuously emboss patterns on thin metal foil and sheet stock.

Drawing

A great variety of sheet metal products, ranging from kitchen sinks and automobile fenders to pots and pans, are made using the manufacturing process called drawing. What is unusual about drawing is that it can be used to produce a three-dimensional part from flat sheet or metal plate. *Drawing* involves both stretching and compressing.

There are several variations of the basic process of drawing. The most common is called deep drawing or shell drawing. Drawing was once used to make large artillery shells. A typical drawn part is shown in **Figure 14-23**.

In *deep drawing*, a female die is pressed into thin sheet stock, stretching the metal over a male

Figure 14-23. An example of parts made by the drawing process. These are shells for automobile oil filters. (The Minster Machine Co.)

Figure 14-24. Mating dies used in the stretch-draw forming process must match perfectly. The workpiece, or blank, is drawn over the lower dies, then stretched tight as the two dies come together to shape the part.

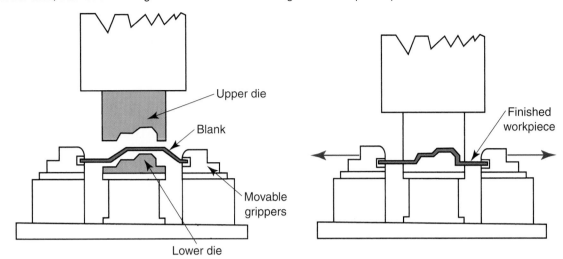

forming punch. Usually the sheet stock is placed on a pressure ring surrounding a bottom punch. An overhead ram then moves the female die down and compresses the stock against the bottom punch. The process is completed when the female die retracts and the pressure ring pushes the part back up to the unload position.

Figure 14-24 illustrates the ***stretch-draw forming*** process, which use mating dies that must match perfectly. Such dies are costly to construct and maintain. These are similar to the types of dies used in stamping processes. However, not all draw forming processes involve mating dies. One

flexible draw forming process that is often a cost-efficient alternative to stretch-draw forming is called ***marforming***. In the marform, no expensive dies are involved. The tooling consists of a thick rubber pad in a metal retainer, which is supported by a platen that is attached to the ram of the press. The forming block is held by a second platen secured to the press bed. Both shallow and deep-drawn parts can be produced with this process.

As shown in **Figure 14-25**, the workpiece is held on a stationary blankholder ring. Drawing begins when the upper platen holding the rubber pad, moves down against the forming punch. The workpiece

Figure 14-25. In the process called marforming, no dies are used. A thick rubber pad in the upper platen forces the blank down over a lower forming punch.

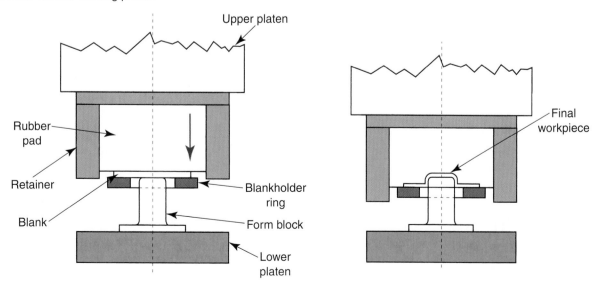

Figure 14-26. Hydroforming is very similar to marforming, but substitutes a forming chamber filled with fluid under pressure for the rubber pad. In this process, the forming chamber is lowered over the workpiece, then the punch moves upward to shape the blank.

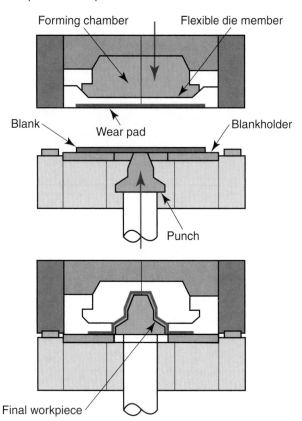

Figure 14-27. Expanding involves driving a punch into the opening of a drawn part to increase its diameter.

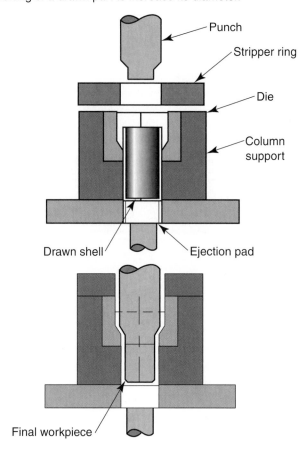

is pressed between the thick rubber pad and the forming punch. Although marforming tooling is inexpensive, it is able to produce wrinkle-free parts.

Hydroforming, is a draw-forming process that uses hydraulic fluid in place of the rubber pad used in marforming. Hydroforming uses a rubber diaphragm that seals the hydraulic fluid in the pressure-forming chamber, **Figure 14-26.** The workpiece is positioned on the bottom die block, and the pressure forming chamber is lowered into place. Forming begins when the punch is forced upward into the workpiece, increasing the pressure in the fluid chamber. This increase in pressure causes the workpiece to be compressed around the punch. When forming is completed, pressure is released and the workpiece is stripped from the punch.

Hydroforming is done using high-speed presses capable of producing parts as large as 25″ in diameter, with a draw depth of 12″. Some companies have hydroformed steel parts as thick as 3/8″ and aluminum up to 1″ in thickness.

Expanding

Expanding is a process that can be used to increase the diameter of tubular parts. **Figure 14-27** shows how a drawn part can be expanded by a forming punch. The punch travels down through a stripper ring, pressing the top of the drawn part against the walls of the die block. When the punch is retracted, the stripper ring helps to remove the expanded part from the forming punch.

A typical application for expanding would be to form a bell on the end of a tube. Expanding can be used to enlarge other portions of drawn work pieces, as well. Specialty machines, called expanders, carry movable shoes on a circular faceplate. The shoes move outward to expand the work. Expanding can be done on other machines as well.

Before the workpiece is expanded it should be softened, or annealed. Methods for annealing metals will be discussed in the chapter on conditioning metallic materials. If the part is to be expanded 30 percent or less of the diameter of the workpiece, the process can be conducted with one stroke. If the expansion is more than 30 percent, several strokes are required, with the part being annealed before each stroke.

Bulging

Another process that can be used to expand tubular shapes such as corrugated tubes, bellows or musical instruments is called *bulging*. The rubber core method is shown in **Figure 14-28**. Here is how it works: the tube to be expanded is placed in the die. The die, which locates and holds the tube, is machined to the desired final shape of the workpiece. The process begins when the bulging punch moves down into the tube. The end of the punch, which is rubber, bottoms out on the base of the die. As the ram continues to move down, the rubber portion of the punch is compressed exerting a force on the sidewalls of the tube causing it to expand into open areas of the die. Under high pressure, rubber provides equal hydraulic pressure in all directions. To achieve this pressure it is essential that the rubber portion of the plunger is completely inside the tube (workpiece). Once the part has been expanded (bulged), the die has to be opened in order to remove the part. Depending on the contour, the die may be split axially or horizontally.

Necking

The forming process called *necking* is used to reduce the diameter of the end of a tubular part.

Figure 14-28. A rubber core is placed inside the part to be shaped by bulging, then a plunger presses down on the rubber. This causes the rubber core to exert outward force.

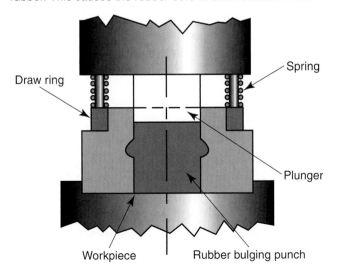

Draw ring

Spring

Plunger

Workpiece

Rubber bulging punch

Figure 14-29. Necking is a die-reduction process that achieves progressive reduction of the part's diameter.

Figure 14-29 shows a typical necked part, a small canister used for carbon dioxide gas. With one operation, the end of the canister is reduced to about 20 percent of the diameter of the tube.

Necking is a die-reduction process that stretches the relatively soft and ductile metal part at the same time that it reduces its cross-sectional area. At first glance, it might appear that the process is used only to reduce the diameter of the workpiece, but there is more to it than simple reduction. As they stretch or elongate the workpiece, the dies turn. The diameter of the stock is reduced in the area where the dies are working. The process results in some inconsistency in wall thickness when the cross-sectional area is reduced and force brings the material to the point of plastic deformation. The actual point where this occurs depends upon the ductility and hardness of the metal.

An important point to remember about necking is that the tensile strength of the part is increased when the specimen is stretched or elongated. This increase in tensile strength makes the part better able to support heavy loads.

The necking process must be used with care; there is a point at which increasing strain on the part will result in it breaking. This is called the *breaking point* or *fracture strength* of the material.

Nosing

Nosing is similar to necking, but it is used to partially close the end of a tube, rather than just reduce the diameter in one section. Nosing can be used to taper or round the end of tubing. This is a process that is particularly useful in making rifle and pistol cartridges, **Figure 14-30**.

Nosing is normally done by using dies to close the end of the tubing. In principle, it is somewhat like the process of rotary swaging. Nosing can be extended to its limits, stretching the metal in order to cause it to close. This type of nosing, called *spinning*, will be discussed later in this chapter.

Figure 14-30. Tubing is first slightly tapered, top. Additional pressure while rotating causes the diameter to be reduced, bottom.

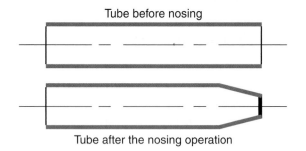

Tube before nosing

Tube after the nosing operation

Electromagnetic Forming

Electromagnetic forming, sometimes called magnetic pulse forming, is a process that forms a workpiece by using intense pulsating magnetic forces. Energy is stored in capacitors that discharge short pulses (measured in microseconds) to a forming coil, **Figure 14-31**. The coil produces a magnetic field, which is passed to the conductor (workpiece). Eddy currents are produced in the workpiece. This current creates its own magnetic field. When the two magnetic fields oppose each other, a repelling force is created between the coil and workpiece. This force pushes the workpiece against a forming die. Since the metal is stressed beyond its yield strength, permanent deformation occurs.

The process can also be used for internal forming by placing the coil inside a tube. When the pulsating electrical current travels through the coil, the force causes the tube to expand.

Magnetic pulse forming is capable of generating pressures as high as 50,000 psi (345,000 kPa) on the workpiece. The process is used extensively as an

Figure 14-31. A forming coil wrapped around a tube generates opposing magnetic fields, with a resulting reduction in diameter of the workpiece. The process can also be used to expand a tube by placing the coil inside it.

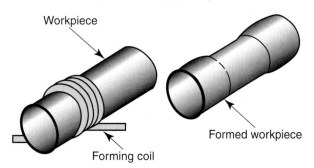

Workpiece

Formed workpiece

Forming coil

assembly technique to join tubular parts to other components. Magnetic pulse forming can be used to reduce the diameter of a section of a tube or to join tubes to rods. In such an application, the tube is inserted into the coil and then a rod is inserted in the tube. The magnetic force created around the coil compresses the tube, pressing it forcefully against the rod.

In the automotive industry, magnetic pulse forming is used to assemble steering gears, ball joints, and shock absorbers. The electrical equipment industry uses the process for assembling coaxial cable, electric motors, potentiometers, and various parts that would be difficult to assemble by other methods.

Joints can be assembled using magnetic pulse forming, instead of welding or brazing. Metal rings can even be assembled to ceramic, thermoset plastic or phenolic parts, since the magnetic field will pass through materials that are electrical insulators.

Contouring

Magnetic pulse forming can also be used for shaping flat parts, a process called *contouring*. In this application, the forming coil looks much like a burner on an electric stove. The coil is placed against the sheet, which presses the flat stock into a forming die.

It is easiest to use magnetic pulse forming with conductive metals. However, metals that are poor conductors can be formed, but a conductive material (called a driver) must be placed between the forming coil and the workpiece. An aluminum driver is used to form stainless steel parts, for example.

This process cannot be used to form complicated shapes. The forming rate is consistent around the coil, so you cannot produce high pressure in one area and low pressure in an adjacent area. Another limitation is that the process does not work well with parts that have many holes, notches, or slots. These voids interfere with the flow of current across the part.

Peen Forming

Shot peening is a cold-working process that is accomplished by bombarding the surface of a part with small spheres called *shot*. Shot can be made of cast steel, glass, or ceramic particles. **Figure 14-32** shows a part being shot peened. During the process, each piece of shot that strikes the metal acts like a tiny peening hammer creating a pattern of small

Figure 14-32. The many thousands of tiny impacts caused by shot striking the metal surface strengthens the part and makes it less subject to surface cracking. (Metal Improvement Company)

overlapping surface indentions. Since surface deformation is not consistent throughout the thickness of the part, some residual compression stress is created on the surface. Cracks will not occur in a compressively stressed zone. Since most cracks caused by fatigue originate on the surface of a part, the compressive stresses produced by shot peening can actually increase a part's life.

Shot peening is extensively used on parts such as oil-well drilling equipment, turbine compressor blades, shafts, and gears. Coil springs (compression) are probably the best known and most widely used parts that are shot peened. All automobile engines use valve springs that are shot peened. Springs as small as 0.005″ in diameter and as large as 3″ in diameter have been shot peened to increase their fatigue resistance. Shot peening can also be used to form or contour parts. See **Figure 14-33**.

Often parts are too large or difficult to transport from the field to the plant for shot peening. Units such as chemical storage tanks and pressure vessels, gas and steam turbines, tubing in heat exchangers and steam generators, and wood processing digesters

have been shot peened on-site using portable equipment. Many different types of machines for spraying shot are commercially available.

Explosive Forming

Some forming processes, such as stretch-draw forming, matched-die method, and coining, require both a male and female die to perform the forming operation. Other forming methods such as hydroforming and flexible die forming use only one die and a rubber forming block. All of these forming processes are used to shape relatively small parts.

Explosive forming uses only a female die form. The sheet metal workpiece is clamped to the die and the assembly is placed in the bottom of a water-filled tank. A vacuum line attached to the die cavity removes air from the void between the workpiece and the die. Next, an explosive charge is suspended in the water above the workpiece. When the charge is detonated, a shock wave forces the sheet stock against the forming die. As shown in **Figure 14-34**, the same principle is used to expand a tubular workpiece, except that the dies enclose the workpiece and the explosive charge is placed inside it. A benefit of this process is that the limit to the size of the workpiece is restricted only by the size of the water tank.

HERF

A variation of the explosive forming process, called electrohydraulic, electrospark, or *high-energy-rate forming (HERF)*, uses a spark-generated shock wave,

Figure 14-33. Shot peening is being used to form the contour on this aircraft wing skin. (Metal Improvement Company)

Figure 14-34. Arrangement for explosive forming used to expand a tubular workpiece.

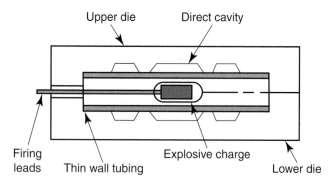

Upper die Direct cavity

Firing leads Thin wall tubing Explosive charge Lower die

rather than a chemical explosion. See **Figure 14-35**. With HERF, a capacitor bank stores electricity until a switch closes a circuit to the forming tank. The tank is normally filled with water, glycerin, or light oil. When the energy is released, an aluminum or magnesium bridge wire connecting the electrodes is vaporized. This creates a plasma channel in the liquid for the spark to cross. The spark generates a shock wave that propagates radially. The concussion forces the stock into a forming die. Sometimes a bridge wire is not used. In such a case, the capacitor voltage is increased from about 4500 to 20,000 volts. With high-energy-rate forming, pressures approaching 175,000 psi can be created.

Figure 14-35. The energy released in HERF causes a shock wave through the transfer medium, forcing the workpiece into the cavity of the forming die.

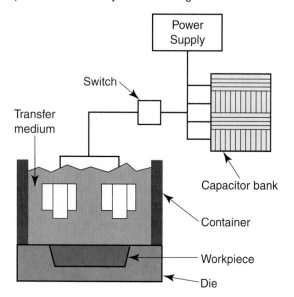

Power Supply

Switch

Transfer medium

Capacitor bank

Container

Workpiece

Die

Spinning

Spinning is a process that involves stretching sheet stock over a rotating male or female mold. The mold is attached to a lathe or spinning machine. Spinning can be either a cold-working or hot-working process, but it is most often done without heating the workpiece.

The beginning stage in the production of a microwave reflector by the spinning process is shown in **Figure 14-36**. A flat disk is snugged up against a male *forming block* (also called a mandrel or mold) that is attached to the headstock of the spinning lathe.

Let us take a closer look at what is happening in the photo. The forming block is fastened to the faceplate, which is attached to the headstock spindle of the lathe. The headstock is located to the left in the photograph. The tailstock, located on the right end of the lathe in the photo, can be loosened and slid along the horizontal rails toward the headstock. The tailstock holds a *follower* (a freewheeling spindle) that is pressed against the stock to be spun. After the tailstock is secured, the pressure between the forming block and follower is sufficient to hold the stock in place.

However, there is more to the photograph than this. First, the machine shown is a lathe operating under *computer numerical control (CNC)*. This enables the operator to run a program stored in the controller. The program is responsible for conducting the actual spinning operations.

To continue with the spinning, the lathe would be turned on and the program would be initiated. This would cause the forming block, workpiece, and follower to rotate. The CNC program would then

Figure 14-36. Using spinning to stretch metal over a rotating mold to form a microwave reflector. (Lake Geneva Spindustries, Inc.)

transmit information to a *forming tool*, and the tool would automatically apply pressure against the face of the disk. This tool pressure stretches the metal over the forming block.

When spinning is accomplished without the aid of CNC, considerable operator skill is required to properly stretch the metal without making it too thin or causing the workpiece to buckle. **Figure 14-37** shows an operator spinning a large aluminum canister by hand. Roller tools like those in the foreground are used to stretch the metal over the form.

Because of their softness, copper and aluminum are the most common materials used in spinning operations. Spinning does not require expensive tooling so production of low volume products is economically feasible. Parts such as flood lamp reflectors, bowls, and bells for musical instruments are normally spun using CNC equipment.

Casting and Molding

The foundry industry is directly concerned with the casting and molding of metals. In many instances, firms in this industry can be classified as primary manufacturers. They are concerned with producing stock for use by secondary manufacturing establishments that, in turn, create the finished product using cast metal parts. Cast parts are made from iron, aluminum, and other metals or alloys.

Cast Iron

The term *cast iron* is unique in that the material and the manufacturing process are essentially inseparable. Initially, cast iron was produced in a high-temperature

Figure 14-37. Large canisters and other products may be spun by hand, using roller tools to stretch the metal over the forming block. Some typical roller tools are visible in the foreground. (Lake Geneva Spindustries, Inc.)

furnace called a *cupola*. The materials loaded into the cupola were usually pig iron combined with scraps of high-carbon-content steel along with coke. Large quantities of forced air raised the temperature of the coke-fired cupolas to the point where the iron and steel became white hot and could pour like water from a ladle. Besides being a fuel, the coke supplies the necessary carbon—cast iron contains from 2 to 4 percent carbon. However, cupola furnaces are very polluting and have been replaced with electric furnaces. In addition to being less polluting, electric furnaces also provide better control over alloying and are more productive than cupolas.

There are many types of cast iron, including ductile, gray, high-alloy, malleable, and white. Gray and ductile cast irons are most common. Of all of the cast ferrous alloys, gray iron has the lowest melting temperature and the best overall castability. It also has tremendous compressive strength. For this reason, machine beds and frames are usually made of gray cast iron. The major limitation of gray cast iron is that it is brittle and the high carbon content makes it difficult to weld.

Ductile, or nodular, iron is used to cast axles, brackets, and crankshafts. Ductile iron is made by adding elements such as magnesium, nickel, or silicon in small amounts to the molten alloy. These elements improve the strength, castability, and machinability of cast iron.

Some foundries are set up as small job shops that contract with customers to produce limited quantities of cast products. Many large companies have foundries in their own facilities.

The raw material to be cast is often purchased as an *ingot* (a brick of stock). Firms that produce large quantities of cast iron products usually have sophisticated facilities for melting castable steel or iron alloys. **Figure 14-38** shows the molten metal being released from a furnace into a heavy metal container, or ladle. This process is called *tapping a heat*. Furnaces typically produce heats ranging from 500 lbs. to 35,000 lbs.

Casting Processes

It is common to think about casting as the process of pouring molten metals. However, not all casting involves pouring liquids. Powders are also cast through compacting and heat-setting processes.

Casting processes can be studied according to the type of mold used in the process. Molds are either

Figure 14-38. Molten steel pours from an electric furnace into a large ladle that will be used to transport the metal to an area where it will be poured into molds to form cast products. (ACIPCO Steel Products)

permanent (reusable) or consumable (expendable). The most significant metal casting processes that use a permanent mold (in terms of industrial use) are powder metallurgy, die casting, and permanent mold casting. Major casting processes that use consumable molds are sand casting and investment casting.

Powder metallurgy (P/M) is a process that involves compacting metallic powders in a permanent (reusable) mold. It can be used to cast both ferrous and nonferrous parts. For many product applications, it is a highly cost-effective process. Powder metallurgy is a near-net-shape manufacturing process and typically uses more than 97 percent of the initially supplied raw material in the finished part. *Near-net-shape* means that parts are produced close to the desired final size, requiring machining only if the part needs threads, holes, or unusual design features. Little shrinkage occurs during firing and virtually no other machining is necessary. Machinability of P/M parts is similar to that of poured castings. An advantage of this process is that additives such as lead, copper, or graphite can be mixed with the powder prior to casting to improve machinability. However, when P/M parts are machined, carbide-tipped cutting tools are recommended.

Powder metallurgy is generally a high-volume production process, capable of producing thousands of parts per hour. In low-volume production (fewer than 1000 pieces, for example), powder metallurgy is

not generally a cost-effective process. The cost of the tooling required would be too expensive to justify its use. A typical powder metallurgy die set is shown in **Figure 14-39**.

Powder metallurgy parts are particularly interesting. Their ability to work in unusual applications is greatly influenced by the mix of powders used and the various processes which they undergo once they are pressed into shape. There are three steps involved in manufacturing a powdered metal part: powder mixing, compacting, and sintering. After sintering, the P/M part is normally ready to use. However, like other metal-formed parts, secondary heat treating and processing operations may be necessary to improve product life in specific environments. Like other metal parts, P/M parts can be annealed, hardened, and tempered.

Very fine powders are used to make P/M parts. Mixing of alloys can be done by the powder producer or the parts manufacturer. The powders from which parts are pressed can be made to assume the characteristics of other metals through a process called infiltration. Self-lubricating bearing surfaces can even be created through the process of impregnation. These and other unusual processes will be discussed in the following paragraphs.

Once the powders are mixed as desired, a metered amount of powder is automatically gravity-fed into a precision die, **Figure 14-40**. After the die is filled, the upper and lower punches close to compact the powder in the die. The densely compacted shape

Figure 14-39. A typical powder metallurgy die set consists of the die itself, an upper punch, a lower punch, and a core.

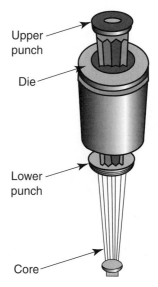

Upper punch

Die

Lower punch

Core

Figure 14-40. To make a part in the powder metallurgy process, the die is filled with blended powder. The upper and lower punches then close to compact the powder. The upper punch retracts, and the lower punch ejects the compact from the die. Finally, the lower die returns to its original position, ready for the next cycle.

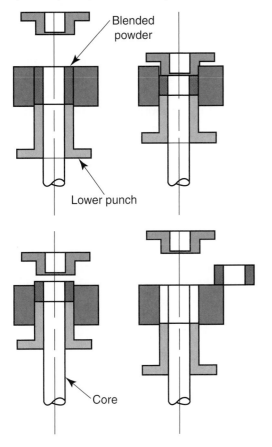

is then ejected and the die is refilled so that another part can be pressed.

The pressure required to press a P/M part varies from 10 to 60 tons per square inch. The higher the pressure, the greater will be the density of the part. Pressing of the powder is done at room temperature.

When the compressed part is ejected from the die, it is referred to as a *green compact*. At this point, the part can be handled, but it is extremely fragile and can be fractured if dropped or carelessly handled. In most production environments, the green compact is transported by conveyor to the sintering furnace, where it travels slowly through a temperature-controlled environment.

Sintering takes place at a temperature just below the melting point of the base metal. Parts are held at this temperature for a short length of time and then cooled. The purpose of sintering is to create a metallurgical bond between the particles of powder. After sintering, the parts can be pressed again to improve detail, or can be plated, heat-treated, impregnated, or machined as desired.

Impregnation is the process of immersing porous P/M parts in heated oil or resin after they are removed from the sintering oven. The parts are allowed to soak long enough so that the pores are saturated by oil. The process has been used by the automobile industry since the 1920s. Parts can also be impregnated with oil using a process called vacuum impregnation.

Care must be taken in handling oil-impregnated parts. Such parts should never be allowed to rest for any length of time on cardboard or porous paper or the impregnated oil will be drawn off by a process called wicking. To avoid wicking, place parts on coated paper or a nonporous surface.

What does impregnation do for the P/M part? If the part is a bearing, for example, it provides a reservoir of lubricating oil. If the bearing surface heats up due to friction, the increase in temperature causes the oil to expand and flow to the surface of the bearing. The oil at the surface reduces the friction, and the bearing cools. As the bearing is cooled, the oil is drawn back into the pores of the metal by capillary action.

It is possible to improve the mechanical properties of P/M parts through another process called *infiltration*. Infiltration is accomplished by placing pieces of the infiltrant metal on or below the green compact before sintering.

The infiltrant metal must have a lower melting point than the porous P/M part. When the part is sintered, the infiltrant is drawn into the part's pores by capillary action. This enables the formation of a composite metal. Not only does the composite provide improved mechanical properties, but the infiltrant also seals the pores of the part. This is particularly helpful when the part is to be electroplated or machined.

When maximum density of P/M parts is desired, a process called hot isostatic pressing can be used. This compacting process is discussed in detail in the section on ceramic material processing. Commercial products, such as cylindrical billets up to 9′ in length and hollow cylinders with diameters as large as 24″, have been produced by hot isostatic pressing.

Powder metallurgy is also used in conjunction with injection molding, a process that is commonly used for making plastic parts. In *P/M injection molding*, very finely ground powders are coated with thermoplastic resin. The mixture is then used in the conventional manner in an injection molding press.

Modified thermoplastic injection molding machines are used at approximately 300°F and 1,000 psi (150°C and 7,000 kPa). After molding, the part is sintered. This burns off the polymer and improves the density of the metal.

Powdered metal parts are found in many products, ranging from automobiles, home appliances, and lawn mowers to farm equipment and sporting goods. Gears and cams, support mounts, racks, and an infinite array of specialty products are being produced each day using powder metallurgy.

Die casting

Like powder metallurgy, die casting is a process that also uses a permanent, or reusable, mold. In *die casting*, molten metal is forced under high pressure into a cavity in the die. Low-melting-point metals such as zinc, tin, or lead are frequently die cast. Die casting has been used in industry since the early 1900s, and is still widely used because it is one of the fastest of all of the casting processes.

Dies used in die casting are constructed of steel alloys. Normally, dies are in two parts: the cover half and the ejector half. The die set, with its die halves, forms a block that is locked into the die casting machine. Dies may be constructed with single or multiple cavities. These dies are expensive, often costing from $10,000 to $100,000 or more to design and machine.

There are two types of die casting machines: hot chamber and cold chamber. Both types operate by forcing melted metal into a forming chamber. The hot chamber machine is faster. It is capable of handling more than 900 shots of metal into the chamber per hour when using zinc. With small parts like zipper teeth, 18,000 shots per hour can be achieved.

Molten metal is normally forced into the die cavity of the hot chamber die casting machine at 2,000 psi. This pressure is maintained until the metal solidifies. The dies are then cooled by water or oil circulating through water jackets. Cold chamber machines also inject molten metal into the dies, but solidification is done in a cold chamber.

One of the disadvantages of die casting is the time needed to develop and produce die sets for a new product. Before a new product can be produced, the casting dies must be made by the tool and die maker. To shorten this waiting time and to reduce changeover time for removing and replacing dies, a number of suppliers offer quick-change die systems. One proprietary system, called the *master unit die method*, was developed by Master Unit Die Products, Inc. of Greenville, Michigan. The system is frequently referred to as a *Mud frame*. In operation, the system uses quick-change inserts (the actual die set) that can be mounted in a permanent frame or fixture that is attached to the press. Thus, the Mud frame stays in the machine. To change a die, the operator loosens four clamps, disconnects heating or cooling lines, then removes one insert and slides another into place. With this die mounting method, tooling costs are reduced because only the insert must be created, not the entire mold base. Additional cost savings are achieved through an increase in productivity resulting from less time spent changing dies. Master unit dies are available for both die casting and injection molding machines. Mud frames have become an essential part of the SMED (Single Minute Exchange of Dies) concept.

Die casting is an efficient process where high production rates and high strength products are desired. Products such as components for appliances, carburetors, and hand tools are often die cast. The process is also effective for combining parts to eliminate a later assembly step, saving time and cost. Visualize a product shaped like an ice cream bar, with two components, a body and a stick. A die could be designed to permit casting the body and stick together as a single unit.

Permanent mold casting

Permanent mold casting, also known as gravity die casting, is used with molten metal. Permanent molds are filled much like the die is filled in powder metallurgy. The major difference is in terms of the material that is used—in permanent mold casting, molten metal is poured into the mold, reaching the cavity through a gating system. When the part solidifies, the mold is opened and the part is removed. Often molds are hinged to permit easy removal of the cast part.

Since the molds used for permanent mold casting are used again and again, they must be constructed of materials that can withstand continuous heating and cooling. Molds are normally made of cast iron, steel, bronze, graphite, ceramic materials, or refractory metal alloys.

A simplified mold of the type used for permanent mold casting is shown in **Figure 14-41**. The metal is poured into the mold through the pouring

Figure 14-41. A mold used for permanent mold casting. This is a simplified cutaway example to emphasize its major parts.

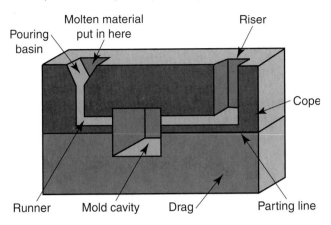

basin. The molten metal flows on into the mold cavity. When the cavity is filled, extra metal flows back out of the riser. This helps to ensure that the metal completely fills the cavity.

When permanent mold casting is used in industry, it is normally an automated process. This makes it cost-efficient to use as many mold cavities as possible. A typical production setup for permanent mold casting might consist of a circular turntable with ten or twelve individual stations. The casting machine would automatically blow out each mold to clean it and then pour in molten metal. The machine would automatically cool each mold, unlock the dies, and eject the casting.

Permanent mold casting is often used to cast parts from iron and nonferrous alloys. The process produces a good surface finish and maintains close tolerances. It is also fast and is suitable to high-volume production.

Permanent mold casting can also be used to make hollow castings. This adaptation of the permanent mold casting process is called slush casting. *Slush casting* is done by removing the casting from the mold when it has just begun to solidify. The part is inverted to pour out the metal that is still molten. The molten metal solidifies when it touches the cooler mold surfaces. The only metal that remains is the outer shell. This process is often used to make ornamental parts and decorative items.

Sand casting

Although sand casting dates back to ancient times, it is still used more frequently than any other casting process. In the United States alone, sand casting accounts for more that 15 million tons of metal poured each year. Typical products manufactured using sand castings include engine blocks and cylinder heads.

Sand casting uses a mold that is expendable, rather than permanent. The mold that is created can be used only once, because it is destroyed when the part is removed after casting. However, the sand is recycled to create another mold. With sand casting, a pattern slightly larger than the final product is placed in a rectangular box called a *flask*. The wood or metal pattern is made slightly oversize, since the casting will shrink during cooling. Moist molding sand with a binder is then packed into the flask around the pattern to create the shape to be cast. Entry and exit holes for the molten metal and pathways to carry it to the cavity are cut in the sand to complete the mold. The patterns are removed before the mold is used. If an opening through the cast part is needed (for a shaft, for example), a molded sand *core* is inserted in the cavity. The core will be broken out after the casting cools.

A typical mold configuration for sand casting is shown in **Figure 14-42**. The flask consists of two frames that fit together, called a *cope* and a *drag*. For many of the parts cast using this method, a *split pattern* makes mold preparation easier.

Figure 14-42. After the cope and drag mold halves have been prepared, the pattern is removed. When the mold is reassembled, its center contains a cavity in the shape of the pattern. If the part is pierced by a hole or other opening, a molded sand core is inserted. Molten metal poured in through the risers will fill the mold. When metal appears in the sprue, the cavity has been filled. Once the metal cools, the mold halves can be opened and the metal part removed. The sand mold is destroyed in the process; a new one must be made for each part that is cast. If a core is used, it is broken out after the part is removed from the mold.

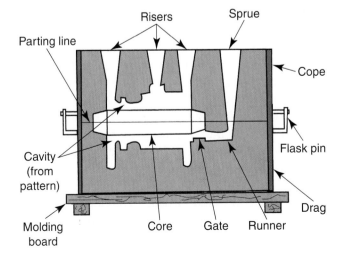

Here is how a typical sand-cast part is made:

First, the lower half of the split pattern is placed flat side down on a molding board. The drag is placed over the pattern, with the pins pointing downward. A fine powdered parting compound is dusted onto the pattern to aid in its removal from the sand. A screened sifter, called a *riddle*, is used to shake sand over the pattern until it is well-covered.

The drag is filled completely by pouring sand to the top, and the sand is firmly packed with a wooden rammer. A bottom board is then placed on top of the drag. The entire sandwich—bottom board, drag, and molding board—is turned completely over so that the molding board is on top.

Once the molding board is removed, the upper half of the pattern is set on top of the lower half that is already embedded in the drag. Pins or other devices are often used to align the two pattern halves. Then, the cope is stacked on top of the drag. Parting compound is dusted on the pattern to prevent the sand from sticking.

The next step is to insert sprue and riser pins in the sand about one inch from each end of the pattern. These pins are tubes that will make holes in the sand filling the cope to serve as pathways for the metal to enter and exit the mold cavity. Metal will be poured in through the sprue and the excess will flow out of the mold cavity through the riser.

Sand is riddled into the cope to cover the top half of the pattern. Additional sand is used to fill the cope and then compacted with the rammer. The sprue and riser pins are then removed. The cope is lifted off the drag and turned pattern-side-up. The pattern is carefully removed. On the surface of the drag, a one-inch wide gate, or pathway, is cut in the sand from the sprue to the cavity created when the pattern was removed. A similar gate is cut from the riser to the pattern cavity. The gates permit molten metal gases to flow into and out of the mold cavity.

At this point, the cope is once again placed atop the drag and the two clamped together so that the mold can be poured. Molten metal is poured down the sprue until it begins to exit the mold at the riser. This ensures complete filling of the mold cavity. After cooling, the sand mold is broken away and the casting removed.

There are many different types of patterns that can be used for sand casting. The split (two-part) pattern already described is popular. An interesting variation is the one-piece polystyrene pattern. This

type of pattern can be left in the mold sand because it is burned up when the molten metal enters the mold cavity.

Investment casting

Similar to sand casting, investment casting also uses a consumable mold. *Investment casting* is a precision casting method used with materials that are difficult to work or expensive to machine.

In investment casting, the cavity created by the pattern is destroyed when the casting is removed. Investment casting is also known as the *lost wax process*. The process was first used around 3000 BC, and is still in use today as a means of producing intricate jewelry, artwork, and ornate metal products.

A pattern is typically carved from wax or expanded polystyrene. Often a number of patterns are attached to each other, much like limbs on a tree. The finished pattern is then dipped in a ceramic refractory slurry.

The refractory coating is allowed to dry, and the process is repeated until the desired wall thickness is achieved. The entire assembly can now be viewed as a mold.

Next, the mold is placed in an oven to melt the wax out of it. The refractory shell remains intact after the wax is melted. Now the mold is ready to be poured. The metal is poured and allowed to solidify, then the refractory mold materials are broken off and the casting removed.

Dentists are one of the most extensive users of investment casting. They use the process to cast silver, gold, stainless steel, and other alloys for making dental bridges and crowns.

Centrifugal casting

Centrifugal casting is a process that is used to mold cylindrical parts from plastics and metals. Centrifugal casting is basically the same process when used with either of these materials.

This casting method relies on centrifugal forces created by rotating a cylindrical part to distribute molten metal against the walls of the mold and into the mold cavities. The process is popular for producing tubular parts such as pipes, tanks, and bearing rings.

The process of centrifugally casting a long hollow tube that is partly closed at each end begins by lining the tube with a slurry of ceramic refractory material. Once the refractory coating is in place, it is dried and baked. This is similar to investment casting. Next, the lined mold (tube) is spun rapidly and the molten metal is poured into it. The spinning process continues until the metal solidifies.

Figure 14-43 illustrates centrifugal casting of tubular parts using a process developed by American Cast Iron Pipe Company of Birmingham, Alabama. The photograph illustrates the casting of a steel tube with close coordination of metal temperature, pouring rate, and speed of rotation.

Conclusion

In this chapter, we have seen that forming metals includes processes that stretch, squeeze, or melt metal. Selecting the appropriate forming process will depend on the metal being used, the number of parts to be made, and the rate at which they need to be made. A part such as a roofing nail could be made by several of the processes described in this chapter. However, the volume of nails and the rate of production will narrow down the field of competing processes to just a few. In addition to production rate and volume, cost is an important consideration. As an example, someone might consider producing roofing nails using die casting. A die casting machine could produce a functional roofing nail. However, it would not be able to compete effectively based on production rates.

Another consideration is the effect the forming process will have on the mechanical properties of the resulting part. Some of the processes can improve the strength of a part while some would make the part brittle. A support made from cast iron will have good compressive strength and natural lubricity but poor tensile strength. However, a part that is forged from steel and looks identical will have different properties that will serve the purpose well. Although the type of material makes a difference, the process has a major role in determining a parts ultimate performance. Therefore, when selecting a forming process, people involved in part design need to work closely with their counterparts who design and operate the manufacturing facilities.

Important Terms

blank
breaking point
bulging
bullnose forming
centrifugal casting
closed-die forging
coining
cold forming
cold heading
cold-roll forming
computer numerical control (CNC)
contouring
contour roll forming
cope
core
cupola
deep drawing
die block
die casting
die set
direct extrusion
drag
drawing
electromagnetic forming
embossing
expanding
explosive forming
extrusion
flash
flask
follower
forming block
forming tool
fracture strength
green compact
hand forging

Figure 14-43. Centrifugal casting is used to create tubular parts. (ACIPCO Steel Products)

headers
high-energy-rate forming (HERF)
hot extrusion
hydroforming
hydrostatic extrusion
impact extrusion
impregnation
infiltration
ingot
investment casting
knurling
lost wax process
marforming
master unit die method
Mud frame
near-net-shape
necking
nosing
open-die forging
P/M injection molding
parting line
permanent mold casting
plastic flow
powder metallurgy (P/M)
punch
punch press
radial forging
riddle
roll bender
roll bending
roller forming
roll-on marking
rotary coining

rotary swaging
sand casting
shot
shot peening
sintering
sizing
slush casting
spinning
split pattern
stamping
stationary mandrel method
stretch draw forming
stretch forming
tapping a heat
thread rolling
three-roll double-pitch machine
two-roll machine
upsetting

Questions for Review and Discussion

1. What are some of the advantages of rolled threads versus machined threads?

2. Explain the process of upsetting, as used to make rivets.

3. Describe the major press-working processes. What do they all have in common?

4. Does the selection of a forming process influence the performance of a part even if the process can produce a part that appears to be identical? Explain your answer.

Processes Used to Form Plastic Materials

Chapter Highlights

- There are two major types of forming processes used with plastics, casting and molding.
- The basic closed molding processes are injection molding, compression molding, rotational molding, blow molding, and extrusion.
- Open molding processes are used to make products such as boat hulls and shower enclosures.
- Blow molding processes are often used for high-volume production of bottles and other containers.
- Lamination and calendaring processes are used to create polymeric "sandwiches" for thin-walled and sheet products.

Introduction

One of the advantages of using an engineered material such as plastic is that stock can be purchased in many forms: sheets, pellets, powder, granules, and rods. There is enough variation in types of raw materials to accommodate almost any manufacturing processes. Another important reason for considering plastics is that the process that forms the part produces a near-net-shape part.

With many materials, shrinkage is a condition that must be taken into consideration. Molds or dies often must be constructed oversize to compensate for this shrinkage during curing or drying. Sometimes, the estimate of shrinkage by the design engineer or manufacturing engineer is inaccurate, resulting in rejected parts.

With plastics, there is normally no scrap generated. When a part is injection molded, the *runners* (pathways that carry plastic to the mold cavity) must be trimmed off. In most cases, however, this material is ground up and reused. Trimmed material and rejected thermoplastic parts are melted and recycled. Rejected thermoset parts or scrap can also be ground and recycled for use as filler. This is particularly useful when constructing composite matrices.

Forming Processes

There are hundreds of processes used in industry to form plastic, but most of these are slight variations or hybrid combinations of eight basic forming processes. These are:

- compression molding
- injection molding
- rotational molding
- blow molding
- thermoforming
- extrusion
- hand layup
- casting

Plastics are extensively used to form composites. Polymeric matrix composites (PMC's) are the most common type of composite structure. Refer to the chapter on forming composites for additional information.

Each of the first seven processes listed above generate parts using some form of molding. Clarification of terminology may be helpful here to differentiate between molding and casting processes. The term *casting* is normally used to describe a process in which a liquid resin is poured into a mold and solidifies. In *molding*, the material is normally in a softened, semi-solid but not liquid state. The product is then formed into the desired shape in a mold or die, using pressure and sometimes, heat.

Closed Molding

Many molding processes utilize a two-piece closed mold. Some molds are more complicated to enable production of large and unusual shapes, and are constructed with more than two parts. There are also open-molding processes that do not require a conventional mold at all.

Closed molding processes are those which compress stock, in a variety of different forms, normally between the halves of a two-piece mold. In the plastics industry, both mold halves are referred to as the die, or the tool. **Figure 15-1** shows a plastic injection molding die used to make a Frisbee. The die half on the right-hand side of the photo shows the sunken part surface used to form the top of the Frisbee®.

When a closed molding process is involved, the mold halves are opened to expedite filling and part removal. The closed mold halves are normally held in a press, which provides pressure on the part during cooling or curing (depending on the type of plastic).

Figure 15-1. Closed Molding die for plastic injection molding used by Pitt Community College Manufacturing Engineering Program, Greenville, NC to make a Frisbee.

Once a thermoplastic part has cooled, or polymerization of a thermoset has occurred, the mold halves open and the part is removed.

In many cases, two-part matched molds are used to compress sheet molding compound, sheets, or preforms into the space left between the mating male and female dies. This is a general classification that is called *matched-mold forming*. Sometimes this is referred to as *compression molding*.

When matched molds are used, the plastic stock is heated while being held tightly in a horizontal position between the two mold halves. Holes through the mold permit heat to circulate in the die halves. After forming is complete, these holes also provide channels for circulating water to hasten cooling. **Figure 15-2.**

Figure 15-2. Matched mold forming is the process of choice when fine surface detail and close product tolerances are involved.

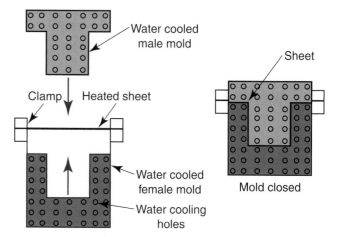

Matched mold forming is an excellent process when fine detail and close parts tolerance are desired. Grained surfaces and lettering can be produced with this process. Conventional metal stamping presses are often used for matched mold forming.

A mold with matched male and female halves is used. The female mold has a sunken surface (*die cavity*), while the male mold has a raised surface that functions as the core for forming the part. Other closed molding techniques incorporate the use of two-piece molds with halves that are identical. Each type of mold will be described in more detail in conjunction with the molding processes with which it is used.

There are many ways to classify the basic forming processes for further study. One approach is to group all of the closed-molding processes according to the way that the stock is applied to the surfaces of the mold. The five major closed-molding processes are:

- Compression molding
- Injection molding
- Rotational molding
- Blow molding
- Extrusion

Each of these processes uses some type of press or support device to compact the mold halves tightly against the raw stock to produce the part. A variety of different types of stock may be used with closed molding processes, including:

- Roving.
- Pellets or tablets, or shapes compressed from powdered polymers. They are usually referred to as *preforms*.
- Cores of partially cured resin with reinforcement, known as *prepregs*.
- Sheets of resin, usually referred to as *sheet molding compounds (SMC)*.
- Materials with the consistency of putty called *bulk molding compounds (BMC)*.

It is important to understand a bit more about the different types of stock used with plastic forming processes. Each type of stock has its own special characteristics, advantages, and weaknesses, and must be properly selected for use with a particular manufacturing process.

Fiberglass roving consists of twisted or woven glass fiber strands. Each of the individual strands is prepared by either slight twisting or straight drawing. There are two methods for producing these strands. They are either woven together to make woven roving, or fed into a chopper gun to be sprayed in chopped form onto resin. The machine shown in **Figure 15-3**

Figure 15-3. This machine weaves almost 300 strands of fiberglass roving at one time. Roving is used as reinforcement for many types of plastic products, including boat hulls, vehicle parts, containers, and sporting goods. (PPG Industries)

is producing nearly 300 strands of fiberglass roving simultaneously. These strands may later be woven into roving fabric, made into a loose reinforcing mat, or fed to a chopper gun for spray application.

Roving is different from the yarns used primarily for weaving cloth. Fiberglass roving is used in plastic fabrication processes because it has higher yield than would be possible with woven yarns of other materials. The yield of a roving is normally from 197 to 980 yards per pound. Fiberglass roving is able to generate more yards of product and still maintain strength requirements than yarns made of most other materials.

Woven fiberglass roving is stronger than conventional yarns, and thus requires less weight than solid materials to provide the desired strength. This makes woven roving a good choice for many fabrication applications involving polymeric and composite materials.

Sometimes roving is applied around the edges of large products to provide extra edge strength. In addition to being the primary material used in hand layup or open molding, woven roving is also used in processes such as filament winding and pultrusion. Roving is also used in mat or woven cloth forms, with molding compounds, in the production of filaments for winding, and in the production of rod stock.

Preforms are the raw materials used as stock in the injection molding machine. They are used with compression and extrusion molding processes as well. The polymer pellets are simple to store and flow easily though equipment. These preforms do not have any special resin coating.

A *coated preform* is considerably different in size and composition than the pellet that would be used in an injection molding press. A coated preform is made by spraying chopped fiberglass and a binder material onto a form. Once the resin has set, the coated preform is rigid enough to be moved to a press for molding. Large parts, such as boat hulls or vehicle body panels, can be produced using coated preforms.

A prepreg is another type of preform. Prepregs consist of a reinforcement material that has been impregnated with liquid thermosetting resin and cured to what is called the *B-stage*. At this stage, the preform is dry, but still slightly tacky. Since prepregs are only partially cured, they can easily be re-formed in a mold. At this point, the resin will behave like a thermoplastic until sufficient heat and pressure have been applied to cause cross-linking. The resulting product will be a thermoset.

Prepregs are sometimes made from continuous-strand woven fiberglass roving. Other materials, such as paper, asbestos, aramid, carbon fiber, and melamine are also used to produce prepregs. Integrated, fiberglass-reinforced composite prepregs can even be made from sheet molding compounds.

Sheet molding compounds (SMC) are sometimes called flow mat or resin sheets. The compound is made by combining layers of resin sheets, reinforcement, fillers, fabric, and additives on automated, continuous-flow machines. The sheets are then rolled into coils, with each layer separated with plastic film to prevent **autoadhesion**. Autoadhesion, or self-adhesion, refers to a strong bond that some surfaces have when they come into contact with each other.

Sheet molding compound is produced by chopping roving into strands up to 2″ long. These strands are then deposited onto a paste coating of polyester resin spread on the lower film carrier. After the first layer of coating is applied, a second film carrier spread with more resin paste is applied, creating a laminated sandwich of fiberglass and resin. The semi-tacky SMC sandwich, with protective film top and bottom, is squeezed to the desired thickness between rolls on the winding station of the machine. The rolled SMC is stored on the film until it is needed.

Composite prepregs are also made using bulk molding compounds. BMC prepregs are produced in the form of a putty consisting of resin, fillers, reinforcement, and other additives. The prepreg is normally produced in the shape of a slug or continuous rope. Sometimes the prepreg emerges in the form of H-beam-shaped slugs that can be automatically fed into forming dies. Bulk molding compounds are used with injection molding, compression molding, and extrusion processes.

Compression molding

Closed molding processes using matched molds can be classified as compression molding. Compression molding is one of the oldest and simplest of all of the closed molding processes. Compression molding can be accomplished using either thermosetting or thermoplastic polymers, but thermosets are popular.

There are two types of compression molding—hot and cold. However, this descriptive terminology may be deceptive. Both forms utilize heat to speed the rate of curing.

With *cold compression molding*, also referred to as *cold press molding*, the material to be formed is placed between the matching opened halves of unheated male and female molds. The material might be roving and catalyzed resin, sheet molding compound, or bulk molding compound. **Figure 15-4** illustrates a typical setup for cold press molding. In a typical setup for cold press molding, the mold halves are opened for loading then closed for forming and curing. Sometimes the mold is opened slightly during curing to permit gases to escape and reduce the formation of air bubbles in the part.

With cold press molding, the part remains in the closed mold and is transferred to an oven for curing.

Figure 15-4. In cold compression molding, the unheated mold closes on molding compound or other material to form it into the desired shape. Curing is sometimes done by moving the still-closed mold into an oven.

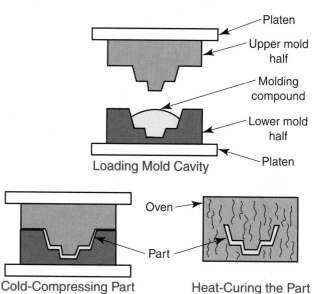

One advantage of cold press molding is that the part that is produced has two smooth, dimensionally accurate surfaces. If the color of the stock needs to be changed, pigment is mixed with the resin before molding.

Since heat speeds the action of the catalyst, many manufacturers prefer to use *hot press compression molding*. In this process, a plastic mixture of resin, reinforcement, filler, and additives is placed between die halves that are heated to between 225°F and 325°F. The die is then closed by pressure that may range from 100 psi to 2000 psi. Curing is accomplished while the mixture is held in the mold, and takes from a few seconds to five minutes, depending on the material. With hot press compression molding, the die halves are constructed with heating and cooling vents to first activate the catalyst and then increase the rate of cooling.

Compression molding is also done using preforms made from sheet molding compounds. This is a popular approach when complex parts with features such as ribs or bosses, or parts with intricate details, are desired. Pressures of 800 psi to 2000 psi (5516 kPa and 13,790 kPa) and temperatures of 275°F to 350°F (135°C to 177°C) are used as SMC material is squeezed between the die halves for several minutes. When compared to cold press molding, pressure and heat improve the flowability of SMC into small channels or voids in complex parts. Hot press compression molding is a process that can produce parts with good surface appearance and close dimensional control.

Compression molding is often used when parts are needed in medium- or high-volume quantities. Compression molding is used to make car body parts, appliance components, truck liner panels, office machine housings, tote boxes, building panels, dinnerware, and transformer cases.

Injection molding

Injection molding is a high-volume production process used primarily with thermoplastic materials, but it can also be used with thermosets. Injection molding is an important process for the plastics industry. If it were possible to retrieve a sample of each plastic product made today throughout the world and classify the products according to the process used to manufacture them, it is likely that injection molding would be listed as the top producer.

With the injection molding process, a measured amount (shot) of liquid plastic is forced into a heated die cavity. While both compression molding and injection molding use a closed mold, the molds are quite different. **Figure 15-5** shows how the injection molding process works.

As shown in the illustration, pellets of thermoplastic stock flow from a funnel-shaped feed hopper into a heated compression cylinder. Once the pellets are melted to a liquid state, a plunger forces a controlled quantity of material through the injection molding nozzle into the closed mold. When the viscous liquid is injected into the mold cavity, it forms a part of the desired shape. Often, the mold will be cooled by water running through cooling channels. After the part cools (often a matter of only seconds), the mold is opened and the part is ejected or removed by robots or some form of simple automation.

Robots are gaining popularity for removing materials from injection molding presses. Robot unloaders are spinoffs of early devices called mold sweeps, which were used to push the runner out of *three-plate molds*. The simplest type of mold consists of two plates. The runner system is located on the parting plane, where the molds separate when they are pulled apart. With a three-plate mold, there are two parting planes between each of two plates. This means that the runner system can be located on one parting plane, and the part on the other. 3-plate molds provide more flexibility in the type of gating location. The part can be located just about anywhere on its surface.

Sprue-picking robots are also used today to remove sprues and runners (trimmed recyclable waste) from opened molds.

Figure 15-5. Thermoplastic materials that have been heated and softened are forced into the mold cavity by the injection molding plunger. After the part cools, the mold opens and the solidified part is ejected or removed. The molding cycle is repeated.

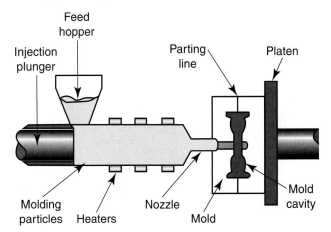

Both thermoplastics and thermosets can be injection-molded. The molding of thermoplastics is more common. With thermosets, molding compound is squeezed from a low-temperature barrel into a closed mold. The matched halves of the mold are then heated to polymerize the thermoset material. After the part has cured, the mold is opened and the part ejected.

When thermosets are heated they soften, then harden to an infusible state. Because of this, it is critical that no material remains in the heating chamber of the injection molding press or it will set. It is important to liquefy the material just before it is squirted into the mold cavity. Because of these concerns, it is much simpler to use thermoplastics with injection molding. See **Figure 15-6**.

Some injection molding machines have a reciprocating screw-type barrel that transports pellets through heating stages before the material is injected into the mold. Other systems use a plunger that forces the stock around a heated mandrel called a torpedo. The reciprocating screw machine is the most popular.

Injection molding machines are rated according to the maximum number of ounces of heated polymer that can be shot, or injected, into the mold with one stroke of the injection ram on the press. Machines are also classified by the force that holds the molds closed. This is often several hundred tons.

Injection molding is a high-speed production process that can quickly produce large quantities of small parts. It is frequently used to manufacture electrical equipment housings, cell phone cases, appliance parts, automotive components, microwaveable dishes, and thousands of other parts of all shapes and descriptions. In most cases, parts ejected from the mold require no finishing, other than trimming off sprues and runners.

Sometimes two or more materials are injected into the mold to make an individual part. This is called *coinjection molding*. Coinjection molding is suited to manufacturing applications for "soft-touch" products where a core must be made of one type of material and the outer shell from another. Products such as plastic cases for some cordless hand tools and armrests for many automobiles are made with this process. The case may be one color and type of material, and the handgrip may be another color and type of material.

Coinjection is a cost-saving process enabling use of lower-cost materials such as recyclable, unpigmented, or foamed resins, into the core of a two-material part. The process involves the use of a solid screw with a second hollow inner screw. Two separate melt streams are produced, and these materials are then layered one on top of the other, at the end of the barrel.

Rotational molding

Rotational molding, or *rotomolding*, is used primarily to make seamless hollow products such as balls, containers, picnic coolers, floats, and toys. Other product applications are storage and feed tanks, agricultural sprayers, automotive dashboards, chemical storage tanks, hot tubs, trash containers, and fuel tanks. Products of almost any size can be made with this process. Historically, rotomolding has been used with thermoplastics, but the process is now being used with some thermosets, as well.

According to *Plastics Technology Online* (1999) rotational molding is the fastest growing process in the United States plastics industry, having grown at a rate of nearly 10 percent per year during the past decade.

Web Site
www.rotomolding.org

The major trade organization for rotational molding is the Association of Rotational Molders (ARM) in Oak Brook, Illinois.

In the early days of rotational molding, the process was limited to using polymer in either powder or a liquid form known as *plastisol*, in a two-piece

Figure 15-6. This injection molding machine forces material around a heated mandrel or "torpedo." Note the heating bands around the cylinder, and the cooling channels through the mold. Use of a cooled mold allows increased production rates.

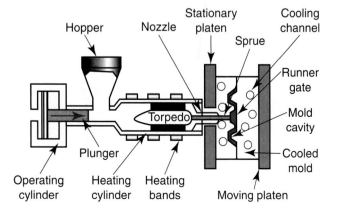

aluminum mold. The mold was then closed and heated, usually to between 500°F and 800 °F (260°C and 427°C), while being rotated simultaneously around two perpendicular axes. See **Figure 15-7**.

With rotational molding the powder or liquid coats the mold's inside surface, where it forms a thick skin or uniform layer. After the part is formed, the mold is indexed to a cooling station where air or water is used to cool it. The mold is then opened, the finished part removed, and another charge of powder or liquid is deposited for the next part.

Production equipment for rotational molding normally consists of a set of three arms. At any given time while the machine is operating, one arm will be in the load/unload cycle, another in the heating/rotation cycle, and the third in the cooling cycle. Some machines are capable of producing tanks as large as 500 gallons (1893 liters) in size. The latest shuttle-type machines have produced tanks with capacities exceeding 22,000 gallons (83 277 liters). With this type of machine, a large motorized shuttle cart travels on tracks from a heating oven to a cooling chamber.

One of the advantages of rotational molding is that several molds can be placed in the machine at the same time. Plastic resin is automatically loaded into each mold, and the molds are moved to an oven where they are slowly rotated. The melted resin sticks to the hot mold and coats the surface consistently. When the parts become cool, they are easily released from the mold.

There are several advantages of rotational molding over other molding processes to produce thin-walled hollow products. An important advantage is that the process produces a one-piece part, rather than requiring an assembly. Another advantage is that tooling is less costly than would be required for injection or blow molding. This is because there is no internal core to manufacture. Also, since no pressure is involved, less expensive materials can be used to produce the tooling. Rotational molding can be used to make just about any size product.

One innovative boat manufacturer, Triumph Boats, of Durham, NC is using a variation of rotational molding called Roplene™ to make boats which are so durable they will resist the impact of a sledge-hammer, and not chip or crack. The process produces the same color throughout the thickness of the wall and it won't blister or fade. Since Triumph boats are made of polyethylene, if they are damaged, they can be repaired with the heat of a propane torch.

Blow molding

Blow molding is a process used for producing hollow thin-walled and medium-walled thermoplastic products. Most blow molding is used to make packaging products such as household bottles and containers for cosmetics, pharmaceuticals, chemicals, foods, and toiletries. Blow molding is also a popular process for making toys and other hollow objects. Many unusual products are made with blow molding. One of the fastest growing applications is the use of blow molding for flexible containers to hold liquids. Flexible containers for juice, milk, and other beverages are now made by most manufacturers.

Blow molding offers tremendous advantages for the production of thin-walled products. It is a process that can produce 60,000 half-liter water bottles per hour (Sidel, Inc.). This makes it possible for blow molding to keep up with the fastest filling machines in soft-drink bottling plants.

Blow molding is a growth industry with leading companies manufacturing equipment in many countries: SIG Corpoplast (Germany), SIPA (Italy), All-Right Machinery (China), Mag-Plastic (Italy), Beutelspacher (Mexico), Chum Power Machinery (Taiwan), and Luxber (Spain). All of these companies are pushing the limit for high-speed production

Figure 15-7. For hollow objects such as balls or storage tanks, rotational molding is ideal. The polymer, in liquid (plastisol) or dry powder form, is placed in the mold. As the mold is heated and rotated in two axes, the polymer forms a thick skin on the inside of the mold.

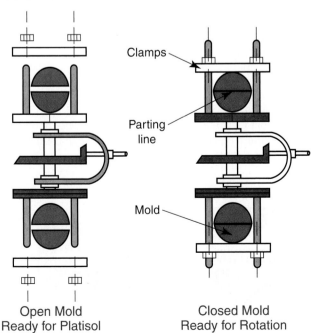

Clamps

Parting line

Mold

Open Mold
Ready for Platisol

Closed Mold
Ready for Rotation

using blow molding. As an example, SIPA manufactures a dual-sided SFR reheat machine, Model SFR 16/21, which can run 20-oz. soda bottles at a rate of 40,000 per hour. That means that approximately eleven bottles are produced per second.

Three major blow molding processes are now commercially used in the United States: extrusion blow molding, coextrusion blow molding, and injection blow molding. Extrusion blow molding is the most common of the three.

Each of the blow molding processes starts with a preformed *parison* (a hollow tube of heat-softened resin) that droops down and through an opening into a convex mold. With extrusion blow molding, an extruder feeding a parison head is used to produce the parison. When the parison is in location between the open mold halves, the mold is closed, crimping the bottom end of the parison. The top of the die also closes, but leaves enough of an opening for air to enter the mold. Air is blown through this hole into the parison, expanding the softened tube against the walls of the mold. Finally, the part is cooled and the mold is opened. See **Figure 15-8**.

Blow molding is a versatile process, used to make a variety of types of products. **Figure 15-9** shows how blow molding can even be used to form two sheets inside a mold into an irregularly shaped

Figure 15-8. Operation of a typical blow molding machine. (A) The heated parison is inserted between the halves of the opened mold. (B) The mold halves close, pinching off the top and bottom of the parison. (C) Air is blown into the parison, stretching its walls to fill the mold. (D) After the part cools, the mold halves open to allow removal of the blown part.

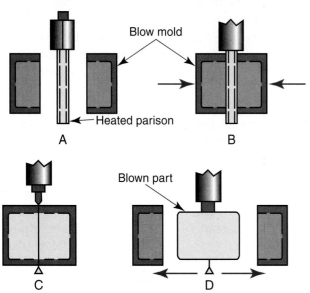

Figure 15-9. The indirect two-sheet blow molding process is similar to normal blow molding, but uses two sheets of material, rather than the parison. As shown, the heated sheets and an air tube are clamped between the mold halves (step 1), then air is injected to force the sheets to take the shape of the mold cavity (step 2). After cooling, the mold will open and the part will be ejected.

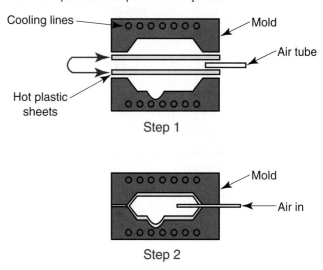

hollow container. This process is called *indirect two-sheet blow molding*.

With this process, an air tube is placed between two heated sheets in an open mold. The mold closes on the sheets, pinching off the ends. Air is blown into the mold through the tube, expanding the sheets against the walls of the mold. The same process can be used with plastic tubing instead of sheet stock. See **Figure 15-10**.

Another type of blow molding using heat is called *free blowing*. This process is popular for producing optically clear products such as acrylic

Figure 15-10. Indirect blow molding can also be used with plastic tubing, expanding it into a hollow container or other blown shape. The sequence is the same as that used with the two-sheet method.

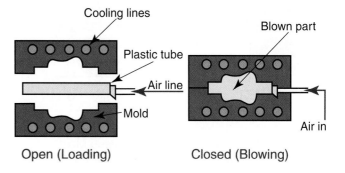

Figure 15-11. Free blowing is used to produce domed shapes of optically clear plastics. As shown, the heated sheet is clamped between the platen and a forming ring, air is then introduced from below to create the bubble or dome shape.

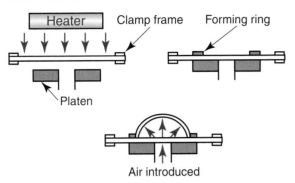

domed windowpanes, serving tray covers, and viewing panels. Free blowing does not use a mold. Instead, the heated plastic sheet is clamped between an upper forming ring and lower metal platen. Air is blown upward from a pressure box through a hole in the platen. The pressure forces the plastic upward into a bubble. See **Figure 15-11**.

A variation of the basic blow molding processes, called *foamed blow molding*, was developed in 1998 by the German company, Boehringer Ingelheim, and later sold to Clairant Corporation of Winchester, Virginia. There are still very few commercial applications of this process, but a variety of packaging and industrial projects are being developed. Most of these are in Europe.

Clairant developers attest to the fact that it can result in a savings of 10 to 30 percent in weight, resin usage, and total cycle time. Foaming of the resin is accomplished through the use of a nucleated foaming agent, based on reactions of baking soda and citric acid. This combination results in a release of carbon dioxide and creation of tiny voids in the part. Each of these results in less material needed for the completed part.

The developer, Boehringer Ingelheim, claims that the process can result in a 60 percent cost savings in material costs and improved cycle time. So far, the greatest potential for the free blowing process seems to be with thin-walled parts, where the walls cool much faster.

Companies that are leaders in the development and application of this process include: Trexel Inc. (Woburn, Mass.); Reedy International Corp. (Keyport, New Jersey); Clariant Additive Masterbatches (Winchester, Virginia); Inkutec GmbH (Gotha, Germany); and Wella AG (Darmstadt, Germany).

Injection blow molding

Injection blow molding is another hybrid process, combining the best features of injection molding and blow molding. The process is used to make bottles from thermoplastic resins.

One of the major advantages of injection blow molding is that parts are produced with no flash (rough edges where the forming seam occurs). This is not the case with either injection molding or blow molding—both these processes typically generate parts with flash that must be trimmed or final finished. When thermoset materials are involved, this results in scrap which is often of little value. Injection blow molding does not produce a seam where the mold goes together. The stock is injected into the mold surrounding a mandrel. Air is then introduced to force the stock against the mandrel and outside mold walls. Let us take a closer look at how this works.

Injection blow molding is a two-step process. First, a tubular parison is formed by injecting resin into a metal die cavity, as was the case with injection molding. The major difference here is that it encircles the blow stem (mandrel). The blow stem is used to create the neck of the bottle. See **Figure 15-12**. The parison is kept hot until it is needed for blow molding. The blow molding phase of this process begins when the machine indexes the preform from the parison station to the blow molding station. There, it is enclosed by the mold and blow stem. Air is then injected, and the desired shape is formed. The final step is removal of the blow stem, cooling, and ejection

Figure 15-12. The injection blow molding process begins by forming the parison using injection molding (top left). The fixture holding the parison then rotates, and the halves of the blowing station mold close over it. Air is introduced to form the blown object. High production rates can be achieved with this process.

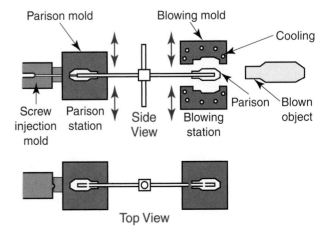

of the part. Injection blow molding is normally used to make smaller containers, with no handles.

There are many different types of injection blow molding machines. They differ primarily in the number of stations that they utilize. Three- and four-station rotary designs are common. Costs for injection blow molding machines range from $100,000 to more than $400,000. Today, 60 percent of all plastic bottles up to 16 oz. (0.47 liter) capacity are produced with injection blow molding.

Many producers are shifting from extrusion to injection blow molding because of its superior quality and productivity. Often, the polished mold cavities produce smoother bottle surfaces. Parts do not have damage caused by removing flash, because there is no flash. Another advantage of this process is the uniformity of wall thickness.

The major disadvantages of injection blow molding are the limitations of product size and design. Normally, only bottles with capacities smaller than 16 oz. are produced. Also, the process is not practical for making bottles with handles. Another disadvantage is that the tooling needed is more expensive than that required for other blow molding processes.

Thermoforming

There are many different thermoforming processes. However, all involve heating a thermoplastic sheet to its softening point and forcing the material into or over the contours of a one-piece mold. Large parts can be produced with thermoforming. See **Figure 15-13**. Some final trimming or material removal is needed to remove unwanted material on the outer edges of the part.

Figure 15-13. Thermoforming processes all involve forcing a heat-softened sheet of material into or over a mold, then allowing it to cool and assume the mold shape. The mold for this complex thermoformed part is cooled internally to speed production. The pipes around the edges of the mold base are tubes carrying coolant. (Aristech Chemical Corporation)

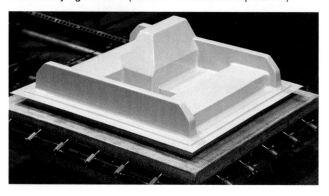

Figure 15-14. Vacuum forming is the most widely used of the thermoforming processes. As shown, a heated sheet of resin stock is clamped to the top of the mold, then a vacuum is drawn from below to permit the pressure of the atmosphere to force the material into the recesses of the mold.

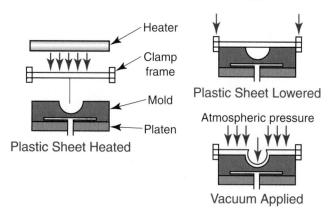

Vacuum forming

A plastic sheet is clamped and heated when using the process of vacuum forming. Then a vacuum is applied beneath the sheet, causing the atmospheric pressure to draw the sheet down against the cavity of the mold. See **Figure 15-14**. The plastic cools when it touches the walls of the mold. In vacuum forming, a male (convex) or female (concave) mold can be used. Refer to **Figure 15-15**.

Softening sheet stock for vacuum forming can be accomplished by heating the sheet in an external oven, then transferring it to the vacuum press. The material can also be heated on the vacuum press with infrared lamps.

The press used for vacuum forming consists of a box that holds the mold, with vacuum openings at the bottom. Clamps hold the sheet tightly against the top of the box. When the vacuum is applied, the sheet

Figure 15-15. Vacuum forming can be used effectively, even for fairly large products, such as this boat hull. (Aristech Chemical Corporation)

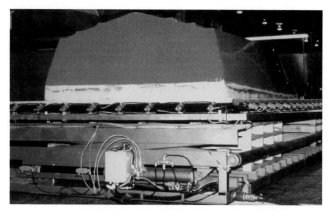

is pulled down over the mold, or the mold is pressed into the sheet. With vacuum forming, the side walls of the product are thicker in the edges, the area that touches the mold last.

One variation of the basic process of thermoforming is called *plug-assist thermoforming*. See **Figure 15-16**. As is the case with most thermoforming processes, the plastic sheet stock is heated to permit it to drape down into a female mold. In the plug-assist process, vacuum is used in conjunction with a cylinder-activated plug to depress the heated sheet into the mold. Plugs are made of materials with low thermal conductivity such as wood, syntactic foam, and cast thermoplastics. They can also be heated, in which case they are made of aluminum.

The plug-assist process is useful for producing parts that have deep draws (a high depth-to-width ratio). Luggage, containers, and some auto parts are products that are made using the plug-assist thermoforming process.

Another variation of the thermoforming process is shown in **Figure 15-17**. This process is called *mechanical stretch forming*. It is very similar to plug-assist thermoforming, except that no vacuum is applied. The plug is used to depress and stretch the stock into the female mold.

Figure 15-16. A variation of vacuum forming, plug-assist thermoforming, uses an air-powered plug or plunger to reinforce the vacuum action when manufacturing a deep-drawn part.

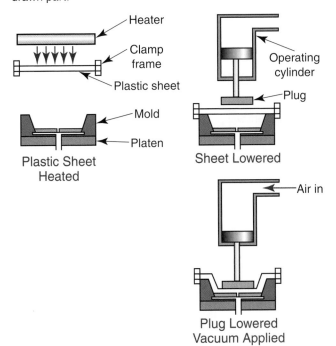

Figure 15-17. In mechanical stretch forming, the air-powered plug is the primary force for shaping the part. No vacuum is used.

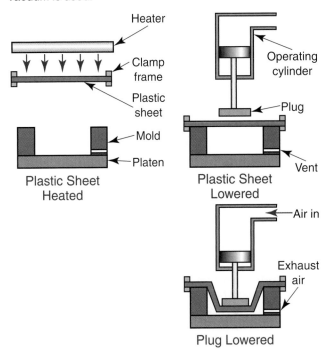

There are many different variations of the basic process of thermoforming. **Figure 15-18** shows how thermoforming is accomplished without vacuum, by using a pressurized chamber. This process is referred to as *pressure forming*. Here, the sheet is softened with heat and placed in the pressure chamber. Air is blown in, providing the pressure to force the stock against the walls of the female mold. See **Figure 15-19**.

Molds for thermoforming can be made of virtually any material. Popular materials are metal, phenolic paper laminate, plaster, wood, and ceramics.

Drape forming

Drape forming is a process that was designed to eliminate one of the weaknesses of vacuum forming: parts with inconsistent wall thickness. You will recall that the areas that touch the edges of the mold are thinner than other areas throughout the parts. With drape forming, the sheet is normally clamped, heated, and then drawn down over the mold. Sometimes the tool or mold is forced into the softened sheet. When the sheet droops, a vacuum is applied.

The unique aspect of drape forming is that after the sheet is softened, the outer edges are draped, or pulled down, over the mold. The use of both draping and vacuum results in more consistent wall thicknesses than other thermoforming processes.

Figure 15-18. Pressure forming uses air pressure to force the softened sheet stock into the female mold. Vents in the mold allow air to be exhausted from beneath the sheet as pressure is applied from above.

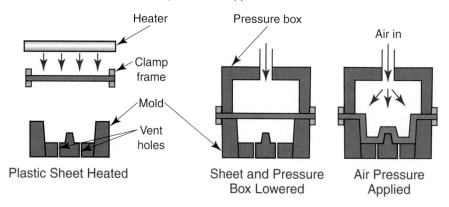

Plastic Sheet Heated

Sheet and Pressure Box Lowered

Air Pressure Applied

Figure 15-19. Pressure-formed products. (A) The surfaces of these electrical panels have been formed with varied textures for appearance and mechanical reasons. (B) Large parts, such as the dash and instrument console of this street sweeper, can be pressure formed. (PMW Products, Inc.)

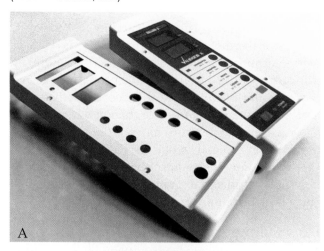

However, the points where the walls fold are still thinner than the other areas of the product. See **Figure 15-20**.

Extrusion

Extrusion is another important process in the plastics industry. It is responsible for the production of more plastic products than any other process except injection molding. In principle, extrusion is a process that is similar to squeezing toothpaste or construction adhesive out of a tube.

Figure 15-20. Drape forming is used to produce a thermoformed product that has a consistent wall thickness.

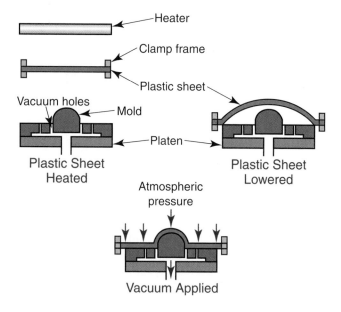

Plastic Sheet Heated

Plastic Sheet Lowered

Vacuum Applied

In this process, a machine called an *extruder* converts thermoplastic powder, pellets, or granules into a continuous melt. The melt is then forced through a die opening to produce long shapes. Typical products manufactured with the extrusion process include auto trim, house siding, garden hoses, rods, soft drink straws, and profile molding. Plastic insulation is also applied to electrical wire and pipe with the extrusion process.

The *single-screw extruder* is the most common type of machine used for extrusion today. Extruders force plastic stock down a long barrel. Machines are sized by the barrel's inner diameter, and range from 1/2″ to 12″ (13 mm to 305 mm) or more.

The process begins with stock being fed by gravity into a hopper on the extruder. The granules, pellets, or powder flow down into a vertical screw which rotates feedstock through the heated barrel of the extruder and into the die. The barrel is heated and cooled by heating/cooling jackets surrounding its outer wall. Microprocessor controllers regulate the temperature throughout different heat zones in the barrel.

After the melt is forced through the die, it must be carefully handled and cooled to minimize distortion. The method used to cool the end product depends on the shape of the material. Sheet products are cooled on carefully polished liquid-cooled rollers. **Figure 15-21** shows how pipes are made by extruding molten plastic through a die, then pulling the workpiece through a cold water bath.

Figure 15-21. Pipe is extruded by forming it over a mandrel as it exits the die. The mandrel is hollow, allowing air to be blown in to keep the soft plastic walls of the pipe from collapsing until it can be hardened by passing through a cold water bath.

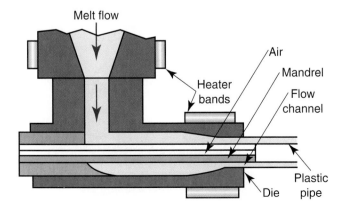

Figure 15-22. A mat of continuous fiberglass roving is being impregnated with resin and emerging as thick bundles or strands. These strands will be further processed into reinforced plastic products. (Creative Pultrusions, Inc.)

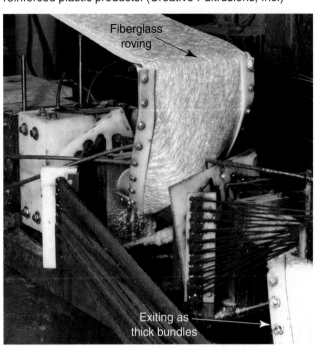

Pultrusion

Pultrusion consists of pulling continuous roving through three basic processing stages. The photograph in **Figure 15-22** shows how raw materials in the form of continuous roving are transformed into strands, or bundles, or pultruded material. The bundles are prepared by pulling the roving through a resin bath or impregnator. Next, the bundles are pulled into preforming fixtures where they are partially shaped and excess resin and air are removed. Finally, the bundles are pulled through heated dies for forming then into an oven for curing. The photograph in **Figure 15-23** shows pultruded bundles that have been processed into continuous lengths of plastic rod.

The manufacturing work cell used for pultrusion consists of a pultrusion machine, the forming dies, a pulling system, and a cut-off saw. The basic machine consists of a creel, resin tank, preforming fixture with a heated die, and a puller. The creel is a shelf-like arrangement holding spools of roving. If mat reinforcement is used, it is dispensed above the machine.

The pultrusion process starts when roving and mat enter the resin tank, where they are impregnated with polyester or epoxy resin. The resin bath,

Figure 15-23. A fluted rod of reinforced plastic is shown at the center of this photo, emerging from a pultruding machine. The large cutoff saw is used to trim the rod to uniform lengths. Inside the machine, bundles of resin-impregnated roving are pulled through heated dies to form the rod. After curing in an in-line oven, the finished rod emerges as shown, ready for trimming. (Creative Pultrusions, Inc.)

or "wet-out tank," is a trough containing rollers that force the reinforcement beneath the surface of the resin mixture. The impregnated resin leaves the wet-out tank through alignment slots in its end, and is pulled through a preform fixture to partially shape it. Preforming fixtures are often made of fluorocarbon or polyethylene because these materials are easy to form and to clean. Once the reinforcement leaves the preforming fixture, it enters a heated matched-metal die for final shaping. The shaped part can be cured in the die or in an oven. The part is then pulled into the pulling station. There are three different variations in design of pulling mechanisms: intermittent-pull reciprocating clamp, continuous-pull reciprocating clamp, and continuous belt or cleated chain. After the stock leaves the puller, it is cut to length with a conventional cut-off saw equipped with an abrasive wheel.

Many products made with the pultrusion process go to the sporting goods market. Fishing rods, hockey sticks, bike flags, tent poles, CB antennas, skateboards, arrows, and golf club shafts are pultruded. However, the bulk of the market for pultrusions (since introduction of the process in the United States in the mid-1950s) has been for electrical applications. Pultruded products in this area include ladders, switch actuators, fuse tubes, transformer

air duct spacers, and pole line hardware. One of the fastest growth areas for pultrusion is for corrosion-resistant products such as bridges and platforms, floor gratings, handrails, and structural supports.

Film blowing

Most polymer films are made using the *film blowing*, or *blown film*, extrusion process. The polymer is melted in a single screw extruder and pumped into a tubular die. Air is then forced in through a die mandrel to blow the tube into a bubble. This thins the walls to the desired final thickness, normally from 0.015" to 0.025". After the tube is cooled by chilled air, it is collapsed by *pinch rolls*. The tube is then slit to make thin sheets and is wound up or made into bags. Film blowing is used most frequently with polyethylene and polypropylene.

Extruded film can be differentiated from sheet stock in that anything less that 0.10" is classified as film. Film is produced in either tubular (blown) form, as just described, or in cast form. Cast film is extruded through a linear slot die and cooled when it comes into contact with chilled metal nip rolls. Most film produced is blown (tubular) film, which is usually stronger than cast flat film.

Wire coating

Extrusion is also used to coat wire. The wire coating process uses a die with a tapered mandrel in the center to keep the inside of the product hollow. Wire is pulled horizontally through the mandrel and the liquid plastic surrounds it. After the coating is extruded, it is cooled and inspected; the coated wire is then wound with a coil-winding machine. See **Figure 15-24**. Hollow tubing is made the same way, using a mandrel that extends through the die.

Figure 15-24. Extrusion can be used to add a coating (electrical insulation) to wire.

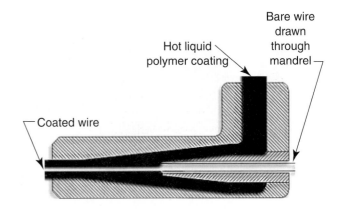

Extrusion blow molding

A process that combines features of extrusion and blow molding, called *extrusion blow molding*, utilizes pellets instead of sheets or tubes. With this process, heated plastic pellets, powders, colorants, and additives are mixed in an extruder to produce a uniform melt. The plastic melt is forced through an extrusion die, which forms a tubular parison. The parison is then pinched together by the top and bottom of the closing mold. The parison is expanded by air pressure to force the plastic against the surface of the mold. The part is cooled and the mold is opened.

Most extrusion blow-molded containers have flash. *Flash* is produced when excess material squeezes out of the seams of the mold. After the product is molded and cooled, the flash is trimmed and the product is complete. Automated flash removal systems are used with high-production blow molding machines. Bottles with necks are normally trimmed in the mold or machine.

Two different types of extrusion blow molding processes, continuous and intermittent, are popular. *Continuous extrusion blow molding* can be used with all blow-molded resins, but it is most often used with polyvinyl chloride (PVC) and other heat-sensitive thermoplastic resins. The parison in continuous extrusion blow molding is continuously formed. This method is used to produce large containers up to one gallon in size.

With the continuous extrusion blow molding process, the parison clamp mechanism, which squeezes the parison shut at the top and bottom of the bottle, is moved back and forth between the blowing station and extruder. Shuttle and rotary clamping systems are widely used. With the *shuttle clamping system*, once the parison reaches the desired length, the clamping mechanism shuttles from the blowing station to a position under the die head, where it surrounds the parison, trims it, and returns to the blowing station. The *rotary wheel system* provides as many as twenty clamping stations. This system enables a parison to be captured from the extruder while some parts are being removed and others are being cooled.

The *intermittent extrusion blow molding* process is used with polyolefins and other plastics that are not sensitive to heat. There are three basic types of intermittent extrusion blow molding machines: accumulator head, ram accumulator, and reciprocating screw. In this process, the parison is quickly formed by a clamp and part ejection system immediately after the product is ejected from the mold.

A reciprocating screw machine pushes the melt backward and forward by hydraulic pressure. This causes the melt to be forced through into the die head for forming the parison. These types of machines are used to produce bottles smaller than three gallons in capacity. Usually, the shot capability in these machines is limited to less than 5 lbs.

The ram accumulator machine accumulates the melt in an auxiliary ram cylinder located beside the extruder. The disadvantage of this machine is that the melt that first enters the cylinder remains in the cylinder for the longest amount of time. Since the melt is heated for different lengths of time, there is lack of uniformity in the material. The ram accumulator system can produce any size parison, but normally it is used for parts from 5 lbs. to 50 lbs. (2.2 kg to 22.7 kg).

The accumulator head system is the most popular intermittent process for extrusion blow molding. The extrusion die head, called an accumulator head, is filled by the extruder. The accumulator head has a tubular plunger, which forces the melt out of the cylinder. With this process, the melt that enters first exits the cylinder first. In a typical operation, melt would fill the cylinder from the side, and would then flow around the inside of a mandrel, reuniting on the opposite side. Some accumulator head systems are capable of extruding shots of up to 300 lb.

Open Molding

All of the processes discussed above use some form of closed molding where the die cavity completely surrounds the workpiece that is being formed. With open molding processes a different kind of mold is used.

There are two types of open molding, hand layup and sprayup ("spray layup"). Basic materials used in open molding include thermosetting resins (usually polyesters), glass fiber reinforcement (roving), and a catalyst (usually methylethylketone peroxide or benzol peroxide). Roving is used in woven or mat form for hand layup, or in strips several inches wide for use in chopper guns for spray layup. Open molding is an important process for constructing composite materials.

With both methods of open molding, the glass fiber reinforcement roving is encapsulated in resin. It is the roving that provides structural strength to the product. When the catalyst and resin are applied to the reinforcement, cross-linking takes place and a structural laminate is formed.

Open molding is a process that is widely used for making prototypes, pools, tanks, boats, ducts, truck and bus components, housings, and corrugated and flat sheet stock. Virtually any size product can be open-molded. It is one of the simplest of the polymeric processes. A minimal amount of equipment is needed, and the molds can easily be fabricated of wood. The major drawback of open molding is that it is time-consuming, resulting in high per-unit labor costs.

Hand layup

The *hand layup* process involves placing a layer of roving into a mold and then saturating the roving with resin. The process may be repeated with additional layers to provide the desired thickness and strength. Molds used for hand layup are often made of wood or plastic. The tool may be complex, incorporating prepreg fabrics, foam plastic cores, or honeycomb cores.

To better understand how the tool works, imagine how a typical product such as a shower enclosure would be made. The final product has a smooth inside surface (the part you see) and a rough outside surface (the part that would face the wall of your house). Imagine that the male mold for making the enclosure is resting on supports on the floor before you, and that the mold extends upward to form the shape of the enclosure. The exposed surface of the mold is highly polished so that a smooth exposed interior surface will result.

Hand layup requires the application of two distinct material surfaces—an aesthetic surface and an interior structural laminate. The process of making a shower enclosure would begin with cleaning and coating the mold with a mold release compound. A thin film material, such as cellophane, might be applied to the mold to achieve a smoother finish. A specially pigmented resin, called gel-coat, could be mixed with a catalyst and sprayed on to form the smooth finished surface of the enclosure.

The gel-coat layer is applied to achieve a thickness of 15 to 20 mil (about the thickness of heavy paper). The gel-coat surface eliminates the need for any final surface treatment or finishing.

Next, fiberglass mat is cut and placed in a single layer to cover the entire surface of the mold. Polyester resin and methylethylketone peroxide (MEKP) catalyst are mixed together and applied to the mat. The resin mixture is brushed or troweled on. In corners and other areas when saturation is difficult, a roller is used.

The gel-coat, fiberglass reinforcement, and resin cross-links and permanently bonds, forming a structural laminate. After a short curing time, the finished product is removed from the mold. The smooth-finished shower enclosure, an exact duplicate of the mold, is ready for trimming and shipping.

Hand layup is preferred because the resin is actually forced into the mat material by roller pressure. This results in a higher resin content, and a heavier and more durable product.

Sprayup

The *sprayup* process uses a chopper gun to spray strands of roving and catalyzed resin into the mold. There are four basic components in the sprayup system: the resin pump, the catalyst dispensing unit, the spray gun, and the glass-fiber cutter. See **Figure 15-25**.

The resin pump feeds resin to the spray gun. The catalyst dispensing unit can either be a catalyst injector or a pump designed for use with MEKP. The resin pump and catalyst dispensing unit are connected so that both cycle at the same rate. There are several different types of spray guns: common types are airless, air-atomized, internal mixing, external mixing, air assist, or airless.

A glass fiber cutter is mounted on top of the spray gun. The cutter chops continuous strips of roving into desired lengths (typically 1″ to 2″). Chopped strands are directed into the resin and catalyst stream. This allows all three materials to be deposited on the mold simultaneously.

Sprayup equipment is available with various sizes of tips and resin pump air delivery systems.

Figure 15-25. The sprayup method of open molding uses a gun that employs the principle illustrated here: streams of catalyst and resin are combined with short chopped lengths of roving and blown with pressure onto the mold to build up a structural layer.

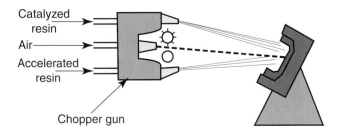

Catalyzed resin

Air

Accelerated resin

Chopper gun

Typical outputs of chopper gun systems range from 3 to 25 lb./min. The sprayup process is sometimes automated to maintain consistency of coverage. Sprayup is particularly popular for producing large tanks and pressure vessels.

There are other closed and open molding processes, but they are variations of the basic processes that have been discussed. There are also other plastic-forming processes that do not use a conventional mold like the ones used with these molding processes. Instead, they depend either on a mold for casting or setting up the material, or on the shape of the product to serve as a dish or boat holding plastic resin. These are called *casting* processes. Sometimes the process of classifying processes may appear to be arbitrary and confusing. It is true that processes used to cast sheets are similar to those used with extrusion. However, the behavior of the material being extruded or cast is completely different.

Casting

Some thermoplastics (nylons and acrylics) and thermosetting plastics (epoxies, polyurethanes, phenolics, and polyesters) can be cast in rigid or flexible molds. Sheets, rods, tubes, or other shapes are produced. Thermoplastics are more frequently used than thermosets for casting.

In order to cast thermoplastics, the monomer, catalyst, and necessary additives are heated and poured directly into the mold. With casting, no pressure is required in the mold. After polymerization occurs, the part is ejected from the mold. One of the advantages of the casting process is that intricate shapes can be made using flexible molds. The molds are peeled off after polymerization takes place.

Sometimes, castings are made by pouring a partially polymerized syrup into the mold, and then heating the mold to complete the polymerization process. One of the disadvantages of this method is that normally *shrinkage* (a reduction in overall size) takes place during the cure. Shrinkage can be as great as 20 percent. One way to reduce the extent of shrinkage is to add finely dissolved polymer to the syrup.

Sheet casting

The most common casting application for acrylics is the manufacture of sheet stock. Either cell casting or continuous casting is normally used to produce plastic sheets. Acrylic, or polymethyl metacrylate

sheets are used to make skylights, greenhouses, picture frames, surveillance windows and mirrors, sight glasses, animal habitats, bus shelters, and boat parts.

Sheets are produced in batches through *continuous casting*. Highly viscous syrup is poured in one end of two high-polished stainless steel belts. Heating and curing takes place while this material is slowly transported down the belt. When it reaches the end of its cycle, it is in the form of a solid sheet. The continuous casting process produces a product with higher molecular weight than extrusion. This results in harder surfaces, making acrylic less likely to be scratched with better impact resistance.

Width of the sheet is determined by the placement of flexible gaskets on the belts. Thickness of the sheet is dependent on the distance between the belts. Continuous casting is used for the production of thin sheets up to 0.375".

The sheet shrinks when it is cooled, but shrinkage is consistent in each direction. The amount of shrinkage depends on thickness and how the process is being performed. The thickness of the sheet also influences impact resistance. Thicker sheets are more difficult to form and do not have the same purity and weatherability characteristics of thinner sheets. Heated acrylic sheets have thickness limitations when it comes to stamping, shearing, and punching.

One interesting application of continuous casting is in the production of film (thin, usually flexible sheets of material). Film is cast by continuously drawing thermoplastic resin through a narrow opening in a die. Once it has left the die opening, the resin is cooled by passing through quenching and chill rolls. Film thickness is controlled by varying the feed control rate and line speed. Film is produced in thicknesses up to 0.005". Thicknesses greater than this are classified as sheet stock.

There are two major pieces of equipment used in producing thermoplastic film: an extruder and a film-casting machine. Supporting process control systems, filtering, trimming, and material handling systems are also required.

Cell casting is normally accomplished using plate glass sheets held together with spring clips to create a *cell*, or forming chamber. The glass exerts pressure on the plastic sheet being formed, and provides tension during cure. A flexible gasket of PVC tubing that separates the glass sheets determines the thickness of the sheet.

Many different types of plastics can be cell cast, but acrylic is the most common. The cell casting process is more versatile than continuous casting and color changes are easy. Cell-cast sheets have better optical properties and smoother surfaces.

However, cell casting is more labor intensive than continuous casting, making acrylic sheets more expensive than extruded sheets. Cell casting is normally used for small batches of 15 to 20 sheets.

Cell casting begins with construction of the sandwich, or cell. A catalyst is mixed with methyl-methacrylate resin and pumped into a blender. Here, ingredients such as plasticizers, modifiers, release agents, colorants, ultraviolet absorbers, and flame retardants and pigments are mixed together. This mixture is then poured into the cell, consisting of two vertical sheets of glass clamped together and separated by a gasket. When the cell is filled, the gasket joint is sealed and the mold is transported in a horizontal position to an oven for curing.

Thin cell-cast sheets are slowly cooled in the mold, using a forced-draft oven, for typically 12 to 16 hours. Thicker sheets are usually formed in a liquid bath, under pressure, in an autoclave. This keeps the monomer from boiling and reduces distortion.

The cell casting process does produce unique characteristics in the products. Gravity results in the pigments settling to the lower face of the mold. This results in poorer surface quality in the material at the top of the mold. A printed-paper protective sheet is normally adhered to the "good face" of a sheet to enable the user to identify the good side. When viewed from a low angle, there is less distortion, and the sheet looks much smoother. This means that the sheet can be viewed from about any angle with undistorted view, making it an ideal product for display cases.

Nylon casting

Nylon is a good material for use in applications requiring abrasion resistance and low weight. Nylon casting is often more economical than injection molding or extrusion. Nylon cast parts can be created in almost any size and thickness. There are four steps in nylon casting: melting the nylon lactam monomer, adding the catalyst and activator, mixing, and pouring.

Melting and mixing is conducted in a temperature- and humidity-controlled environment, since the ingredients are extremely *hygroscopic* (water-absorbing). Water absorbed from the mixture would cause the catalyst to decompose.

The lactam flakes are melted under very precise temperature and environmental conditions. Any additives to be introduced into the melt must be carefully dried to eliminate moisture. The melt is then transferred by gravity or mechanical means to the mold. Parts that will require machining after casting are often annealed to reduce brittleness. Plastics are annealed by slowly cooling the parts in circulating-air ovens or in tanks of mineral oil.

Potting and encapsulating

Potting and encapsulating are casting processes that are often used in the production of electrical and electronic devices and components. Both processes are used to protect electrical or electronic devices from moisture and damage by mechanical abrasion through use. There is often confusion in the way that the terms *casting, potting,* and *encapsulation* are used in the industry. They are similar but do have unique characteristics and differences.

Sometimes chopped fibers are combined with catalyzed resin and poured into molds to cast the desired product. In some cases the mold is reusable. In other instances, the mold stays with the product and serves as the actual case or housing. Potting and encapsulation processes are ideal for use with products containing wires, connections, or components that do not need to be accessible at some later time—once potted, or encapsulated, the components are sealed for life.

Potting can be done by applying potting compound using a dispensing gun, or through automation in a continuous production situation. When the process is automated, liquid polyester thermosetting resin is poured into the housing of the product and the resin and case become an integral part of the product.

Most often, potting is used to completely cover an electrical device. The device is placed in the mold or pot, and resin poured in until it reaches the top of the mold. Electrical transformers are often made with the potting process.

The distinguishing characteristic about potting is that the mold, or pot, usually stays with the finished product. The plastic resin serves as a dielectric (nonconductor). In some cases, chopped fillers are added to the resin to provide additional strength. Potting provides strength and protection to products containing wires or electrical components. Terminal blocks and electrical casings are among products that use potting to improve their durability.

The plastic shells or cases used with potted products are usually made of ABS, nylon, or PVC. There are three common types of resins used for potting: epoxy, urethane, and silicone. Acrylic potting compounds that cure through exposure to UV and heat may be used. Hot melt materials are also popular, but are not very resistant to heat and chemicals.

Encapsulating involves covering the component or part with resin, normally a potting compound. There are two different types of encapsulation. The simplest method of encapsulation is to just coat the area that needs to be protected with potting compound. This is normally done using a squirt application container, like a catsup container, or with an applicator gun.

When an entire component, such as a capacitor, is to be encapsulated, a process called *dip coating* may be used. This is frequently used to coat or seal devices and connections, and embed coils, windings, transformers, chokes, resistors, transistors, diodes, and other electrical components. Dip coating is accomplished by dipping the component in resin. When the potting compound sets up, the gel that is left serves as a permanent cover around the product. Sometimes the entire component is covered with plastic. In other applications, only the connections where the wires or leads attach to the component are dipped or encapsulated.

Another method of encapsulation is accomplished using a mold that is removed from the product when then product is cast. Some manufacturers use a two-part die of aluminum, coated with Teflon® to aid in removal of the part. Parts made from molds have a flash where the inside edges of the molds meet.

Calendering

Calendering is a process that was used more than 100 years ago to process natural rubber. Now, it is widely used in the plastics industry to manufacture polyvinyl chloride (PVC) film and sheeting, and single ply roofing membranes. This is also a process that is important to the papermaking industry.

Calendering involves pulling and squeezing pliable thermoplastic stock in a one or two pass operation between a series of turning rollers to produce the desired film thickness. Calendering is normally conducted in three stages: mixing, feeding, and calendering, followed by a post-calendering treatment. In some cases, a polyester fabric or glass fiber mat is used as reinforcement between two applications of raw material.

If PVC film is being produced, both solid and liquid raw materials may be required. Solids are conveyed to a high-speed mixer from silos or bins. Liquids are metered to the mixer. Mixers usually consist of planetary gear extruders. The mixing process is computer-controlled and fully automated.

Additives are introduced to the mixture, which is then heated to promote blending. The mixing of this dry blend is called *fluxing*.

When the mixture has reached a consistency of soft clay, it is fed from the extruder through a metal detector or strainer to filter out contaminants from the stock. It is then conveyed onto the calender rolls. Calendering is normally accomplished on machines with from four to seven cast-iron rolls, depending on the design configuration of the machine. The rolls are heated to keep the sheet pliable during squeezing.

After calendering, the film that has been produced is conveyed by pick-off or stripper rolls in a vertical position to a unit called the embosser. The embosser consists of three rolls: the embossing roll, a contact roll, and a chilled rubber roll. It is the embosser that imparts texture or pattern to the film or sheet stock.

Next, the film travels through sets of rollers for tempering (hardening) and cooling. These rolls are carefully temperature-controlled. At this point, the product is nearly completed. The only operation left is entering the winding machine, where the completed film is trimmed, cut, and spooled. The winding machine has its own computer and programmable controllers that regulate the roll size according to the desired weight.

Centrifugal casting

Pipe, tubing, and other round objects can be produced using the centrifugal casting process. Centrifugal casting is a process that is used to form ceramics, metals, plastics, and composites. The process is particularly useful with ceramics.

Fiberglass strands are saturated with thermoplastic polyester resin inside a hollow mandrel (tapered cylinder) that serves as the mold. Metal tubing is normally used as the mandrel. The mandrel is heated in an oven and rotated. Centrifugal forces throw the mix of resin and reinforcement against the walls of the mandrel. The mandrel continues to turn during the mixing and curing cycles. Curing is accelerated by pumping hot air through the oven. Products manufactured with this process include large tanks for agricultural chemicals, water softener tanks, pipe, tubing, and liquid storage tanks. Centrifugal casting can even be used to manufacture long cylinders with external threads.

The centrifugal casting process requires a minimum of labor and can be automated if high volumes are desired. It is a resourceful process with low tooling costs and little waste.

The major disadvantage of centrifugal casting is that it is limited to the production of cylinders with uniform thicknesses. Irregularly shaped or tapered cylinders with different wall thicknesses are difficult to produce.

Important Terms

autoadhesion
blown film
B-stage
bulk molding compound (BMC)
calendering
casting
cell
cell casting
closed molding
coated preform
coinjection molding
cold compression molding
compression molding
continuous casting
continuous extrusion blow molding
die cavity
drape forming
encapsulating
extruder
extrusion blow molding
film blowing
flash
fluxing
foamed blow molding
free blowing
hand layup
hot press compression molding
hygroscopic

indirect two-sheet blow molding
injection blow molding
intermittent extrusion blow molding
matched-mold forming
mechanical stretch forming
molding
parison
pinch rolls
plastisol
plug-assist thermoforming
potting
preforms
prepregs
pressure forming
pultrusion
rotary wheel system
rotational molding
rotomolding
runners
sheet molding compounds (SMC)
shrinkage
shuttle clamping system
sprayup
three-plate molds

Questions for Review and Discussion

1. What are the similarities and differences between hot and cold compression molding?

2. Select a plastic container such as a parts bin. What steps would you perform if you wanted to use this container as your mold for producing a part using hand layup? When and how would you add color, if desired?

3. What are the advantages and disadvantages of hand layup, compared to sprayup?

4. How is wire coating accomplished using the extrusion process?

5. Sheets of thin film are produced from tubular stock. What keeps the walls of the tubing from sticking together?

16 Processes Used to Form Wood Materials

Chapter Highlights

- There are two basic wood forming processes: bonding and bending.
- Hardboard, insulation board, and particleboard can be referred to as composition boards.
- Particleboards of various types are formed from pieces of wood and a synthetic resin binder.
- Reconstituted wood is one of the newest forms of engineered wood materials.
- Steam or soaking is often used to plasticize wood that is to be bent.

Introduction

There are three major types of wood forming processes—bonding, laminating, and bending. Bonding processes are used primarily to form sheet-type materials that are used in construction, to provide structural support in manufactured products, or to manufacture industrial stock. Laminating processes are used to construct sandwiches of layers of wood or wood and other materials. Bending processes involve the use of chemicals, heat, moisture, and tension to force solid or laminated wood materials into desired curvatures and irregular shapes. Wood can even be made so elastic that it can be tied into knots. See **Figure 16-1** for a systems model showing the major processes and products that will be discussed in this chapter.

Figure 16-1. Major wood forming processes and products.

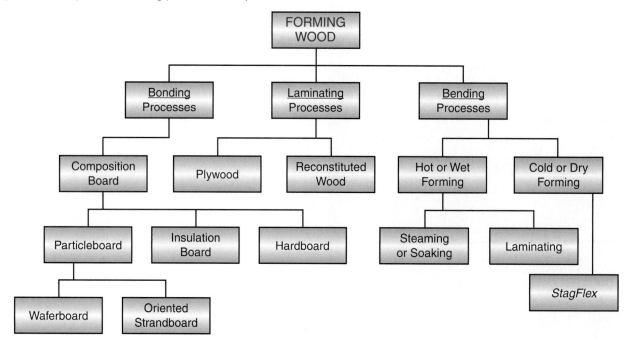

Bonding Processes

Bonding processes using heat and pressure to compact particles or chips into sheet stock are used in the production of wood composition board. *Composition board* is made from wood that has been broken down into particles or fibers and reconstituted to form products such as insulation board, hardboard, waferboard, particleboard, and oriented strandboard. Commercial development of composition board began as a means of using waste byproducts from the manufacture of paper. The manufacture of wood composition board takes advantage of many parts of the tree, even those parts that would have been considered unusable for making other types of wood products. Sawdust, planer shavings, and other wood residues are also used.

Composition Boards

There are three major types of composition board: hardboard, insulation board, and particleboard. In New Zealand and Canada, hardboard and insulation board are referred to as *fibreboard*. The medium density fiberboard (MDF) market has grown rapidly in the past ten years, with New Zealand being the world's leading producer of radiata pine-based MDF.

Both hardboard and insulation board are made by compacting and bonding fibers. The fibers in hardboard are so fine and tightly compacted that it is difficult to see them when looking at a piece of stock that has been sawn to make the interior composition visible. Insulation board is more loosely compacted, and thus the fibers are easy to see. Particleboard is not made from fibers. It is made using larger particles of wood and synthetic resins. Other types of stock are also made from particles and pieces of wood pressed together. The most popular types are waferboard and oriented strandboard.

Hardboard

Hardboard was first produced in 1924 by William H. Mason, founder of the Masonite Corporation. Mason discovered that small wood chips could be blown apart in a digester with high-pressure steam. The fibers that were produced were perfect for producing a continuous mat of pulp that could be pressed into a sheet of strong, hard material. The basic process developed by Mason is still used today. See **Figure 16-2**.

The U.S. Department of Commerce, in *Commercial Standard CS 251-63*, defines **hardboard** as a panel "manufactured from interfelted lignocellulosic fibers consolidated under heat and pressure in a hot press to a density of at least 31 pounds per cubic foot." Hardboard has high tensile strength, high density, and low water absorption. Often, the board has greater hardness and weight per cubic inch than the wood from which it was made.

Figure 16-2. The processing stages used in manufacturing hardboard. Wood is broken down into fibers, which are mixed with water to form a pulp that is pressed into sheet form. After further dehumidifying, the sheets are cut to size.

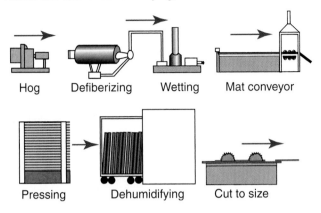

Hog   Defiberizing   Wetting   Mat conveyor

Pressing   Dehumidifying   Cut to size

Hardboard can be purchased in either tempered or standard form. *Tempered hardboard* has been impregnated with a resin and oil blend, which is stabilized during drying. Tempered hardboard has better strength, stiffness, and resistance to water and abrasion than standard hardboard. Tempered hardboard has a density of 60 to 75 pounds per cubic foot. Hardboard is used for floor underlayment, door panels, cabinet shelves, and furniture backs. Hardboard is also used as a core material that is overlaid with veneer. See **Figure 16-3**.

Hardboard has several properties that give it an advantage over many other products:

- Smooth surface finish, enabling wood grain to be printed directly on the material
- Dense composition, resulting in smooth edges when machined or cut
- Superior strength, permitting removal of screws from the material without damage to the material

A process called *wet felting* is used to make hardboard and insulation board. First, the wood material is broken up into chips in a gigantic chipper called a *hog*. Then the chips are softened with heat prior to refining. Refining is accomplished with a pair of large discs that rotate against each other at speeds up to 1800 rpm. Steam pressure and centrifugal force carry the chips outward, where cutting knives break the chips into fibers. The wet felting process uses a large amount of water. Plants producing volumes of 300 tons of hardboard per day use as much as 94 gallons of water (360 liters) of water per second in preparing the mat material. Some plants produce as much as 720 tons of hardboard each day.

The next step is to *defiberize*, or dry out, the particles so that most of the chemicals are removed. Water is then added to the dry particles so that they can be mixed further and form a pulp that is easier to handle. Once all of the pulp material is in dispersion, it is run off onto a moving conveyor. Here, some of the water is removed through absorption. At this point the material is dried to remove most of the water. Finally, the material is placed in a dehumidifying oven to dry it further. The final stage in the process is cutting the hardboard to size, usually $4' \times 8'$ panels.

A similar process for making hardboard is called *dry felting*. With this process, the wood fibers are produced in the same way as with wet felting, but the fibers are blown directly into the dryer where the moisture is evaporated. Resin is added to the fibers before they enter the dryer. Due to the high heat and mixture of the material, any small spark from a foreign object may result in spontaneous combustion. Dryers are constructed in such a way as to limit damage to the facility if this occurs.

Insulation board

Another type of composition board is called *insulation board*. This material has been used

Figure 16-3. This door is covered with a thin molded facing of hardboard that can be painted or stained for a natural wood look. As shown in the cutaway view, the facing is applied over a solid core of particleboard. Solid wood edging is used for additional rigidity. (Masonite)

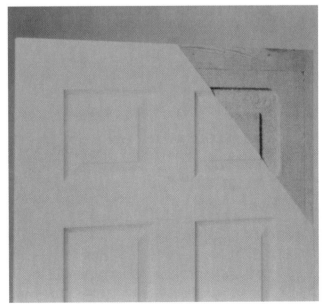

since 1914, when it was developed by Carl Muench, a refrigerator manufacturer who was working to find a use for residue from a friend's pulp mill. The machine designed by Muench was an adaptation of the papermaking machine and was capable of producing 3,000 square feet of insulation board per day. His process went through many refinements. The first plant producing Muench's Celotex insulation board was built in 1920 near Marrero, Louisiana. It is still the largest producer of insulation board in the U.S. Today, insulation board plants have achieved daily capacities of more than 3 million feet.

Insulation board has high resistance to heat transfer and is also low in density. Small air pockets in the board slow down the passage of hot or cold air. Insulation board is also good for reducing noise. As an acoustical barrier, it can absorb more than 70 percent of the noise that strikes it. Billions of feet of insulation board are produced for use by the housing industry each year. See **Figure 16-4**.

Insulation board is manufactured using pulpwood and *bagasse*, a cellulosic byproduct produced during the pressing of sugar cane. The process of making insulation board begins with feeding logs into a chipper, where they are first reduced to chips, then to fibers. Water and chemicals are added to turn the fibers into a continuous mat. The mats are fed into a drier, then cut and trimmed to size, in sheets from 16 to 26′ long.

Particleboard

Today there are many different types of wood polymer composite boards that are commonly called

Figure 16-4. The many small air spaces between the fibers of insulation board make it an efficient sound-absorbing material. Ceiling tiles and suspended ceiling panels like those in this photo are forms of insulation board. The exposed surfaces are often painted or laminated with a plastic material for a finished appearance. (USG Interiors, Inc.)

particleboard. Most particleboard is made from non-chemically processed dry wood particles in various shapes and sizes. See **Figure 16-5**.

Particleboard is popular in the construction and furniture-making industries. *Particleboard* was developed in Germany in 1943; it was first sold in the U.S. for construction use in the mid-1950s. As a manufacturing material, particleboard met with only

Figure 16-5. Many shapes and sizes of chips are used for particleboard.

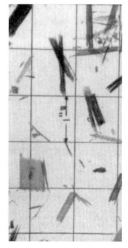   

Drum Cut Flakes   Planer Shavings   Sawmill Dust   Ring Cut Flakes

limited success until the 1960s, when users began to learn how to work with this new material. Today, more than one billion square feet of particleboard is produced annually.

About 40 percent of the particleboard used today is purchased by the construction industry for flooring **underlayment**. Most of the remaining 60 percent is used as core material by the furniture industry.

The U.S. Department of Commerce, under the *Commercial Standard CS 236-66*, defines particleboard as "a panel material composed of small discrete pieces of wood bonded together in the presence of heat and pressure by an extraneous binder." Particleboard can also be defined in terms of the method by which it is pressed. When the pressure is applied in a direction perpendicular to the faces, as in a conventional multi-platen hot press, the board is defined as *flat-platen pressed*. Most of the particleboard produced today is made using flat-plated pressing.

The process of making flat-platen pressed particleboard usually starts with the chips being prepared in the form desired and mixed with resin in a blender. The material is then transported to a mat forming station. In some cases the mat is prepressed. Then it is placed in a series of hot presses where the individual mats are compacted into a single board. The final step includes trimming the sheets to the final 4' × 8' sheet and sanding.

Particleboard is also produced through continuous extrusion. When the applied pressure is parallel to the faces, the board is defined as *extruded*. Extruded board is weaker across the length of the board and exhibits more lengthwise swelling during pressing. Most extruded hardboard is used as core material supporting veneered surfaces. When continuous extrusion is used to produce particleboard, small wood chips are mixed with resin. The mass is then forced through the opening in the extrusion die. With extruded particleboard, the chips and fibers are oriented in a plane perpendicular to the horizontal surface. This results in a stiffer product with decreased bending strength. In some cases, extruded boards are laminated with veneers or plastic to compensate for this.

Particleboard can be treated with chemicals to resist attack by insects or fungi. It can even be embossed or stamped with special surface treatments. At the same time, it should be remembered that this engineered product has the ability to absorb water readily, and thus can swell and warp. Without

surface treatment, it is not very useful in high-humidity environments.

About 90 percent of the materials used to manufacture particleboard come from wood-processing mills. The ingredients consist of planer shavings, sawmill chips, veneer wastes, and other residues from preparing wood stock. Sometimes, logs that are too small to be useful for other purposes are also used to produce wood flakes. After the wood flakes are reduced to the desired size and shape, they are mixed in a blender with synthetic **binders** (normally phenolic resins) and other chemicals. The resin-coated chips are then fed into a forming machine, which deposits them directly on a metal belt to form a mat. Mats are made in layers, using different-sized particles. Large particles are positioned in the middle to provide strength, and the small particles are layered on the outside to produce smoothness. The forming machine produces a dry mat. The mat is then conveyed to a hydraulic press, where heat and pressure are applied. This is normally about 400°F at 1,000 psi.

The Indian Plywood Research Institute, a subsidiary of the National Research Development Corporation (New Delhi, India), has developed an interesting new type of particleboard using rice instead of wood. The product is claimed to have superior characteristics over wood: termite resistance, high decay resistance, improved water resistance, and high durability. With their process, an adhesive bond is created between the rice husks, and then the mats are formed and pressed using heat, much like the conventional process of manufacturing particleboard. The process is currently being licensed to firms in India and Malaysia.

Waferboard and oriented strandboard

Waferboard (WB) and oriented strandboard (OSB) are also types of reconstituted wood. Sometimes these are referred to a *flakeboards*. Both WB and OSB are purchased in 4' × 8' sheets. The only difference between the manufacturing process of WB and OSB is in the forming processes used to distribute material on the final screen and prepare the final mat. See **Figure 16-6**.

Waferboard, or waferwood, is made using high-quality wood flakes that are about 0.028" thick, 1.5" wide, and 3" to 6" long. The flakes are bonded together under heat and pressure with phenolic resin that serves as a binder and waterproof adhesive. Both sides of waferboard have the same type of textured

Figure 16-6. Three of the common forms of composition board. From top to bottom: waferboard, oriented strandboard, particleboard. All are widely used in both building construction and in manufacturing. (American Plywood Association)

surface appearance. Waferboard is mainly used in the construction and furniture-making industries. Waferboard has a density of about 40 lbs. per ft^2. **Figure 16-7** provides a detailed explanation of this process.

Oriented strandboard is made from wood fibers, rather than particles or flakes. The fibers are large and irregularly shaped. Consequently, they are easy to see after the mat material is pressed into a sheet. As is the case with all particleboards, the materials are bonded together with resins and glues. Each successive layer of fibers is arranged at right angles to the preceding one. As is the case with other types of composition boards, waferboard and oriented

strandboard can be cut using common woodworking machines and tools.

Lamination Processes

Lamination

One of the most important processes to the woods industry is lamination. It is the same process that was used with plastic materials, but in this case, involves sandwiching sheets or pieces of wood together to make larger pieces of wood. Lamination is used extensively for the construction of plywood. It is also used to produce reconstituted wood. This is a relatively new engineered-wood product that can be purchased either as a veneer or in solid blocks. Veneer comes in very thin sheets cut from expensive wood. These sheets are used as the exposed (top, bottom, or side) layer that is glued over a core of less expensive wood (such as particleboard). Veneer is used when a fault-free, decorative wood with a particular color or grain is desired. It is less expensive than using solid hardwood. Most lower-end furniture manufactured today is made using veneer.

Reconstituted wood is made from plywood that is cut apart in pieces and reassembled. The strips that are glued together are often smaller than 0.030″. Strips this size are so small that they can scarcely be seen with the naked eye. When the wood is glued together, it is hard to tell that it is not "real" wood.

Figure 16-7. Steps in the production of waferboard and oriented strandboard.

Step	Operation	Description
1	Waferizing	Feeding of debarked log into a waferizer (slicer) to create wafers.
2	Wet Wafer Storage	Wafers are screened to separate core from surface material.
3	Drying	Wafers are dried to low moisture content (4–10%).
4	Screening	Wafers are conveyed pneumatically from the dryer, and then separated from the gas stream at the primary cyclone according to surface area and weight.
5	Blending	Dried wafers are conveyed to the blender, where they are mixed with resin, wax, and other additives.
6	Mat Forming	Resinated wafers are conveyed to the *former*, where they are metered out on a continuously moving screen.With WB production, the wafers are dropped randomly. For OSB production the wafers are oriented in one direction with the next application layer being deposited at right angles to the first.
7	Trimming	Boards are trimmed to finish 4′ x 8′ sheet.
8	Hot Pressing	Boards are compacted and hot pressed to create adhesive bond between wafers.
9	Final Finishing	Waferboard is trimmed, sawed, and edge painted.

Plywood

Plywood is made by gluing a number of layers, called *plies*, together at right angles to each other. The plies on the surface (top and bottom) are arranged with the grain approximately parallel. An odd number of plies (3, 5, 7) is used so that the material will be balanced, with the same number of plies on either side of the center core. This makes it possible to run the grain on both sides in the same direction. See **Figure 16-8**.

Plywood is manufactured in accordance with the *U.S. Product Standard PS 1-74/ANSI A199.1*. This standard, established by the American National Standards Institute, designates the species, strength, type of adhesive, and appearance of plywood.

Lamination is also used for processes other than making plywood. It is used, for example, to produce bent or irregularly shaped parts in the furniture industry. Typical applications include bent forms for tables and chairs.

Advantages of lamination

When shapes such as huge beams or arches are desired, there are really only two ways that they can be constructed: by sawing them to the proper shape, or by *laminating* them (building up the desired shape in layers). Lamination is preferred, since cutting is usually more wasteful of material and the grain does not run parallel with the flow of the part's surface. Another advantage of laminating is that it produces a much stronger product than solid stock.

Figure 16-8. Plywood is laminated from 3 to 7 layers of veneer for strength and dimensional stability. The top sheet is plywood with a brushed surface to emphasize the wood grain. The bottom sheet is a variation of plywood called "composite board." It has facings of veneer applied to a core of particleboard, and can be used for many of the same applications as traditional veneer plywood. (American Plywood Association)

There are other reasons why lamination may be the ideal process for forming many wood products. With laminated parts, the highest quality materials can be used for the face surface and material of lower quality or appearance for the inside. This results in a better-looking product for less money.

Another advantage of lamination is that the process enables manufacture of more-complex and more-intricately shaped larger pieces. There is little waste with lamination, since smaller pieces can be glued together. The process is often used in the boat manufacturing industry for making wood molds. Thousands of individual wood pieces are glued together in contact with each other to form the mold. Since the mold is often large (60' or more in size), a great deal of precision work is necessary to fit all of the pieces together in perfect alignment.

The sequence of operations normally followed for laminating irregularly shaped parts, such as large architectural beams, might be something like this:

First, the stock (board) is cut to length and planed to remove the rough surface from both faces. The ends of each board are beveled to facilitate joining end-to-end. *Beveling* is the process of changing the sharp (90°) angle where vertical and horizontal surfaces join into a (usually) 45° angle.

After beveling, each piece of stock is run through a glue-spreading machine, where an adhesive is applied to both sides. The glued pieces are joined end-to-end and placed in plies in a jig that conforms to the desired shape. The jig with glued plies is placed in a forming press where heat and pressure are applied. After the adhesive cures, the piece is removed from the press. Finally, the product is machined on a specialty-type planer designed to handle the desired shape.

Thick pieces of wood tend to crack and split over time. Lamination is often a useful method for eliminating these types of worries—products can be made in any size, and they will retain their structural integrity without cracking.

Bending Processes

Bending processes are used to form curved shapes and can be done with either solid or laminated materials. Hardwoods such as white oak, elm, hickory, ash, birch, maple, and walnut are particularly suited to the bending process. Softwoods are difficult to bend.

Wood can be bent either across the grain or with the grain. On a typical piece of lumber 6′ long and 10″ wide, the grain runs parallel to the length of the board. Wood is easier to bend going with the grain. Bending of solid wood can be done either when the wood is dry or when it is wet. Laminated plies are bent when they are wet.

Some woods are bent when they are wet because they are more prone to fracture (breaking) than others. In most cases, moisture and heat are applied to the dry wood, and it is then bent and clamped over a form until it cools and dries. It will then retain its new form.

Wet or Hot Bending

There are two basic methods of wet bending: steaming or soaking, and laminating in a two-piece mold. Steaming or soaking will soften (plasticize) the wood so that it can more easily be formed. Laminating is accomplished using adhesives in a forming jig. Plywood, solid, or laminated wood can be bent using either method. It is more difficult to bend solid pieces of wood than laminated stock. See **Figure 16-9**.

Steaming or *soaking* is the better way to bend wood. Here is how the process works:

First, the stock is subjected to steam or is soaked in boiling water until it reaches a moisture content of about 20 percent. Dry wood normally has to be soaked or steamed about one hour for each inch of thickness. After it is soaked or steamed, the stock is quickly wrapped around a solid form or placed in a forming jig that conforms to the desired shape. Dowels are inserted in the base plate to hold the steamed part tightly against the form. The part is kept in this position until it cools and dries. Once it has dried, the part will hold its desired shape.

Sometimes, sheet metal is clamped to both faces of the stock during bending to provide additional support and help in holding the desired shape. The inside radius of the wooden mass can be compressed as much as 25 percent, but the exterior surface cannot be stretched more than one or two percent. For this reason, bending must be done slowly, or the outside layers will split. After the stock is bent, it should be allowed to dry on the form for at least 24 hours before it is removed.

In a high-volume production environment, large hot or cold presses are used to press and form wood. The process is the same as with bending of wood using a simple fixture. In some cases the wood is softened and then conveyed to the press. In other cases, the wood is placed in the press and steam or oil is applied to soften the wood prior to forming.

Cold or Dry Bending

Normally, large hydraulic cold presses are used to prefabricate plywood, rather than for forming. Moisture, steam, and heat are usually required for bending. However, some firms do manufacture flexible plywood designed especially for bending. Clarks Wood Company Ltd., manufactures and distributes flexible plywood called *StagFlex*, that is suitable for cold forming small radii. StagFlex is unique because it can easily be formed to take on any desired shape. When it is glued, laminated or veneered, StagFlex will maintain the desired shape. StagFlex is made of thin sheets of plywood sandwiched together. There is no need for heat or water treatment. Thicker layers can be formed by gluing layers together while they are held in a fixed position. In this application, hot pressing may be used with thermosetting glues. See **Figure 16-10**.

Figure 16-9. The curvature in the back and seat of this Charles Eames molded plywood chair, which was first introduced in 1946. The chair is still made today of five-ply molded plywood using a forming jig. (Herman Miller, Inc.)

Figure 16-10. StagFlex flexible bending plywood available in long grain and cross grain, and in thicknesses of 6 mm and 8 mm. These veneers can be laminated to produce greater thicknesses. (Clarks Wood Company Ltd., UK)

Important Terms

bagasse	laminating
bending	oriented strandboard
beveling	particleboard
binders	plies
bonding	plywood
composition board	soaking
defiberize	StagFlex
dry felting	steaming
flat-platen-pressed	tempered hardboard
hardboard	underlayment
hog	waferboard
insulation board	wet felting

Questions for Review and Discussion

1. What are the major steps you would follow to plasticize wood prior to bending?

2. What are the differences between hardboard and hardwood?

3. Which of the major types of composition boards would you specify for use in a very wet and humid environment?

4. How would you be able to recognize a sheet of reconstituted wood from "real" wood?

This 150-ton press uses high-frequency radio waves to heat plywood while bending it into various forms. (Nemeth Engineering)

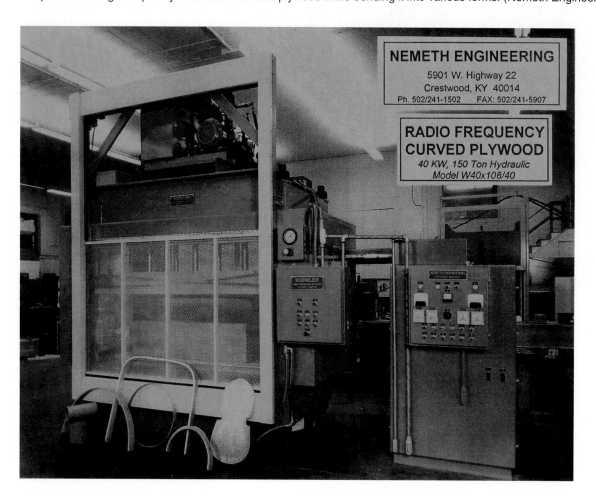

17

Processes Used to Form Ceramic Materials

Chapter Highlights

- There are four major types of ceramic forming processes.
- Most industrial ceramic parts are formed by the dry pressing method.
- Material preparation, forming, densification, and finishing are the four process actions involved in ceramics.
- Deflocculants are used to reduce the percentage of water in slip and to maintain formability of the material.

Introduction

There are many different types of forming processes used to manufacture products made of ceramic materials. These can be divided into four broad groups. See **Figure 17-1**.

The first group of materials—lime, cement, and plaster—uses water to turn finely ground powder into fluid slurry. Water causes a chemical process called *hydration* that hardens the cement and holds the mixture together.

The second group of materials, glass products, involves processes that require no pressing. The product is created by melting stock in solid or particle form until it becomes a thick liquid. The liquid is shaped as it cools and solidifies.

The third group, ceramic clay products, uses forming processes with heat and pressure to compact dry powders mixed with binders. *Binders* are chemicals that are added to reduce the *viscosity* (thickness of the mixture) and hold the mixture together.

Figure 17-1. Ceramic materials are organized into these four classifications.

Group	Description	Products
1	Pulverized products where the raw materials are heated to create cement-like properties	Cements, limes, plasters
2	Products that are heated until they become fluid and are formed while still in a fluid state	Glass
3	Products that are formed through compaction of finely ground powders, followed by heat treatment to create a sintered bond and strength	Ceramic clay products
4	Products made from a liquid slurry or slip that is compacted under pressure	Contoured clay products; often high volume production

The fourth group mixes dry powders, binders, water, and other chemicals called lubricants. Lubricants are added to improve the shaping characteristics of the material and to aid in removal of the product from the mold. Once the thorough mixing is complete, the liquefied ceramic material is referred to as *slurry*. When binders are added to the slurry, the amount of water that must be used to produce a viscous mixture can be reduced. Many of the ceramic processes in this category are used to produce cast products by pouring slurry directly into a mold.

Clay mixtures are engineered so that they can be used with a particular process, as well as conform to the design requirements of the product. Many of the processes that use either compacted dry powders or binders and lubricants in a slurry also use heat to improve the binding characteristics of the ingredients. Other processes involve cold-pressing. Many different types of clay mixtures can be created, and the nature of the material dictates what type of temperature is appropriate. This is discussed in greater detail later in this chapter.

The process of manufacturing metal, wood, or plastic parts normally begins with the purchase of raw material in the form of powders, pellets, sheets, or bars. This stock is then transformed into the desired shape using a manufacturing process such as drawing, forging, pressing, or bending. Stock may be usable in the form in which it is received. However, in the field of ceramics the stock that is purchased has to be further modified before it can be formed or shaped.

By now, you should be developing an appreciation for the tremendous diversity of product applications in ceramics. There are hundreds of unique products and process applications with glass products, traditional clay products, electronic and high-temperature ceramics, and ceramic glasses. In spite of this diversity, ceramic processes do have some commonalities.

Manufacturing engineers must pay careful attention to three different types of process actions influencing the behavior and performance of the ceramic stock and workpiece: preparation of materials, densification, and finishing. The greatest process in the world will lead to products that fail quality assurance tests if the material is not prepared properly, if it is too wet, or if it is improperly glazed and fired. The problem is further complicated by the many stages of processing required before stock can be used with a particular process.

Preparing Ceramic Materials for Processing

Each particle of finely ground clay powder that is used in a dry mixture contains many individual ceramic crystals. When these crystals are crushed or deformed under pressure, they retain their new shape when the pressure is removed. At the same time, the crushing process must be conducted with care. Unlike metals, ceramic crystals are brittle and will fracture, rather than flow, under applied pressure. Clay stock can be made less brittle by reducing its viscosity using water, binders, lubricants, and other additives.

When clay must be reduced to a semiliquid plastic state, a process called *doping* is often used. This process is also used by industrial ceramicists to create a plastic condition in non-clay materials. Doping is done by adding organic binders and lubricants to the mix. Binders provide wet strength during forming and dry strength after drying. Lubricants decrease friction during forming, thus enabling forming to be accomplished with lower pressures.

Adding water to doped ingredients improves the plasticizing action of the clay mass. Binders that are used with water are polyvinyl alcohol, various starches, methylcellulose, and protein. Waxes and water-soluble oils are lubricants used with water.

For injection molding and other processes that use ceramic stock in a liquid state, synthetic thermo-plastic resins can be used with organic solvents or *plasticizers*. In this situation, the resin acts as a lubricant as well as a binder.

Most ceramic products are formed and air-dried. They are then fired in an oven called a *kiln*. Firing heats the product until it is completely dry. Firing also binds the clay molecules together to create a usable product. Before they are fired, parts are referred to as *greenware*. After being fired, the parts are in the *leather-hard* state.

Before being fired, ceramic parts can be machined using conventional metal-cutting tools. However, because of the fragile nature of these parts, extreme care must be taken when securing the workpiece and using the tool. Often, parts are fired before they are machined. However, when this is done, the machining is limited mainly to grinding, lapping, and honing processes using specialty tools or abrasives.

Many ceramic pieces, particularly those made from slurry or viscous plastic material, contain a high percentage of water after forming. To control the strength and size of the product, it is particularly important to remove most of the water from the product before it is fired. This is normally done through air-drying or controlled-heat drying of the part.

Mixing the clay powder with the correct amount of water, binders, and lubricant is important. Clay that is mixed with too much water, binder, or lubricant retains a high percentage of moisture after the initial drying cycle. If that part is then placed in the kiln and fired, heat will cause it to quickly release water. This results in drastic shrinkage, and could cause the part to crack or explode in the kiln.

The proper degree of dryness of ceramic parts is also important. If the part is too dry, it will be excessively fragile and will tend to crumble when it is transported to the kiln. Excessively dry parts also require greater forming pressures. Now that you have a basic understanding of ceramic forming, you can take a closer look at some of the processes.

Dry Forming Processes

As noted at the beginning of this chapter, one group of forming processes for ceramics uses pressure to compact dry powders that are then fired. There are a number of variations of the dry forming process.

Dry Pressing

Most industrial ceramic parts are formed using very little water in a process called ***dry pressing***. Automated dry pressing is normally accomplished using ceramic powders with 2 percent or less moisture content. Stock used for dry pressing consists of the dry powder, binders, and lubricants.

The dry-pressing process was first used in Germany in the late 1800s. The first use in the United States was in the 1930s when this process was used to make electrical insulators. Dry pressing has a number of major advantages over other ceramic processes. Dry-pressed parts have less flash. The most significant advantage, however, is that dry pressing is a near-net-shape process.

Dry pressing and other processes that use pressing to create near-net-shape parts are particularly attractive in applications with stringent electrical requirements. The process is used extensively for producing computer memory cores, refractories, tiles, spark plug insulators, and many electronic components. Capacitor dielectrics as thin as 0.020″ and thick- and thin-walled insulator boxes for protecting miniature electronic circuits are also produced with this process. A variation of dry pressing, called ***tablet pressing***, is used primarily in the pharmaceutical industry.

Another major market for near-net-shape dry pressing is the nuclear power industry. Today, more than 100 nuclear power plants in the United States supply about 20 percent of the electricity used in this country. Nuclear fuel fabricating plants convert enriched uranium hexafluoride to uranium dioxide powder and then press this into fuel pellets using dry pressing. See **Figure 17-2**. Once the pellets are pressed, they are inserted into fuel rods made of zirconium alloy.

Some of the first electrical products made by dry pressing used a type of ceramic material called steatite. ***Steatite*** is a natural crystalline form of magnesium silicate or talc. This was mixed with clay and barium carbonate flux. In composition, the steatite

Figure 17-2. The ability to produce near-net-shape parts is one of the advantages of dry pressing. This rotary press is turning out nuclear fuel pellets. (General Electric)

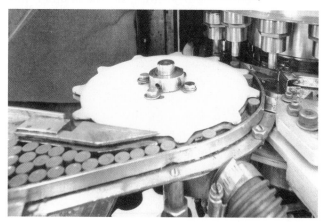

contains a small amount of clay mixed with artificial plasticizers, such as wax or polyvinyl alcohol. When they are fired at around 2460°F (1750°C), a glassy bond is formed that holds all of the particles together. Fired steatite insulators are as strong as porcelain.

Dry pressing is one of the simplest high production processes used today for producing dense parts. Production speeds often reach 5,000 parts per minute (ppm). The dies are compacted in a press using pressure from 3,000 psi to 30,000 psi.

The term "dry pressing" is not technically accurate because, most of the time, the stock contains from 3 percent to 15 percent water. Preparation of the powder for dry pressing takes place by spray-drying the slip to produce solid particles, or by screening the slip through a fine mesh. Once the fine powders are separated and mixed with binders and lubricants, they are ready to be compacted in hardened steel or tungsten carbide dies. The amount of pressure required for pressing depends on the materials being compacted, the shape of the piece, and the end use for the product. In most cases pressures vary from several thousand psi to 100,000 psi. Dry pressing of tile often requires as much as 2,500 tons.

The selection of binders is a critical factor in the dry-pressing process. The ideal binder improves flowability of the powder, reduces abrasion, improves bonding strength, increases internal lubrication, and aids in removal of the part from the die. The binder burns off during the initial stages of the firing cycle, without disturbing the structural density of the compacted part. The compacting process causes the powdered particles to shift and deform, reducing

the porosity of the material and thus improving the density of the part. After the granulated powder is pressed, elastic compression of the granules continues through compaction. On ejection, this stored elastic energy produces an increase in the dimensions of the compacted part. This is referred to as *springback*.

In addition to this deviation in the size of the part, there are other limitations to the dry-pressing process. A major concern is a variation in the amount of shrinkage throughout the part during firing and a loss of original tolerances. That is, not all areas of the part undergo the same amount of shrinkage. Despite these limitations, dimensional tolerances of plus or minus 1 percent are common in most applications.

Steps in dry pressing

There are four steps involved in the dry pressing production cycle:

1. Filling the die.
2. Compacting the powder.
3. Ejecting the part.
4. Firing or sintering.

When the die set opens, the matched dies are filled with free-flowing granules of fine powder. Powder flows from a hopper to a feed shoe, which rests on top of the lower die. The shoe slides across the top of the die body and distributes powder into the lower die cavity. The dies then close and the punches compact the part.

Once the part is compacted, the lower punch is raised and the part is ejected from the die. The process continues when the die shoe, filled with a new load of powder, pushes the part out of the cavity and on a conveyor for transportation to the kiln for firing. The firing of dense ceramic compacts is referred to as *sintering*.

Normally, firing causes ceramic products to shrink to about two-thirds of their green volume. This shrinkage makes it difficult to produce parts at the desired size. One method for obtaining near-net-shape products is to machine the part prior to sintering.

One method of improving the performance of dry-pressed parts is called *hot forging*. In this process, a dry-pressed part is heated and then pressed in a cold steel die. Hot forging improves the density and strength of the part. Unfortunately, this additional pressing creates stresses in the part that must be removed by annealing.

Refractory materials

Hot forging is often used to produce *refractories*, ceramics that can withstand continued exposure to high temperatures. Refractories can withstand temperatures that often approach 5000°F (2760°C) while resisting chemicals, thermal shock, and physical impact. Because of their ability to survive in high temperature environments, refractory materials are particularly useful for bricks to line the inside of kilns and for crucibles to carry molten metal.

There are many different types of ceramic materials used for making refractories. The most common are fire clay and silica, but kaolin, magnesite, bauxite, alumina, and limestone are also used.

Refractory products are usually made in brick or tile form. The manufacturing process for refractories begins with stock preparation: separating, crushing, grinding, and screening. The powder is then mixed and compacted in the same manner as it is in other dry press applications.

Refractory coatings are also produced for use on metals and ceramics. In this application, the coating is sprayed directly onto the refractory brick and then fired. Ceramic coatings are being used effectively in high-temperature applications to provide energy savings and extend the service life of components. Ceramic coatings prevent oxidation and fluxing (burnout) of refractories at elevated temperatures. In **Figure 17-3**, a black-body ceramic coating is being sprayed on the refractory lining of a reheat furnace at a steel mill.

Figure 17-3. A ceramic coating being sprayed onto the refractory lining of a reheat furnace. Such applications are used in the steel industry and others where high-temperature furnaces are involved. (Ceramic-Refractory Corp.)

Dust Pressing

The raw materials used for dust pressing are different from those used for dry pressing. *Dust pressing* is a popular process for making white-body, glazed ceramic tiles. The stock for these tiles is referred to as *dust*, which is composed of 62 percent talc and 38 percent clay. These materials are mixed in a ribbon blender thoroughly before being transported to a mix muller. A *mix muller* is a machine with heavy steel wheels that mulls or kneads the mixture. During the mulling process water is added to bring the moisture content to approximately 8 percent by volume. The addition of water also causes the talc and clay to adhere to each other. After mulling, the dust mixture is placed into a *pulverizer*, a machine that breaks down the globs of dry clay. The stock is placed into containers and stored until it is needed.

The stock is transported to the press room, where hydraulically-driven presses are used to press the dust into a solid body. Pressures of approximately 2,400 psi are common. Tiles that have as much as 8 psi tensile strength are pressed. This green body is then carried through three drying cycles over a 36-hour period. The first cycle is at 100°F (38°C), cycle two at 160° F (71°C), and cycle three at 210°F (99°C). What is interesting to note is that at the end of the third cycle, the tile contains no moisture and has a tensile strength of 16 psi.

The tile is conveyed through a spray booth where glaze is applied. Glaze is a mixture of minerals, water, and oxides or carbonates that are applied to ceramics and fired to produce a glassy finish. After application of glaze, the tiles are fired in a kiln at approximately 2000°F (1093°C). The heat from the kiln causes the glaze to fuse together and bond to the body of the tile. After firing, the tile is taken on pallets to the sorting department where it is grouped according to grade and shade. The matching of shades is an exacting process requiring an individual to evaluate each batch in accordance with specifications established by the U.S. Bureau of Standards and the U.S. Department of Commerce.

Experimental programs are presently underway at the U.S. Army Materials Technology Laboratory and the Ballistic Missile Research Laboratory to investigate the feasibility of combining the dust pressing process with explosive forming to produce tiles with an increased density. Results indicate that powder can be pressed into a green compact with approximately 60 percent density. After pressing, the compact is then placed in a forming chamber, similar to the type used for dust pressing. An electrically-detonated charge forces punches against the compact and compresses it against the walls of the metal die. The gas produced during compaction escapes through vertical grooves on the internal surface of the die.

As noted, in both dry and dust pressing, water and binders are added to the mix. Some dry pressing processes require only pressure; other variations of the process require both heat and pressure.

Cold Isostatic Pressing

Isostatic pressing is conducted by applying pressure to the powder from three directions, rather than the single direction used in linear pressing. Isostatic pressing is also known as *uniaxial pressing, hydrostatic pressing,* or *hydrostatic molding.* Higher pressures are required for isostatic pressing than for conventional cold or hot pressing. Isostatic pressing is used for parts that, because of intricate configuration or extreme density requirements, cannot be produced with conventional hot or cold pressing methods.

Isostatic pressing can be done either cold or hot. **Cold isostatic pressing** at room temperature, using pressures ranging from 5000 psi to 20,000 psi (35,000 kPa to 240,000 kPa), is the process most often used in industry. Molds are normally filled with a free-flowing spray-dried powder. Vibration is used to eliminate voids (holes) in the powder mass inside the mold.

Isostatic pressing is based on Pascal's Law. This law states that when pressure on a liquid in a closed container is increased or decreased, the resulting change in pressure occurs uniformly throughout the liquid. As applied to powder compaction, this means that powdered ceramic particles in a flexible, air-tight container, when placed in a closed vessel filled with pressurized liquid, will receive an applied force uniformly from all sides. The result is that material compacted by isostatic pressing will be uniformly pressed in every direction.

Dry bag isopressing

There are two basic types of cold isostatic pressing: compacting powders in a two-piece matched mold or using fluid and a rubber liner to create pressure. The first method was discussed in conjunction with other pressing methods. The

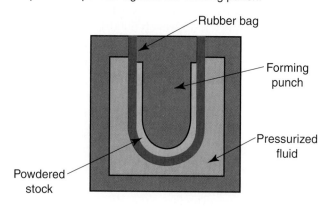

Figure 17-4. The ceramic powder and a forming punch are sealed inside a bag in the dry bag isopressing method. Pressure is exerted on the bag from the outside by fluid to compress the powder against the forming punch.

Rubber bag

Forming punch

Pressurized fluid

Powdered stock

second method, called *dry bag isopressing*, requires additional discussion. See **Figure 17-4.**

In dry bag isopressing, a male die punch is used with an open-cavity lower die block. A rubber bag is inserted into the lower die block, which has no female impression, only an open cavity containing fluid. The fluid used is usually hydraulic oil, water, or glycerin. The bag is filled with powder and a forming punch, and sealed. After the bag is submerged in the liquid, pressure is applied. The ceramic powder in the rubber bag is compacted as the liquid in the mold transmits the pressure.

Hot Isostatic Pressing

Hot isostatic pressing is accomplished by pressing ceramic powder at 15,000 psi (103,000 kPa), while heating the compact to approximately 3000° F (1650°C). Hot isostatic pressing is also called *gas pressure bonding.* Heating the pressed compact improves the quality of the part that is produced.

The heating action, called sintering, causes the loosely bonded powder to be transformed into a dense ceramic body. During sintering, moisture and organic materials are removed by burning them out of the greenware body. A process called **diffusion** then takes place, causing material to move from the particles into the void spaces between them. This improves the density of the part and causes less shrinkage during firing.

In hot isostatic pressing, the container is placed in a pressure vessel inside a high-temperature furnace. The pressure vessel is purged with gas and then evacuated to remove air and moisture. Helium is injected into the vessel at the desired pressure, and the furnace temperature is raised and held at the final bonding level until the part is properly sintered.

Explosive Forming

Explosive forming is used more frequently to shape refractory metals and carbides than to shape ceramics, but it can be used with ceramics, as well. When ceramics are to be explosively formed, the material used is in the form of granulated powder.

The powder is poured around a forming mandrel, which is encased in a pliable envelope. Often, this envelope is made of a thin-gage metal. Explosives are placed around the envelope, and in some instances, the entire assembly is placed under water. The powder is compacted against the mandrel when the explosive is detonated.

A new field of study, called *dynamic compaction*, has evolved from the process of explosive forming. The Lawrence Livermore National Laboratory, Lewis Research Center (NASA), Los Alamos National Laboratory, and Defense Advanced Research Projects Agency are conducting research on this technology.

Dynamic compaction is a process that can be conducted either at room temperature or with heat. The amount of heat required for dynamic compaction is much lower than is required for hot pressing or sintering of ceramics. Firing at lower temperatures results in a part with fewer contaminants. Dynamic compaction can be used in conjunction with cold and hot pressing, reaction bonding, hot isostatic pressing, and injection molding.

Wet Forming Processes

A second large group of forming processes used for ceramics utilizes powdered clay suspended in water. The liquid clay is mixed with binders and lubricants to improve shaping characteristics and aid in removal from the mold. Depending upon the water content, the liquefied ceramic material may be referred to as slurry or *slip*.

Wet Pressing

Relatively dry mixes are used with both dry and dust pressing. In comparison to these processes, the *wet pressing* process adds water to the powdered stock before the mixture is compacted in a die. One would infer from the process name that the clay mixture must be viscous, or at least plastic, in nature.

It is true that the mixture for wet pressing is fluid, but it is not necessarily wet.

There are two variations of wet pressing. The first method uses powdered stock, with a moisture content of approximately 15 percent. The water creates plastic deformation during compacting while the mixture is forced against the contour of the die cavity. Compacted parts will have flash in the areas where the mold halves come together.

Wet pressing is not suitable to automation, because parts are easily damaged and are difficult to transport after pressing. However, the process does have one major advantage over dry pressing: because of the higher moisture content, less pressure is involved.

The most significant disadvantage of the process is the dimensional accuracy of wet-pressed products. Wet processing is not as good a near-net-shape process as dry pressing, since dimensional tolerances can vary by as much as 2 percent.

A second form of wet pressing, called **ram pressing**, was developed and patented by Ram, Inc., of Springfield, Ohio. Ram pressing requires the use of a plastic (pliable) mass, rather than powder. The material is mixed and extruded into a storage canister until it is needed at the machine.

The material to be formed is pressed in a steel or plaster die. As pressure is applied, the plastic mass flows into the various contours of a two-piece mold. Air entrapped in the clay mass is removed by applying a vacuum through holes in the mold. As in the previously discussed method of wet pressing, flash is also created at the parting line of the mold. Ram pressing is often used for making oval-shaped dishes and contoured parts.

Hot Pressing

In the hot pressing process, powders are pressed at high temperatures in heated graphite dies. Hot-pressed parts exhibit strength and density superior to products made by cold dry-pressing.

Hot-pressing dies are usually heated by an induction or resistance coil. See **Figure 17-5**. The heating of the dies makes it possible to use lower pressures (approximately 3000 psi) to compact the part. This results in cost savings for the press.

The main expense of the process is the graphite dies, which are soft and must be discarded after a single pressing.

Hot pressing can also be accomplished using a more plastic ceramic body. In this method, a heated

Figure 17-5. The water-cooled induction coil of this 100-ton hot press heats the graphite die to a temperature of approximately 3600°F (2000°C). (Ceradyne, Inc.)

die lowers to sear the surface of the ceramic material, causing a cushion of steam to form between the ceramic body and the die surface. This cushion acts as a lubricant to keep the part from sticking to the mold. The heated pressing ram is made of metal, but the molds themselves are normally made of plaster.

One interesting application of hot pressing is in the field of electro-optic ceramics, where solid-state (no moving parts) switches are being made. A unique type of ferroelectric material, lead zirconate-titanate doped with lanthanum, is used. This material, called **PLZT**, has the ability to shift its polarity under the influence of an electric field, permitting the solid ceramic part to act as a switch.

The manufacturing process for producing the switch begins by placing a prepressed slug of PLZT in an alumina-lined, silicon carbide mold inside a pressurized chamber. The mold rests on an alumina plate and is surrounded with a thin layer of refractory material. Oxygen is backfilled into the pressurized chamber, which is heated to approximately 1700°F (700°C). The temperature and pressure are maintained for 18 hours while a pushrod presses the part and holds it in alignment in the mold.

After the hot-pressed slug is cooled and extracted from the mold, it is cleaned of refractory grain, annealed, and polished for optical testing and evaluation. This hot pressing application has allowed a very significant reduction in the size of switches while increasing the performance capabilities. **Figure 17-6** illustrates hot pressing of PLZT.

Throwing

Any discussion of contemporary manufacturing processes must include mention of one of the foundational processes in ceramics, called *throwing*. While throwing dates back to antiquity, it is still used today by firms producing round objects such as dinnerware, pottery, and production prototypes. In addition, many of today's applications still depend on variations of this basic process.

Throwing involves hurling a body, or mass, of clay that is in a plastic state onto a revolving potter's wheel. Once the clay is attached to the wheel, alignment (centering) of the mass is done to ensure a symmetrical product.

Shaping of the product is accomplished by hand. The body is formed by pulling the clay upward and outward with moistened fingers, thumbs, and palms. This ancient process is still used today to make large electrical insulators that are later joined together in sections.

Jiggering

Jiggering is a ceramic production process that is used to make flatware, such as plates, saucers, and oval dishes. The process begins with a slice of clay being placed on a revolving convex mold that will form the inside of the plate. The mold typically is rotating at a speed of about 400 revolutions per minute (rpm). A template or forming tool then forces the clay over the mold. The process can be accomplished with or without automation.

Figure 17-6. Solid-state switches can be dry pressed from the ferroelectric ceramic material PLZT by applying pressure and heat for a lengthy period. A typical pressing setup for PLZT is shown.

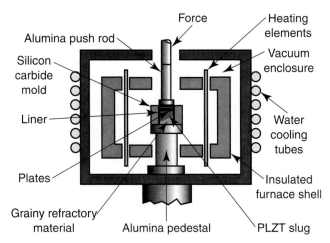

Figure 17-7. Jiggering machines like this one are used to form greenware. (Lenox China)

In manual jiggering, the operator wets the plaster mold with a sponge, places the clay, and then uses a forming template to press the plastic mass down over the contour of the rotating mold. When the process is automated, the forming template is normally a dome-shaped profile tool that automatically contacts the surface to compress the clay and form the second side of the product. See **Figure 17-7**.

Jollying

The difference between jiggering and jollying is that in *jollying*, the plaster mold forms the outside surface of the product and a forming template shapes the inside. A simple jollying operation would involve placing the plastic clay, in the form of a ball, onto the spinning mold. The sides would then be drawn upward, as in throwing and shaping a cup. The template is pushed down into the center of the ball of clay to shape the inside of the product. China and porcelain cups are made this way.

Today, many jiggering and jollying operations are performed by a roller machine. The major difference here is that instead of using a forming template, a heated die, called a "bomb," is used to perform the shaping operations. **Figure 17-8** illustrates how a roller machine works.

Extrusion

Extrusion is a process that pushes metals, plastics, and composites through the profile of a die to form continuous lengths of stock. Examples of extruded products include moldings and door casings. As you may recall from earlier chapters of this book, resin is heated to a pliable form in the barrel

Figure 17-8. This roller machine is used to form a ceramic product using the process called "jiggering." The ceramic material, in the plastic clay state, is pressed into shape between two rotating surfaces. After forming, the product can be removed from the jigger head.

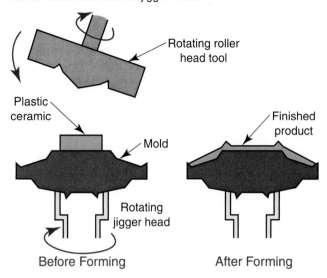

of a machine called an extruder, and is then forced through an opening in a rigid metal die.

The process of extruding ceramics does not require heat to make the material pliable. Ceramic stock requires only water and additives to make material that can be forced through the extrusion die. See **Figure 17-9**.

Sometimes extrusion is used to *de-air* (remove the air from) highly viscous ceramic stock. This airless stock is used to produce slugs, called **blanks** or **pugs**, that will be used with pressing and jiggering processes. **Figure 17-10** shows extruded body pugs that have been stacked prior to being sliced into individual blanks. When high-temperature refractories are produced, extruded bricks are made and then re-pressed in dies before drying.

Extrusion is a useful process for making control rods and ceramic fuel for use in the nuclear

Figure 17-9. Clay, in a pliable plastic state, can be forced through a die to form an extruded shape. This simplified drawing shows a piston extruder.

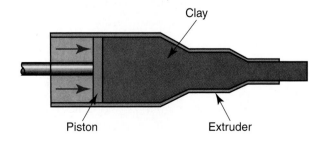

Figure 17-10. These body pugs are being stacked on a pallet by automated equipment prior to being sliced into disk-shaped blanks. (Lenox China)

fuel industry. See **Figure 17-11**. Before the advent of ceramic nuclear fuel, the pure isotope uranium-235 had to be extracted from uranium metal. Earlier reactors used this type of fuel.

Ceramic fuels have become popular because of their stability, high heat-resistance, and superior ability to ward off corrosion. Ceramic forms of uranium are made from oxides or carbides. Uranium dioxide (UO_2) is popular. UO_2 has a melting temperature of 5070°F (2800°C), making it one of the most refractory (heat-resistant) materials used today. Uranium dioxide must be enclosed in a container to keep it from further oxidizing.

Uranium carbide is more frequently used as a uranium substitute. It has a thermal conductivity higher than uranium dioxide, and has no pores. This makes it a very useful material for use as a fuel rod

Figure 17-11. A technician monitors production of ceramic pellets that will be used to assemble nuclear fuel rods. (General Electric)

Figure 17-12. Ceramic fuel rods are assembled into bundles like these for installation in nuclear reactors. Uranium dioxide and uranium carbide are both used for such applications. (General Electric)

in nuclear reactors. See **Figure 17-12**. Any increase in the temperature of the rod is consistent throughout the rod.

Another interesting application is the use of ceramic extrusion to manufacture nuclear fuel with cermets (ceramic/metal composites) made of uranium oxide, and boron carbide powders that are suspended in a stainless steel matrix. Experimentation has also been conducted using mounds of extruded strands of *morillonite* clay to absorb radioactive waste. Waste is mixed with other clay to reduce the extent of contamination and the clay bricks are fired to permanently lock in the radioactive atoms. According to a report from *The West Australian,* (February 1, 2002), bricks made from radioactive by-products (bauxite residue from Alcoa's alumina processing plant at Kwinana, Western Australia) have been used to build homes in Australia and Jamaica.

Injection Molding

Injection molding is a process used extensively in the plastics industry. The same process is used by the ceramics industry for making spark plug insulators and electronic parts. Although the basic concept of squeezing material into a cavity is similar, there are several details that are unique to the ceramics injection molding process.

Injection molding of ceramics requires mixing of special plasticizers and resins with the dry ceramic powder. The clay mix is heated and formed into pellets by extrusion. The pellet mix is loaded into a hopper that feeds the mix into a heated cylinder called a barrel.

Hydraulic pressure then forces the material through an opening in the cylinder and into the cooled metal mold. With ceramics, the dies must be made of harder, more wear-resistant metal alloys than the dies used for injection molding of plastics.

Often, thermoplastic or thermosetting polymers are mixed with the ceramic powder to act as a filler; the polymers fill the pores of the ceramic body. When the resins are heated, they soften and are injected into a cold die. The plastic additives are then removed from the product by thermal treatment before firing. Sometimes, the removal of the resin takes several days. Kiln firing is normally the final stage of the process.

Tape (Band) Casting

Band casting, or **tape casting**, is a continuous production process used to produce thin strips of ceramics for use as electrical substrates and heat-exchanger devices. Some of the most frequent uses of this process are for production of multilayer titanate capacitors, piezoelectric devices, thin film insulators, ferrite memory chips, and multilayer electronic packaging. Electrical substrates, called *fabrication boards*, are normally green in color. Resistors and metal circuit patterns are usually printed onto this substrate and *co-fired*.

Co-firing is an important process in the production of integrated circuit chips, multilayered RF (radio frequency) modules, and microwave components. Co-fired products are those that contain a sandwich of several materials, such as metals and ceramics, and are fired together (co-fired). Co-firing not only joins the materials together, but also permits the formation of signal connections and buried pathways. There are two major types of co-fired ceramics: low-temperature co-fired ceramics (LTCC) and high-temperature co-fired ceramics (HTCC).

In LTCC, many thin layers of ceramic materials and conductors are combined. This permits the construction of a multilayered product with different types and shapes of interconnects and components. Co-firing enables the manufacture of a unique structure that would not have been possible using conventional alumina or other soft substrates. LTCC has been used in the industry for about fifteen years. The temperature required for LTCC is normally around 1470°F (800°C), depending on the base materials. The temperature must be high enough to *densify* (make hard and bind particles) the ceramic materials, but must not exceed the melting point of the electrode metal.

Before the layers can be formed, a ceramic/glass frit is held together with a binder and pressed into a sheet carried on a delivery roll. This is similar to the process of *tape casting* that will be discussed later in this chapter. The rolled product is often referred to as **green tape**. The next step in the process is to punch holes, called **via holes,** into the material where vertical interconnects will be required. Conductors are then screened onto the layers to form horizontal interconnects called **groundplanes** and **striplines**. These layers are then stacked up on alignment pins and pressed together to remove air pockets. The final step in the LTCC process is firing. Here, the binder is driven from the material and the glass frit melts and joins the layers.

High-temperature co-fired ceramic integrated circuits are made in layers that are fired at the same time under high temperatures, approximately 2910°F (1600°C). One of the most popular applications today is for single-chip packaging of high input/output (I/O) integrated circuits. The base conductor metal for these IC's is normally tungsten. Few other metals can survive this type of temperature. The process to create HTCCs is similar to LTCCs. The major difference is not only in the temperature, but also in the method used to produce conductor patterns. Dielectric tape is added by screen printing after the via holes are created. This is necessary because of the shrinkage that occurs due to high temperature firing.

The tape casting process is a method for forming powdered slurry into thin flat sheets. Controlled-thickness film is produced by flowing ceramic slurry down an inclined surface, or passing it through forming rolls. **Figure 17-13** shows two different types of tape casting processes. In the top illustration, thin film is cast on low-ash paper by pulling the paper through ceramic slurry. Other types of carriers can also be used. The film and carrier is then dried and collected on a take-up reel.

The lower illustration shows thin film casting using rollers to regulate the thickness of the film.

Figure 17-13. Two of the several types of tape casting processes.

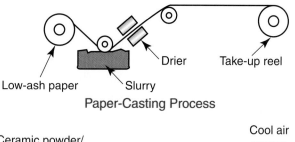

Paper-Casting Process

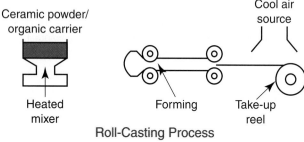

Roll-Casting Process

With this variation, the ceramic powder is mixed with organic binders.

Another type of tape casting, called the ***doctor blade casting process***, involves the use of a doctor blade, or metal squeegee, to regulate the amount of ceramic slurry that is allowed to flow on a metal belt. In doctor blade casting, organic binders are added to the powder. A solvent makes the mixture fluid so that it can be cast on a continuous steel ribbon.

The basic difference between doctor blade and tape casting is motion: When the blade is held stationary and the surface moves, the process is called tape casting. When the doctor blade moves across a stationary surface holding the slurry, it is referred to as doctor blade casting.

With either type of casting, a flexible, smooth-surfaced, white leathery film is produced in thicknesses of approximately 0.010″ to 0.060″ (0.25 mm to 1.52 mm). The process is continuous and automated.

The most critical processing step in tape or doctor blade casting is the formulation of the ceramic slip. Slip consists of the ceramic powder, a dispersant to stabilize the powder, a solvent to reduce the mix viscosity to facilitate casting, a binder to provide strength prior to firing, and a plasticizer to modify the properties of the mixture.

Let us take a closer look at how the tape casting process works. First, the casting slurry is prepared by mixing powdered ceramic stock with binders, plasticizers, and dispersants. Mixing is usually accomplished by milling, a process similar to the barrel tumbling used to finish metallic materials.

After milling, the slurry mixture is heated, filtered, and de-aired to remove air bubbles. The de-airing process is accomplished in a vessel where the slip is under vacuum for several minutes.

Slurry is then cast on a very clean Mylar®, Teflon®, or acetate film. The film is continuously supplied from a spool at one end of the machine. The ceramic tape is applied to the top surface of the film, where its thickness is regulated by the size of the opening between the bottom of the doctor blade and the film carrier. The film dries slowly as it is being transported through the machine on the film carrier.

The tape that is produced is flexible and does not permanently bond to the carrier belt. When the tape reaches the end of the machine, it can be cut or stored in a reel with the film carrier.

When it is time for further processing, the tape can be removed from the carrier and stamped, slit, or scored to the desired size and configuration. Parts are produced approximately 16 percent oversize to compensate for shrinkage during firing.

Calendering

Calendering is a process that can be used in place of extrusion or tape casting to prepare thin ceramic plates or sheets. In the calendering process, powder is mixed with a polymer in a paddle mixer. Agitation causes the solvent to be removed through evaporation. Stirring continues until the solvent evaporates. The clay is then transferred to mixing rollers. Next, the clay mix is squeezed between the two mixing rollers and a corrugated calendering roller. The rollers function as a cutter, turning in opposite directions and producing a shearing action on the material being pinched between them.

The rollers squeeze the clay powder and polymer until the film is the desired thickness. The film is then collected on a paper or polymer reel and stored for further processing.

One of the major advantages of calendering is that the final product is solvent-free. In the band or tape casting process, the evaporation of solvent creates pores in the final product.

Slip Casting

You will recall from its description earlier in this chapter that slip is a very liquid clay and water mixture. Slip often involves adding as much as 25 percent water to dry or nearly dry powder. The water content is kept as low as possible to reduce the extent of shrinkage that occurs during drying.

Slip casting, also called *drain casting*, is a process that has been used to make figurines and intricately designed products since the early 1700s. See **Figure 17-14**. The process begins by pouring slip into a hollow two-piece plaster mold. When the slip has been drawn by capillary action into the porous walls of the mold, the mold is inverted and the undried slip is poured out. This results in solid layers of clay being cast, or layered, on top of other layers to form the walls of the product. The walls become thicker with each pour.

Theoretically, the walls could keep getting thicker and thicker until a solid cast product is developed. Actually, however, as the walls become thicker, it takes longer for the liquid slip to be absorbed into the clay. Trying to cast a solid clay body with this process would require further modification of the slip material. The desired thickness produced with conventional slip normally would not exceed 0.050″ (1.27 mm). After the product has dried, the mold is separated and the product is further dried before it is fired.

The formation of slip is often a complicated process. Generally, it is not practical to produce slip by just adding water to clay. Such an approach would require excessive proportions of water, perhaps as

great as 50 or 60 percent. This would result in slip that would quickly saturate the pores of dry plaster molds. The wet molds would then have to be dried before more castings could be made. Time would be wasted and mold life would be shortened.

An even more critical factor, however, is the high percentage of shrinkage. An excessive amount of water results in distortion and cracking during the drying and firing processes.

For these reasons, it is necessary to use a *deflocculant*, a chemical that helps to make slip fluid while keeping the water content low. Deflocculants can reduce the water content of slip to 25 percent or less. Sodium salts of oxalic and tannic acid are common deflocculants.

The effect of a deflocculant is often dramatic. If clay is mixed with just enough water to create a stiff paste and a tiny amount of deflocculant is added, a highly liquid slip will be immediately produced upon stirring. Too much deflocculant, as expected, results in a thickening of the slip.

Sometimes a slip consists of coarse-grained materials that cannot be dispersed effectively with water and additives. In this case, a process called *freeze casting* is sometimes helpful. In freeze casting, thick slip is poured into a smooth-walled, rigid-rubber mold. The mold is then frozen to force the clay to expand and press against the walls of the mold. Let us take a closer look at this process.

Freeze casting is also called *freeze gelation*. It is a *sol gel* process, where a suspension of silicon dioxide, or sand (a sol) is gelled to form a solid. With freeze drying, ceramic workpieces can be made without the need for high-temperature firing.

A typical application is to mix aluminum oxide slurry with an aluminum filler powder. Then, a wetting agent is added to disperse the filler powder in the slurry. The mixture becomes doughy and stiff, but still remains thixotropic when vibrated. *Thixotropic* means that the material remains in a solid condition at rest but becomes fluid when agitated.

The material is vibrated to turn it into a liquid and remove trapped air bubbles. The liquid that is created is used to fill a mold. The filled mold is frozen to cause the silica to precipate from the mixture (sol) to form a solid (gel). The new material is held together much like a sintered green form. The final step is to dry the part in a furnace.

The major advantages of freeze casting over other casting processes for making simple products relates to speed and cost. Freeze casting does not

Figure 17-14. Slip casting is a process that has been used for hundreds of years to produce hollow ceramic products. Wall thickness can be built up in several layers.

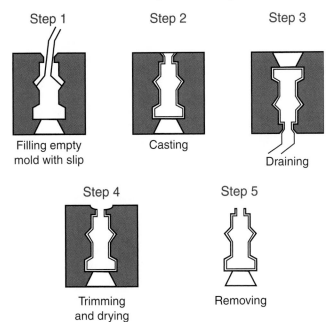

Step 1

Filling empty mold with slip

Step 2

Casting

Step 3

Draining

Step 4

Trimming and drying

Step 5

Removing

require high pressure equipment or an expensive furnace. Low temperatures, just above the boiling point of water, can be used to make products that accurately portray the details created by the mold.

Another unusual variation of the slip-casting process is called **ultrasonic casting**. With this method, low-amplitude audio frequency vibrations of 15–20 kilohertz are transmitted to the plaster mold during the casting process. Some of the heat generated in the plaster mold by the ultrasonic vibrations is transferred to the workpiece. A number of firms are using this process to slip-cast cermet and alumina plates. Cermets are composite materials consisting of ceramic and metal particles held in suspension in a metal matrix. Heat causes the cermet to be sintered, bonding the particles together.

Casting

Casting is a relatively simple process that is often used to create solid ceramic objects. It is a variation of the process of slip casting. A liquid suspension of clay (slip) is poured at room temperature into a porous plaster-of-paris mold. The dry mold absorbs moisture into its pores, causing a hard layer to build on its walls. The process is continued, building layer by layer, until the interior of the mold is completely filled. In this case, it is referred to as *solid casting*. Slip with a high percentage of deflocculants is used in solid casting. The addition of deflocculants creates a fluid suspension with as little as 20 percent liquid. If the slip was produced with a high percentage of water, excessive shrinkage would occur during drying, and the part would be distorted and cracked.

A less-common method of casting involves using molten ceramics to manufacture high-density refractories, grinding wheels, and other forms of abrasives. *Molten ceramic casting* is done by pouring the melt into cooled metal molds to accomplish rapid cooling, or *quenching*. The quenching process causes stresses that result in the formation of fine-grained crystals, which results in a very hard product.

Glass Forming Processes

Processes used to form glass usually do not involve compacting or pressing solid material in a mold. Glass is created by melting stock until it becomes a viscous (thick) liquid and shaping it as it cools and solidifies. These basic steps have been refined into specialized manufacturing processes for creating specific products. These processes include drawing, floating, fiber drawing, tube drawing, centrifugal casting, glassblowing, glass rolling, glass pressing, and sagging.

Drawing

Drawing is a process that was used for many years to produce sheet glass. The first unsuccessful attempts to draw sheet glass directly from the furnace date back to the 1850s. It was not until 1905 that a Belgian named Fourcault improved the process enough to make it practical for manufacturing. The Fourcault process, first used in industry in 1914, involved pulling molten glass upward from an oven and through rollers to a horizontal conveyor. After being transported through an *annealing lehr* (oven), the glass is cooled and cut.

Since the early days of glassmaking, manufacturers have developed their own unique processes. Popular at various times were the Colburn process and the Pennvernon process. Today, most sheet glass is made using the float glass process. Drawing is more frequently used for producing glass fibers and tubes.

Floating

The **float glass process** produces sheet glass with the qualities of thicker plate glass. Today, there are approximately 260 float-glass plants worldwide that make around 95 percent of the world's supply of flat glass. Float glass is the most popular process for making safety glass, mirrors, and transparent glazing for commercial buildings. The patent for this process was obtained in 1848 by Henry Bessemer, but the UK firm Pilkington Brothers is credited for much of the early development. A schematic illustration of this process is shown in **Figure 17-15**.

The raw materials, sand, calcium, oxide, soda, and magnesium are weighed, mixed, and then introduced into a furnace where they are melted at temperatures between 1830°F (1000°C) and 2730°F (1500°C). The molten glass flows in a continuous ribbon from the furnace into a bath of molten tin. The glass is highly viscous and the tin is very fluid. These two materials do not mix and the contact surface between these two materials stays perfectly flat. Once the glass is formed, the next step is to add metal oxides directly to the glass. When the glass leaves the bath of molten tin, it has cooled down enough to be passed through an annealing lehr. In the lehr, the glass is cooled

Figure 17-15. Much of the glass produced today is made by the float glass method, where the raw materials: sand, calcium, oxide, soda, and magnesium are introduced into a furnace and melted at a temperature of 1000°C to 1500°C. The molten glass, which is highly viscous, flows in a continuous ribbon from the furnace into a bath of molten tin. The tin and glass do not mix and the contact surface makes the glass perfectly flat.

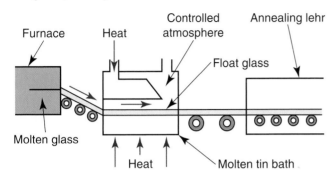

Figure 17-17. These workers, with the aid of an overhead crane, are moving a bundle of float glass sheets weighing four tons. The sheets will be fabricated into architectural glass panels. (PPG Industries)

at controlled temperatures, until it is reaches room temperature. A float line in a float glass plant can be nearly half a kilometer long. See **Figure 17-16**.

After floating on top of the molten tin, the glass cools to its exit point of 1100°F (600°C). Tin is used to float the glass because it is liquid at this temperature range and does not vaporize and burn at 1800°F (1000°C).

To reduce oxidation of the tin, the bath is enclosed in a protective gas. With the float glass process, the entire operation is almost totally automated. Today, throughout the world, almost all traditional plate glass manufacturing has been eliminated and float glass manufacturing has taken its place. See **Figure 17-17**.

Figure 17-16. This quality control technician is performing a visual inspection of a wide ribbon of float glass as it is conveyed through a cooling section of the processing operation. The glass ribbon runs back through the annealing lehr (shed-like structure in background) to the molten tin bath where molten glass first assumes the flat ribbon form. (PPG Industries)

Fiber Drawing

Glass fibers are made by drawing heat-softened glass marbles through small orifices, called *bushings*, in the bottom of a tank on a remelting furnace. Drawing fibers is an ancient process that is actually older than blowing glass. Today, the process has been automated and upgraded. See **Figure 17-18**.

Today, the production of glass fiber is still a relatively simple process. Here is how the process of fiber drawing works. A cylindrical, optically pure glass rod called a **preform** is heated to its softening point inside a furnace, normally around 2912° F (1600°C). The preform is then drawn through holes in a melting tank. The fiber that is produced is pulled through a cooling tank. Here, convective gases reduce the temperature to about 300°F (150°C). Finally, a plastic coating or *sizing* is applied, and the plastic-bound fiber is cured in an ultraviolet oven before it is spooled on a take-up drum. In some cases, the fibers are not plastic coated and are coated with sizing by a roller to protect them from becoming scratched and damaged.

Figure 17-18. This droplet of molten glass will be drawn into a fine optical fiber. (Corning Incorporated)

Once the fibers are sized, they are gathered together on a spinning drum. After collecting, they are further processed into yarns and roving material using conventional textile manufacturing processes.

Two basic types of fibers are **staple fibers** (6″–15″ in length) and **continuous fibers**. The fibers are sometimes sprayed with binders to hold them together in a continuous mat.

Tube Drawing

Glass tubes are made using the **Danner process.** With this process, molten glass flows down over a hollow mandrel. The mandrel is positioned at an angle to the glass, so the molten material runs down its length and off the lower end. Air is blown in through the center of the mandrel to keep the walls of the glass tubing from collapsing. The tubing is carried by rollers into an annealing lehr. The diameter of the tubing depends on the temperature of the glass, the amount of pressure blown through the tubing, and the speed at which the tubing is pulled off of the mandrel.

Centrifugal Casting

You will recall from an earlier chapter that the centrifugal casting process was used with metallic materials. The same process is also used to form ceramic pieces, particularly cone-shaped glassware. **Figure 17-19** illustrates how this process works.

Centrifugal casting is used with many different types of materials. When used with glass, the process begins by filling the rotating, funnel-shaped mold with molten glass. Centrifugal forces created by the rapidly turning mold throw the mass of molten glass, called a **gob**, against the walls of the mold, causing it to creep upward as the mold continues to turn. A distribution rod is then inserted into the mass of molten glass, helping to force the glass against the walls of the mold. Once the walls are coated to the desired thickness, the glass funnel is trimmed to height with the trimming wheel. When the product is cooled, the finished part is ejected from the mold.

Glassblowing

You are probably familiar in a general way with the ancient process of glassblowing. In this process, a gob of molten glass, called a *gather,* is formed into the desired shape using air pressure. Much scientific glassware is still produced in this way.

Most glassblowing in industry today, however, is done on automated high-production glassmaking

Figure 17-19. Centrifugal force causes molten glass to flow outward and upward along the walls of a rotating mold, forming a cone shape. As the glass cools, it is first trimmed, then removed from the mold.

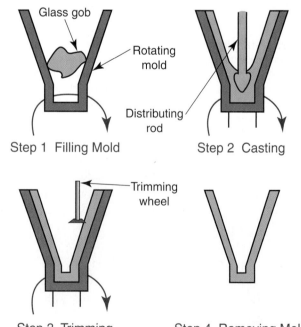

machines. Michael Owens invented the first automated blowing machine in 1903.

There are three major types of automated glass-blowing machines in use today: the press and blow machine, the blow and blow machine, and the ribbon blow molding machine. Some machines are used primarily for the manufacture of bottles, jars, or containers. Others are used to manufacture such products as lightbulbs and Christmas tree ornaments.

Press and blow machine

The *press and blow machine* is used primarily for container production. This process is illustrated in **Figure 17-20**.

Container production is accomplished in two major operations. The neck of the container is made by pressing or blowing, then the rest of the container is formed by blowing.

After the part is blown, the mold is opened and the product is transported to the annealing lehr. The annealing process permits the glass product to gradually cool down, thus relieving stresses and strains created during forming. The press and blow process is ideal for producing containers with thin walls, such as disposable jars and bottles. See **Figure 17-21**.

Blow and blow machine

The *blow and blow process* is a two-mold process that creates a blow-molded bottle by forming the neck first and then the rest of the bottle. See **Figure 17-22**. This process begins with molten glass being placed into the parison mold. When the blow head is closed, a blast of air blows the molten glass down into the mold. This forms the neck of the bottle. At this point, an air blast is introduced from the bottom, blowing the glass upward and against the walls of the mold. The basic shape of the parison has now been created.

The parison is removed from the first mold, inverted, and placed in the second mold. The parison is heated, then a blast of air into the mouth of the bottle expands it to its final shape.

Figure 17-20. The press and blow process.

Step 1
Molten glass is
placed in a mold

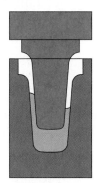

Step 2
A forming tool is inserted in the mold causing
the molten glass to fill the mold cavity

Step 3
The forming tool is removed
and the preform is complete

Step 4
The preform is placed
in the blow mold

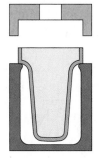

Step 5
The neck of the
preform is sealed

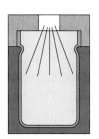

Step 6
Air is blown in, forcing
the molten glass against
the walls of the mold

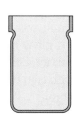

Step 7
The finished glass
product is removed
from the mold

Figure 17-21. Glass bottles that will eventually hold salad dressing are shown emerging from their molds in this blow and press operation. Note the rapid cooling of the glass, as indicated by the color difference between the bottles in foreground and those just removed from the molds. (Ball Corporation)

Ribbon blow molding machine

In the *ribbon blow molding process*, a ribbon of molten glass, approximately 3″ wide, flows continuously from an overhead melting tank through two water-cooled forming rollers. As the ribbon travels along a steel track, air heads from above contact the ribbon and blow the glass to shape the parison. At the same time, split molds come up to surround the molten glass. Air blows the parison against the walls of the mold to shape the final product. After the mold halves separate, the completed bulbs are cracked off the ribbon and fall onto a transporting conveyor. Ribbon blow molding machines are used to manufacture incandescent lightbulb blanks (envelopes), vacuum bottles, and clock domes.

Glass Rolling

If flat glass does not need to be totally transparent, it can be poured and rolled. *Rolled glass* is translucent, with a light transmission capability of 50 to 80 percent. Rolled glass is often used in skylights, bathrooms, and interior lights.

The molten glass flows out of the furnace over a refractory barrier called a *weir*, onto a machine slab, and through two water-cooled rollers. It is these rollers that establish the thickness of the plate. A refractory gate permits glass to flow onto the slab. If wire-reinforced glass is being made, the wire is introduced into the hot glass using a locating roller. If the glass is to be decorated with a design, shaping rollers are used to emboss the glass, creating the desired surface relief.

Once the glass is shaped, it is carried by rollers to the annealing lehr. When it enters the lehr, the cast glass is reheated to about 1470°F (800°C). When the glass comes out of the lehr, it is cool enough to handle. Finally, it is cut to size and packed for shipment. The rolling process is used to manufacture wire reinforced safety glass, colored plate glass, greenhouse glass, and opaque glass.

Glass Pressing

Automatic presses are also used to press molten glass into metal molds. Once the molds are loaded with glass, the desired gob is pressed using a metal plunger (*follower*), to form the desired shape. While this is happening, another molten gob of glass is being placed in the next mold. After each piece is pressed, it is air-cooled, ejected, and transported by a conveyor to the next work area. The piece that is *mold pressed* is unique, in that the shape or form of the interior wall is independent of the exterior. With pieces that are blown, the outside and inside walls have a corresponding shape. Pressing glass was first mechanized in the United States in the mid 1820s.

Figure 17-22. The blow and blow molding process for glass containers involves the use of two molds. The first mold is used to blow the neck, then the parison is transferred to a second mold to be blown into final form.

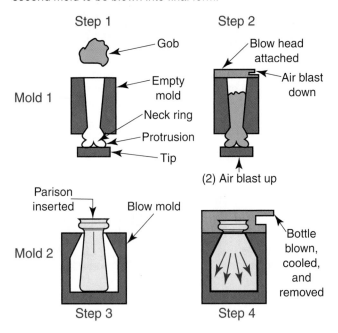

Slumping

Slumping is a technique that has been used commercially for many years for shaping glass. Slumping is accomplished by placing sheet glass over a mold, called a *slump mold*, and applying heat until the glass softens and drapes downward. The glass conforms to the shape of the mold and takes on the texture of the mold surface. With this process, the temperature, normally from 1200°F to 1700°F (650°C to 930°C), is maintained to minimize stretching and maintain consistency in the cross-sectional thickness of the walls.

The process of slumping is simple, but not all shapes can be effectively slumped. The mold must have a gradual slope and not be too deep. It is also important that slump molds have holes in the bottom so air can escape during cooling. Molds for slumping glass can be made of metal or soft firebrick. Often *grog mix* (clay mixed with fine particles of previously fired clay) is used to make the slump mold.

Important Terms

annealing lehr
band casting
binders
blanks
blow and blow process
calendering
co-fired
cold isostatic pressing
continuous fibers
Danner process
de-air
deflocculant
diffusion
doctor blade casting process
doping
drain casting
drawing
dry bag isopressing
dry pressing
dust
dust pressing
dynamic compaction
extrusion

follower
float glass process
freeze casting
gob
greenware
green tape
grog mix
groundplanes
hot forging
hot isostatic pressing
hydration
injection molding
isostatic pressing
jiggering
jollying
mix muller
molten ceramic casting
morillonite
plasticizers
PLZT
preform
press and blow machine
pugs
pulverizer
ram pressing
refractories
ribbon blow molding process
rolled glass
sintering
slip casting
slumping
slump mold
slurry
springback
staple fibers
steatite
striplines
tablet pressing
tape casting
thixotropic
throwing
ultrasonic casting
via holes
viscosity
weir
wet pressing

Questions for Review and Discussion

1. What causes shrinkage in a clay product? Describe several methods that can be used to reduce the amount of shrinkage.

2. Distinguish between the processes of dust press and hot forging.

3. What is the advantage of hot pressing in terms of consistency and strength of material?

4. Why does calendering produce a better quality product than tape casting?

18

Processes Used to Form Composite Materials

Chapter Highlights

- One of the most common processes used to form composites is open molding.
- Closed-molding systems are more expensive but they offer some advantages that cannot be attained in open molding.
- Many of the forming processes for composites are tailored to specific applications to provide the necessary performance features desired for that product.

Introduction

Generally, a forming process changes the shape of a material without changing its physical or chemical properties. In addition, many forming processes are a relatively uncomplicated step in the overall manufacturing process. However, composite forming is the exception being one of the most involved and critical processes in manufacturing. With most materials, it is relatively easy to classify a processing method as being either a primary and secondary process. As you may recall, a primary process manufactures the industrial stock while a secondary process uses the stock to manufacture the final product. In the field of composites, however, it is difficult to distinguish between primary and secondary processing. A composite fabricator is making the composite material or stock (primary processing) as it is being formed into the final product—secondary processing. In most cases, a composite manufacturer combines raw materials through a chemical reaction to form the part.

There are many ways to form a composite. The selection of a forming process is influenced by the type of matrix and the reinforcement that is used to create the composite. You may recall that there are three types of reinforcement materials—fillers, fibers, and solids—that can be used to create a composite. In addition, the polymeric-matrix (resin) used in the composite influences the manufacturing process, including cycle time and lot size. All of these factors must be considered when selecting a forming process.

Some of the major processes used to form composites are open molding, vacuum-bag molding, infusion molding, resin transfer molding (RTM), thermal expansion molding, pultrusion, and filament winding. Each of these forming processes has one or more methods for applying the matrix material. In some cases, such as infusion molding, the distribution and infusion of the matrix material is the most complex part of the process.

Open Molding

The simplest way to form a polymeric-matrix composite is *open molding*. In its most basic form, the reinforcement is saturated with the matrix resin and laid against a form (the mold) that creates the outside finished surface of the part. The resin's function is to transfer the load to the reinforcement.

Generally, the resin is applied to the reinforcement materials with a flow coater. A *flow coater* looks like a spray gun except the nozzle has a series of holes which send out a fan of droplets. Flow coaters, unlike spray guns, do not atomize the resin. Consequently, styrene emissions and overspray are greatly reduced.

In some cases, chopped fiber is mixed with the resin as it flows onto the mold surface to speed the lay-up process. Once the resin and fiber are applied, they are rolled out to remove trapped air and redistribute surplus resin to areas that are not thoroughly saturated. After this part of the process is completed, the composite is allowed to cure. See **Figure 18-1**.

Curing is a chemical reaction that begins when the resin is mixed with a small amount of an *initiator*, a substance that causes the chemical reaction (curing). When an initiator is mixed with these resins, the initiator is consumed in the curing process. The initiator is often incorrectly referred to as a catalyst. In scientific terms, a *catalyst* begins a chemical reaction but is not consumed by the reaction. In flow coaters the initiator is added just before the resin reaches the nozzle. The most commonly used matrix materials for fiberglass composites are polyester and vinylester resins.

One of the big drawbacks to open molding is the emission of styrene, which is a component in

Figure 18-1. Chopped fiberglass and resin are being applied to an open mold using a chopper gun. The mold is for a fiberglass skiff that is popular for fishing on lakes and rivers.

A High Volume Low Pressure (HVLP) internal mix spray gun sends resin out of the nozzle in a fan along with chopped fiberglass woven roving.

Woven roving is pulled into the gun from a container mounted over the mold.

Resin, initiator, and air lines are supported by a mechanism that gives the operator full mobility to move the gun around the mold.

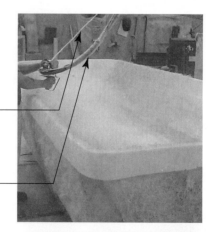

The completed boat's high gloss exterior finish is due to the smooth surface of the open mold.

the resin. The styrene serves two functions in the forming process. First, it acts as a solvent to improve the resin's ability to saturate the reinforcement. Second, it is a component in the chemical reaction that occurs when the resin cures.

Composite manufacturers have made several changes in their processing over the past decade to reduce styrene emissions. First, they began replacing spray guns with flow coaters. Improvements in resin formulations and the use of flow coaters made it possible to saturate glass fiber reinforcement with lower levels of styrene. However, open molding leaves a large surface area exposed to the atmosphere, which allows styrene to escape until the resin is finally cured.

Styrene is a clear colorless oily liquid that plays an important role in the application and curing of the resin. At room temperature, styrene is very volatile and has a strong, but sweet, odor. The human nose is able to detect styrene at levels well below what the United States Occupational Safety and Health Administration (OSHA) has established as the airborne permissible exposure limit (PEL). There are actually two PEL limits for styrene. The first is 50 parts per million (ppm) averaged over an 8 hour work shift and the second is 100 ppm not to be exceeded

during any 15 minute work period. Exposure above these levels can irritate the eyes, nose, throat and skin. Very high levels of exposure can cause a person to feel dizzy, lightheaded, and even become unconscious. Continued high-level exposure could cause brain and liver damage.

One of the methods developed to prevent styrene from escaping to the atmosphere during curing is to cover the surface while the chemical reaction is taking place. One of the more popular methods uses a simple plastic sheet as the covering.

Vacuum Bagging

When open molding, *vacuum bagging* can be used to cover the surface of the composite once it has been laid up in the mold. The bag is a flexible plastic sheet that conforms to the composite. Once the desired matrix and reinforcement materials are applied, the bagging film is placed over the mold. A fitting on the film enables a vacuum pump to remove the air between the film and the mold, which compresses the composite during curing and squeezes out trapped air. See **Figure 18-2**.

Besides providing a barrier to prevent styrene from escaping, squeezing the composite increases

Figure 18-2. This diagram compares vacuum bagging and infusion molding processes.

The **vacuum bagging process** begins with an open mold. In this example, a section of a boat hull is shown.

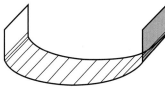

The **infusion process** begins with an open mold. In this example, a section of a boat hull is shown.

The first step in vacuum bagging is to coat the mold with gel coat.

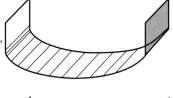

The first step in the infusion process is to place all the reinforcement in the mold.

Next, the fiberglass reinforcement is placed, saturated with resin, and rolled out.

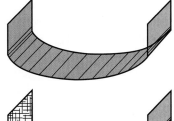

A flexible cover fitted with a resin distribution system is placed over the reinforcement and sealed at the edges. A vacuum is drawn once the resin supply lines are connected to the cover. Next, a valve is opened allowing resin to infuse the reinforcement. The vacuum is maintained until the resin is cured.

Finally, a plastic cover fitted with a vacuum hose is placed over the uncured composite. The cover is sealed at the edges, a vacuum is drawn, and it remains in place until the resin is cured.

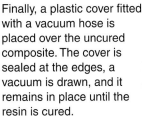

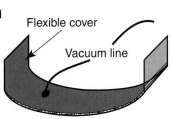

Flexible cover

Vacuum line

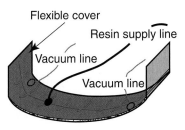

Flexible cover

Resin supply line

Vacuum line

Vacuum line

the density and improves some physical properties. Vacuum bagging, however, is often not practical on large molds that are being progressively laid up. In situations like this, the mold is continuously curing as the reinforcement and resin is being applied.

Infusion Molding or Closed System Molding

Infusion, also known as the Seemann Composite Resin Infusion Molding Process (SCRIMP), reduces styrene emissions by wetting out and curing the laminate in a closed system. The key to the process is the resin distribution system patented by Seemann Composite Systems, Inc., of Gulfport, Mississippi. In general, the patent centers on a flexible cover incorporating a medium for resin distribution that enable a builder to get repeatable properties from a closed molding system. Infusion molding uses atmospheric pressure to squeeze the resin into the reinforcement fibers, similar to the vacuum bagging process.

The infusion process provides some structural benefits that the developers of the process say rival the material and mechanical properties obtained in highly controlled autoclave processes. In addition, the infusion process creates a laminate in "one shot", eliminating secondary bonding problems. The process also provides opportunities to achieve fiber to resin ratios as high as 70:30 along with the virtual elimination of air entrapment and voids. The infusion system is suitable for low-volume production.

The process requires a mold similar to any open molding process and a flexible covering similar to a vacuum bag. The process, as described in the patents, begins by fabricating a bag from silicone rubber to conform to the mold. The silicone rubber starts out as a brushable liquid so the bag can be made by applying several coats over a completed lay-up that has been left in the mold. Once the silicone rubber is cured, it can be peeled from the mold. The result is a tough, conforming mold cover that can be reused many times. The cover also incorporates some other features such as resin distribution manifold with branch conduits to provide resin paths to all parts of the laminate. At the edges of the bag are vacuum manifolds. The bag may also include a pattern of small pillars, cones, or pyramid shapes formed on the inner surface of the bag to hold the bag off the laminate. This provides local paths for the resin to flow into the laminate. An alternative to this approach is the inclusion of a mesh or a resin distribution medium that serves to hold the bag off the laminate.

Some of the commercial processes use disposable bags, which are used only once. This approach however creates a significant amount of solid waste and is justified only for "one off" production.

The sequence of operation begins with laying up the dry fiberglass against the mold in the desired amount and orientation. Some builders use a spray adhesive compatible with the resin to hold the dry fiberglass in place. The mold cover with the resin distribution medium is put in place over the dry laminate. Once the system is sealed, a vacuum is applied and the encapsulated mold is checked for leaks. After the system has been evacuated, the resin is introduced until runoff in the resin channels indicates that the laminate has been completely impregnated. The mold remains sealed until the resin is cured. After curing, the bag is carefully peeled away and cleaned for reuse.

The key objective in the infusion process is to create a closed molding system using a conforming bag structure that can be reused. This closed system, along with a carefully planned and placed reinforcement system, creates a high strength composite with a minimum of waste and emissions.

Prepregs

Most open-mold fiberglass fabricators utilize resin spray or flow coaters for applying gel coat and resin to a mold. Conventional gun-type resin application systems use compressed air, high fluid pressures, or combinations of fluid pressure and compressed air to spray resin materials. Initiators for the resin systems are normally introduced inside the spray apparatus or outside the nozzle in the turbulent resin stream. Spray guns and flow coaters may also be equipped with glass choppers, which cut fiberglass roving into short lengths and propel it onto the molding surface.

Spray guns and flow coaters are considered very productive in transferring resins to the work surface for low and medium volume applications. However, for a number of years manufacturers that make composite parts for the aircraft, automotive, and military markets have relied on the use of fiber reinforcements that are presaturated with resins. These materials, referred to as "prepregs", offer a number of advantages over conventional resin application techniques. Prepregs enable manufacturers to closely control resin to fiber ratios, and atomization of pollutants (styrene) is practically eliminated. A further benefit is that cleanup and waste (no over spray) is greatly reduced. These advantages are, however, not enough to make prepregs widely accepted by most fiberglass fabricators.

Prepregs are generally formulated with more expensive epoxy based resins, which require heat to complete the cure cycle. These more expensive resins are normally combined with exotic, high strength reinforcing materials, such as graphite fibers. Storage and shelf life is also a problem since the prepreg must remain refrigerated until the lay-up process begins. Prepregs appear to be best suited for applications where extremely high strength-to-weight-ratios are required and cost factors are secondary.

Resin Transfer Molding Processes

Resin transfer molding (RTM) production systems can be used to replace many conventional open molding processes. RTM molds can be produced using the same materials and techniques used to fabricate conventional open molds. The molding resins and reinforcement materials used are similar to the materials used to produce similar components in open molds. Even the gel coat finishes that provide the color and protective surface for a composite are the same as those used in open molding. The major difference between open molding and RTM is the use of an "inside" mold. RTM requires this inside mold to finish the part, similar to the two halves of an injection mold. See **Figure 18-3**. Open molding has a rough inside surface because it has been rolled out by hand. With RTM, the inside surface can also be a finished surface, providing sufficient material and resin has been placed to fill the cavity. The need to completely fill the RTM mold requires a great deal of technical skill to produce uniform good quality parts. However, for parts such as truck hoods, which require a good surface finish on the outside and inside surfaces, the extra care that RTM requires is worth the investment.

RTM processing begins with the application of a gel coat to one or both sides of the mold, depending on requirements. Glass reinforcing and other materials, such as core stock, are placed in the bottom half of the mold. Once the reinforcement is in place, the mold halves are closed and securely clamped.

Figure 18-3. Resin transfer molding (RTM) provides a finished surface on both sides of the part.

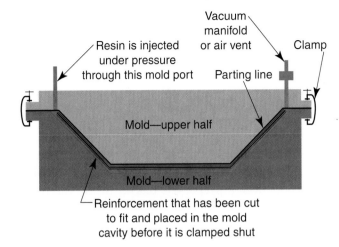

Next, resin that has been mixed with an initiator is injected through one or more strategically located ports. Inlet ports and vents are normally located in the top half of the mold. Molding pressures typically are between 30 and 75 psi and the molding process is usually carried out at room temperature.

Making a set of matched molds for RTM is not difficult. The two halves of the mold are laid-up over a pattern that is identical in size and finish to a completed part. The mold is generally made from the same types of materials (a fiberglass composite) used to produce a mold for open molding. Some specialized tooling is required to ensure that the two halves maintain alignment and are able to withstand the pressures exerted during operation. The tooling must also reinforce the molds to prevent flexing during the injection and curing cycles. Providing the correct number and location of vents is essential to ensure that air is not trapped in the mold. Any air that is trapped can cause voids or blisters in the finished part. Careful design and workmanship is even more critical in RTM molding than in open molding.

Thermal Expansion Resin Transfer Molding

A variation of the basic process of resin transfer molding is *thermal expansion resin transfer molding* (TERTM), where a preformed rigid cellular (foam) core is covered with reinforcement material. Core materials are usually polyvinyl chloride, polyurethane, or polyamide rigid foams. Reinforcement materials such as fiberglass, carbon fibers, or Kevlar®

are also popular. Once the core is covered, it is placed inside a matched die and impregnated with epoxy resin. At this stage, the mold is only partially filled. Heat is applied and the foam expands. This creates pressure that squeezes the reinforcement and matrix materials together.

Expansion Molding

A two-part flexible mold is used in the process known as *expansion molding*. The mold is usually made from an elastic material such as silicone rubber. In production, a prepreg is placed inside the mold. The mold is then closed and placed between two platens in a press. The platens provide support and keep the exterior walls of the mold from expanding. The entire assembly is then placed in an oven for curing. The silicone rubber of the mold will expand until it reaches the curing temperature. The mold then stops expanding and provides forming pressure necessary to compress the prepreg.

Pultrusion

Pultrusion, also referred to as *continuous extrusion*, is a specialized application of the basic process of extrusion. Continuous extrusion is particularly useful for constructing composites, since it orients and controls the matrix and reinforcement materials much better than either the open molding or liquid infiltration processes. Continuous extrusion also permits the blending of different types of materials.

In the pultrusion process, the matrix and reinforcement materials (generally fibers) are pulled through an impregnator or resin bath. See **Figure 18-4**. As they

pass through the bath the reinforcement materials are saturated with the matrix material (resin or a molten polymer). Immediately after the bath, there is an aperture or preformer that serves to remove any excess matrix material. This device also provides a means to organize and shape the saturated materials before they enter the die. The die that is used in continuous extrusion may include several different pathways or channels. These channels permit the process to combine multiple sources of matrix/fibers, and in some cases fillers and adhesives. The die squeezes these materials together to form them into a multilayered composite. The matrix begins to cure in the die and continues as the pultruded shape moves through a curing oven. Some of the most common profiles produced are rounds, squares, U-shapes, or stars.

Almost any reinforcement material can be adapted for pultrusion to produce products such as the side rails for ladders, construction panels, and roofing panels.

Filament Winding

In the *filament winding* process, a spinning mandrel is covered with a resin-coated filament to form a tube-shaped product. See **Figure 18-5**. Products such as tubes, struts, drive shafts, underground storage tanks, and pressure vessels are produced by filament winding. The *filament* can be made from strands of carbon fiber, fiberglass, or aramid. Filament winding is divided into wet-winding and dry-winding. In *wet winding*, the filament is drawn through a bath of liquid resin before being wound around the mandrel. In *dry winding*, the filament is dry when it is wound on the mandrel and cures with the application of heat.

Figure 18-4. Pultrusion, also referred to as *continuous extrusion,* is a specialized application of the basic process of extrusion.

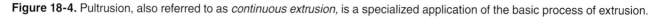

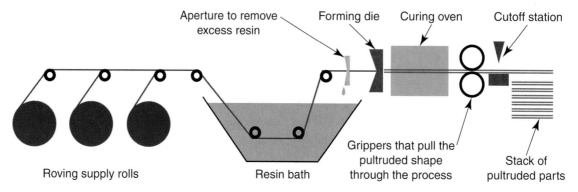

Figure 18-5. Filament winding uses a spinning mandrel that is covered with a resin-coated fiber or filament to form tube-shaped products.

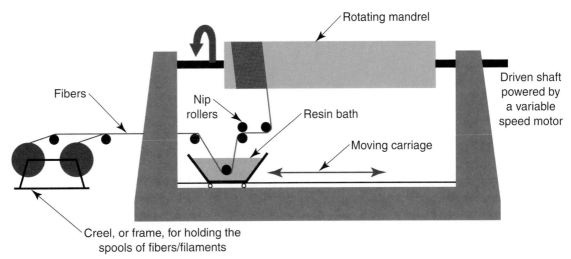

Here are the detailed steps for wet winding.

1. The resin is mixed with an initiator and flows into a dip bath.
2. The fiber is saturated by running it through the dip bath.
3. The mandrel is coated with a mold release compound.
4. The resin-coated fiber is wound around the rotating mandrel.
5. The composite is permitted to gel and cure.
6. The mandrel is removed.
7. The product is finished and inspected.

Wet winding is a rapid process. As an example, one manufacturer makes a tube 59" long by 2.6" in diameter (1500 mm × 65 mm) with a wall thickness of 0.1" (3 mm) in approximately one minute.

Tubular products are made on winding machines that vary in construction from simple lathe-type winders to complex computer-controlled production machines. Portable equipment has also been designed for manufacturing large storage tanks on location.

Dry winding uses prepregged *fiber tows* (bundles) or tape. Dry winding is a popular process for manufacturing fishing rods and golf club shafts. The process usually involves winding the prepreg onto a warmed mandrel that begins the curing process.

Mandrels serve as the molds in both the wet and dry winding processes. Some mandrels are designed to make only tubes, but it is possible to wind parts with closed ends. Irregular shaped parts are also routinely wound. To aid in removing the mandrel from a cured part, a mandrel can be made to be collapsible. In some instances, mandrels are made of low-melting-point metals. These types of mandrels are melted so that they can be poured out of the completed part. The molten metal is used to cast another mandrel. Other types of mandrels are inflatable. Once they are wound and the product is cured, the mandrel is deflated and removed. Even cardboard or wooden tubes are used, in some cases, to wind parts as long as 30' or 40'.

Important Terms

catalyst
curing
dry winding
expansion molding
fiber tows
filament
filament winding
flow coater
infusion
initiator
mandrel
open molding
pultrusion
resin transfer molding
thermal expansion resin transfer molding (TERTM)
vacuum bagging
wet winding

Questions for Review and Discussion

1. Explain the difference between open molding and vacuum bag molding.

2. How does infusion molding differ from resin transfer molding?

3. What is the difference between a catalyst and an initiator?

4. Which forming process would you use to make the following composite parts?

   A. A 12″ diameter pipe with a 1/4″ wall thickness to be used in a chemical plant.

   B. A roof panel for a road grader. Production is one grader per day.

   C. Tool chests for contractors. The company produces five units a day.

   D. Overhead storage bin doors for a passenger jet. A hundred parts are needed for each plane and production is four planes per month.

19

Processes Used to Separate Metallic Materials

Chapter Highlights

- Shearing is a method of separation that is used in several processes to cut out parts without making chips.
- Chip-making machine tools are extremely versatile and automated.
- Cutting tools are highly specialized and enable machining and turning centers to remove high volumes of metal quickly and accurately.
- Drilling is a common operation that requires cutting tools that have been designed exactly for that purpose.
- Grinding can finish parts to exact dimensions with a fine surface finish.
- There are several processes that have been developed to cut or remove material in difficult situations with precision and with a minimum of stress on the part.

Introduction

There are many manufacturing processes for separating material from a metal workpiece. Some separating processes shear or punch parts from flat stock while others cut material from bars and plate. The processes that use a cutting tool turn the unwanted material into chips. Some examples of separating processes that generate chips are turning, drilling, and milling. The chips become scrap that may be recycled, but nevertheless the chips are wasted material. Therefore, conversion efficiency (reducing the amount of waste generated) is an important factor for selecting a separating process.

There are other types of separating processes that do not produce chips. These processes punch or shear the stock cleanly. Although there are no chips produced, there is still waste. The material separated from the part is the *drop* or *punching*, which becomes waste unless other uses can be found for this material. Other categories of separation processes use chemical, heat, or abrasion.

Major Separating Process Categories

The first two separation processes described here are the most universal of all. You have probably used them in shaping paper. Common tools for separating paper are scissors, hole punches, and knives. The tools used to accomplish the separation process in metal are similar in principle to the paper tools but much more robust to handle the forces needed to separate metallic materials.

Shearing

Shearing is a mechanical separating process that is used to cut sheet metal or plate. There are three forms of shearing, which are based on the type of blade or cutter used. They are straight shearing, punch-and-die shearing, and rotary shearing.

All shearing processes involve the use of opposing surfaces. See **Figure 19-1**. The upper blade moves down into a workpiece that is held securely by the bottom fixed blade. The upper blade pushes the stock down into the opposing bottom blade, severing the metal.

Straight shearing, or cutting of sheet metal to size, is usually done with a machine called a squaring shear. Sheet metal is placed on the horizontal machine table of the shear, where a clamping

device secures the sheet metal. The operator then steps on a foot treadle, engages a lever, or pushes a button to power the blade.

Opposing surfaces can also function as a punch-and-die. The action is similar to that of a scissors. The blade edge is angled so that when it travels down, the blade starts shearing on one side and continues across the sheet metal to the opposite side. The path that the shearing follows is called the *shear line*. This action concentrates the force at a point and minimizes the distortion of the material along the shear line. **Figure 19-2** shows a CNC-controlled hydraulic shear that can cut steel sheets up to 3/8" thick and 161" wide.

Punch-and-die shearing requires a punch to push into the sheet metal and a die to support the sheet metal and provide another shearing edge. Punches may be round and produce circular holes or they may be machined to almost any custom shape. The matching die must have a hole that is the exact shape of the punch. The shear line follows the profile of the punch and may travel in two directions at the same time. An example of this type of shearing is seen in a standard paper hole punch. By examining the curved profile of the punch while it is piercing a piece of paper, you will see how the shearing starts on two opposite sides and travels around the punch's shape.

Rotary shearing is done with a blade that is a rotary wheel. See **Figure 19-3**. Rotary shearing can be used to cut either straight or circular shapes. This allows the operator to cut small radii and irregular pieces. Rotary shearing is a hand process that is not practical for high-volume production applications.

Figure 19-1. Shearing is a process that cuts material by pinching it between a moving blade and a fixed blade.

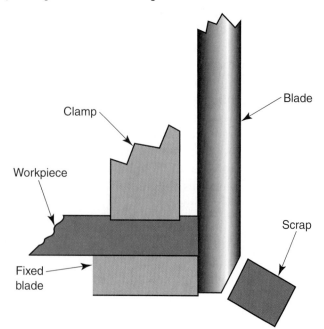

Figure 19-2. This hydraulic shear can cut steel sheets up to 3/8" thick and 161" wide. (Photo courtesy of Dreis & Krump)

Figure 19-3. Rotary shearing uses opposed wheels to make straight and curved cuts.

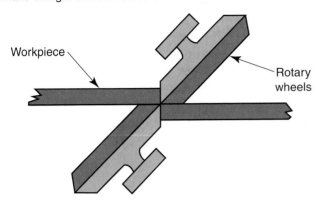

Workpiece

Rotary wheels

Specialized types of shearing

In addition to the three major types of shearing described, there are specialty processes based on the principle of a shearing or punching action. These include lancing, slitting, notching, perforating, punching, and nibbling. Each of these processes use a shearing action, but is done with a different type of shearing machine.

Perforating is a punching process used to produce a number of closely and regularly spaced holes in a straight line across a sheet metal section. This is often done to facilitate bending of metal in a particular area or to expand the metal. See **Figure 19-4**. Very small holes are difficult to punch and seldom justified in terms of being cost-effective.

Both lancing and slitting are accomplished without removing any material. *Lancing* is a punching process that is used to make a tab without removing any material, **Figure 19-5**. The tab that is punched is normally bent to provide an area for gripping or moving a part. *Slitting*, **Figure 19-6**, is

Figure 19-4. The process of perforating results in a line of evenly spaced holes across a surface to allow the metal to be expanded.

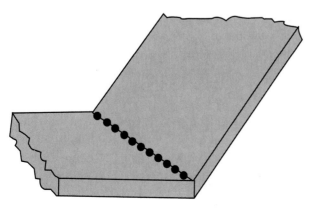

Figure 19-5. The lancing process shears metal to make an L-shaped tab for assembly or other purposes. No metal is removed.

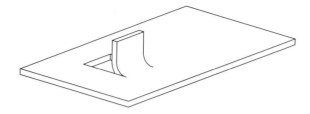

Figure 19-6. Slitting is a punching process often used to prepare sheet metal parts for assembly. As shown in the cross section, the material between the parallel slits is pressed upward or downward to allow a tab to be inserted.

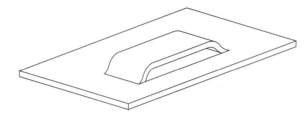

done to create an area where another part or device can be inserted. Both lancing and slitting are variations of the basic process of shearing.

Blanking and Punching

Blanking is used to punch a flat metal part (blank) from sheet metal. This process produces no chips and often generates very little waste. Blanks are produced with each stroke of a punch press. Special punches and dies are used.

Blanking is often confused with punching. In the process of punching, the punch moves down into a mating die. The stock to be punched is fed between the punch and die. As shown in **Figure 19-7**, the

Figure 19-7. When punching is used to produce a part, the punched out area is usually waste.

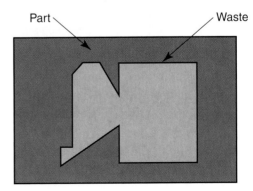

Part

Waste

Figure 19-8. When blanking is used to produce parts, the punched out areas are usually the parts, while the surrounding material is waste.

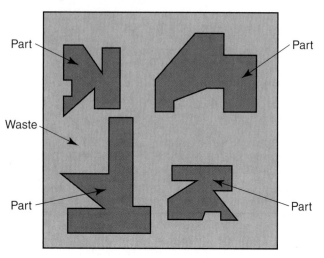

stock that is stamped out with the punch is generally waste, and the part is what is left. In blanking, the part is the area that is stamped out. The stock that is outside the blanked part is generally waste. See **Figure 19-8**.

In blanking, the punch travels only about halfway through the thickness of the stock. Usually, several blanks are stamped at the same time. These are carefully positioned so that multiple parts can be cut from one workpiece with a minimum amount of waste. Positioning of parts in this manner is called

nesting. Care must be taken in designing parts to make them nest effectively in the work blank. See **Figure 19-9**. The purpose of nesting parts, in addition to minimizing waste, is to simplify handling and subsequent processing. Several parts can be handled or processed at one time. After all operations have been performed on a part, it is broken off the sheet.

When individual parts are needed in low volume, other process options may be more practical than blanking. Nibbling and notching are two processes that are used to cut or punch out individual parts. Both of these processes differ from blanking in that they generate chips.

Nibbling

Nibbling is a simple process that requires no custom tooling or unique fixturing. It is a practical method for cutting out limited numbers of flat parts with complex shapes. The process is done on a *nibbling machine,* which looks something like a foot-powered stapler. The machine supports a round or triangular punch that moves up and down in a mating die at a rate of up to 900 strokes per minute. The process may be cost-effective when limited numbers of unique parts must be cut from sheet metal, but it is time-consuming.

Nibbling is done by inserting sheet metal into the opening between the punch and die. The metal is

Figure 19-9. To minimize waste, parts must be carefully positioned or nested. Industry uses computer programs to accomplish part nesting.

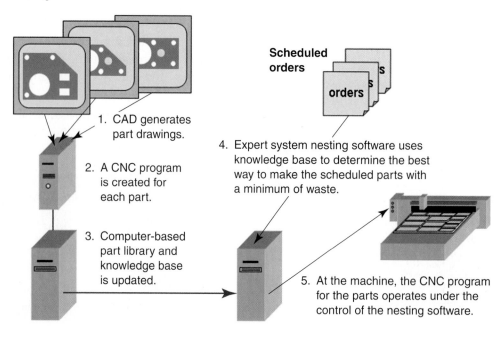

held and manually or automatically guided into the machine, using a template. The sheet that is nibbled is held down in the machine by a stripper. This prevents the part from being bent or distorted when the punch pierces the metal. Nibbling is able to cut metal using many overlapping strokes of a small punch. The material on the inside of the punched line is the finished part, which generally has a smooth edge.

A process similar to nibbling is called *notching* or piercing. See **Figure 19-10**. Notching is also a punching process, and is typically done using a punch press. The material that is removed by the punch is scrap.

All shearing, punching, and stamping processes must be carefully selected, depending on the type and gage of metal that must be processed. When a stamping process is involved, the frame and the mechanical components that hold the die and punches must be extremely rigid to prevent flexing. Consequently, punch presses are massive and heavy compared to the size of the metal stock being processed. Stock is generally classified according to gage. Aluminum, stainless steel, and steel sheets are all classified by gage numbers that indicate the thickness of the sheet. Surprisingly, the thickness for the same gage number is not the same for each of the three materials. As an example, the thickness of 18 gage steel is .048", stainless steel is .050" while aluminum is .052". There is also a more general way to classify the thickness of these materials by referring to the sheet as being light, medium, or heavy gage. Light gage stock covers thicknesses up to 0.031" (0.79 mm) or 22 gage. Medium gage stock consists of sheet from 0.031" to 0.109" (0.79 mm to 2.77 mm) or 12 gage. Heavy gage stock is any material that is above 12 gage.

Turning

All the processes discussed so far in this chapter are used to shear, cut, or press parts from flat stock. It is often necessary to reduce the diameter or change the profile of round stock as well. *Turning processes* are used to machine rotating parts. The most common machine used to perform machining operations on round parts is the *lathe*, which is more commonly referred to as a *turning center*. The primary function of a turning center is to place a stationary cutting edge against a rotating bar of metal to reduce the diameter of the bar. However, the turning center has expanded into one of the most versatile machining

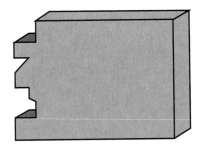

Figure 19-10. Notching to produce irregular contours on the edge of a part can be done on a CNC punching machine or a punch press.

tools in use today. Some of the tasks routinely performed by the turning center include:

- **Drilling and boring.** A workpiece is held in a rotating chuck while a cutting tool or drill bit carves an accurate hole in the end of the workpiece.
- **Facing.** As the workpiece is rotated, a cutting tool moves across the face to square the end of the workpiece and reduce it to the desired length.
- **Knurling.** A special tool is pressed into the circumference of a slowly rotating workpiece to form diamond-shaped indentations. Knurling is used to provide a gripping surface on a part, or to expand the part's diameter.
- **Parting.** A cutting tool is pressed into a rotating workpiece to cut off a portion of the material.
- **Threading.** To form internal or external threads, a cutting tool is pressed into the surface of a slowly rotating round bar while the tool slides along the bar's axis at a controlled rate.
- **Turning.** Rotating a length of round stock while a cutting tool moves toward and along the stock's axis forms a part with various diameters or a tapered profile.

Many complex parts, such as a fuel injection body, can be completely machined in a turning center. **Figure 19-11** shows a turning center and the workpiece made on the machine. Turning centers are normally classified according to two factors. The first is the maximum diameter of workpiece that can be turned, which is referred to as *swing*. The second is the maximum length that can be turned between centers. **Figure 19-12** shows how these factors are measured.

Work-holding devices and methods

The rotational forces and cutting loads that a workpiece receives in a turning center can be

Figure 19-11. This is a turning center ready to machine a workpiece.

This turret holds the tools needed to machine the fitting being made by this turning center

The workpiece is held in place by a three-jaw chuck

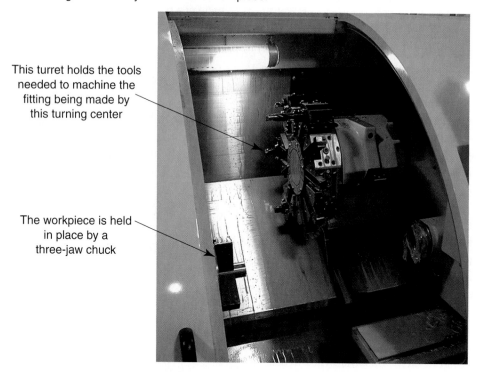

Figure 19-12. Turning center factors that determine the size of the workpiece that can be machined.

Note the turret holding cutting tools

The swing is the maximum diameter of a workpiece that can be held in a turning center

The distance between the chuck and the tailstock determines the length of the work area

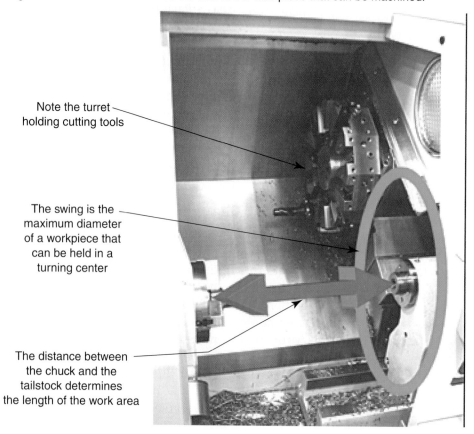

tremendous. The method of securing work in the machine must be considered before machining can begin. Since the workpiece rotates, it is subject to rotational forces and cutting loads. Therefore, the work-holding devices have to be able to securely grip the part but not interfere with the machining operations. The most widely used work-holding device is the ***universal chuck***. The most common version of the universal chuck has three jaws and is like the chuck found on electric drills in most home workshops. The three jaws move simultaneously so that the item, whether it is a drill bit or a workpiece, is gripped and centered along the rotating axis of the chuck. Therefore, a universal chuck is a self-centering chuck.

A ***collet*** functions much like a universal chuck and is used to grip and center smaller round, square, hexagonal, or octagonal stock. A collet is a thin steel cylinder with an accurately machined bore that is slightly larger than the workpiece. The cylinder also has three or four lengthwise slots that run about two-thirds the length of the cylinder creating fingers. The outside of the collet is machined to a taper so that when it is drawn or pushed into a similarly tapered sleeve the fingers will move in to securely grip the

workpiece. Since they are made for a workpiece of a specific size and shape, collets offer better accuracy and repeatability than a universal chuck.

A *faceplate*, which is attached to a lathe's spindle may also be used to hold workpieces. In application, a workpiece is clamped or bolted to the faceplate. Generally, a faceplate is used with workpieces that are too large or irregular in shape to be held in a chuck. A casting is a common item that is frequently mounted on a faceplate. The primary drawback in using a faceplate is the time required to attach the workpiece and position it for machining. **Figure 19-13** shows several work-holding devices for a turning center.

Turning tools

Years ago, cutting tools were made from high-speed steel and ground by hand to the desired shape. Hand-ground cutting tools are now in limited use mainly for low-volume specialty machining. Today, cutting tools use disposable inserts made from engineered materials. These inserts are designed to control tool loads, heat, and surface finish. Tool holders provide the needed support and rigidity these inserts require.

Cemented carbide is one of the most common engineered materials used in tool inserts. This insert

Figure 19-13. A variety of work-holding devices for a turning center are shown here. (Positrol Inc.)

A drawdown chuck

A second operation collet chuck

A collet chuck

A gear chuck

can withstand cutting speeds three to four times faster than high-speed cutting tools. Cemented carbide is made from tungsten carbide sintered into a cobalt matrix. Other metals, such as tungsten, may be added to the matrix to improve the tools properties. The carbide cutting tools may also be coated with a wear resistant material such as titanium nitride. This thin, fused layer of material improves the lubricity of the tool at the cutting edge, and therefore increases wear resistance several hundred percent.

Another popular cutting tool is made from a ceramic material called *cermet*. Cermet is an extremely hard and heat resistant material made from carbide and sintered oxides. Cermet cutting tools permit even higher cutting speeds and a better surface finish than cemented carbide. However, these tools are weaker than carbide tools. To work around its fragility, cermet inserts are limited to shock-free or low-shock applications.

Figure 19-14 shows some milling bits with cemented carbide tips. In **Figure 19-15**, a variety of turning and milling tools are shown. Notice the diamond- and triangular-shaped inserts mounted in the tool holders and scattered around the bottom of the photo. The inserts provide two or three cutting surfaces. Therefore, when one side becomes dull, the insert can be removed and turned to expose another cutting edge. When all sides have been used, the insert is disposed of and a new one is inserted. In some cases, the carbide material is brazed to the tool

Figure 19-14. Cemented carbide tips are brazed on drills used for high-speed cutting. (Mitsubishi Carbide)

Figure 19-15. Cermet inserts and tools used for turning, milling, and drilling. The inserts are screwed into place on the tools, and can be replaced easily when worn. (Mitsubishi Carbide)

holder. Brazed tools are suitable for low volume or custom machining.

Productivity factors

Minimizing the amount of material that must be removed saves time and minimizes waste. However, once a design has been established and the metallic stock selected the next task is to set up the machine to remove the unwanted material as quickly as possible. The factors that govern how fast material can be removed are depth of cut, feed, and speed. The distance that the cutting tool penetrates the workpiece is the *depth of cut*. *Feed* is the distance the cutting tool travels in one minute. The feed is usually specified as inches per minute or millimeters per minute. *Speed* refers to the revolutions per minute (rpm) that the workpiece or cutting tool is turning. The tool spins in a milling machine and the workpiece spins in a lathe. **Figure 19-16** illustrates how these factors determine the volume of material that can be removed per minute.

These three factors determine the volume of metal that is removed by a cutting tool. The settings for each of these factors depend on machine and part characteristics. The key part characteristic is the material type (aluminum, brass, or steel) and its physical properties, such as hardness and machinability. The next characteristic that influences material removal is part design. Some parts are very fragile or are difficult to grip securely, which reduces the cutting load that the part can withstand and consequently reduces the rate of material removal.

The machine tool being used also influences the rate of material removal. Larger robust machine

Figure 19-16. These are the factors that determine the volume of material that can be removed per minute.

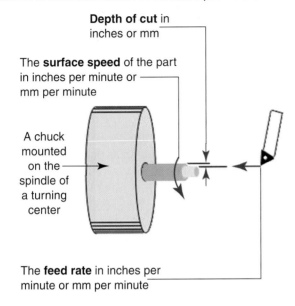

Depth of cut in inches or mm

The **surface speed** of the part in inches per minute or mm per minute

A chuck mounted on the spindle of a turning center

The **feed rate** in inches per minute or mm per minute

tools can withstand higher tool loads, and therefore remove material at a faster rate. The condition of the machine is also critical. A worn machine will not be able to maintain accuracy at higher loads.

Another factor to consider when determining the correct volume of material to be removed is the ability to control heat. Production machining operations are done with the cutting tool being flooded with a coolant. The coolant washes away the chips, but more importantly it cools the workpiece and the cutting tool. It was once thought that the cutting tool should be lubricated, however, investigations indicated that there was no void for the oil to enter between the cutting surface of the tool and the metal being removed. A void is essential to create a lubricating film. Therefore, the oil just flowed over the cutting tool. However, the oil did provide a valuable function in cooling the cutting tool and carrying away the chip debris. There are more effective and less expensive cooling agents than petroleum-based fluids. The fluids most commonly used are water-based coolants with rust inhibitors. These coolants do have lubricating properties to assist in lubricating the moving parts of the machine tool.

Planing

Planing is an operation used to remove large amounts of material from horizontal, vertical, or angular flat surfaces. In some applications, the material is removed to reduce the size of the workpiece.

In other instances, the planing process is done to produce slots or angular grooves in the material.

There are four types of planing machines that are found in industry. These are planers, shapers, slotters, and broaches. The first two types, the planer and the shaper, are seldom used anymore. These machines are slow and have been replaced by other, more specialized machines such as slotters, broaches, and machining centers (mills).

Planing processes generate chips while separating material over the length of the workpiece. In the planer, work is moved back and forth into a fixed, overhead cutter. Planers are huge machines useful for handling workpieces as large as 40′ wide by 80′ long.

Like the planer, the shaper is seldom used anymore in industry. However, a description of how it works should prove useful in understanding the operation of many other planing processes.

In shaping, the work is held stationary in a vise as an overhead tool removes stock in forward passes over the workpiece. Shapers are able to index the work sideways to set up a different tool path for the next forward cutting stroke. Shapers have an overhead ram that moves forward and backward, carrying the cutting tool.

Slotting

Many smaller job shops use the type of planer called a *slotter*. Much of the work once completed by slotters is now being done by another process, called milling.

A *slotter* is a vertical shaper that is sometimes used to cut both internal and external slots and keyways. To visualize how the process is done, consider a workpiece 3″ in diameter with a 1″ hole drilled through the center. The slotting machine looks somewhat like a large drill press. The cutting tool would be powered down inside of the hole, removing material and creating a slot or keyway the length of the hole. Keyways are used to keep pulleys from turning on rotating shafts.

Broaching is a process that is faster than slotting and produces a finer surface finish. Like slotting, broaching can be used to do internal or external planing.

Broaching

Broaching is a process usually done using a hydraulic-powered broaching machine, **Figure 19-17**. The process is ideal for internal machining of keyways, splines, and irregularly shaped openings. Note

Figure 19-17. A broaching machine with two broaches ready to be pulled down through workpieces. (National Broach & Machine Co.)

Broaches

the shape of the hole shown on the bottom of **Figure 19-18**. Broaching can also be used to remove surface material on metal workpieces.

The cutting tool used in broaching is called a *broach*. Broaches create a planing action when they are pushed or pulled across or through a workpiece.

Figure 19-18. Configuration of cutting teeth on a pull broach. This broach produces the irregularly shaped hole shown below it.

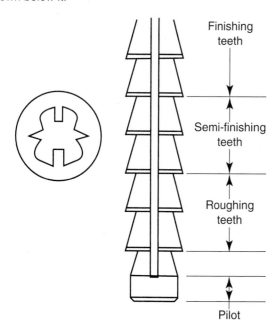

Finishing teeth

Semi-finishing teeth

Roughing teeth

Pilot

The high-speed steel broach is a long rod or bar with many individual teeth. Each tooth is just a few thousandths of an inch longer than the one before. When the tool first enters the work, only a little bit of stock is removed. As the broach continues along the workpiece, each tooth cuts a little deeper. By the time it is at the end of its pass, the maximum depth is achieved. Broaching is normally done using cutting fluid to lubricate and cool the tool. Broaching is a rapid process that can remove as much as 0.250″ (6.35 mm) of material in one pass of the cutting tool. The final finish that is achieved with broaching is comparable with other fine machining processes. The type of material being broached will determine how fast the broach can move through the material. Speeds can be as high as 50 fpm (feet per minute) for aluminum and soft alloys.

There are several different types of broaches. **Figure 19-18** shows the tooth configuration that would be found on a *pull broach* used for internal holes. This broach is tapered, with the first teeth being the smallest. The broach starts with chip breakers, expands with roughing teeth. These teeth remove the largest amount of material. The semi-finishing and finishing teeth then follow to produce a finer finish. The broaching machine pulls the broach on the pull end, which extends beyond the pilot area of the broach. Next comes a section with roughing teeth. **Figure 19-19** shows an actual pull broach. The pull end is located on the left side of

Figure 19-19. A pull broach being assembled. The ring-like cutting tooth units (note different tooth sizes) are assembled in order on the mandrel in the foreground. The pull end, which enters the workpiece first, is at left in the photo. (National Broach & Machine Co.)

the illustration. On the right side is the end that is last to cut the workpiece. This is called the rear pilot end. The pull broach is ideal for internal openings through a workpiece. **Figure 19-20** shows two pull broaches that are about to be pulled down through two steel workpieces.

When material is to be removed on the outside of a workpiece, another type of broach, called a *slab broach*, is needed. The ***slab broach*** consists of a steel blank that may be up to several inches wide and is usually about the same thickness. Slab broaches can be just about any length. This type of broach has cutting teeth arranged on the cutting face. Slab broaches are often constructed in sections joined end-to-end. Several slab broaches are shown at left in **Figure 19-21**. The broaches are installed in slots in the center of the fixture. After the fixture is assembled and locked together, a workpiece can be pulled through the opening in the center of the fixture. The slab broaches will then cut slots on the outside of the workpiece.

Another type of internal broach is called the ***shell broach***. This broach is constructed in three parts: the broach body, an arbor section holding the shell, and the finishing section. Helical splines, such as the rifling in gun barrels or the teeth of gears, are often formed by rotating these broaches as they are pulled through the part.

Machining of parts such as crankcases, cylinder heads, and transmission components is often done by broaching. The gears shown in **Figure 19-22** were

Figure 19-21. Slab broaches, like those at lower left, are held in a fixture and used to cut slots on the outside of a workpiece. (National Broach & Machine Co.)

produced with a process called ***pot broaching***. This type of broaching is also considered an economical method for producing precision external spur gears and automotive front-wheel-drive transmission gears. In some situations, gears can be machined two at a time, achieving production speeds of more than 500 pieces per hour.

In the pot broaching process, the tool holder (called a *pot*) is held stationary, and the parts to be machined are pushed or pulled upward through the

Figure 19-20. Two large pull broaches positioned above the clamping fixtures used to hold workpieces in the machine. (National Broach & Machine Co.)

Figure 19-22. The internal teeth of these ring gears were cut by the pot-broaching process. (National Broach & Machine Co.)

broach. Broaching can be done at speeds up to 25 fpm. An important advantage of the pot broaching process is chip removal. Since the broach is inverted, chips fall away from the tool and part during the machining process. This keeps the teeth clean and aids in producing an improved product.

There are three basic types of broaching tools used with this process: stick, ring, and ring-and-stick. Each has similarities to the slab broach. The *stick broach* has a split, two-piece broach holder that supports broach bushings. The bushings mount high-speed-steel stick slab broaches.

The *ring broach* has a split, two-piece holder that mounts a series of high-speed steel rings. Parts with an outside diameter (OD) in excess of 10″ can be pot broached with this type of tool.

Combination *ring-and-stick broaches* are used to broach teeth on gear blanks while other operations are being performed. This type of tool is used when complex precision gear forms are needed.

Milling

Milling is a process that provides for the controlled movement of a workpiece in conjunction with a rotating multi-toothed cutter to produce slots, grooves, pockets, and contoured surfaces. Eli Whitney developed the first true milling machine in 1818. Over the past two centuries, the milling machine has evolved into a truly sophisticated and versatile production machine that can make a diverse range of parts without the constant attention of a skilled operator. Because of their versatility, production milling machines are now referred to as *machining centers*.

Like the lathe, production machining centers are computer controlled and sophisticated. Machining centers are capable of producing virtually any part a designer can imagine. To achieve this capability, machining centers are able to move in all three dimensions and on an axis around each dimension. Machining centers are classified according to their physical structure. The most popular are the *column-and-knee-type machine* and the *bed-type machine*. In the column-and-knee-type machine, a worktable is mounted on a knee that moves up and down the column. The bed-type machine does not have a knee. Instead, it has a worktable that is mounted directly onto the machine bed. The bed-type machine is capable of handling workpieces that can weigh tons.

Of the two machine configurations, the column-and-knee-type is the most versatile. Both types can be purchased with either vertical or horizontal spindles. Thus, a milling machine is often referred to as a horizontal milling machine or a vertical milling machine.

A *horizontal milling machine* has the cutting tool carried on an *arbor* (spindle) that travels along an axis parallel to the worktable. In a *vertical milling machine*, the cutter is positioned perpendicular to the worktable. **Figure 19-23** shows an example of each type of machine.

Machining centers can have either horizontal or vertical spindles. In the horizontal configuration, the spindle can be either fixed or traveling. The fixed

Figure 19-23. A horizontal and vertical milling machines are shown here. (Haas Automation, Inc.)

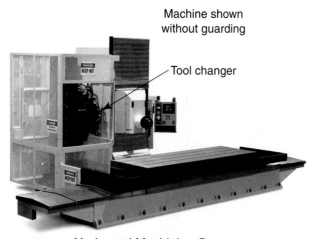

Horizontal Machining Center

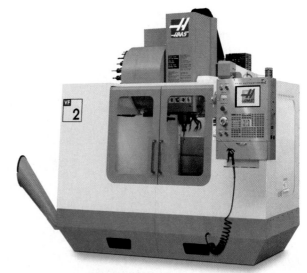

Vertical Machining Center

column provides greater rigidity but offers less flexibility. There are two other features that machining centers require. The first is a tool changer that can store dozens of cutting tools. These various cutting tools are automatically selected, loaded, and unloaded as required. This enables the machine to run unattended for periods of time. The tool changer is visible on the horizontal milling machine in Figure 19-23.

The second feature is a workpiece loading system that allows finished and unfinished parts to be unloaded and loaded while the machining center is working. This feature reduces downtime to the few seconds it takes for a shuttle to move the completed part out and swing in the next part to be machined. In order for the parts to be accurately located, they are preloaded on a pallet. A pallet is a fixture that holds one or more parts securely and attaches to the machine's worktable quickly and accurately. Some machining centers use index tables or transfer lines to transport the workpiece. An index table is able to rotate intermittently, coming to a stop precisely at the same point every time. The table will remain fixed until it receives a signal to rotate again. At a minimum, an index table will rotate 180° creating two fixed points or stations: a loading/unloading and a machining station. Transfer lines also move intermittently, however, the transfer line conveys workpieces in a straight line. A transfer line generally has many stations so several machining or ancillary operations can be done sequentially.

Milling cutters

Almost all milling cutters are multi-tooth tools. Most are made from high-speed steel or tungsten carbide. For high volume production, tools with inserts are available. There is a variety of cutter forms in use. They are:

- Plain milling cutters
- Side milling cutters
- Face milling cutters
- End milling cutters
- Angle milling cutters
- Formed milling cutters
- Slot milling cutters

Figure 19-14 and 19-15 showed examples of many of these tools.

When only one side of a cutter is in contact with the workpiece, the operator must decide whether to use conventional (up) milling or climb (down) milling. With *up milling*, the workpiece is fed in the direction opposite the rotation of the cutter. See **Figure 19-24**.

Up milling is the method used most frequently in manufacturing. In this method, the teeth of the milling cutter approach the work with a gradual sliding action that eventually leads to the tooth biting into the work. This produces thin chips and creates milling marks on the workpiece when each tooth begins its cut. Up milling is the preferred method for machining castings or parts with hard surfaces.

Down milling is sometimes called *climb milling*, because the cutter rides down on top of the workpiece. In this process, the cutter turns in the same direction that the workpiece is traveling. This means that full downward pressure is placed on each tooth as it contacts the workpiece. There is no gradual slivering as in up milling—the cutter bites into the work with a thick cut that tapers off at the end of the stroke. The downward pressure has the added

Figure 19-24. Conventional (up) milling compared to climb (down) milling. Note the direction of table travel in each system.

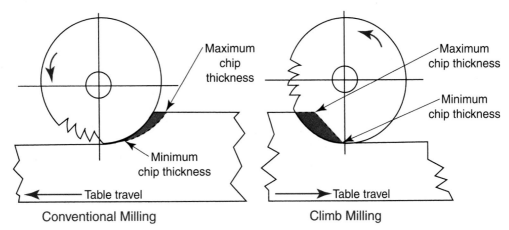

Conventional Milling

Climb Milling

benefit of helping to hold the stock down under the cutter. Down milling provides a smoother surface finish than up milling. Down milling, however, is a more dangerous approach than up milling and should not be attempted with machines that have a great deal of play in the table.

The goal of the designer or technician setting up a part to be made in a machining center is to remove material at the fastest rate possible. The volume of material that can be turned into chips each minute is dependent on the cutting speed, depth of cut, and the feed rate of the tool. The type of material being machined, the strength of the part and how well it can be held, the characteristics of the cutting tool and the machining center condition also influence the rate at which metal can be removed. Generally, the deeper the cut, the slower the cutting speed, and the rougher the surface finish.

The manufacturers of machining centers have developed strategies designed to provide good productivity, smooth surface finishes, and dimensional accuracy. The approach is called high-speed machining. The benefits of high-speed machining include the following:

- Lighter cuts
- Higher speeds and feeds
- Lower cutting forces
- Reduced vibration
- Cooler parts

The combination of machining centers and high-speed machining enables small- and medium-sized companies to be competitive with large corporations and low-wage offshore manufacturers. The productive capability of machining centers is also due to their ability to produce a wider range of parts without the need for secondary finishing operations.

Machining centers provide a range of spindle speeds from 7500 to 30,000 rpm with feed rates up to 850″ per minute. Machines also move quickly to get parts into position. Speed is not the only feature that makes these machines productive. The computer-based control systems and sensors provide capabilities that are not possible with manually controlled machines. In addition to selecting and guiding the cutting tool, the control system is able to compensate for temperature by adjusting the cutting tool and can control spindle and headstock temperatures with a water chiller. Machining centers also share another characteristic with turning centers—they are very inexpensive.

Drilling

Creating a round hole is one of the most fundamental and common operations in manufacturing. The most common operation used to create holes is drilling. Drilling operations can be accomplished with equipment ranging from a battery-powered hand drill to a machining center. A specialty tool called a drill press is frequently used for nonproduction drilling work. The basic process of drilling, however, is defined more by the cutting tool (drill) than the machine that rotates the drill.

When drilling, a hole is created by turning a drill in the spindle of a machine tool. The drilling operation produces chips that are carried out of the hole through *flutes*—special channels in the drill. The most common type of drill used for metals is called a *twist drill*. It is made by forging the basic form and then twisting the blank while it is hot to achieve a spiral shape. Twist drills are normally made of high-speed steel or carbon steel. The life of a drill can be extended by coating it with titanium nitride.

The shank of the twist drill must be gripped by the spindle of the machine. Most machines have a self-centering (three jaw) chuck to grasp straight-shank drills. Larger drills (greater than 3/4″ in diameter) have a tapered shank designed to fit the internal taper in the spindle of the machine. A variety of straight and taper shank drills are shown in **Figure 19-25**.

The angle at the tip of the drill (the tip angle) varies, depending on the material drilled. The *tip angle* is the included angle formed when measuring from one side of the tip of a drill to the other. Most

Figure 19-25. Twist drills and other rotational cutting devices. (National Twist Drill)

metals require an included angle of from 90° to 118°. Plastics require a 60°–90° angle.

Drills can be used to make holes as small as 0.005″ (0.127 mm) and as large as 4″ (102 mm) in diameter. Holes larger than 2″ are seldom drilled—they are more likely to be milled. Drills can be purchased in many different lengths. Long drills are used in a process referred to as *gun drilling* or deep-hole drilling. *Gun drilling*, as the name indicates, was originally used to drill gun barrels. As drilling proceeds, the long drill shank is supported by pads that slide along the inside of the hole. This enables the gun drill to be self-centering. Cutting fluid is forced under pressure through the center of the drill to aid in cooling and to flush chips from the hole. **Figure 19-26** shows a series of deep holes drilled through a thick steel plate.

Drill speeds vary according to drill size—smaller drills can be run faster than larger drills. However, drills are easily broken, so care must be taken in powering the drill into the workpiece. Binding is one of the main reasons drills break. Binding is caused by chips that pack along the flute and are unable to rise out of the hole. To minimize this problem, the drill must be retracted from the hole and allowed to spin off the chips that are packed in the flutes. The empty hole should be flooded with cutting fluid so that the chips can be flushed out when the drill reenters the hole. This sequence should be followed when manually drilling or during a drilling operation under program control.

Drill sizing systems

Twist drill sizes are designated by one of four systems: numbers, letters, fractions, or millimeters. The largest number-sized drill is number 1, which is 0.228″ (5.8 mm) in diameter. The smallest number-sized drill,

number 80, is 0.0135″ (0.342 mm) in diameter. Letter-sized drills are classified from A to Z. The smallest letter-sized drill is the A drill, which is 0.234″ (6 mm) in diameter. The Z drill is 0.413″ (10.4 mm) in diameter. Fractional drills are available in sizes from 1/64″ to 3 1/2″ (0.397 mm to 89 mm). Metric drills are available in sizes from 3.0 mm to 76.0 mm (0.118″ to 2.992″).

Reaming

Most drilled holes have a relatively rough surface finish and are seldom round. They are often "egg" shaped or in thin material, they may even be shaped like a "bow tie." If the hole is to act as a bearing or a bearing guide, it will need to be reamed. *Reaming* is a final finishing process that improves the dimensional accuracy and surface finish of a drilled hole. Reaming produces the smoother finish required for the insertion of bushings or bearings.

Reamers are long multiple-edged cutting tools made of carbon tool steel or high-speed steel. There are many types of reamers. Some are turned in the drilled hole by hand. Others can be inserted in the hole using a drill press, milling machine, or one of the tools mouthed in the turret of a turning center.

The *hand reamer* has a tip that is flat, rather than being angled like a drill, and has straight or helical flutes that run several inches up the shank. Hand reamers cut on their peripheral edge. They are turned with a T-handle wrench.

Machine reamers have either tapered shanks or straight shanks. They are used when greater cutting action is desired. Machine reamers are constructed with a beveled end that cuts somewhat like a drill when entering a hole.

The speed of reaming depends on the type of reamer, the material, and the method. Machine reaming is normally done at about two-thirds the cutting speed (rpm) that would be used with a drill of the same diameter. The feed rate for machine reaming is about three times the rate used in drilling. Cutting fluids should be applied when reaming holes. Holes that are to be reamed must be drilled slightly undersize.

Tapping

Tapping is the process used to cut threads inside a hole *(internal threads)*. Tapping is done with a *tap*, a cutting tool with rows of cutting teeth separated by flutes. Hand taps are made of carbon steels or high-speed steels. See **Figure 19-27**.

Figure 19-26. The series of holes running the length of this steel plate were made with the process known as gun drilling, or deep-hole drilling. (Betar, Inc.)

Figure 19-27. A tap can be used manually or on automated machinery to cut internal threads. (National Twist Drill)

Tapping can be done by hand, with the tool held in a T-handle tap wrench, or with a machine such as a drill press, automatic screw machine, or lathe. Automated tapping heads can produce as many as 500 threaded nuts per hour.

There are three common types of hand taps: taper, plug, and bottoming. The taper tap has a long "lead" allowing it to enter the pilot hole. This makes it easier to start cutting the thread. Often the three taps are used consecutively: taper, then plug, then bottoming. The three differ in the degree of taper (lead) on the start of the tap. A bottoming tap has no lead and is used for tapping *blind holes* (holes that do not run through the workpiece).

To produce a tapped hole, you would first drill an undersized hole, according to the dimension specified on a tap drill chart. Tapping is done by starting the tap into the hole carefully, taking care to hold the tap perpendicular to the workpiece. The tap should be turned two revolutions, then backed 1/4 revolution to break the chip being formed. After several revolutions, the tap should be backed out to the top of the hole to remove the chips that have been produced. Most taps move the chips back out to the top of the hole using their flutes. A *rake angle* on the end of each row of teeth provides wider flutes at the bottom of the tap. This helps to remove chips when the tap is backed out of the hole. If the chips are not removed, the tap may bind and break. **Figure 19-28** shows the rake angle of two different taps.

Figure 19-28. The end views show two different rake angles used for taps. (National Twist Drill)

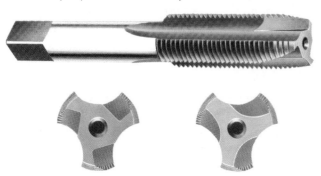

Grinding

Grinding is a cutting process that uses abrasive particles to perform the cutting action. Often, these particles are bonded together and shaped into cylindrical grinding wheels. Abrasives may also be bonded to a cloth backing to make sheets, cylinders, disks, or belts. Grinding can even be accomplished by using a paste compound with a binder that clings to the separate grains. Abrasives are also used in such processes as abrasive jet machining and water jet machining.

There are two types of grinding: rough and precision. *Rough grinding*, also called *offhand grinding* or *snag grinding*, is used for rapid material removal on castings, forgings, and welded parts. It is used to remove parting lines or flash on castings. Rough grinding is normally done with a bench or tool grinder.

Precision grinding is a process often used on materials that are too hard to cut with conventional tools. Precision grinding machines are able to grind flat surfaces, cylinders, threads, or cutting tool blades. The gears shown in **Figure 19-29** were precision-ground.

Figure 19-29. Precision grinding provided the fine finish on these two gears, which were ground to very close tolerances. (National Broach & Machine Co.)

Figure 19-30. This CNC external grinding machine is used to precision-grind the teeth on large gears. (National Broach & Machine Co.)

Precision grinding requires working to close tolerances and is an effective process for producing a fine finish on the workpiece. The major forms of precision grinding are surface grinding, internal cylindrical grinding, and external cylindrical grinding. **Figure 19-30** shows a CNC gear grinder used for external grinding. **Figure 19-31** shows a finished gear that has cycled out of the machine. The 17″ diameter aircraft gear in **Figure 19-32** is being ground using a six-axis CNC gear grinding machine that is capable of grinding up to 20″ diameter gears.

Grinding wheels

Grinding wheels are used on both rough and precision grinders. Coarse wheels are used for roughing, or rapid material removal. The wheels used for precision

Figure 19-31. A finished gear that has cycled out of the grinder for unloading. Note the gear blank positioned inside the machine, ready for grinding. (National Broach & Machine Co.)

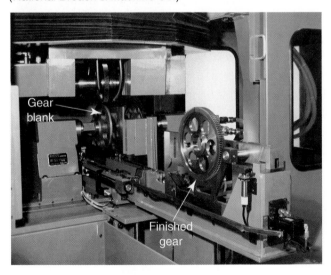

Figure 19-32. This large aircraft gear is being ground to close tolerances on a six-axis CNC machine. Note the thin grinding wheel removing material from teeth at the top of the gear. (National Broach & Machine Co.)

grinding are more dense, which makes them stronger and more durable. Abrasives used to manufacture the wheels are either natural or synthetic.

Natural abrasives include sandstone, quartz, emery, corundum, garnet, and diamond. Natural abrasives are seldom used in production grinding operations. Synthetic abrasives include silicon carbide, aluminum oxide, diamond, and boron nitride.

Grinding wheels are made from abrasive grains held together with a binding agent. There are several major types of binders used today: vitrified, silicate, resinoid, metal, rubber, and shellac.

Vitrified-bonded wheels are the type most frequently used. They are strong and rigid, and are not affected by water, oils, or acids. Vitrified bonded wheels are designed to operate at speeds approaching 6500 sfpm (surface feet per minute).

Silicate-bonded wheels use sodium silicate as a binder. The mix of binder and abrasives is compressed in a metal mold and then heated for a prolonged length of time. The abrasive grains of silicate-bonded wheels are released faster than vitrified wheels because their abrasive grains are released more quickly. Silicate-bonded wheels are recommended for use in such applications as sharpening cutting tools.

Resinoid-bonded wheels use Bakelite™ or Resino™ as a binder. These wheels are very strong and tough. Resinoid-bonded wheels can operate at high speeds.

Wheels with metal or rubber binders are used for specialty applications. Metal binders are used with diamond abrasive, for grinding hard materials, or for electrolytic grinding. The rubber binder is embedded with abrasive and rolled out into sheets. The wheels are die-cut from the sheets and vulcanized. Rubber-bound wheels are made as thin as 0.005" (0.127 mm). These wheels are designed for use at high speeds (often exceeding 15,000 sfpm). High-speed rubber wheels are designed primarily for use with materials such as plastic, porcelain, and tile. They are also used on cutoff saws for cutting metal.

When a fine finish is desired, shellac-bonded wheels are recommended. Shellac wheels are made by compacting shellac and abrasives in steel molds. Shellac wheels are thin and flexible.

All metal-grinding processes generate chips and produce heat. Grinding fluids are used to cool the workpiece and carry the chips away from the cutting area. This helps to keep the wheel clean so that the open pores between the abrasives can continue to perform their cutting action.

Surface grinding

Surface grinding is a form of precision grinding that is done on flat workpieces. It is a good process for producing fine finishes to close tolerances. Surface grinding is also used for cutting hard materials that are not machinable by conventional methods. Examples include ceramics, tile, glass, and composite materials.

This process is done using a machine called a surface grinder, which is constructed somewhat like a milling machine. Surface grinders consist of a machine base, worktable, and arbor. The spindles are arranged in either a vertical or horizontal position. Surface grinders are sized by the length of workpiece that can be ground. Horizontal machines, which position the grinding wheel above and parallel to the worktable, are common. See **Figure 19-33**.

The grinding operation usually begins by securing the workpiece to the worktable of the machine with a magnetic chuck. The worktable moves longitudinally (from left to right), while the grinding wheel removes stock the length of the workpiece. The table position is adjusted laterally (front to back) after each stroke. Grinding is done with cutting fluids to keep the wheel cool and clean, **Figure 19-34**. The coolant helps to extend the life of the wheel and keeps the working surface cool. Water-soluble coolants and water-soluble oils are preferred by most manufacturers.

Workpieces may be secured with a magnetic

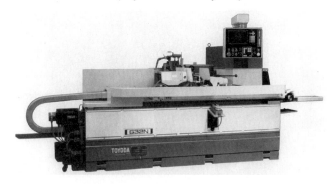

Figure 19-33. This CNC cylindrical grinder can be programmed with up to nine different grinding cycle selections. It can be used effectively for single-part grinding or continuous runs. (Toyoda Machinery Co.)

chuck during the grinding process. The iron workpiece shown in **Figure 19-35** was cast with a flat base, making it a perfect candidate for the magnetic chuck.

With the help of fixtures, surface grinders can also be used to grind workpieces with irregular shapes. A *fixture* is a device that holds a workpiece securely while processes, such as grinding, are performed. The fixture is then clamped to the table or held in a magnetic chuck.

Some applications require the grinding of nonferrous metals, such as brass or aluminum, or nonmetallics like plastic. Sometimes this is done using fixtures or steel braces to prevent movement. In other instances, particularly when the material is very thin, an operator may use double-faced tape to secure the workpiece to the table.

Figure 19-34. Cutting fluids are used in grinding to cool the wheel and wash away particles of ground-away stock. (DoALL Company)

Figure 19-35. A surface grinder can be used on workpieces with irregular shapes. This workpiece is clamped in a magnetic chuck that holds it tightly in place on the worktable. (DoALL Company)

Magnetic chuck   Workpiece

Cylindrical grinding

Cylindrical grinding is used on workpieces with curved surfaces (cylindrical shapes). Cylindrical grinding can be done on the outside of a workpiece (external), or on the inside (internal). **Figure 19-36** illustrates some of the different types of cylindrical grinding applications.

External cylindrical grinding uses a grinding wheel which runs over the outside surface of a workpiece. The work is secured between centers or in a chuck.

Internal cylindrical grinding is done when the surface of a hole must be accurately smoothed. This

Figure 19-37. The basic arrangement used for external centerless grinding.

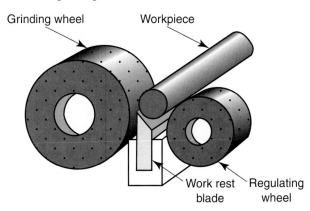

Grinding wheel   Workpiece

Work rest   Regulating
blade   wheel

can be accomplished with a tool post grinder attached to a lathe or with a specialty grinding machine.

Another type of cylindrical grinding is called *centerless grinding*. This form of grinding can be used for external or internal applications. **Figure 19-37** illustrates the method used for external centerless grinding: the workpiece is held on a work rest blade positioned between a grinding wheel and a regulating wheel. The regulating wheel presses against the workpiece, causing it to rotate. The grinding wheel grinds the surface. Products such as lathe centers and roller bearings are often ground using this process.

Internal centerless grinding is used to finish the inside of a hole in a cylindrical part. There are several variations of this process. The most common method involves positioning the tubular workpiece between

Figure 19-36. Various grinding applications can be performed on cylindrical, flat, or hollow workpieces.

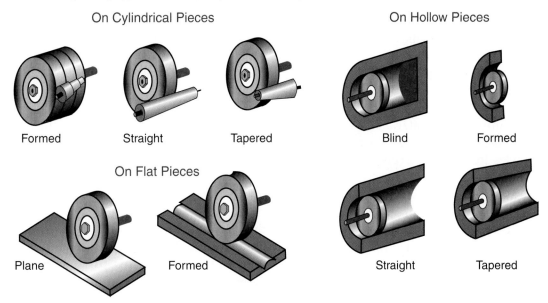

On Cylindrical Pieces

Formed   Straight   Tapered

On Flat Pieces

Plane   Formed

On Hollow Pieces

Blind   Formed

Straight   Tapered

Figure 19-38. Three rolls—pressure, support, and regulating—hold the tubular workpiece for internal cylindrical grinding.

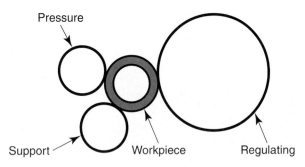

three rolls: a pressure roll, a support roll, and a larger regulating roll. The regulating roll turns the stock while the other two rolls hold it tightly against the regulating roll. The grinding wheel is attached to a shaft that travels inside the tubular part. **Figure 19-38** shows how the three rolls work together to hold the tubular part.

Abrasive Jet Machining

Abrasive jet machining is a process that appears, at first glance, to be much like sandblasting. While there are some similarities, there are also major differences.

Sandblasting is a process that uses air pressure to blast sand particles against the surface of a workpiece to clean it or dull its finish. *Abrasive jet machining* is a grinding process that suspends tiny particles of abrasive material in a low-pressure stream of gas (dry air, carbon dioxide, or nitrogen) sprayed through a sapphire nozzle. This process is used to etch, scribe, groove, and cut holes or slots in hard metals and nonmetallic materials such as ceramics and glass. See **Figure 19-39**.

Many different types of abrasives are used. When very hard materials are to be machined, typical abrasive choices would be silicon carbide or aluminum oxide. For machining a soft material, a soft abrasive such as bicarbonate of soda would be used. Since the abrasive particles lose their sharpness during use, they cannot be reused.

The pressure that is used for abrasive jet machining is low enough that an operator can safely pass a hand through the abrasive stream. Velocity of the gas stream ranges from nearly nothing up to 1000 fps. The abrasive stream is directed toward the workpiece by guiding the spray nozzle.

Figure 19-39. Abrasive jet machining is a versatile process with many uses. Here, it is being used to etch and clean the surface of a part. (S.S. White Technologies, Inc.)

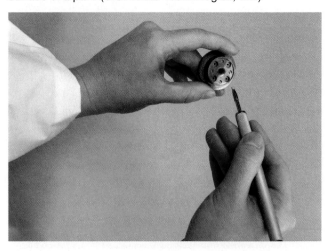

Abrasive jet machining is ideal for materials that are heat-sensitive. Conventional grinding processes create heat, so they cannot be used with such parts. Because of the low pressure employed, however, abrasive jet machining is a relatively slow process. Cutting rates are about 0.001" (0.025 mm) per minute.

Applications of abrasive jet machining include drilling and slicing thin wafers of hardened metal, etching numbers on parts, removing broken taps from holes, and deburring.

Waterjet Machining

Waterjet machining is a process that uses a high-velocity stream of water to cut materials ranging from paper to stone or metals. The pressurized water stream can cut and slit porous materials such as wood, paper, leather, brick, and foam. When abrasives are introduced to the stream of water, the process is effective for cutting hard materials. Abrasive waterjet systems have been used commercially since 1985 for cutting rocks, metals, glass, ceramics, and composites with great speed. See **Figure 19-40**. A waterjet system can cut hard materials in thicknesses from 0.003" (0.08 mm) to 1" (25 mm) or more, and softer solids up to 10" (25.4 cm) in thickness.

Typically, the process uses focused jets of water 0.006" to 0.008" (0.15 mm to 0.20 mm) in diameter, pressurized to as high as 55,000 psi (380000 kPa). In the automotive industry, materials such as carpet, fiber-reinforced plastics, sheet molding compound, reaction injection molded parts, thermoplastics, and fiberglass are being cut with waterjet equipment.

Figure 19-40. Abrasive waterjet machining can rapidly cut through hard materials, such as this steel plate being cut for use as a saw blade. The tooth configuration being cut is shown on the blade segment in the background. (Ingersoll-Rand Waterjet Cutting Systems)

Figure 19-41. This abrasive waterjet nozzle design has a rated life of 500 hours. (Ingersoll-Rand Waterjet Cutting Systems)

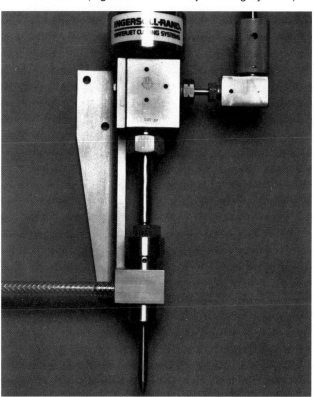

Computer-controlled waterjet cutting systems are being used all over the world to cut printed circuit boards.

Waterjet machining is being used on metals for deburring, descaling, and degreasing. It is also used to automatically strip coatings from wire at low pressure, which avoids damaging the wire. For large-diameter cables, stripping takes about 5 to 10 seconds per cable section. This is about five times faster than traditional thermal or mechanical stripping techniques.

The principle of waterjet machining—removing material through erosion from a high-velocity, small-diameter jet of water—was discovered by accident. Workers in steam plants observed that pinhole leaks in pressure steam lines had enough power to cut quickly and cleanly through a wooden broomstick. The technology was perfected and patented as waterjet machining in 1960, and was first used in industry in the early 1970s.

An advantage of waterjet machining is that the cutting tool never breaks and never needs sharpening; however, the abrasive action causes nozzles to wear quickly. The sapphire nozzles are rated at a useful life of between 50 and 500 hours, **Figure 19-41**.

Waterjet machining produces little or no dust and generates no heat during cutting, making it ideal for certain types of applications. An interesting feature of this process is that the cut can begin at any location in the part. Not many cutting processes can accomplish this.

The equipment required for waterjet machining consists of a large electric motor driving an oil pump. An intensifier is used to increase the water pressure about forty-fold. Additional parts of the system consist of an accumulator and the waterjet nozzle. The water is adjusted by regulating low-pressure oil actuators. Often, the nozzle is used as an ***end effector*** (tool on the end of the arm) of an industrial robot. For safety reasons, human operators must be kept out of the work area. A production setup for waterjet machining is shown in **Figure 19-42**.

Major industries that use waterjet cutting include:
- Aerospace
- Automotive
- Electronics
- Nonwoven textiles
- Food products
- Paper and corrugated board
- Shoe and garment products
- Building products

Figure 19-42. A production setup for waterjet cutting from large workpieces. (Ingersoll-Rand Waterjet Cutting Systems)

Laser Cutting

A *laser* is used for a variety of processes, including cutting, drilling, welding, heat treating, soldering, and wire stripping. Metal cutting is the single largest application of the laser in manufacturing, and is normally done using a carbon dioxide (CO_2) gas laser, **Figure 19-43**.

In a typical gas laser, a mixture of carbon dioxide and nitrogen is stored in a glass *lasing tube* at low pressure. This mixture is referred to as the *lasing medium*. When high voltage is transmitted from a power source to electrodes in the tube, the discharge excites the nitrogen molecules. These excited molecules, in turn, cause the single carbon atom in each CO_2 molecule to vibrate back and forth between the two oxygen atoms. The vibrating CO_2 molecule gives up energy as it changes from one energy vibration pattern to another. Light is then emitted in the form of photons. As each photon moves along to the

Figure 19-43. A gas laser can cut metals quickly and cleanly. Lasers are also used for drilling, welding, and a variety of other operations. (Rofin-Sinar, Inc.)

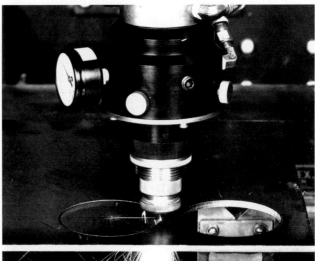

next excited molecule, it stimulates that molecule to give up a photon as it changes to a lower energy vibrational state. This causes a stimulated emission of still more photons.

Figure 19-44 illustrates what happens when mirrors are placed at both ends of the lasing tube. As light is reflected back and forth, the mirrors cause the photons to form into a highly *collimated* (focused)

Figure 19-44. Basic laser setup. The mirror at the right side of the illustration is not fully reflective, so it allows the beam of laser light to pass through.

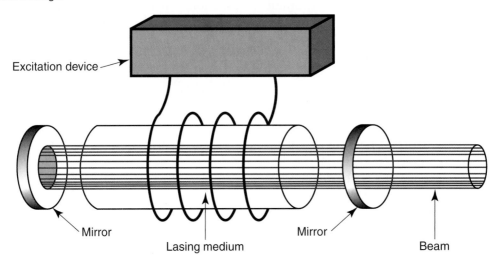

Excitation device

Mirror

Lasing medium

Mirror

Beam

beam. One of the mirrors does not reflect all of the light, so some passes through (is *transmitted*) in the form of a laser beam. The beam is not continuous; it is produced in short, rapidly repeated pulses.

The laser beam for cutting is only 0.005″ to 0.012″ (0.127 mm to 0.30 mm) in width. This means that a minimal amount of heat will be generated in the area of the *kerf* (cut). This makes the laser a good tool for cutting small holes, narrow slots, and closely spaced patterns.

The narrower the kerf, the faster the laser can cut. Narrower kerfs require less energy from the laser and result in less heat on the workpiece. **Figure 19-45** illustrates a typical laser cutting system.

One currently available CO_2 laser cutting system can be used to make a high-quality cut in steel up to 0.375″ (9.5 mm) in thickness. A low-quality cut can be made in steel as thick as 0.5″ (12.7 mm). Aluminum and brass up to 0.125″ (3.17 mm) thick can be cut.

With laser cutting, there are no cutting tools to break or wear. Power capacities above 105 watts/in². are necessary for laser cutting of metal. The most common power range for commercial carbon dioxide lasers is 400 to 1500 watts.

Lasers are often built as part of a turnkey (ready to operate) system that includes CNC and automated tooling. Like waterjet cutting systems, lasers can be dangerous if used carelessly. Lasers should always be operated in a controlled environment.

Figure 19-45. Parts of a typical laser cutting system. The lens holder is water-cooled to prevent heat buildup that would otherwise damage it.

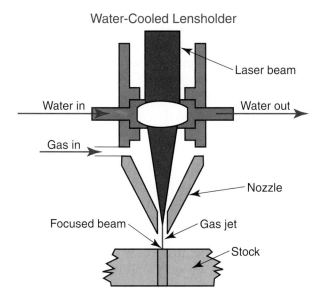

Water-Cooled Lensholder

Chemical Milling

Chemical milling is also known by a variety of other names: chemical machining, photofabrication, photoforming, photoetching, and photochemical machining. The terminology that is applied depends on the end use. When the process is used to etch or form metallic parts, it is usually called chemical machining, photoforming, or chemical milling. When it is used to prepare electronic substrates, such as printed circuit boards, it is referred to as photofabrication.

There are two different types of chemical milling processes: chemical blanking and contour machining. *Chemical blanking* involves forming a part by etching metal away completely through the workpiece. *Contour machining* selectively etches a desired area to some specified depth.

Chemical milling can be used with virtually all metals and alloys, including stainless steel, beryllium-copper, aluminum copper-nickel, molybdenum, tin, titanium, and many other magnetic and special purpose alloys. The process is most effective when used with parts up to 0.020″ (0.51 mm) in thickness. Chemical milling is the most effective method for producing exacting detail in parts from foil under 0.001″ (0.025 mm).

Chemical machining is basically a photographic process. The first step in the process is to thoroughly degrease and clean the workpiece. Next, a light-sensitive coating called a *photoresist* is applied to the surface.

There are two different types of photoresist: wet film and dry film. The wet film resist is a liquid applied by spraying, roller coating, or dipping the metal workpiece. Once the wet resist is applied and dried, a photographic master (negative or positive) is placed over the coated metal plate. The master and plate are then exposed to ultraviolet light. The light goes through the clear areas in the master and strikes the light-sensitive resist coated on the plate. The light hardens the resist in this area. The plate is then etched by running it through an acid spray bath. *Etchants* are normally sodium hydroxide (for aluminum), a hydrochloric/nitric acid blend (for steels), and iron chloride (for stainless steel). When the etching is completed, the part is removed and washed.

Since the wet method involves solvent-based materials, it is gradually being displaced by the dry method because of environmental and hazardous waste considerations. Despite their limitations, the wet films are preferred for microelectronic applications because better resolution and image clarity are produced.

Most companies involved in chemical milling now use the dry film method, since these water-based resists are easier to apply and more environment-friendly. The dry film is a light-sensitive polymer sheet that is laminated to the metal workpiece. As in the wet-film process, a photographic master is applied and the sandwich is exposed and etched. After etching, the polymer coating is chemically removed from the metal part.

Chemical machining can also be done by applying a *maskant* (another term for resist) that is not light sensitive to the metal workpiece. The maskant can be a sprayed or brushed coating, a tape, or other type of material that will adhere to the surface. With this method, the maskant is scribed away in the area to be etched. Where the maskant remains, the etchant (acid) cannot penetrate. Where it has been removed, the acid can attack the metal and etch it away.

Chemical milling is sometimes done to remove surface material or an entire section of a part. This might be done for ornamental purposes or primarily to reduce the weight of the part. Chemical milling can also be used for deburring (removing rough edges or burrs on the edges of the machined parts). When it is used in this way, the process is often referred to as chemical blanking.

Contouring is a method used to remove metal from surfaces of irregularly shaped parts. The part is selectively etched to a desired depth. Contouring can be done using a brushed-on mask. With this method, the steps of masking, scribing, and etching are the same. The difference is that the part is removed before it is completely etched.

Chemical machining has several advantages over producing parts by conventional stamping or punching methods. Since chemically machined parts do not undergo stresses and strains in forming, the parts are stronger and more consistent. There is no expensive tooling involved, and no retooling costs. Changes in design can be made quickly without the long lead times needed for tooling revisions.

The aviation industry is the largest user of chemically milled parts, particularly parts that are contour-machined. Other products include custom hardware, decorative panels, plaques, and instrument dials.

Ultrasonic Machining

Ultrasonic machining is a process that removes material by eroding it using vibrations generated by high-frequency sound waves. The sound waves are amplified in a funnel-shaped horn to create the desired vibrations. Attached directly to the horn is a cutting tool made in the shape of the hole or cavity that is to be cut in the workpiece. Ultrasonic machining has many similarities to ultrasonic welding, a process that is used extensively to join plastics.

Ultrasonic machining can be used to drill and shape very hard, small, high-tolerance parts made from materials such as germanium, ceramics, or glass. This method of machining is a slow process, but has the advantage of not generating any heat or stress, which could break or cause the distortion of very thin parts.

There are two methods of ultrasonic machining. Both use a vibrating metal tool. In the first method, sometimes called *impact machining*, the tool is covered with an abrasive slurry mixture. The tool itself never touches the workpiece. Cutting takes place through a constant vibratory action of the tool, which causes the abrasive particles to gradually erode away the workpiece, creating the shape pattern that is desired. The abrasive slurry is produced by mixing water with iron carbide, silicon carbide, or aluminum oxide powder.

The tool used for impact machining oscillates about 25,000 times each second. At the end of the tool, where it is held close to the workpiece, the actual movement due to vibration is about 0.003" (0.08 mm).

The second method of ultrasonic machining uses a rotating, diamond-tipped vibratory cutting tool. The cutting action is produced by the tool itself; no abrasive slurry is required. There are several types of tools used in this method. Some resemble conventional grinding wheels, others look like twist drills or end mills. The cutting tool has the same shape as the hole or cavity that is to be produced in the workpiece.

An important consideration in ultrasonic machining is the specific frequency at which the tool must vibrate to be effective. This is called the tool's *resonance point*. Cutting takes place when the tool reaches the frequency that causes it to resonate. Experienced machine operators can tell when the tool has reached this point by listening to the sound of the machine.

Ultrasonic machining systems resemble ultrasonic welders. The workpiece is placed in a slurry pot, which is secured to the table. The vibrating tool is attached to an acoustic tool head that moves vertically. Other components of the ultrasonic machining system are the circulating pump and power supply.

Electrochemical Machining

Electrochemical machining (ECM) could really be called reverse electroplating. ECM is used to shape a workpiece to a desired external contour, produce odd-shaped blind holes, and machine almost any shape in the cavity of a part to a specified depth. The process often provides advantages over conventional methods used to machine complicated shapes in hard and tough materials. Electrochemical machining is most cost-effective when used for high-volume production.

Normally, ECM is selected only for those applications where conventional machining processes cannot be used. An example is the production run of aircraft struts as shown in **Figure 19-46**. The struts have pockets that must be machined on both sides of the arm. Due to an overhanging flange, however, the bottom pocket is obstructed, preventing the use of conventional milling. The solution was to machine the pockets electrochemically.

Electrochemical machining first gained popularity in the aircraft industry, where it was used to produce

Figure 19-47. This large turbine rotor was produced with the aid of electrochemical machining. ECM was first used in the aircraft industry. (Anocut, Inc.).

gas turbine engine components, **Figure 19-47**. The aircraft industry is still the major user of ECM, but the process is also being used in the making of dies and machine parts. The process is particularly attractive to these industries because the finished part is burr-free, eliminating the need for further processing. Parts produced by ECM retain their structural integrity, since no heat is involved in the machining process. A limitation of electrochemical machining is that the workpiece must conduct electricity. Cast iron, for example, does not work well with ECM.

Here is how the ECM process works. The workpiece is the positive electrode, called the *anode*. The hollow cutting tool is the negative electrode, or *cathode*. **Figure 19-48** shows the anode (an aircraft

Figure 19-46. Aircraft struts machined with ECM. The pocket next to the left hand of the worker holding the part could not have been made with conventional machining methods. (Anocut, Inc.)

Figure 19-48. An ECM machine ready to produce an aircraft turbine rotor. (Anocut, Inc.)

turbine rotor) and cathode in an ECM machine ready for forming.

The machine is closed and a neutral salt electrolyte flows continuously between the anode and cathode. Low-voltage DC current is transmitted across the electrodes, removing electrons from the surface atoms of the workpiece. These *ions* (free electrons) are drawn toward the hollow cutting tool. The rapidly flowing electrolyte moves the ions out of the gap between the tool and the workpiece. This prevents *plating* (deposition of metal) from taking place on the cutting tool. This process continues as the tool is moved automatically across the workpiece.

ECM equipment is available in both vertical and horizontal types. **Figure 19-49** shows vertical ECM machines that are used to produce small beryllium parts.

An important advantage of ECM is that there is no tool wear. Once the machine is set, the tool can produce thousands of parts.

While electrochemical machining has many advantages, it also carries with it a number of disadvantages. Setting up ECM systems for production is a time-consuming process, requiring a great deal of experimentation and testing. A more troublesome concern is the corrosive mist that ECM produces, particularly when sodium chloride is used in the electrolytic solution. Fumes must be carefully handled. However, systems for detecting leakage and handling exhaust vapors are standard on most machines. Approved safety procedures must be carefully followed for disposing of waste material. One of the waste materials, sodium chloride–soaked cloth or paper, is a fire hazard.

Figure 19-49. The vertical ECM units in the row at right are used to produce small parts from the metal beryllium. (Anocut, Inc.)

Electron Beam Machining

Electron beam machining (EBM) is a thermoelectric process that focuses a high-speed beam of electrons on the workpiece. The heat produced by EBM is sufficient to melt superalloys or any known material that can exist in a high vacuum. EBM is particularly useful for drilling small holes and for perforating or slotting materials that are difficult to machine, such as superalloys, ceramic oxides, carbides, and diamonds. The material removal rate is very slow.

Figure 19-50 shows the components of the electron beam machining system. The heart of the system is the electron beam gun. This generates *thermal energy* in the form of a high-velocity stream of electrons.

Material is instantly removed by the melting and vaporizing action caused by concentrating the high-velocity electron beam on the workpiece. The

Figure 19-50. Components of an electron beam machining system.

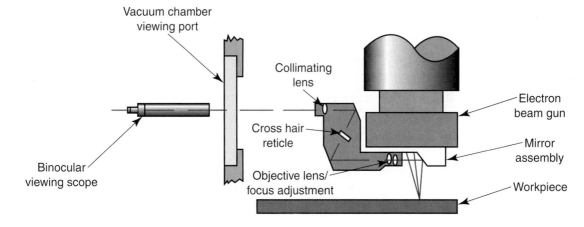

electron beam strikes the workpiece with on-and-off pulsations several milliseconds in duration. Temperatures in excess of 12,000°F (6650°C) can be generated.

Electron beam machining is normally conducted in a vacuum. When electrons strike the workpiece, their kinetic energy is converted to heat energy. This causes vaporization. The heat is pulsed on and off, and it is this pulsing action that causes the cutting. The equipment used for electron beam machining is the same as that used for electron beam welding.

Product applications include micromachining to manufacture computer memories and integrated circuits. Holes or slots as tiny as 0.005″ (0.127 mm) in diameter can be drilled using EBM. Drilling time for a 0.005″ hole would be instantaneous. Larger holes are produced by rotating the workpiece.

One drawback of EBM is that work is limited to the size of the vacuum chamber. Another drawback is that the process generates radiation; therefore, X-ray shielding around the work area is necessary.

Electrodischarge Machining

Electrodischarge machining (EDM) is a process that uses electrical energy to remove stock from metal workpieces. Electrodischarge machining bombards the workpiece with 20,000-30,000 electrical sparks per second, and thus is sometimes called *spark erosion machining.*

EDM can be used to cut hard metals and can form deep internal shapes or irregularly-shaped holes. It is an ideal process for removing material from hard-to-reach areas of parts. It is often used to produce intricate shapes on hardened materials that cannot be machined with other processes. However, EDM is a slow process. Thus, it is not cost-effective for conventional machining applications.

Like electrochemical machining, electrodischarge machining can be used only with electrically conductive stock. Here is how it works:

The tool, or electrode, has the same shape as the hole that will be produced. A male electrode could be constructed in relief to resemble an intricately carved sculptured cameo. If the tool is moved into close proximity with the workpiece, the same intricate detail would be created, but in reverse, as a sunken-surface female form. The workpiece is placed in a dielectric fluid, such as mineral oil or a kerosene and distilled water mixture. The tool is brought close to the surface of the workpiece. The dielectric fluid

flows between the tool and the workpiece. When dc current is transmitted to the electrode in short pulses, an electric arc is created between the tool and workpiece.

The sparks strike the surface of the workpiece and remove small bits of metal. The higher the energy in the pulse, the greater the amount of material that will be removed.

Sometimes the tool is moved downward into the workpiece; at other times, the workpiece is moved upward. This depends a great deal on the design of the electrodischarge machining equipment. Typical machines look like a vertical end mill, with a ram or quill replacing the cutter spindle. A tank is attached to the table to hold the dielectric fluid.

One of the major uses for EDM is the machining of matched molds used for injection molding. EDM is also a popular process for shaping carbide tools and for machining complex shapes in exotic metals.

The process can easily be adapted to computer numerical control (CNC) for automated operation. A number of CNC electrodischarge machines are commercially available. They can be programmed to automatically handle electrode changes, table movement, changes in power requirements, and dielectric flushing. The CNC operator can recall from memory a multitude of preset patterns and programs.

Traveling Wire EDM

The major difference between conventional EDM and traveling wire EDM is the type of electrode used. In *traveling wire EDM*, cutting takes place using a round wire that travels through the workpiece. This type of EDM is also referred to as *electrical-discharge wire cutting.* EDM is sometimes compared to band saw cutting, because both processes can be used to cut intricate shapes from a metal workpiece. However, the two processes really have few similarities. The wire used in traveling wire EDM never touches the workpiece, and its movement is directed by computer numerical control. Band saw cutting is a process in which the saw blade remains in constant contact with the cutting surface, and the direction of the blade is controlled by an operator.

In traveling wire EDM, the wire moves down through the workpiece. As in conventional EDM, the sparks arcing to the workpiece from the electrode act like small teeth, each removing a small amount of metal. This results in a narrow kerf being machined in the workpiece; the wire travels inside this kerf.

Figure 19-51. This traveling wire EDM machine is cutting a thick piece of stock. The EDM creates a narrow kerf, resulting in minimal stock loss during cutting. (Elox Corporation, an AGIE Group Company)

Figure 19-51 shows a traveling wire EDM system cutting thick metal stock.

The position of the wire in a traveling wire EDM system is determined by computer numerical control. A typical CNC traveling wire system is shown in **Figure 19-52**. Many machines are equipped with an automatic wirefeed mechanism that will rethread the wire if a break occurs. The wire used is usually only 0.010" (0.254 mm) in diameter. For many years, copper was the only type of wire available. Now, other types of wire are gaining popularity including brass, molybdenum, tungsten, or copper coated with zinc. Previously, wire was only used once because its surface was worn away during cutting. Wires that can be reused are now being introduced.

Traveling wire EDM is a popular process for making blanking dies for sheet metal stamping. Any shape can be cut, in any direction. One of the advantages of the process is that the die metal can be hardened before it is cut to size with the traveling wire machine. This permits construction of both a punch and a die from the same die block.

One of the unique characteristics of traveling wire EDM is that cutting thicker stock does not increase cutting time. As an example, a piece of stock 1" thick (25.4 mm) might be cut at a rate of about 10 linear inches per hour. This cutting rate is expressed as 10 in./hr. As a comparison, a 7" thick (178 mm) workpiece might be cut at a rate of 11.6 linear inches

per hour (11.6 in./hr.). Since the wire is discharging along the entire length that is facing the workpiece, the rate at which the metal is cut is not actually a function of the thickness. Therefore, the cutting rate (inches per hour) for the much thicker workpiece can be the same or even slightly faster than for the 1" piece.

Plates more than a foot in thickness have been cut with traveling wire EDM. Traveling wire EDM is a popular process in tool and die work, and for gang cutting of stacked sheet metal parts. *Gang cutting* refers to the use of multiple heads that simultaneously cut single or multiple workpieces.

CNC

A turning center is a computer-controlled machine, or CNC machine. CNC stands for *Computer Numerically Controlled*. The programs to operate these machines can be developed by the machine operator at the machine. For simple parts, this is very effective. However, the program for the machining center is often developed along with a CAD drawing that was created when the part was designed. After the design has been checked and approved, the designer answers a series of questions posed by the CAM (computer aided machining) portion of the CAD/CAM software program. These questions are used to identify the type of machine tool that will be used and some particulars about the part and how it will be machined. Next, the designer will identify

Figure 19-52. A traveling-wire EDM machine operating under computer numerical control. (Elox Corporation, an AGIE Group Company)

the surfaces to be machined and the tools that will be used. When this work is completed, the designer will *post process* the computer file that was created by the CAM portion of the software. The post processing generates the program for the machine tool. This program is transmitted to the machine tool or stored to a disk or CD and carried out to the machine tool. There is a variety of CAD software available that includes a CAM package. However, most CAM software can import standard CAD files

The ability for CAD/CAM software to generate machine code for CNC machines has simplified and expanded the use of CNC machines. These machines are very productive and surprisingly inexpensive.

Important Terms

abrasive jet machining
anode
arbor
bed-type machine
blanking
blind holes
broach
cathode
cemented carbide
centerless grinding
cermet
chemical blanking
chemical milling
climb milling
collet
collimated
column-and-knee-type machine
contouring
contour machining
cylindrical grinding
depth of cut
drop
electrochemical machining (ECM)
electrodischarge machining (EDM)
electron beam machining (EBM)
end effector
etchants
external cylindrical grinding
faceplate
feed
fixture
flutes
gang cutting
grinding
gun drilling
horizontal milling machine
internal cylindrical grinding
internal threads
ions
kerf
lancing
laser
lasing medium
lasing tube
lathe
maskant
milling
nesting
nibbling
notching
perforating
photoresist
planing
plating
post process
pot broaching
precision grinding
pull broach
punch-and-die shearing
punching
rake angle
reaming
resonance point
ring-and-stick broach
ring broach
rotary shearing
rough grinding
shearing
shear line
shell broach
slab broach
slitting
slotter
speed
stick broach
straight shearing
surface grinding
swing
tap
tapping
thermal energy
tip angle
traveling wire EDM
turning center

turning processes
twist drill
ultrasonic machining
universal chuck
up milling
vertical milling machine
waterjet machining

Questions for Review and Discussion

1. Explain how electrodischarge machining would be used to remove material from the inside of a hollow part.

2. How should the speed (rpm) of a turning center be changed to maintain the same rate of metal removal if the diameter of the workpiece is increased? Also, how should the depth of cut and the feed be adjusted if the rpm is not changed? Explain your answers.

3. When would up-milling be preferred to down-milling?

4. Why is tool wear not a consideration with electrochemical machining?

5. What type of grinding wheel would be best for use on cutoff saws that are used to cut metal? Why?

20 Processes Used to Separate Plastic Materials

Chapter Highlights

- Proper selection of cutting speed and sawtooth shape is important for efficient material removal when cutting plastics.
- Cutting speed is calculated in feet per minute, as measured along the circumference of the cutting tool.
- Twist drill bits for plastic are ground to a sharper angle than those used for metal.
- Cutting speeds must be calculated carefully to avoid material breakage.
- Plastic materials are cut using a variety of processes, including die cutting, extrusion cutting, laser cutting, waterjet cutting, and hot wire cutting.

Background

Most of the processes that are used to cut metal and wood materials are also used to cut plastics, including circular sawing and drilling. The circular saw and drill used to cut plastics may be the same as those used on wood or metal. However, the saw blade and drill bit may vary slightly to match the properties of the specific plastic.

Plastic sheets, rods, and tubes are usually cut using fine-tooth blades, with approximately twelve teeth per inch. Specialty blades can also be purchased for cutting plastic. Band saw blades with four to six teeth per inch are ideal for use with many plastics.

Most plastics can also be cut using traditional metal removal processes such as blanking, shearing, or die cutting. The removal process called routing, widely used with wood materials, can also be used with plastics. Routing is particularly useful in applications where the cutting tool does not go completely through the workpiece.

Plastics can be drilled with the same conventional twist drills used for metals and woods. However, since plastics are more brittle and easier to fracture than metals, a drill may be modified to prevent it from grabbing into the stock. If the bit is correctly sharpened and operated at the correct speed, two continuous spiral ribbons of plastic will emerge from the hole that is being drilled.

Twist drill bits for plastics should be ground so that the cutting edges have a zero rake angle (is perpendicular to the material being drilled). In most cases, drills used for plastics have a sharper point than those used for metals. Blunter point angles may work better when drilling nylon workpieces less than 3/8″ thick. See **Figure 20-1**.

Typically, a drill used with plastics should be ground to a point angle of 60° to 90°. A drill for metals will have a 118° point angle. Slower speeds of 500–1,000 rpm are best for drilling plastics. This speed should be further reduced immediately before removing the bit from the workpiece. The bit should always be removed from the stock before it is stopped. Otherwise, the heat generated by the drilling action will melt the plastic around the bit, cause the material to stick to the bit, and possibly crack the workpiece. A good practice is to retract the drill at least once for every three diameters the drill penetrates.

The challenge of picking the proper type of drill or cutting bit for plastic can be complicated because of the many different types of materials that are available. The International Association of Plastics Distributors (IAPD), in cooperation with Onsrud Cutter, provides an online database to help manufacturers select the best cutting parameters for specific materials and tools.

Figure 20-1. Twist drill bits for plastics should be ground so that the cutting edge has a zero rake angle (is perpendicular to the material being drilled). Drills used for plastics have a sharper point than those used for metals. Typically, a drill used with plastics should be ground to a point angle of 60° to 90°. A drill for metals will normally have a 118° point angle.

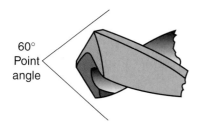

60° Point angle

Machining Plastics

Nearly all plastics can be machined—milled, cut, turned, planed, shaped, routed, stamped, drilled, and tapped. However, the approach taken with a particular manufacturing process must be adjusted to fit the unique characteristics of each material. When turning plastics, tools with a carbide insert will be helpful. Most often, these differences in approach relate to the types and speed of the cutting tool, and the use or absence of a lubricant or coolant.

Coolants are normally not required for machining operations other than drilling and parting. However, improved surface finishes and closer tolerances can be achieved by using non-aromatic, water-soluble coolants.

The cutting area can be cooled by using spray mists and pressurized air on the cutting tool. Conventional petroleum-based cutting fluids may work with some plastics, but may cause stress cracking when used with some polycarbonates and polysulfones.

Cutting Speeds

Saw blades are classified according to the number of teeth per inch. A fine-tooth cutter has more teeth per inch than a coarse-tooth cutter. In most cases, a fine-tooth blade is used to cut plastics. The tooth shape is also important. When cutting with a band saw, the skip-tooth pattern is effective in clearing chips out of the kerf. For some types of thermosets, the hook-tooth blade is preferred. See **Figure 20-2**.

Plastics can also be cut using a circular saw blade with carbide teeth arranged in a "triple chip" tooth design. See **Figure 20-3**. Table or overhead panel saws are often used, operating at a speed of from 6000′ to 8000′ per min. Blades used to cut stock from 3/32″ and thicker normally have from 3 to 5 teeth per inch. Blades used with thin-gage sheet should have a hook or rake angle from 10° to 15°.

Plastics have a machining speed of approximately 200 fpm (feet per minute). Fpm refers to the distance that the cutting edge of the tool travels in one minute (measured along the circumference of the cutting tool). Fpm can be calculated as follows:

$$fpm = \pi D \times rpm\ /\ 12$$
$$rpm = fpm \times 12\ /\ \pi D$$

where:

rpm = revolutions per minute,
fpm = surface feet per minute,
D = diameter of cutting tool in inches,
π = 3.1415.

Figure 20-2. The skip tooth bandsaw blade is preferred for cutting plastics, since it more effectively clears chips from the saw cut, or kerf. The hook-tooth design is used for some reinforced plastics that are harder or denser. Compare the hook- and skip-tooth designs to the conventional tooth design.

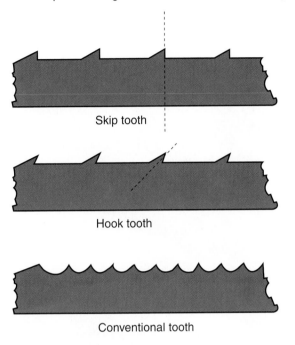

Skip tooth

Hook tooth

Conventional tooth

To fully understand the importance of selecting the proper cutting speed for a particular process, consider the cutting speed (expressed in fpm) of plastics

Figure 20-3. Triple chip tooth design for circular saw blades used to cut plastics.

compared to the speed used with other materials. Hard rubber has a cutting speed of 200 fpm, the same as plastics. Aluminum has a faster cutting speed, 250 fpm. The cutting speed of soft cast iron is 125 fpm. The ultra-hard metal titanium has a cutting speed of only 15 fpm.

Not all plastics are cut in the same way. Normally, thermoset plastics are more abrasive to cut than are thermoplastics. Some high-pressure plastics even require diamond- or carbide-tipped tools. Cutting may also be done in some applications with waterjet or laser tools, **Figure 20-4**. Tools must cut cleanly without burning, since the coefficient of thermal expansion for plastics is nearly ten times greater than it is with metals. When plastic is heated, the workpiece exhibits three thermal effects:

- It absorbs heat
- It expands
- It transmits heat

Expansion is described by the *coefficient of thermal expansion*, which is represented in equations as the Greek letter alpha (α). In plastics, there is a close correlation between the coefficient of thermal expansion and the melting point. The higher the coefficient, the lower the melting point.

When using a manufacturing process such as drilling, remember that the harder the material, the slower the cutting speed. The speed at which the drill spindle turns is described in terms of revolutions per minute (rpm). However, this is not the same thing as

Figure 20-4. Plastics may be separated by many different means, including laser cutting done on a machine like this one. Computer numerical control allows the laser to rapidly perform cutting operations on simple or complex shapes. (Shibuya Kogyo Co.)

cutting speed. The cutting speed of a drill is called its *peripheral speed*. This is the distance in feet per minute that the cutting edge of the drill actually travels. If soft materials, such as plastics, are drilled or cut using too great a cutting speed, breakage of the material is likely to occur.

Cutting Extruded Lengths

Cutting long lengths of extruded plastic is often done with an attachment to the extruder called an *extrusion cutter* or *traveling extrusion saw*. This is a machine that travels with the extrudate and uses a high-velocity rotating knife to cut it to length. Extrusion cutters are used to cut plastic hose, tubing, and various shapes with a diameter of less than 4″ (102 mm). When cutting short lengths of material 1/8″ (3.2 mm) or less in diameter, extrusion cutters can make several hundred cuts per minute. Machines are also designed to cut long lengths, often up to several hundred feet.

Die Cutting

Die cutting is a process that is used to cut plastic sheets, blister packaging, foam for upholstered furniture, and openings in molded parts. The process has long been used in the printing industry to cut packaging and containers using presses that force sharp-edged cutting rules through the paper or cardboard stock. See **Figure 20-5**. Die-cutting equipment ranges in complexity from simple hand-held dies to automated cutting equipment.

Hand-held dies consist of sharp-edged cutting rules bent to the desired configuration, much like a cookie cutter. The die is struck with a mallet to force the cutting rule through the material. Die cutting is a simple stamping process.

Die cutting machines are either single plane or rotary. With the *single-plane die-cutting* machine, a flat die set with multiple patterns is pressed into the stock. Normally the stock to be cut is placed on top of the press bed (a heavy shelf), and the die travels down on top of the stock. Some machines use a traversing roll to apply pressure.

Continuous die cutting can be accomplished with either a flatbed press or a rotary machine. In the rotary configuration, dies are arranged around the circumference of a cylinder. A back-up roll provides pressure for the cutting action; the material to be cut is fed between the die cylinder and the backup roll. **Figure 20-6** shows a computer integrated die

Figure 20-5. Continuous-die cutting. A—Rotary dies are arranged around the circumference of a cylinder allowing high-volume production. Two of the many sizes of cylinders are shown. (The Rotometrics Group) B—This full-head hydraulic cutting press can exert 80 tons of pressure to cut material using the dies shown on the open press bed. (Hudson Machinery Worldwide)

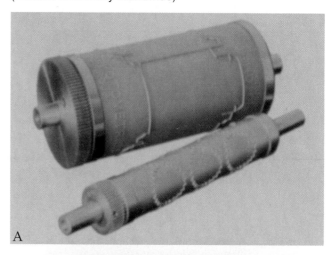

A

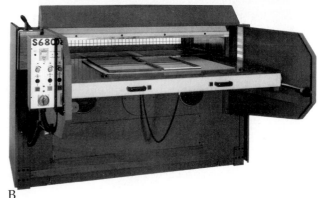

B

cutting system capable of 120 cutting strokes per minute. The system uses PLC (programmable logic control) microprocessor control. The machine shown in **Figure 20-7** incorporates a *traveling-head press* designed to make multiple cuts on wide sheets or roll goods.

Die cutting is used to cut parts such as gaskets and labels from various materials including plastics, rubber, foam, paper, and composites. Precise control of the die cutting process allows it to be used for such applications as *kiss cutting* of self-adhesive materials without penetrating the paper or film backing. See **Figure 20-8**.

Foam Cutting

The most common method for cutting plastic foam is with a hot wire. *Hot-wire cutting* is a simple

Figure 20-6. This computer-integrated flat bed converting system uses a programmable logic controller (PLC) to monitor and control high-speed die cutting operations. (Hudson Machinery Worldwide)

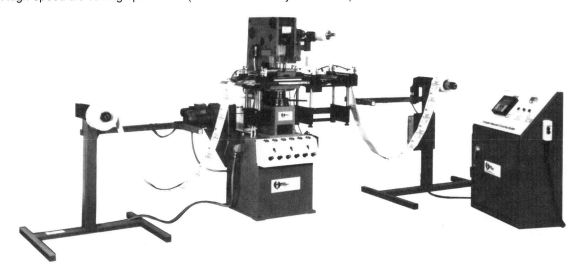

process that is accomplished with a machine built for specific product applications. Both hand-operated and computer-controlled machines are used.

Hot-wire cutting of foam is done with a thin wire that has high electrical resistance. The wire heats up when it is connected to an electrical power supply. The hot wire is used to slice through foam plastic by melting the material. Often, multiple wires are arranged side-by-side, so that several different parts can be cut at the same time. They can also be rolled around pulleys to produce cylindrical parts. Unusual shapes are produced using wire that is preformed to produce custom shapes.

Figure 20-8. Note how the material between the die cut shapes has been removed after cutting, leaving the backing intact. This mechanical platen press performs "kiss cutting," a precision operation in which the die cuts the material but does not penetrate the backing. It is often used for self-adhesive materials that will later be peeled off the backing and attached to another surface. (Hudson Machinery Worldwide)

Figure 20-7. Traveling-head presses like this one are designed to enable multiple-piece cutting from wide sheets or continuous rolls or material. (Hudson Machinery Worldwide)

Degating

To remove gate and runner systems from molded parts, a specialized hand tool called a *degating cutter* is used. A runner is the channel in the mold that conveys the plastic from the barrel of the injection molding machine to the part.

These specially designed pliers are shaped so that the cutting surfaces are able to make a cut flush with the edge of the part. Cutters are available in various shapes to facilitate cutting in locations that normally would be difficult to reach. Degating tools are also used to cut structural foam molded parts. Degating tools can also be attached to the gripper (end effector) on a robot and performed as an automated process.

Deflashing and Deburring

When a two-piece mold is used to produce the part, a thin seam or fine line of material, called *flash*, results at the point where the mold halves come together. The amount of flash that is generated increases over the life of the mold due to wear caused by the abrasive molding compounds rubbing against the mold surface.

Blow molded parts can be identified by looking at the bottom of the container for the fine line and spotting signs of flash. Injection molded parts can be identified by looking for a small bead at the bottom of the part at the point where the material was distributed to the mold.

Burrs are sharp edges on parts that are produced by force, usually when a cutting tool removes material from the workpiece. Both flash and burrs can be removed by several methods. The simplest approach is hand scraping. The processes most frequently used by manufacturing companies to remove flash and burrs from the outside of a part are *tumbling*, or *vibratory finishing*.

Tumbling

Tumbling is a process much like the process of milling, which is used to mix ceramic powders and additives. When tumbling is used to deflash plastic parts, the parts are placed inside a canister or drum. As the drum turns, the parts slide against each other, gradually rubbing away the flash. No other grinding medium is used; the parts produce the abrasive action.

Sometimes flash is produced inside the part or around holes, slots, and grooves. It is more difficult to remove flash or burrs on internal holes or grooves. Heavy parts and those with projections that may tangle or break may be unsuitable for deflashing in a tumbler or vibratory bowl. In these cases, it may be necessary to remove unwanted material by hand, using cutters, grinders, punches, burr knives, and files.

Vibratory deflashing

Another deflashing process, called *vibratory deflashing*, involves the use of a vibratory bowl or high-speed centrifugal units to improve the abrasive rubbing (or *scrubbing*) action between the parts. Many different types and shapes of media are available, depending on the material that is to be removed and the hardness that is necessary to remove the desired amount of material. The abrasives used include powdered mineral material, glass beads, walnut shells, and corncobs. Due to the abrasive cutting action, this process creates a dull matte finish on the parts.

Flash can also be removed using a media gun. The media gun may be used for *sandblasting*, in which abrasive sand particles are carried in a stream of air from a nozzle to abrade away the flash. Many firms transport parts on a conveyor through a media stream. Blast cleaning equipment, ranging in complexity from simple air blast guns to more complex systems with multiple guns and automated material handling conveyors are used to deflash parts. The deflashing media might consist of synthetic polymers, natural organic products, or minerals. Organic and polymeric materials are most frequently used as grinding media with plastics. See **Figure 20-9**.

Figure 20-9. This nonabrasive plastic media's size and irregular shape enable it to clean heavy residue quickly and easily without damaging mold surfaces. (Maxi-Blast, Inc.)

Cryogenic Deflashing

Cryogenic temperatures, below –150°C or 123 K, are created by introducing liquid nitrogen or dry ice to the deflashing environment. This makes the flash brittle and assists in material removal. Cyrogenic deflashing is popular for use with molded rubber parts, but it can be used with other materials as well. Tumblers, vibratory bowls, and high-speed centrifugal units are used in cryogenic deflashing. Other gases, such as liquid oxygen, liquid hydrogen, and liquid helium, are used as cooling agents or propellants in many rocket engines used in space vehicles.

In cryogenic deflashing, liquid nitrogen is injected into the insulated chamber of the vibratory bowl or tumbler. Flash, which is thinner than the parts being finished, is removed by the low temperature and plastic shot which is hurled from a precision throwing wheel. The wheel turns at high speeds approaching 8,000 rpm and breaks off the brittle flash on impact. The temperature of the chamber and speed of the throwing wheel depends on the media, type of material being removed, flash thickness, and part configuration. Silicones can require temperatures of –250°F (157°C) and 15,000 rpm wheel speeds. Other materials can be effectively deflashed at –40°F (-40°C) and 5,000 rpm speeds.

A very durable polycarbonate blast media is preferred for cryogenic deflashing. The ultra-cold temperature and high speeds can be very demanding on the media. Agricultural materials such as walnut shells, apricot pits, and corncobs are sometimes used. However, plastic media provides extremely long life, more consistent results, shorter cycle times, and a cleaner working environment. See **Figure 20-10**.

Figure 20-10. Cryogenic polycarbonate media used for deflashing silicone parts with high-speed centrifugal deflashing machinery at temperatures as low as –250° F. (Maxi-Blast, Inc.)

There are several advantages of cryogenic deflashing over the use of other methods. The use of plastic blast media prevents scratching of parts caused by blasting with abrasives such as steel shot, aluminum oxide, or glass beads. The cryogen can be recycled and there is no problem associated with disposal of chemicals or waste materials. A major advantage is the faster speed for deflashing enabled through cryogenic conditioning.

Grinding

In order to recycle plastics, it is usually necessary to break up rejected parts, runners, or other scrap material into small pieces or granular form. This is accomplished by grinding.

There are four different types of grinders that are used in the plastics industry to reclaim parts and scrap. These are generally referred to as the light-duty grinder, auger-fed grinder, medium-duty grinder, and heavy-duty grinder.

For light-duty grinding, the machine is normally placed in the manufacturing workcell, directly beside the equipment used to form the part. Small grinders, with motors of up to 40 hp, are capable of handling scrap with a wall thickness of approximately 1/4" (6.3 mm). Such light-duty grinders are normally fed and emptied by hand, but can be attached to blowers and hopper loaders.

Another type of grinding machine is referred to as the auger-fed grinder or ***granulator***. These machines are used when the manufacturing process is automated or semiautomated. Auger-fed grinders work in conjunction with other simple automation devices, such as sprue pickers (robot arms), and automatic load/unload systems. The material that is fed into these grinders typically consists of sprue and runner systems, as well as occasional reject parts.

Medium-duty grinders, with motors up to 100 hp in size, are larger and heavier than light-duty grinders. Medium-duty grinders have larger throats and cutters capable of grinding stock up to 1/2" (12.7 mm) thick. In most instances, the particles that these machines produce are loaded into drums, either by gravity feed or a pneumatic system.

Heavy-duty grinders, with motors of up to 400 hp, are used for large parts and tough grinding applications. Due to the noise they make when crushing and grinding parts, these machines usually are installed in a separate room, away from the work area. Heavy-duty machines are automatically loaded and unloaded.

The finished product of each of these grinding operations is called *regrind*. Regrind ranges from a coarse granule to a fine powder. The ground material can be recycled into new products at the injection molding station. Care must be taken not to mix unlike colors, or the ground material will be less desirable. Mixed colors are used to make stock that is used for filler and in applications where the color of the product is not important.

There are many processes that can be used to separate plastics. Many additional separation processes are discussed throughout the text, in conjunction with the separation of metals, woods, ceramics, and composite materials. These processes often can be used successfully with plastics. It is important to remember that each material has its own unique behavioral characteristics. Adjustments in cutting speed, type of tools used, and type of process chosen may be necessary.

Important Terms

coefficient of thermal expansion (α)
continuous die cutting
cryogenic
cutting speed
degating cutter
die cutting
extrusion cutter
flash
granulator
hot-wire cutting
kiss cutting
peripheral speed
regrind
sandblasting
single-plane die-cutting
traveling extrusion saw
traveling-head press
tumbling
vibratory deflashing
vibratory finishing

Questions for Review and Discussion

1. Describe the type of blade that you would purchase if you were interested in cutting plastic sheet stock with a saber saw or reciprocating saw.

2. When drilling plastic, should you use a cutting speed faster or slower than you would when drilling metal? What is the relationship between cutting speed and the coefficient of expansion of plastic?

3. If you were setting up your own plastics manufacturing firm, how important would you consider the purchase of a grinder or granulator? How would you use the recycled material? What would you do with the regrind from thermoset parts?

Processes Used to Separate Wood Materials

Chapter Highlights

- Edged tools are the primary means used to separate wood materials.
- Planers and surfacers are used to provide parallel, smooth-planed surfaces on wood.
- The rotating workpiece in a lathe may be attached to a faceplate or held between centers.
- Sawing is a chip-producing separating process, with each succeeding tooth of the saw cutting away a tiny chip from the wood.
- Drills, boring machines, and mortising machines are used to produce holes in wood.

Background

Most consumers have a working familiarity with many different types of edged tools that are used to separate wood stock. We know that some of these processes, such as drilling or sawing, produce chips. Others can separate the material with no chip being generated at all. The edged tools responsible for all of this may be used in both hand- and power-operated equipment. These tools—generally knives, saw blades, drills, and chisels—are used to cut pieces of wood apart, make holes of various shapes in the wood, smooth the wood's surface, or reduce its dimensions.

There are nine major types of processes used in manufacturing to separate wood:

- Planing
- Jointing
- Shaping
- Routing
- Turning
- Sawing
- Drilling
- Boring
- Mortising and tenoning

When wood is purchased as rough stock, the first step is usually to create a smooth surface on opposing faces by removing a small amount of stock using the process called *planing*. Edges are then smoothed using another process called *jointing.* Let us take a closer look at how this process works.

Planing

Machine planing, or surfacing, is a process that mills wood to a uniform thickness and produces a smooth surface. Planing is used to remove rough mill marks on stock that has not been surfaced at the mill. Planing is also used to **true-up** (eliminate bow in) the wood, but it cannot straighten warped stock.

Most stock purchased directly from the mill needs to be planed to produce smooth faces. Normally, planing is done in the direction of the grain, with a maximum cut of 1/16″ (1.6 mm).

There are three major types of machines used for planing: the knife-blade planer, the abrasive belt planer, and the jointer. Both the knife-blade and abrasive belt planers are used to smooth the face surfaces of stock. The jointer can be used to surface the faces of narrow-width stock (boards up to 6″ or 8″ wide). The jointer is the preferred machine for removing stock from the edges of boards.

Knife-Blade Planer

Figure 21-1 illustrates the design of a knife-blade planer, which works on the same principle as

the hand plane. The knife blades shear off a uniform layer of stock as the material is passed through the machine. Knife-blade planers with cutterheads only on the top are called *single planers* or single surfacers, and are the most common type. *Double planers* have blades on the top and bottom, so that both sides of the stock can be planed at the same time. With either type of planer, stock is pushed into the machine on a bed, and power feed rollers carry it through the machine. Chip breakers and pressure bars help to reduce chattering and keep the stock from jumping during planing.

Planers are sized by the *bed width*, which essentially means the width of stock that can be milled in the machine. Planers range in bed width from 12″ to 52″ (30 cm to 132 cm). Knife-blade cutterheads are normally adjustable to suit the width of the stock to be planed. Cutterheads may carry a number of knives, or may have only one, with a counterbalancing blank on the other side of the planer head. Large planers may carry as many as 20 or 30 jointed knives. **Figure 21-2** shows an operator planning a thick board on a single-surface knife-blade planer.

One of the disadvantages of the knife-blade planer is that the knives must be removed for sharpening and changes in set-up. The practice is

Figure 21-1. A single surfacer is a knife-blade planer with one rotating cutterhead. The knife blades, held in place by plates called gibs, shear off a thin layer of material from the top of the stock as it passes through the machine. Units with two cutterheads, called double planers, surface both top and bottom of the stock at the same time.

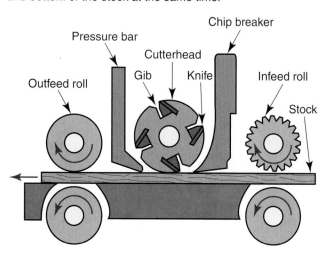

Figure 21-2. This worker is feeding a thick plank through a knife-blade planer. Note the vacuum system used to collect sawdust and chips attached to the top of the machine. It simplifies cleanup and minimizes the amount of dust in the air. (Delta International Machinery Corporation)

time-consuming, and the knives must be handled with extreme care. Another drawback of the knife-blade planer is the danger of *kickback*. This means that it is possible for the cutter to bite into the stock so that the workpiece is thrown out of the machine back toward the operator. Kickback is usually caused by such operator errors as feeding stock that is badly warped or bowed, taking too large a cut in one pass, or feeding stock that is too short to be safely planed. A safety-conscious operator always stands to one side of the machine, out of the path of the moving stock.

Abrasive Belt Planer

The **abrasive belt planer** is a machine that uses a large abrasive belt to remove material from the surface of wood to smooth it and reduce the board's thickness. The abrasive belt planer is a safer machine than the knife-blade planer. The abrasive belt planer eliminates kickback and the problems related to removing and sharpening knives. In addition, abrasive belt planers generally have lower overall maintenance costs and are less noisy than knife-blade planers. Abrasive planing can improve lumber yields up to 20 percent by eliminating product defects caused by knife-blade planers. The machine shown in **Figure 21-3** is a high-production abrasive belt planer used in furniture manufacturing. It can handle large parts at speeds of up to 80 feet per minute. On this planer, the stock is fed into the infeed section of the first unit, which planes the top side of the stock. The wood is then

Figure 21-3. This large abrasive belt planer surfaces the top side of the wood in the first unit and the bottom side in the second unit. It is used for high-speed production planing in a furniture factory. (Timesavers, Inc.)

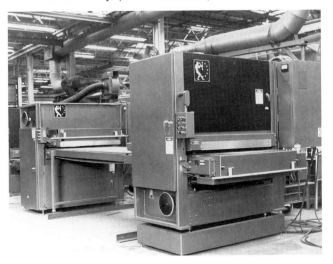

Figure 21-4. This is the abrasive belt planer shown in Figure 21-3 with covers removed to show roller drive mechanisms that carry the abrasive belts. Note that the belt is on top in unit at right, and at bottom in unit at left. This permits surfacing both sides of the stock in a single pass. (Timesavers, Inc.)

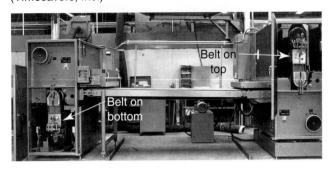

carried on a conveyor to the second head, which planes the bottom of the board.

Abrasive belt planers have heavy sanding belts instead of conventional metal knife-blade cutters used in other planers. Abrasive belt planers can be purchased with belts from 12″ (30 cm) to almost 10′ (3 m) in width. Machines of this type are constructed with any number of planing heads to meet high-volume production requirements.

Figure 21-4 shows the interior of the same machine. Note the vertical planning belt exposed on each head. At first glance, you might think that a sanding belt would wear out quickly in a production environment. If the machine is used properly, this is far from true—the belts are tough and will last for a long time. However, if you try to take too large a cut with one pass, or if you attempt to plane stock with nails or other metal fasteners, the belt will tear. Tears or holes shorten the life of the belt, leading to premature failure.

Planing can also be done using hand planes. Removal of stock on the edges is a relatively simple process. Planing of end grain can result in the blade digging in, and chipping or breaking off the edges of the workpiece. The proper procedure is to hand plane from the edges working toward the center.

Jointing

The *jointer* is a machine that can improve overall quality of a workpiece by effectively complementing the work of the planer and circular saw. For example, even if stock is purchased with both face sides planed from the mill, it is still necessary to true-up one edge on a jointer before proceeding.

Figure 21-5. A small jointer of the type used to true the edges of boards.

Figure 21-6. Cross-sectional view of a typical jointer. Stock is held against the fence and moved over the rotating cutterhead for material removal. The heights of the infeed and outfeed tables are independently adjustable.

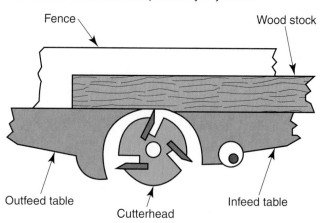

This sequence of operations is often followed when jointing. If the stock is narrower than the jointer's cutterhead, the jointer is normally used to plane the surface and edge of the best face. The opposing face is then planed on the planer, and the board is cut to width on the circular saw.

Jointers are sized by the length of the knives in the cutterhead. A small jointer, as shown in **Figure 21-5**, might have a cutterhead with three or four knives, and run at a speed of approximately 4500 rpm.

When a jointer is used to true an edge, the operator places the face of the stock against a *fence* (adjustable stop running perpendicular to the cutterhead axis) and rests the edge on the infeed table of the machine. This will ensure that the finished edge is perpendicular to the face. The stock is pushed over the rotating cutterhead and on the outfeed table. See **Figure 21-6**.

After the back end of the stock passes the cutterhead, a machine guard closes the opening over the cutterhead. On small jointers, a *push stick* is normally used to hold down the stock and push it through the machine. A larger jointer is shown in **Figure 21-7**. With this machine, the stock is of sufficient thickness and width to permit the operator to push it through the machine without using a push stick.

All of the major parts of the jointer—the infeed table, outfeed table, and fence—can be adjusted to produce the desired cut. The adjustment of the outfeed table is very important. The outfeed table must

be positioned so that it is level with the highest point of the knife edges. If the table is too low, the trailing edge of the board will drop down, producing a dip on that end of the stock. If the outfeed table is too high, it will push the board upward and produce a taper. In order to remove stock on the jointer, the depth of cut is established by adjusting the infeed table to a level that is lower than the outfeed table and the knives on the cutterhead.

If it is not used properly, the jointer can be a dangerous machine. Stock must be at least 12″ (30 cm) long and at least 3/8″ (9.5 mm) thick. End grain should never be planed unless the stock is at least

Figure 21-7. This machine operator is facing one side of a thick board, using a jointer. The dark-colored mechanism pressing against the edge of the workpiece is a spring-loaded guard that swings into place to cover the rotating cutterhead as soon as the workpiece has cleared it. (Delta International Machinery Corp.)

12″ (30 cm) wide. A push stick or push block should be used when removing material from the face surfaces of small- to medium-sized workpieces. When the stock is large and heavy, it may be safer to push it through without a push stick.

Shaping

The *shaper* is a useful and versatile machine for producing the intricate shapes required for molding or window framing. Shaping produces a straight line or design pattern along the length of the stock. Elaborate crown molding and exterior molding is often made by joining together several different pieces of molding.

There are various types of shapers. The most common of these is the *single-spindle shaper*, which has a vertical spindle (shaft) that projects through an opening in a horizontal metal worktable. See **Figure 21-8**. The shaper cutters are mounted on the spindle. Cutters have many different shapes, ranging from straight to elaborate designs with curves and angles. Here is how the process of shaping works:

If you are shaping material where the edge to be grooved is a straight line, the fence is used to gage the depth of the cut that will be made into the edge of the stock by the rotating cutter. After the fence has been set, the stock is placed on the table and pressed against the

Figure 21-8. The single vertical spindle of this shaper carries a cutter that produces a desired edge-shape on stock. The worker in this photo is using the shaper to finish the edge of a door facing. The two diagonal bars in front of the fence are anti-kickback devices. (Delta International Machinery Corp.)

Anti-kickback devices

Figure 21-9. The shaper cutter is secured on the spindle with a nut and lock washer, as shown. The cutter configuration will produce a matching shape on the edge of the workpiece, as shown. When shaping the edge of an irregular workpiece, a template is often used. It rides against the collar to maintain the proper depth of cut.

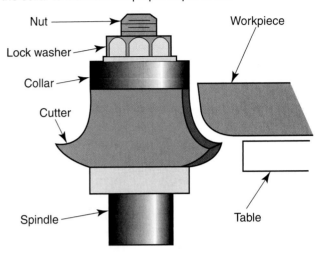

adjustable fence. The workpiece is then pushed carefully into the rotating shaper cutters. A push block or push stick should be used, whenever possible.

Stock must always be fed against the rotation of the cutters in a shaper. The height of the cut is controlled by a handwheel that raises the height of the vertical shaft. The depth of cut is regulated by adjusting the position of the fence. The shaper cutter rotates at high speed (5000 rpm to 10,000 rpm). **Figure 21-9** shows a shaper cutter attached to the spindle, and its relationship to the workpiece.

Not all pieces are easy to shape. If the workpiece has been bandsawed into a curve or irregular shape, the fence cannot be used. Instead, the workpiece is pushed against the top and bottom collars, while the knives work to remove the desired amount of stock.

A third method of guiding the stock against the cutter is to use a shaped pattern or template. The stock is attached to the pattern, which is placed on the shaper table. The edge of the pattern rests against a collar that is above the cutter on the spindle. (Refer again to Figure 21-8.) The guiding edge of the pattern regulates the cut. One of the advantages of this method is that the entire edge of the part can be shaped, since no stock has to be left uncut to run against the collar. Another advantage is that the edge of the stock can be rough, since it is not in contact with the collar.

As in the case of the jointer, stock being processed on the shaper should be held and positioned with push sticks, guards, or hold-down devices when possible. Often, special-purpose wooden guides or fences are fastened to the shaper table. At other times, the stock is held against templates or special forming jigs.

In production, shaping is often performed using *multiple-spindle shapers*, which have a number of spindles arranged side by side. Automated equipment that clamps the stock to the table and feeds it into the cutters is also popular.

Routing

The router has become one of the most important machines in the furniture-making industry because of its great versatility. Routers can be mounted on a table or may be handheld. Large production routers look much like vertical milling machines or drill presses. Production routers may have motors capable of rotating spindles at more than 50,000 rpm.

The router is used to add simple round or decorative shapes to the corners of stock, tables, and countertops. A router that is mounted in a table can shape moldings and cut grooves in the edges of stock. Special templates and fixtures are used with routers to produce items from intricate carvings, such as rosettes and overlays, to simple signs often seen at craft shows. Routers can even be used to produce spiral flutings and rope moldings on lathe-mounted work.

When a *handheld router* is used, it is held securely in both hands, with one hand grasping a handle on each side of the tool. The cutterhead is either plunged into the stock or pushed into it from the side. The best cut is made by using a fairly rapid feed rate and several shallow cuts. Usually a template or guide is used to produce the desired pattern. A bushing or bearing on the end of the cutting tool helps to guide the cutter around the template.

Stationary routers are used to make grooves and cut irregular shapes. A pin is mounted in the table and serves as a guide. The template runs against the pin to shape pieces as desired.

Routers are popular in the furniture-making and boatbuilding industries. Computer numerical control (CNC) routers are often used to cut patterns and designs when a number of identical parts are required. **Figure 21-10** shows an operator loading an uncut piece of wood onto the router table while the routing process is carried out on two other pieces.

Figure 21-10. This CNC router can complete two workpieces at the same time with great precision because of the program stored in the computer controller. The operator is loading an uncut piece of wood on a transfer table in preparation for the next cycle.
(Stanton Manufacturing Co., Inc. Photo by Matt Bentz)

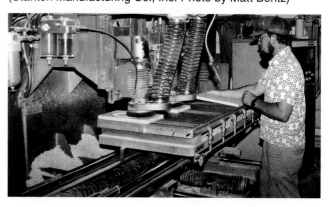

A computer control unit is located to the operator's right. The program for cutting the desired pattern is stored on a computer disk until it is needed. Then, the controller transmits the electronic signals to the router to make the desired cuts.

In industry, it is common practice to use two or more routers attached to the same control system so they will perform identical movements on workpieces. **Figure 21-11** shows an operator waiting for a pair of CNC routers to finish cutting parts. The vacuum hoses connected to each router head are used to draw sawdust away from the router bits, keeping the cutting paths clear and unobstructed.

Figure 21-11. To keep the router path clear, a vacuum dust collection system is attached to each of the router heads of this CNC tandem routing machine. The two hoses carry sawdust away to a central collection point. Note the two pieces of uncut stock on the transfer table at left, ready to move into place beneath the routers.
(Stanton Manufacturing Co., Inc. Photo by Matt Bentz)

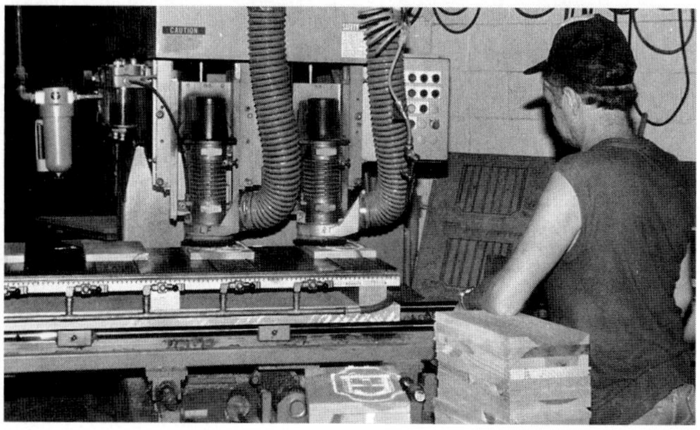

Figure 21-12. Spindle turning on a lathe is done by rotating the workpiece between centers and using a chisel-type tool to scrape material from its surface until the desired contour is achieved. (Delta International Machinery Corp.)

Turning

Turning of cylindrical parts is done on a wood lathe. The workpiece is either held between centers or is attached to a faceplate mounted on the spindle of the lathe's headstock. Turning between centers is referred to as *spindle turning*. Most lathes used for wood turning are similar to metal lathes in concept, but not as heavy and rugged. See **Figure 21-12**.

Spindle Turning

The operator in Figure 21-12 is removing stock using a lathe tool called a chisel. Chisels come in six different shapes: gouge, skew, parting tool, diamond point, round nose, and square nose. See **Figure 21-13**. The *gouge* is a round-nosed, cupped tool for roughing and making cove cuts. The *skew* is a flat tool that is used to smooth cylinders and cut shoulders. The *parting tool* is used to separate material and to cut off stock. The *diamond point, round nose,* and *square nose tools* are used to produce specialized contours.

Lathes are sized by the swing and length of the bed. The *swing* is the dimension that is equal to

the largest diameter piece that can be turned on the lathe. Swing is equal to twice the distance from the center of the spindle to the bed.

Major parts of the wood lathe are the headstock, tailstock, bed, dead center and spur center, and tool rest. A wooden workpiece is secured between centers for turning as follows:

First, the center of each end of the stock to be turned must be determined by drawing intersecting diagonals from the corners. The point where the lines intersect is where the point of the tailstock center is placed to secure the part for turning. The tailstock is loosened and moved to the right to permit insertion of the workpiece, and the stock is tapped against the spur center in the headstock. The tailstock is then moved up against the stock until the point of the dead center can make contact with the point of intersection on the right end of the stock. The tailstock is then locked in place and the center turned snugly into the stock. Finally, the tool rest is moved to the area where stock removal will begin, and is raised to a height of about 1/8″ (3 mm) above the workpiece center.

Once the workpiece is secured in the lathe, it is time to perform the initial turning operation, called *roughing*. It is advisable to use slow speeds of 1000 rpm or less to perform roughing operations. When roughing, the gouge is held firmly on the tool holder and pushed into the work. A cut is taken down the length of the tool holder. During turning, the tools will get hot from friction. They should be

Figure 21-13. Three of the most common lathe tools are the gouge, skew, and parting tool. The diamond point, round nose, and square nose tools are used to create specialized contours. (Robert Larson Company)

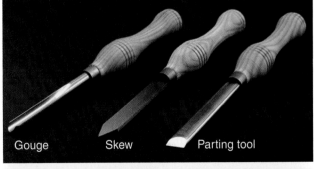

Gouge   Skew   Parting tool

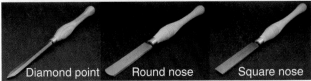

Diamond point   Round nose   Square nose

Figure 21-14. Cutting across the grain is *crosscutting* and with the grain is *ripping*.

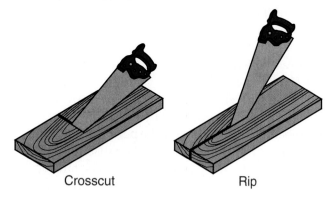

Crosscut   Rip

cooled in water to prevent destroying their temper and also to keep them from burning the wood.

Facing

Facing is also called **faceplate turning**. Facing of wood is normally done on a wood lathe, with the stock screwed directly to the faceplate. Sometimes the stock is glued to a backing block, which is then screwed to the faceplate. This eliminates the problem of holes showing in the work. When the facing operation is completed, the part can be separated from the backing block with a wood chisel.

Faceplate turning can be done using the diamond point, round nose, or square nose tools. Lathe tools are positioned using the tool rest. The rest should be raised slightly above center of the workpiece and aligned parallel to the surface to be turned. Turning is done using a slow speed. A scraping, rather than cutting, action is used. The gouge should never be used for faceplate turning.

Sawing

There are many different types of machine- and hand-powered sawing operations. All are chip-producing separating processes. Chips are produced when the teeth of the saw blade cut into the work. The tooth shape and the type of blades vary from saw to saw, but the principle of sawing is always the same. The teeth cut into the stock, with the blade moving (turning or being pushed or pulled) in the direction that the teeth are pointing. The action of the teeth digging into the stock helps to push the material being sawed down against the table of the saw. Sometimes, stock is fed into the blade, but in other cases the blade is pulled into the stock.

Saw cuts made across the grain are referred to as **crosscut sawing**. Cuts made in the direction of the grain are referred to as *rip sawing*. See **Figure 21-14**. Saw blades can be purchased with either crosscut or rip teeth. Circular power saw blades are also available with teeth that can be used for both types of cutting. They are called **combination blades**.

Circular saw blades are used for most of the machine-powered saws: portable circular saws, radial arm saws, cutoff saws, and panel saws. The major sawing processes are scroll sawing, bandsawing, circular sawing, radial arm sawing, cutoff sawing, and panel sawing. Each of these will be discussed.

Scroll Saws

Scroll saws are often referred to as *jig saws*. Scroll saws are used by patternmakers to produce intricate cuts within the inside dimensions of a workpiece. See **Figure 21-15**. Scroll saws can also produce sharp radii and fancy designs of the outside edges of the material being cut. However, this type of work can be accomplished with other machines. Cutting irregular curves inside of the stock is an operation unique to this machine. When many ornate shapes need to be cut, scroll sawing is often an ideal choice.

Figure 21-15. To cut intricate shapes on the inside of a workpiece, such as the numerals being cut out of a thick block of wood in this photo, the scroll saw is the preferred power tool. (Delta International Machinery Corp.)

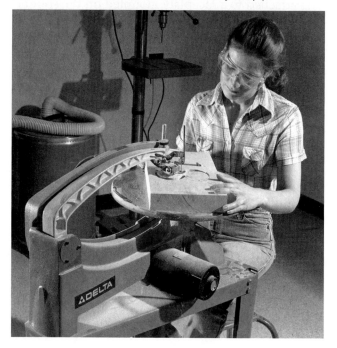

Blades used for scroll sawing are sized in terms of the number of *teeth per inch*. The finer-toothed blades (those with more teeth per inch) can be used to cut sharper curves. Harder material such as aluminum, copper, and plastics can also be cut with scroll saws, but fine-toothed blades and slower speeds are necessary. The blades are about 6″ to 8″ (15 cm to 20 cm) long, and are held in a vertical position with spring tension between upper and lower chucks. It is the up-and-down movement of the lower chuck that moves the blade.

The scroll saw can also be used to produce angled, rather than vertical, cuts. This is accomplished by tilting the table supporting the stock. Typically, tables will tilt up to 15° to the left or 45° to the right.

The scroll saw is ideal for making inside cuts. These cuts can be accomplished without producing an undesired saw cut across the pattern to get to the inside. To do so, drill a relief hole, slightly larger than the width of the blade, in an area where stock will be removed. Then, loosen the blade from the upper chuck of the saw. Move the lower chuck down by rolling the motor pulley by hand. The stock can then be placed on the saw table with the blade projecting through the drilled hole. Fasten the blade in the upper chuck again, and proceed with the inside cut.

While the scroll saw is fine for making small-radius turns and cutting intricate designs, sharp turns must be made slowly or the blade will break. One disadvantage of the scroll saw is that it is not very useful for cutting long straight lines. Because of the very thin blade, it is difficult to cut long lines which are perfectly straight and do not waver from one side to the other.

Bandsaws

The *bandsaw* is used primarily for cutting curved edges. The blade used in a bandsaw is a *band* (continuous loop) with teeth cut into its edge. The blade is held vertically in tension between two rotating wheels. See **Figure 21-16**. The blade will last a long time without needing to be replaced if it is operated at a speed appropriate for the material being cut. If the blade breaks, the ends of the broken blade can be rejoined by welding. In some cases, a butt welding fixture is attached directly to the bandsaw.

Bandsaws are sized by the diameter of the wheels. The smallest machines are normally 10″, while others have wheels as large as seven or eight

Figure 21-16. A bandsaw is useful for cutting circles and arcs with a fairly large radius. The continuous-loop saw blade is carried on large wheels inside the rounded safety covers at the top and bottom of the machine. (Delta International Machinery Corp.)

feet in diameter. These very large saws are used in sawmills to cut logs into timber.

Angled cuts can be made by changing the angle of the table. Straight cuts are made using a fence for accuracy. On long straight cuts, however, the blade sometimes will pull slightly to the left or right. This is referred to as *lead*, and is usually caused by improper tracking of the blade on the wheels. Since it is difficult to avoid this problem, many operators prefer making long cuts using a circular saw. The bandsaw works best for cutting angles and large curves.

It is important to remember that the capability of every saw is limited by the shape and width of its blade. In bandsawing, the blade used is normally from 3/8″ (9.5 mm) up to several inches in width. A circle with a 2″ (5 cm) diameter can easily be cut with a 3/8″ (9.5 mm) blade. However, if you try to cut a 1/4″ (6.3 mm) radius, the blade will not be able to turn this sharply. It will bind and may break. In order to cut such a sharp radius, the cut either must be broken up into a series of tangential cuts that gradually work toward accomplishing the sharp radius desired, or a narrower width blade must be selected.

Circular Saws

When long, straight, or angular cuts are desired, the circular saw is normally the preferred woodworking machine. Circular sawing is a very versatile process. Most wood manufacturing plants have at least one of these machines.

However, the circular saw cannot do everything. For example, circles cannot be made with this saw without a special fixture. It is limited to making straight cuts. This includes ripping, cutting off stock, making dadoes and miters, and producing grooves. Despite this limitation, the circular saw is still the most important machine for separating wood material.

There are many different types of circular saws. The common table saw is the foundation for all of these saws, and will be used here to convey the basic process of circular sawing. Other types of circular sawing are accomplished with machines such as radial arm saws, cutoff saws, and panel saws. Portable circular saws are extensively used in the construction industry. See **Figure 21-17**.

On a table saw, the circular saw blade is mounted on a horizontal shaft, called an *arbor*, located beneath the center of the table. Single-arbor saws are most common, and are often called *variety saws*, because they can perform a number of tasks. Sometimes they are referred to as cutoff saws, or *cut saws*. Another type of circular saw commonly found on construction sites is the cutoff saw. A *cutoff saw* is mounted on a pivot and is popular for cutting angles on framing lumber. See **Figure 21-18**. Saws with two arbors (one with a ripping blade mounted on it, the other with a crosscut blade) are called *universal saws*.

Figure 21-17. Because of their ease of handling on the job site, handheld portable electric circular saws are widely used in the construction industry. Saws are available in several different blade-size configurations, with the 7 1/4″ size shown here among the most widely used. (Milwaukee Electric Tool Corporation)

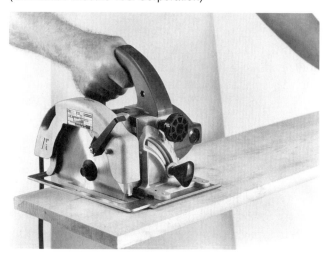

Figure 21-18. The cutoff saw is used on many construction sites to quickly and accurately trim framing lumber to size. For rafters and molding, precise angles can be set and cut repeatedly. (Black and Decker)

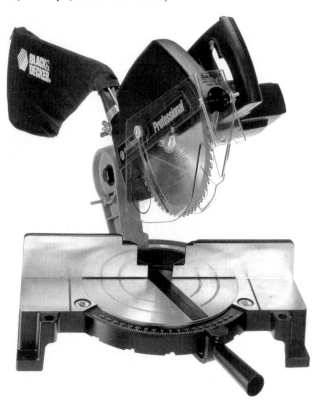

Circular saws are sized by the diameter of the blade. Common sizes for circular saws range from 7 1/4″ (18.4 cm) portable units and 8″ (20.3 cm) table-top models to production machines with blades 16″ (40.6 cm) or more in diameter.

Before using the circular saw, the operator must make a decision about the type of cut. If the cut involves ripping the stock, this is normally accomplished using a fence. In **Figure 21-19**, the fence can be seen to the operator's right, with the board snugged up securely against it.

When setting up the machine, the blade is raised to the level necessary by turning a handwheel. If the cut is to be made completely through the board, the blade should be raised to a height just above the top of the stock.

The fence is adjusted by releasing a locking lever, then sliding it toward or away from the blade. If the fence is moved closer to the blade, a narrower width cut will be made; if moved away, a wider cut will result. Once the fence is positioned for the proper width of cut, the lever is used to lock it in place.

Figure 21-19. Table saws can be used to rip or crosscut stock. Many have a tilting arbor carrying the blade that allows making angled cuts. (Delta International Machinery Corp.)

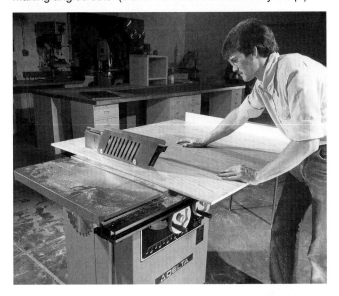

The stock is fed by hand into the blade. During ripping, the stock is pushed against the fence, and forward into the blade. If the stock is too narrow to be safely guided through by hand, a push stick should be used to keep the operator's hands away from the rotating blade.

When using the table saw to make cuts across the grain, a miter gauge is required. The fence is removed or moved out of the work area. A miter gauge is then placed in a slot in the table. The board is placed against the miter gauge, so that it runs across the table (perpendicular to the blade). The stock is shifted left or right, so that the correct amount of material will be cut off by the blade.

Whether crosscutting or ripping, be sure to avoid standing directly behind the stock that is being removed. Sometimes kickback occurs, propelling loose stock violently backward. The stock being kicked back could cause a serious injury. Kickback most often occurs when ripping. Most saws are equipped with an *anti-kickback device* to eliminate this problem.

Radial Arm Saws

The radial arm saw is another type of circular sawing machine. While the table saw has its blade mounted beneath the table, the blade on the *radial arm saw* is carried on a horizontal arbor suspended on an arm above the worktable. See **Figure 21-20.**

Radial arm saws are normally used for crosscutting, particularly trimming boards to length. Most of the cutting processes performed on the table saw can also be done on the radial arm saw. The radial arm saw is particularly handy when crosscutting long stock that would be unwieldy to work with on a table saw.

Another interesting feature of the radial arm saw is its capability of being used for purposes other than sawing. The blade can be replaced with a sanding disc for sanding or an abrasive wheel for grinding.

When cutting on the radial arm saw, the stock is held securely against a stop-type fence located at the rear of the saw table. The saw head is then pulled across the stock, toward the operator. Radial arm saws are classified by the size of the saw blade and the horsepower rating of the motor. Radial saws normally are from 8″ to 14″ (20.3 cm to 35.5 cm) in diameter, with 10″ (25.4 cm) the most common size.

Miter Saws

The *miter saw* is similar to the cutoff saw, but has the capability of cutting stock (especially trim and molding) at an angle from the vertical, as well as moving through a 90° arc horizontally. See **Figure 21-21.** The miter saw is attached to a spring-loaded arm, which raises the saw until the operator pulls it down. Most cutoff and miter saws use 10″ (25.4 cm) blades; some use 8 1/2″ (21.6 cm) blades.

Figure 21-20. In addition to duplicating most of the functions of a table saw, the radial arm saw can also be used as a sander or grinder by replacing the blade with the appropriate accessory. (Black and Decker)

Figure 21-21. The miter saw, or power miter box, can be used to cut trim and molding stock at compound (both horizontal and vertical) angles. A spring loaded arm keeps the saw off the work until it is pulled down by the operator. (Black and Decker)

Panel Saws

Sometimes, large sheets of plywood or other polymeric material must be cut into sections. While this can be done on the table saw, it is often easier to use another type of saw called a *panel saw*. See **Figure 21-22**. The panel saw consists of a circular saw attached to steel rails that are suspended over a rack that supports the material to be cut. The saw slides up and down easily on the rails, and can be turned to cut the sheet either vertically or horizontally.

Figure 21-22. Panel saws can quickly and accurately cut large sheets of plywood or similar materials into smaller pieces. The saw can be rotated to cut either horizontally or vertically. (Black and Decker)

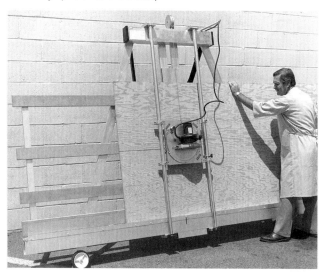

Drilling

Most readers will be familiar with the process of drilling holes with drill bits using hand drills or machine drill presses. However, there are many different ways to make holes in wood.

Machine drilling, or boring, of holes in wood is normally done with a drill press or a boring machine. The tools that are used to produce holes are called ***bits***. They are held in place in the drill press by clamping them in a chuck or by inserting them in a tapered sleeve.

In addition to common drill bits (usually referred to as "twist drills"), other cutting tools are available, such as plug cutters, Forstner bits, hole saws, auger bits, and spur machine bits. Holes 1/4" (6.3 mm) or smaller in diameter are normally drilled with a handheld power drill, a manually operated rotary drill, or a push drill. Any of these tools will drill small holes quickly. Larger holes are usually made with a drill press, **Figure 21-23**. The operator in the photo is checking the positioning of the workpiece before starting to drill the hole. Note the large bit, which has a tapered shank that is inserted in a sleeve attached to the drill press spindle. To ensure accurate and safe operation, the workpiece is held securely in a fixture bolted to the table of the drill press.

Figure 21-23. A drill press is used with larger drill bits. This operator is checking for proper positioning of the drill over a workpiece that is held in the fixture bolted to the drill press table. (Delta International Machinery Corp.)

Boring

In the furniture-making industry, holes are often bored for inserting *dowels* (wooden connecting pegs) in parts such as drawers or shelves. In such applications, perfect alignment is necessary. This is normally accomplished using a boring machine. While boring machines look something like conventional drill presses, they are much heavier and often have several spindles. This enables more than one hole to be bored at a time.

With some single-spindle boring machines, the table moves upward toward the tool. In other instances, the table is stationary and the spindle moves toward it. Often, the travel of the tool is regulated by a foot pedal, as is the case with mortisers and punch presses.

Multi-spindle boring machines are normally available in sizes up to 6′ long. There may be as many as twenty spindles on large production machines.

Mortising and Tenoning

Sometimes, an odd-shaped joint called a *mortise-and-tenon joint* is required to secure drawer components or other pieces of fine furniture. The vertical mortising machine cuts the rectangular opening in wood for a mortise-and-tenon joint. See **Figure 21-24**. The mortise-and-tenon joint provides much more holding strength than a simple *right angle butt joint*, where two pieces of wood are pressed together in a 90° angle, glued, and nailed.

Most furniture joints are secured with screws, dowels, or *biscuits* (flat discs). These approaches are inexpensive but eventually result in wobbly chairs or weak products. Screws provide the weakest joint, and dowels create a loose joint through expansion and contraction. Mortise-and-tenon joints, with one piece chiseled to fit into a slot cut in another, are the mark of fine furniture.

Figure 21-24 Mortise and tenon joint. The protruding tenon, or tongue is inserted in the mortise. The mortise can be cut all the way through or partially through the wood.

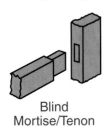

Blind
Mortise/Tenon

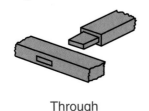

Through
Mortise/Tenon

Figure 21-25. Mortising chisel for use in a drill press. The square cutter cuts the mortise and the drill inside removes the chips. (American Machine & Tool Co.

Mortising Machines

Only the *mortise* is cut on the mortising machine. The mortising machine looks like a large vertical drill press or boring machine. The major difference is that the stock is advanced into the chisel, rather than the chisel being advanced into the work. The hollow *mortising chisel* is carried in a vertical spindle, somewhat like the drill press. The outside rectangular shape of the mortise is cut with the chisel, while the inside is cleaned out with the mortising bit. The bit turns inside of the chisel, making the hole and removing chips. The chips travel up the spirals of the bit and flow away from the bit through an opening in the front of the chisel. See **Figure 21-25**.

To cut a mortise, the mortising chisel and bit are placed in a bushing, and then locked in place in the spindle with setscrews. The stock is clamped to the table, and a mark showing where the first mortise is to be cut is lined up properly beneath the tool. After the depth of cut is set, the machine is turned on. When the operator depresses a foot pedal, the table moves up, and the mortising tool immediately cuts the mortise. When the pedal is released, the table retracts. The process is repeated until the mortise is the correct size.

Oscillating-chisel mortising machines are available in single- or multiple-head models. Some machines can operate either vertically or horizontally. Multiple-head machines can be equipped with as many as twenty mortising heads.

Tenoning Machines

The square mortise is of little value without a machine to cut a tenon. The *tenon* is cut on a tenoning machine, often called a "tenoner." Single-end and double-end tenoners are used in industry to cut tenons for products such as window sash rails. They are also used to shape shoulders and to cut corner joints for cabinets.

A typical setup with a single-end tenoning machine might consist of several tenoning heads, coping heads, a cutoff saw, and a movable carriage. When using a machine such as this, the operator clamps the stock to the carriage. The role of the carriage is to transport the stock to the tenoning heads. The tenoning heads make the face and shoulder cuts. The stock is moved to the coping heads, which cut contours on the shoulders. The stock is then carried to the cutoff saw, which trims the tenon to the proper length.

Important Terms

abrasive belt planer
anti-kickback device
arbor
band
bandsaw
bed width
biscuits
bits
combination blades
crosscut sawing
cutoff saw
diamond point tool
double planers
dowels
faceplate turning
fence
gouge
handheld router
jointer
kickback
lead
machine planing
miter saw
mortise

mortising chisel
mortise-and-tenon joint
multiple-spindle shapers
panel saw
parting tool
push stick
radial arm saw
right angle butt joint
rip sawing
roughing
round nose tool
scroll saws
shaper
single planers
single-spindle shaper
skew
spindle turning
square nose tool
stationary routers
swing
teeth per inch
tenon
true-up
turning
universal saw

Questions for Review and Discussion

1. What is the advantage of using a mortise-and-tenon joint, rather than a right-angle butt joint?

2. How would you go about removing the warp in a 1″ × 10″ (2.5 cm × 25.4 cm) board?

3. What is the difference between a knife-blade planer and a jointer?

4. What causes kickback on a circular saw? How can it be avoided?

22 Processes Used to Separate Ceramic Materials

Chapter Highlights

- Glass, ceramic, or steel media are used in the various types of mills to grind clay into extremely fine particles for mixing as a slurry.
- Filter pressing, spray drying, and other methods are used to remove excess water from bulk ceramic material.
- Stock for dry pressing is ground and milled with a variety of processes.
- Removal of excess or unwanted material from sintered workpieces is typically done by grinding.

Introduction

Raw clay must be ground and refined many times before it is useful in manufacturing. The initial stages of refining are done by primary manufacturing firms (processing plants) that mine mineral deposits from the earth. When the material is first mined, pieces can be as large as a yard in diameter or as fine as a granule of powder. The raw materials are often *beneficiated*, which means that, to reduce transportation costs, the raw material is refined by plants located near the mine. The purpose of the refining is to separate or remove impurities from the clay. This material is then packaged in the form of blocks of clay. The blocks are purchased by secondary manufacturing industries for the creation of hard good manufactured products, the emphasis of this textbook.

Grinding Clay to Unify the Mixture

When raw material is purchased from the processing plant, it usually must undergo additional blending and crushing before use. Often, a pulverizing machine called a *mix muller* is used. The muller consists of a large circular pan in which two large steel wheels revolve. Some models use rollers suspended in the middle of the pan from a revolving pivot arm, rather than wheels.

Clay placed in the muller pan may be in chunks up to several inches in size, or as small as a grain of sand. The wheels or rollers grind and crush the material until it is reduced to a size where it can drop through openings in the bottom of the pan. After the clay particles are reduced to the desired fineness, they are passed through a series of screens to eliminate those that are oversize or undersize.

In many cases, a manufacturer purchases stock that has already been crushed, screened, and refined. One of the preferred methods is to receive dry, powdered stock in railroad tank cars or enclosed tractor-trailers that can be unloaded by vacuum. Sometimes, the clay is mixed with water, with the resulting slurry shipped by tank truck.

If the stock is received in powdered form, the first step in making it useful for manufacturing is usually tempering. *Tempering* is the process of mixing and kneading liquids into a dry material to produce stock that is pliable enough for forming. In most cases, tempering consists of mixing water with the dry clay stock while continuously cutting and kneading the mixture. The clay mixture produced can vary from almost dry to a paste suitable for plastic forming.

When the clay emerges from initial mixing, it has the desired water content, but is not compact. At this point, the clay mixture still contains air bubbles. When the clay is to be used for extrusion, these bubbles must be removed to increase the density and strength of the part. This is done in a machine called a *de-airing pug mill*. The pug mill consists of knives on a rotating shaft. The knives cut and fold the stock in a shallow mixing chamber. This kneading action traps air bubbles and compresses and compacts the clay. Extruded clay leaves the pug mill in a continuous column.

Milling

Ceramic stock is usually ground in order to produce fine particles for use in making slip or with pressing processes. In this case, the grinding process is usually called *milling*. When hard clay in lump form is used, it is first passed through roller crushers to break it down into smaller particles. The crushing operation is followed by milling to further reduce the size of the particles, if needed.

There are many similarities in milling the basic ingredients for the different types of clay and product applications. Clays exhibit varying degrees of plasticity, and all contain nonplastic elements such as flint or quartz, and mineral fluxes such as feldspar. All of these ingredients must be thoroughly milled and mixed to ensure that the body will have uniform composition throughout.

Compositions will differ, depending on the type of ceramic product to be made. For example, if bone china tableware is being made, the raw material consists of calcined cattle bones mixed with china clay and feldspathic stone. About 50 percent of the mix is bone, 25 percent is clay, and the rest is feldspathic stone. The china clay adds plasticity to the mix, while the feldspathic stone serves as a flux that makes the clay body flow when it is heated. The bones provide a refractory element that helps to reduce the impact of the heat on the clay in the furnace. Nonplastic materials such as bone or stone must be subjected to the process of *calcination* to mix properly with the clay. In this process, the material is first broken down by heating it to a red heat. Then, the material is milled to produce particles that are sufficiently fine. Bone china is normally fired at a temperature of 1200°C. The temperature needed to break down the nonplastic materials is approximately 450°C.

Milling is normally accomplished in a tumbling or rotating *ball mill*. Industrial ball mills normally consist of a steel cylinder, up to 10′ in diameter, which is partially filled with spheres of heavy steel or dense ceramic *grinding media*. Grinding media consists of small pieces of siliceous rock or rubber, usually less than an inch in diameter. Porcelain balls, flint pebbles, or alumina balls are also used.

Conventional Milling

The process of milling can be done wet or dry, but it is easier to pulverize the mixture when the clay is suspended in a slurry.

In conventional milling, the process involves introducing flint, stone, feldspar, or other materials with water as the mill turns. The amount of powder that is added is usually around 25 percent of the total mill volume for dry milling, and about 40 percent for wet milling. The balls or grinding media lift, turn, and crush the mixture as the cylinder rotates. The turning, mixing, and crushing action may take many hours to reduce the particles to the desired size.

Often, the milling process is more complex than just adding the ingredients and grinding them in a ball mill. For example, if color is critical in the finished product, metallic iron contaminants, which would create dark spots during firing, must be removed using powerful magnets. As noted earlier, nonplastic materials may have to be calcined before milling so that they will properly mix with the clay.

A variation of the ball mill that has gained popularity in industry is called the **Hardinge conical mill**. The major advantage of the Hardinge mill is continuous operation—it never has to stop to be filled. The mill turns around a horizontal axis, while the mix is continuously fed in at one end and automatically discharged at the other. Due to the conical shape, the material slides down the walls of the mill and is ground in a graduated fashion. The largest and heaviest particles are pushed to the bottom of the mill, where they are lifted and dropped with the greatest impact at the widest diameter of the cone. The smaller particles move automatically to the narrow discharge end.

Other Wet Milling Processes

Continuous grinding and dispersion of slurry, consisting of solids suspended in water, is also done using a John mill, a Molinex mill, or an attrition mill. All three of these mills accomplish grinding action through the use of glass, ceramic, or steel media. In the *John mill*, short agitator pegs of tungsten carbide are set into the inner and outer walls of a grinding chamber. See **Figure 22-1**. As the grinding cylinder is turned at high speed, agitation is introduced by the protruding pegs, producing a uniformly ground product. To dissipate the heat generated by friction, cooling water is often circulated in the large-diameter agitator shaft and in the chamber's cooling jacket. The John mill is a high-energy mill, requiring only one pass for most grinding operations involving solids and high viscosity slurries. Materials typically ground in a John mill include ferrites, toner for copying machines, printing inks, conductive coatings, and metal oxides.

In the *Molinex mill*, eccentric grinding disks (rather than the agitator pegs of the John mill) are

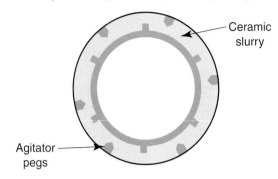

Figure 22-1. This is a John mill, as it would be seen from above. The agitator pegs on the walls help move the grinding media through the slurry to accomplish fine grinding.

Ceramic slurry

Agitator pegs

used to obtain an extremely fine grind. The disks are mounted on a rotating shaft and are staggered to function as an auger. See **Figure 22-2**. This auger arrangement helps move the grinding medium against the flow of material to be ground. The disks are sectioned so that the grinding media can be agitated from both inside and outside. The Molinex mill is used in industry for grinding a wide array of materials including chocolate, microorganisms, zirconium oxide, and clay.

Attrition milling is a batch operation that involves pumping the materials to be ground and the grinding medium into a stationary chamber. A rotating internal shaft with extended arms, **Figure 22-3**, is used to agitate the grinding medium and material to be ground. The beating action and the rubbing of the material against the grinding media produce a uniform particle size. Attrition mills are extensively used to grind food products, cosmetics, graphite dispersions, many oxides, and ferrites.

Figure 22-2. In the Molinex mill, openings in the eccentrically mounted disks permit the grinding media to circulate and reduce the particles of clay to a small size.

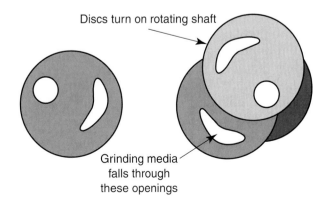

Discs turn on rotating shaft

Grinding media falls through these openings

Figure 22-3. The attrition mill uses a rotating shaft with arms to continually agitate the mixture of grinding media and clay. A high-pressure pump helps to keep the mixture moving.

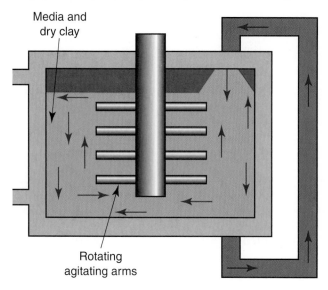

Dry Milling

Dry milling is normally accomplished with a different type of milling process, such as the fluid energy mill. Commonly known as the *jet mill*, the **Trost fluid energy mill**, **Figure 22-4**, has the advantage of producing very fine powders that are uniform, and often of submicron size.

In the Trost mill, a vibratory hopper feeds solid clay into a stream of compressed air that is made turbulent by the action of two opposed jets.

Figure 22-4. In the Trost fluid energy mill, opposing jets of air create a turbulence that causes material particles to collide and fracture. The classification chamber separates materials by particle size.

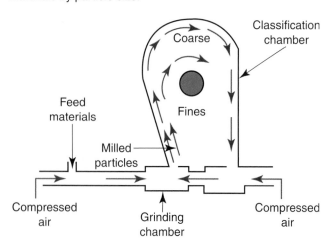

This aggressive action accelerates the particles to sonic velocities. The particles collide with each other, fracturing as a result of the impact. When the particles are reduced, they travel through an upstack channel at reduced velocities and are carried into a flat classifying chamber. The mill has essentially no moving parts. Jets of superheated steam, nitrogen, carbon dioxide, or water are used. This process, producing particles at sizes from micron to submicrons, can mill both hard and soft materials. The output capacity of Trost mills ranges from grams per hour to thousands of pounds per hour. The major advantage of the jet mill is its ability to break dry material into the ultrafine particles that permit dry pressing products with improved strength, hardness, and density.

Gears and bearing inserts, cermets, and heat-resistant devices are typical products that must be tough to continue functioning in hostile environments. To ensure high-quality products, finely ground powders must be compacted using great amounts of pressure to produce dense compacts. Other applications for the fluid energy mill are in making pharmaceuticals and fine chemicals, electronic materials, coatings, and inks used by computer printers. See **Figure 22-5**.

Figure 22-5. The exposed dies shown in this photo are used to apply 40 tons of pressure to form the dense compact. (Gebruder Netzsch Maschinenfabrik GmbH)

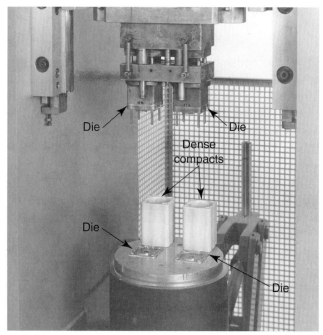

Filter Pressing

We have discussed the grinding of ceramic clay and glass materials with many different viscosities. The process that will be used to shape the product depends on the desired viscosity of the material.

Potteries often require the **ceramic body**, the material that is to be shaped into the form desired, to be in a plastic state. A high percentage of water is added to improve mixing. Clay that is transformed into liquid by adding water is called **slip**. Sometimes, the water content of the slip may be 40 to 50 percent; before it can be used for forming, the prepared slip has to be dewatered. This is accomplished by forcing it through a filter press. If a product is to be made from slurry and shaped plastically, the slip is dewatered only to the point where it can be shaped, but still has a plastic consistency. Tableware is made from this type of material.

Filter presses used to dewater slip consist of an iron frame with nylon filters. Hydraulic pressure is used to close the filter press over the slip. The filters remove part of the water to produce squares or thin slabs of material. What is left after pressing is called a **filter cake**.

When the filter cake is first produced, it is not ready for processing. There is more moisture trapped in the center of the filter cake than on its surface or outside walls. The cake also contains air, which would cause holes in the final product. After curing and inspection, the cake is normally further refined using a process called *pugging*.

Pugging

The filter cake may be stored for some time before it is used. Consequently, a process called **pugging** is often used to remove air bubbles from the cake and further refine the clay body. The equipment used in this process is referred to as a *de-airing pug mill*. See **Figure 22-6**.

The pugging process is used to temper the mixture (give the clay uniform consistency), remove air bubbles, and improve the workability of the mass for use in plastic-state processes such as jiggering or jollying. Often, an extrusion press is attached to the delivery end of the pug mill.

Figure 22-6. The de-airing pug mill tempers a clay mixture by mixing and extruding it; entrapped air is removed by applying a vacuum to the mixture.

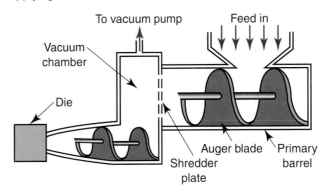

When a cake is fed into the pug mill, it is taken into the barrel of the machine and shredded by rotating auger blades, which then carry the clay to a shredder plate. This plate contains small holes through which the plastic clay is extruded. The "worms" that are produced are cut into short lengths by a rotating blade in the vacuum chamber where *de-airing* takes place. De-airing, or removal of air from the clay can be done in many ways: pinching, pulling, chopping extruding, or cutting processes are examples. The purpose for de-airing is to remove air holes in the clay mass, which would reduce the strength of the completed product. After the clay is de-aired and extruded, it is then cured and inspected and is ready for further processing.

Sometimes a different type of material is needed. Pugging is used to separate the clay mixture and remove air. Other types of processes are used when the clay stock needs to be converted into a damp powder. Let us take a closer look at how this is done.

Spray Drying

Spray drying is another method for reducing the amount of water in slip and reducing the percentage of binder in the mixture. A rule of thumb is to reduce the percentage of binder to not more than 2 percent and the percentage of lubricant to not more than 1 percent. Keeping the amount of water, binder, and lubricant to a minimum is important because these elements are burned out when the product is fired, thereby resulting in shrinkage. Spray drying is used to produce granular particles with a moisture content of 7 percent or less. This is a dewatering method that is particularly useful when dry forming processes are to be used.

The advantage of spray drying is the forming of powders with a very uniform particle size. This uniformity of powders cannot be obtained with any other drying method. The solid granules do not contain any air bubbles or dust. Particles can be produced that are either solid or hollow, depending on the pressure and type of spray nozzle used. Here is how the process works:

There are two basic methods of producing spray-dried particles. The first of these involves pumping slip onto rotating disks located at the top of the spray drier. The disks then throw the collected droplets of slip outward through heated air and onto the walls of the drier. When the slip hits the walls, it falls down through a cloud of hot air. The second type of system uses a spray nozzle at the top of the drier, rather than rotating disks. In both cases, the droplets are completely dry by the time they reach the bottom of the chamber. The particles that are produced are small, free-flowing granules that are perfect for dry pressing. The size of the granules will vary depending on such factors as the type of drier, temperature, binder used, and type of clay. **Figure 22-7** shows the complexity of air-handling systems in a large commercial spray-drying installation.

Grinding for Material Removal

Ceramic workpieces may have to go through a grinding process to remove unwanted material. It may be necessary to remove undesirable material that may have escaped from mold seams. In other cases, supporting structures may have been necessary to hold up elaborate workpieces during firing, and may need to be removed. Efforts are made to reduce or eliminate the need for grinding, but sometime, it may be necessary.

You will recall that grinding was a basic material removal process used with metals and other materials. The basic process used with ceramics is the same. The differences here are in terms of grinding media, and the methods used to support more fragile workpieces. Many different types of grinding wheels are used.

Grinding Fired Clay

It is also possible to grind ceramic parts that have been fired. Ceramic parts can be ground to remove undesirable material. Hard materials such as alumina may require diamond grinding wheels. Softer materials may be ground using silicon carbide or alumina.

Grinding often has to be done on parts with unusual shapes that are supported in the kiln with struts, called **kiln furniture**. Often, there will be imperfections in the surface finish due to furniture interfering with the final coating (glaze). It may be necessary to grind or polish the part after it is fired to remove the imperfections.

Heat generated by grinding is a concern because temperatures as high as 2200°F (1000°C) are produced at the grinding surface. This often results in some microscopic cracking, which can extend into the body of the ceramic part. The use of cutting fluids generally improves the cutting efficiency of the wheel, while helping to prevent overheating.

Nonabrasive methods are also used to remove ceramic material or to polish a ceramic part. These include ion beam, laser beam, flame, and chemical polishing techniques. Most machining of ceramics, however, is done using abrasive grinding processes.

Grinding Glass

Processes used for grinding and shaping glass are different from those used with clay-based ceramics. In most industrial environments, diamond and resin bond abrasive grinding wheels and abrasive belts have replaced traditional grinding wheels because of their extreme efficiency. **Figure 22-8** shows a diamond grinding wheel. The diamond coating is located on the outer circumference of the wheel.

Metal bond diamond wheels are used for high-volume rough grinding. The wheels are normally

Figure 22-7. This is a schematic of a large commercial spray-drying installation with a complex air-handling system. (Gebruder Netzsch Maschinenfabrik GmbH)

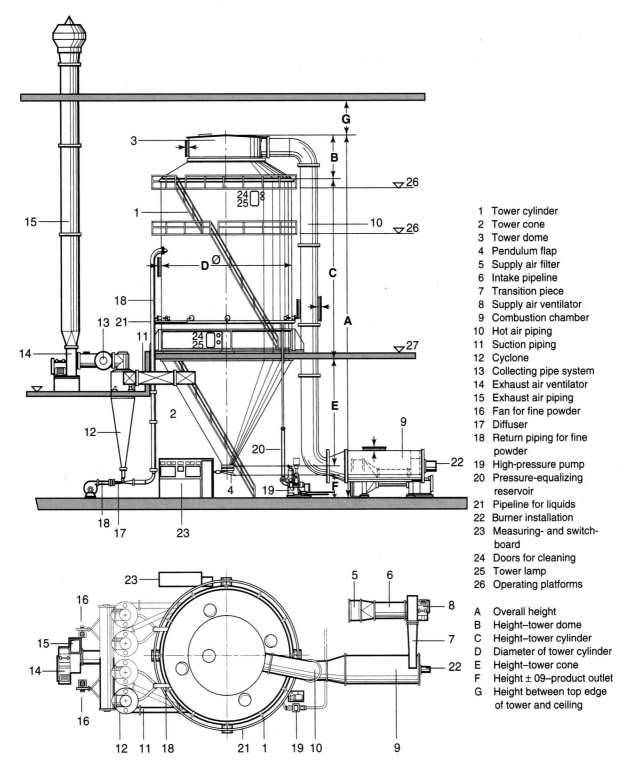

1 Tower cylinder
2 Tower cone
3 Tower dome
4 Pendulum flap
5 Supply air filter
6 Intake pipeline
7 Transition piece
8 Supply air ventilator
9 Combustion chamber
10 Hot air piping
11 Suction piping
12 Cyclone
13 Collecting pipe system
14 Exhaust air ventilator
15 Exhaust air piping
16 Fan for fine powder
17 Diffuser
18 Return piping for fine powder
19 High-pressure pump
20 Pressure-equalizing reservoir
21 Pipeline for liquids
22 Burner installation
23 Measuring- and switchboard
24 Doors for cleaning
25 Tower lamp
26 Operating platforms

A Overall height
B Height–tower dome
C Height–tower cylinder
D Diameter of tower cylinder
E Height–tower cone
F Height ± 09–product outlet
G Height between top edge of tower and ceiling

Figure 22-8. Shown is a vacuum-brazed grinding wheel with a diamond coating on the outer circumference of the wheel. A grinding wheel like this is used to remove hard metal and other materials such as glass. (Alibaba.com Corporation)

made with a 1/8″ thick band of solid diamonds. Metal bond means that the diamonds are embedded in a sintered metal matrix. Standard diamond wheels use nickel to bond the diamonds. Metal bond wheels will normally last six to ten times longer than a standard diamond wheel, and cost about two to three times as much.

Diamond abrasives greatly speed up the rate of cutting, increase productivity, provide a more consistent finish, and reduce damage to the glass. Often, diamond belts are used in place of conventional wheels and abrasives for grinding glass. Belts can be used to grind flat glass, laminated glass, mirrors, decorative glass, crystal, art glass, scientific glassware, and clay-based ceramics.

Workers of cold glass often use diamond cones to produce very small inside curves. Nickel is used to bond the diamonds to the cones, which are available only in a smooth pattern finish. Cones are difficult to use and are used more frequently on horizontal than vertical surfaces. Cones are used on a grinder with a minimum shaft speed of 3600 rpm.

Green Machining

Ceramic parts are often made by compacting powder under high pressure. This produces near-net-shape products. Material removal to size the product to final dimension is accomplished with *green machining* if the material is in a dry-pressed state. It is *green* if it has not been sintered or fired. Machining is done on the loosely compacted powdered part. Since the part has not been sintered, it is quite soft and can easily be machined using conventional metal-cutting equipment. Diamond tooling is not required.

Machining greenware requires skill and careful handling. The formed part, or **compact**, is fragile; care must be taken to avoid collapsing the material when securing the part in the chuck or workholder on the machine tool. In addition, machining must proceed slowly to prevent cracking the part and producing unnecessary stresses. Securing the greenware is often accomplished by fastening the part to fixturing using beeswax.

The machinability of green parts is improved by modifying the composition of the powder mixture. Some firms compress zirconia-toughened alumina (ZTA), with a process called *reaction bonding*. Normally, a mixture of about 20 percent zirconia is used. Higher amounts of zirconia result in higher toughness of the material. ZTA green bodies exhibit very high strength in parts, but are also easily machined. Standard tools made from high-speed steel are recommended.

If final machining is done on parts that have been fired, this is referred to as **hard machining**. Hard machining requires the use of diamond tooling and is very slow and time-consuming.

Etching Processes

Material removal from a glass surface, resulting in a frosted or opaque appearance, is done by the use of etching processes. Makers of decorative glass of the nineteenth and twentieth centuries sometimes used etching to produce a frosted pattern on windowpanes, tavern windows, and other surfaces.

There are several methods that were used for the past 200 years for etching glass. Sandblasting and acid etching were common methods. Today, sandblasting is still used. A more recent method for etching involves the application of etching cream. Cream is simply applied where the glass is to be etched, left on the surface for a few moments, and then washed off with water.

Acid etching is rarely done these days for safety reasons, but it does have a softer, more pleasing appearance than sandblasting. Acid etching was usually done in two stages. First the whole glass was etched to a matt finish. This was followed by a different type of acid-etch to bring back some areas to a clear finish.

If a design is desired on the surface, the glass is covered with either wax, a photo resist, or masking material. The material is then removed where the acid will act on the glass surface. Sometimes, the gas from the acid is used to create the etched pattern; in other instances, the entire plate is submerged in the acid. Glass can be frosted, or made opaque, by *chemical etching*. An entire pane or portions of glass can be frosted with an acid (*etchant*, or etching agent). A coating of wax is applied to the glass on any area that is not to be etched. Designs may be scratched through the hardened wax for detailed work. A weak etching solution made with hydrofluoric acid (or a stronger commercial etching solution) is then applied. The acid eats into the glass wherever the wax was removed. When the etching reaches the desired depth, the acid is drained off and the glass is washed in hot water. This halts the etching action of the acid and melts the wax coating.

Frosted products are popular for applications where obscured vision is desired to ensure privacy, or to produce artistic designs directly on the glass product. Frosted incandescent lamps and door panes are made with this method.

Photographic resists, similar to those used with the process of chemical milling and photo fabrication on metals, can also be used. With these resists, masks are used to block exposure to light. The exposed resist is hardened by light and is unaffected by the acid. The protected areas covered by the mask remain soft, and are washed off with water. When the pane is immersed in the acid bath, the acid acts on the soft material but does not affect the hardened area.

Mechanical etching involves bombarding the glass surface with a very fine abrasive and lead shot mixture. Usually, the pattern in the glass is made by masking off the areas to be left unetched with a protective film. Sandblasting produces a recessed, or sunken surface etching.

With the *abrasive shot method*, a mixture of abrasives and lead shot is placed inside of an agitation frame, along with the sheet of glass. The abrasive particles cling to the lead shot. When the mix is agitated, these coated particles of shot act like individual cutting tools. The abrasive and shot combination can be used over and over again.

The mechanical etching process called *sandblasting* can also be used with many different types of materials. Due to the danger of inhaling poisonous abrasive dust, sand is seldom used today; instead, alumina oxide abrasive is normally used. Unique designs can be produced on sheet glass with frosting methods. The opaque frosted surface can even be shaded with sprayed colors or oil colors. The frosted surface holds the color well and adds a beautiful sheen to the glasswork.

Important Terms

abrasive shot method
attrition milling
ball mill
beneficiated
calcination
ceramic body
chemical etching
compact
de-airing
de-airing pug mill
etchant
filter cake
filter presses
green machining
grinding media
hard machining
Hardinge conical mill
jet mill
John mill
kiln furniture
mechanical etching
milling
mix muller
Molinex mill
pugging
sandblasting
slip
spray drying
tempering
Trost fluid energy mill

Questions for Review and Discussion

1. Think about creating a hollow granule using the process of spray drying. Why does the pressure and size of the spray nozzle affect the solidity of the particle?

2. How is it possible to produce ultra-fine particles from dry stock, using the fluid energy mill? Why do you think that this is called fluid energy milling? Where does the fluid come in?

3. List three processes presented in this chapter that could be used to remove air in clay.

23

Processes Used to Separate Composite Materials

Chapter Highlights

- Abrasive sawing separates composites with a minimum of damage.
- Waterjet cutting is a fast and dust-free method for separating composites.
- Lasers work well in cutting some composites.
- Ultrasonic machining provides a fast, precise, and reliable means to cut many of the reinforcement materials used in composites.

Introduction

Composite products are normally formed to the size and shape desired. Since they are produced in the desired shape, the major reason for separating composite materials is to trim parts. Conventional trimming tools, such as the band saw, router, hacksaw, belt sander, reciprocating saber saw, and circular saw, are used for many trimming applications. A less-conventional trimming tool, a 5-axis robotic router, is shown in **Figure 23-1**. The router is trimming a dashboard unit, for use in a street sweeper, to close tolerances.

There are, however, composites that are made as a stock to be shaped or machined to create a part or product. The processes used to separate these materials must be designed to handle the abrasive, hard, and brittle characteristics of typical composite materials. There are four processes that are well suited to this task. They are diamond wire sawing, waterjet cutting, laser machining, and ultrasonic machining.

Figure 23-1. This composite part, a dashboard unit for a street sweeper, is being trimmed to close tolerances by a robotic router. The 5-axis machine can be programmed to precisely repeat the trimming operation on each workpiece. (PMW Products, Inc.)

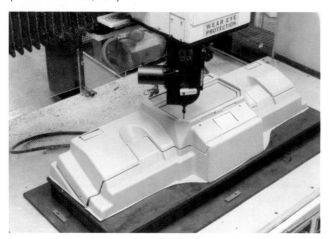

Figure 23-2. A diamond wire saw quickly and easily cut through the high-strength composite materials used for this helicopter rotor blade. (Laser Technology, Inc.)

Diamond Wire Sawing

Wire sawing provides a means to follow a curved path with tight radii while creating a minimum of stress at the cut line. The wire that is used to saw composites is a diamond wire. *Diamond wire* is a round, high-tensile-strength, copper-plated wire with fine particles of industrial diamond impregnated on its surface. These diamond particles cut through extremely hard materials with virtually no scrap loss. Cuts are burr-free, and the sawing action generates very little heat. This makes the process particularly useful for cutting heat-sensitive semiconductor composites. Wire sawing is also a cost-effective solution for cutting expensive materials as well as very small parts since diamond wire sawing produces a saw kerf of only 0.009″ (0.229 mm).

Diamond wire is useful for cutting fragile materials that would be cracked or broken by other cutting methods. Diamond wire saws have been successfully used to cut boron and tungsten composites, carbon and graphite, carbon fibers, ceramics, chromium, crystals, fiber composites, gallium arsenide, garnet, glass, and gold. See **Figure 23-2**.

A diamond wire saw incorporating a table that floats on a cushion of air is shown in **Figure 23-3**. The circular cutting attachment is mounted on top of the table. With this saw, the round diamond wire can cut around a corner, leaving a radius only slightly larger than the small radius of the wire. This makes possible machining of complex parts with a degree of accuracy not possible with most cutting tools.

Diamond wire saws typically use 100′–200′ of continuous length cutting wire wound on a drum. The wire has no joints or welds, and is designed to cut under high tension. In operation, the wire unwinds from one drum and is rewound on another. See **Figure 23-4**.

When small or fragile workpieces are to be cut, wax is used to affix the stock to a chalk-like ceramic support base.

Waterjet Cutting

Using a stream of water to cut or shape a material may not seem to be all that practical. However, in 1968 Norman Franz of Ingersoll-Rand obtained the basic patents for a waterjet technology that could cut materials with the precision of a band saw. The

Figure 23-3. This diamond wire saw can be used for extremely precise cutting of complex shapes, since the round worktable floats on a cushion of air. (Laser Technology, Inc.)

Figure 23-4. This 24″ (61 cm) diamond wire saw is being used to slice the square composite workpiece into several thinner sections. (Laser Technology, Inc.).

first commercial demonstration of the technology was made in 1971, when a waterjet was used to cut 3/8″ (9.5 mm) thick pressed board for the furniture industry. Since that time, the waterjet (a high-pressure water stream) has demonstrated its effectiveness in quickly and cleanly cutting materials including titanium, ceramics, glass, carbon fibers, stainless steel, concrete, polymers, and fiberglass reinforced composites. The basic technology behind the waterjet is both simple and complex. The basic components of a typical waterjet system are a booster pump, an intensifier, and a cutting nozzle. Clean water enters the booster pump, where it passes through filters, then into the intensifier. In the intensifier, the water velocity is increased tremendously. The narrow jet of water exits the cutting nozzle at speeds as high as 3000′ (914 m) per second, or approximately three times the speed of sound! At the nozzle, the water is under a pressure of nearly 60,000 psi (pounds per square inch). Depending on the material, a waterjet can cut at a speed of around 800″ (20 meters) per minute.

The diameter of the stream can range from 0.004″ to 0.014″. This thin stream of pressurized water can quickly cut gaskets, foams, plastics, thin rubber, food, and a variety of other soft materials. When small amounts of abrasive particles, such as garnet, are mixed into the jet stream, the resulting *abrasive jet* can cut virtually any hard material such as metal, composites, stone, or glass. Waterjets provide extremely accurate cuts, with a linear accuracy of ± 0.005″ and a high degree of repeatability. Most abrasive waterjet cutting produces net shaped parts (a finished part), thereby eliminating any secondary finishing.

There are other benefits. A waterjet cuts at room temperature, so there is no thermal stress at the cut line. Since there is no tool-to-part contact, mechanical stress, shredding, crushing and other surface damage is also eliminated. Since waterjets cut with a narrow kerf, parts can be tightly nested or common-line cut so raw material usage is maximized.

For safety, people must be kept away from a waterjet when it is in operation. To use a waterjet safely and to take advantage of its versatility and accuracy, waterjet cutting is generally controlled with computer-based robotic motion systems—not hard tooling. These systems minimize change-over and setup time.

The waterjet process provides a number of advantages for cutting composites. A waterjet can be used to cut, drill, or machine. It generates no dust, requires no additional lubricant to keep the kerf clean, and does not produce heat in the cutting area. In waterjet cutting (sometimes referred to as *waterjet machining*), material is removed by a compressive, shearing action. This makes the process faster and more cost-effective than other techniques when cutting materials such as boron/aluminum honeycomb, aluminum/boron carbide, and graphite composites. The abrasive waterjet cutting process is powerful enough to cut through 3″ (7.6 cm) tool steel at a rate of 1.5″ (3.8 cm) per minute or 10″ (25.4 cm) of reinforced concrete at speeds exceeding 1″ (2.54 cm) per minute. See **Figure 23-5**.

Laser Machining

The term *laser* is an acronym for *light amplification by stimulated emission of radiation*. The amplified light produced by a laser is very different from normal light. A laser light is monochromatic, meaning that it

Figure 23-5. This waterjet cutting system for sheets of flat stock up to 9′ × 14′ (2.7 m × 4.3 m) can cut at speeds of up to 500″ (12.7 m) per minute. Computer controls are used to program movements of the cutting head in the X, Y, or Z axes. (Technicut)

Figure 23-6. A CO_2 laser is used to cut exotic metals and many other types of materials. (Coherent General)

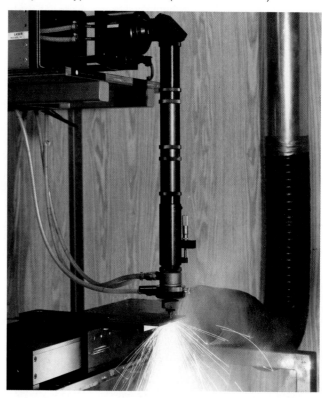

consists of one specific coherent wavelength or color. A coherent wavelength (light) is just another way of saying that the light is organized, producing a strong very directional and concentrated beam. A laser of modest power can be focused to a very high intensity for cutting, burning, or even vaporizing materials. When used for cutting, a laser melts, burns, or vaporizes the material, leaving an edge with a high quality surface finish. Because of these properties, the laser has been used in industry for more than a generation.

Lasers have been used extensively for separating metallics, but are suitable for many composite applications. While lasers are not appropriate for every application, they can beat many traditional manufacturing processes when flexibility, speed, and accuracy are needed. Lasers are now being used to cut honeycomb panels and plastics, **Figure 23-6.**

Since no tool bit is involved, lasers are even useful for drilling holes at an angle into the workpiece. With conventional drilling, if the drill bit is not held perpendicular to the stock, it is difficult to penetrate the workpiece accurately. In addition, the likelihood for breaking the drill bit is very high. Lasers are also useful for cutting and trimming composite parts within ± 0.005″ (0.127 mm) with a high degree of repeatability. Lasers also eliminate the presence of dust during cutting.

The safe operating practice for laser cutting, as with waterjet cutting, requires operators to be kept out of the cutting area. Therefore, lasers are most often guided by computer numerical control (CNC) machine. The CNC system on many of these machines also controls the material handling equipment that moves the parts. These systems can quickly stack or

place completed parts so that the laser can begin cutting the next part. See **Figure 23-7.**

There are three basic types of lasers used in manufacturing today. These are referred to as the CO_2, Nd:YAG, and the excimer lasers. CO_2 lasers are used for processing both metals and nonmetallics.

Figure 23-7. This CNC laser machine provides high-speed cutting and piercing of metal or composite materials, with precise repeatability of actions from one workpiece to the next. It combines the flexibility of a laser cutting machine with the productivity of a traditional turret punch press. (U.S. Amada, Ltd.)

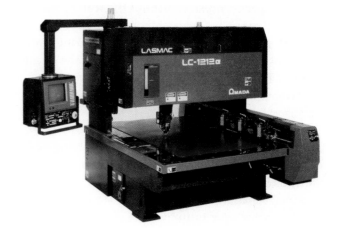

The Nd:YAG lasers are used primarily for marking, scribing, drilling, welding, heat-treating, and cutting of metals. They are also used to transmit electrical signals through fiber optic "light pipes."

Excimer lasers are still essentially in the experimental stages of development, although they have been used in delicate eye surgery and similar applications. These lasers use high-energy gases such as argon fluoride (ArF), krypton fluoride (KrF), and xenon chloride (XeCl).

CO_2 lasers

Low-powered *CO_2 lasers* in the 20 to 500 watt range are good for cutting polymers such as wood and plastic. These types of lasers are also used to cut and drill ceramics. At the low end of the power range, CO_2 lasers can be used for scribing, soldering, and trimming. More powerful units, beyond the 500 watt range, are used to weld, cut, and heat-treat metals. The CO_2 laser is well suited to cutting and heating many types of plastics and composites.

Let us take a closer look at how lasers work. LASER is an acronym for *Light Amplification by Stimulated Emission of Radiation*. As you think about the process, think of the stages involved—generation of light, then amplification, then stimulation and finally emission of radiation.

First, electrical current is transmitted to the *lasing tube*. The tube is either a solid crystalline rod (of aluminum oxide, sapphire, or yttrium-aluminum-garnet) or a hollow tube filled with a gas (CO_2, helium, or nitrogen). When the tube is energized, it lights up and glows, much like a fluorescent light tube. The tube can be energized by using alternating current (AC), direct current (DC), or radio frequency (RF). At this point, the process is very inefficient. The light that is created is very bright (several times brighter than light emitted by the sun), but is *incoherent*. This means that the light spectrum includes many different wavelengths and phases. Without some means of focusing the light and making it *coherent*, the light has little or no value in manufacturing.

The second stage of the process involves passing the light energy back and forth between the cathode and anode at opposite ends of the lasing tube. Movement of the light energy through the lasing medium amplifies the light intensity. The light beam is then released to mirrors that direct it through a prism, where it is focused into a beam of the desired size. The light is now coherent, with a narrow spectrum of wavelengths that are in phase.

Once the desired laser beam is produced, it is ready to be directed toward the workpiece. The beam is normally pulsed; that is, it occurs in short on-off bursts. The power requirements and the quality and size of the beam are controlled to achieve the necessary processing actions: cutting, welding, or marking.

Nd:YAG lasers

In the *Nd:YAG laser*, Nd stands for the element neodymium, an expensive metal that causes the lasing action. The lasing rod—a crystal composed of yttrium (Y), aluminum (A), and garnet (G)—is used as the active medium in the laser. The Nd:YAG laser is smaller than the CO_2 laser, but it is just as powerful. A 450-watt YAG crystal might be no more than 6" (15.2 cm) long by 3/8" (9.5 mm) in diameter.

The Nd:YAG is a solid-state laser that uses a synthetic crystal lasing rod. The beam of this laser is nearly visible and is focused with lenses to improve the output. Green safety lenses are needed to view the Nd:YAG laser beam.

There are two types of Nd:YAG laser beams: continuous and pulsed. Continuous-beam lasers rated at 25 to 50 watts are used for soldering and scribing; those rated at 50 to 100 watts are used for marking. Pulsed Nd:YAG lasers up to 450 watts are good for drilling, cutting, and welding of metals.

The Nd:YAG laser is particularly useful for welding delicate components, such as heart pacemakers. This is accomplished because the laser emits a pulsed, highly focused beam for fractions of a second, resulting in rapid heating.

Excimer lasers

Excimer lasers use exotic gases to create a beam that pulsates on and off for periods of from 10 to 16 nanoseconds (billionths of a second). These lasers actually remove stock, molecule by molecule, rather than vaporizing it.

The excimer laser has found several applications in medicine. One of the most important uses is in laser eye surgery. The industrial applications are principally micromachining plastics, paper, ceramics, glass, crystals, and composites. A cut of several hundred microns (.1 to .3 mm) can be made using an excimer laser. When one of these materials is illuminated with an excimer laser, a pressure rise and subsequent shock wave is generated, breaking the chemical bond holding the material together. The process creates very little heat in the surrounding material. Other excimer laser applications are still in the developmental stages, and will probably not be used very extensively in manufacturing for many years.

Ultrasonic Machining

Ultrasonic machining is a process used extensively with plastics. It is also used to cut prepregs, aramid fiber, graphite, boron epoxy, carbon phenolic, hybrid composites and fiberglass, as well as aluminum honeycomb core, Nomex® fabric, leathers, fabrics, films, and sealants.

With the variety of composite materials in use, ultrasonic machining was developed to provide a fast, precise, and reliable means for cutting and shaping these materials. The rotary ultrasonic machine tool shown in **Figure 23-8** is used to process composites, ferrites, zirconium, beryllium oxide, ruby, sapphire, and other difficult-to-machine materials.

This machine applies axial 20 kHz (20,000 cycles per second) ultrasonic vibrations to a rotating

Figure 23-8. Vibrations at the rate of 20,000 per second, used with a rotating diamond tool, allow this equipment to perform rapid drilling and machining of composites and other materials. (Branson Ultrasonics Corporation)

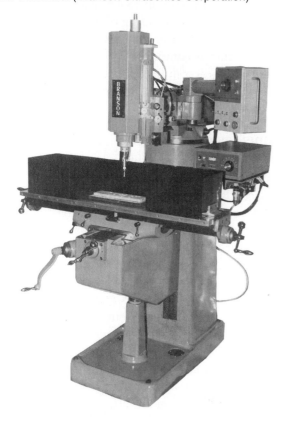

Figure 23-9. The horn assembly of this ultrasonic machining system expands and contracts thousands of times each second, causing the tool to move rapidly up and down, cutting into the workpiece. (Sonic-Mill, Albuquerque Division, Rio Grande Albuquerque, Inc.)

diamond tool. The ultrasonic vibrations reduce friction between the tool and material. This permits drilling and milling of workpieces with less pressure.

Rotary ultrasonic machining processes use a power supply to generate a high-frequency electrical signal that is transmitted to a piezoelectric *transducer* (converter). The transducer converts the electrical signal to mechanical motion. This motion is conveyed to the *horn*, which holds the rotating tool.

Figure 23-9 shows a close-up view of a horn extended down to the workpiece. The horn expands and contracts about 20,000 times each second. This causes the tool to vibrate in a longitudinal (along its length) direction while also rotating. See **Figure 23-10**.

The diamond tools that are used in this process rotate at a speed of up to 4000 rpm. The combined action of the rotation and the longitudinal ultrasonic vibration provides the cutting action. Frictional heat is reduced, eliminating the need for coolant, which could cause workpiece contamination. Ultrasonics enable fast and efficient cutting at lighter tool pressure than with traditional machining.

Ultrasonic drilling and milling has several advantages over conventional machining processes when used with ultra-hard materials. The process is faster, permitting continuous drilling. There is no need to withdraw a drill to flush it. Ultrasonic machine tools can drill deep, straight, small-diameter holes. As noted, minimal heat is generated in the cutting area. This makes ultrasonic machining a

Figure 23-10. In rotary ultrasonic machining, a diamond tool is vibrated up and down, while simultaneously rotating. This provides rapid cutting action. (Sonic-Mill, Albuquerque Division, Rio Grande Albuquerque, Inc.)

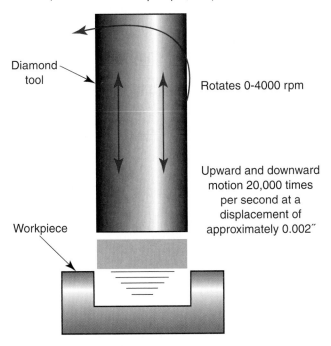

Diamond tool

Rotates 0-4000 rpm

Upward and downward motion 20,000 times per second at a displacement of approximately 0.002″

Workpiece

valuable process for use with fragile materials that are prone to thermal cracking.

Ultrasonic machining is used for a wide array of manufacturing applications, ranging from the fabrication of carbide dies and the machining of small holes in alumina substrates, to the drilling of long holes through blocks for use with lasers. Ultrasonic machining can be used to drill, engrave, broach, and shape silicon, silicon carbide, silicon nitride, sapphire, glass, quartz, ferrite, alumina, alumina nitride, gallium arsenide, and many exotic composites.

Ultrasonic machining can also be done in a stationary manner, without tool rotation. This permits the cutting of holes or pockets of virtually any shape in the workpiece. The cutting tool used has a cross-section that is the size and shape of the desired hole or pocket.

As shown in **Figure 23-11**, a recirculating pump forces an abrasive slurry (abrasive suspended in a liquid medium) into the working area between the vibrating tool and the workpiece. The abrasive particles strike the workpiece at a force equivalent to 150,000 times their weight. The particles chip off microscopic flakes of material. As the tool is advanced into the workpiece, the abrasives grind out

an opposite but perfect match of the tool face. The presence of the cool slurry makes this a cold cutting process.

Ultrasonics are used to drill, machine, engrave, broach, and shape materials. The principle has now been applied to hand-held ultrasonic knives or cutters that can be used to cut or trim composites. See **Figure 23-12**. These ultrasonic drives provide some worthwhile advantages for handheld cutters. The first benefit is lighter weight, compared to motorized reciprocating or round blade cutters that can weigh up to 30 pounds. Many of the ultrasonic knives currently available weigh as little as 2 pounds. This lighter weight makes the use of this device less tiring. But the light weight and compactness of the tool also helps make it easier to maneuver, thereby improving the precision of cutting.

The ultrasonic trim knife has three basic components: a power supply, a transducer, and a horn/blade assembly. The power supply increases the frequency of the electric current from 60 Hz to 20,000 Hz. The high-frequency energy is supplied to the transducer, which changes it into mechanical vibratory energy that powers the blade. This tool is able to cut multiple-ply composite materials precisely and rapidly without fiber disorientation or damage.

Figure 23-11. When a stationary tool is used, a slurry of abrasive suspended in water or another liquid does the actual cutting. Energy is transmitted to the abrasive particles by the rapid longitudinal vibrations of the tool. (Sonic-Mill, Albuquerque Division, Rio Grande Albuquerque, Inc.)

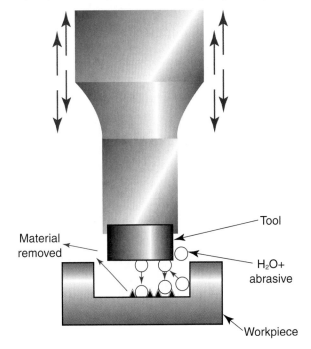

Material removed

Tool

H_2O+ abrasive

Workpiece

Figure 23-12. The blade of this hand-held ultrasonic knife vibrates 20,000 times per second, allowing it to easily cut most composites. Cutting speeds are governed by the material being cut and its thickness. (Branson Ultrasonics Corporation)

Important Terms

abrasive jet
CO_2 lasers
diamond wire
excimer lasers
horn
laser
lasing tube
Nd:YAG laser
transducer
ultrasonic machining

Questions for Review and Discussion

1. What are some of the advantages of diamond sawing compared to conventional sawing when cutting composite workpieces?

2. List the advantages for waterjet cutting.

3. What type of application is suitable for waterjet cutting? Explain why?

4. What is the difference between ultrasonic cutting and ultrasonic machining?

24

Processes Used to Fabricate Metallic Materials

Chapter Highlights

- Mechanical joining includes threaded fasteners, which among their many advantages have the unique ability to generate large clamping forces.
- Welding joins metals by melting adjoining surfaces so that they are able to fuse together.
- Ultrasonic welding uses sound energy to generate heat for joining metals or plastics.
- Brazing and soldering are two of the oldest and most versatile processes used to join metals.

Introduction

The process of joining metal parts is one of the most common activities in manufacturing. The processes used can be categorized as being either permanent or reversible. Some of the fabrication processes permanently adhere materials (gluing) together or provide a permanent bond through cohesion (welding). There are hundreds of metal fabrication processes, both permanent and reversible. The major techniques are presented in this chapter.

Mechanical Joining

Mechanical joining techniques include methods for attaching two or more components together. Threaded fasteners, such as screws or bolts, are among the best methods of assembly for products that must be disassembled. If the method of assembly requires a more permanent linkage, rivets may be a more appropriate alternative. There are hundreds of unique mechanical fasteners that are used for specialized applications. There are reference books available that contain detailed information on most fasteners in use today. These details include thread dimensions, pitch, and load capacity.

A unique advantage that threaded fasteners have is the ability to generate large clamping forces. The thread of the bolt is essentially an inclined plane that is pulling the nut up the distance of the pitch of the thread for every full revolution of the nut. The mechanical advantage can be tremendous. As an example, a coarse thread, half-inch bolt (1/2–13) has 13 threads per inch. This means the nut moves up the bolt 1/13″ or .077″ every time the nut turns one revolution. If a person uses a 6″ long wrench (the lever arm), the distance traveled in one revolution by the point where the force is exerted on the wrench is 37.7″ (pi times the diameter of the circle or 3.1416 × 12″). The total mechanical advantage therefore is 490:1 (37.7″ ÷ .077″ = 490). The high mechanical advantage that can be developed with a simple hand tool explains why threaded fasteners are so widely used even in very critical assemblies.

There are concerns about the strength of fasteners. One of the areas of concern is damage to the thread itself. The depth of a thread on a bolt is small compared to the overall diameter and can be easily damaged, which may severely reduce the fastener's overall strength. However, the design of a thread is surprisingly robust and will retain substantial strength even with what appears to be considerable damage to the threads. The other concern is in "twisting off" a bolt by overtightening it. Since threaded fasteners provide tremendous mechanical advantage, it is important they not be overtightened. In order to achieve the maximum clamping force without damaging the fastener, the "torque" (force times the lever arm, which is the length of the wrench handle) must be limited to the maximum value for the fastener. Torque tables contain this value for each type of fastener. A torque wrench or torque screwdriver must be used in order to know when the desired torque has been reached.

Welding Processes

Another common method for joining metals is welding. *Welding* processes join metallic parts by locally heating the metal of two adjacent parts until that area becomes molten, permitting the metal to fuse together. In most welding processes, the heat is created from a flame, an electrical current, or a chemical reaction. Some welding processes generate heat by a combination of these methods.

Welding can involve joining components so small that they must be viewed through a microscope. Welding also is used to fabricate structures that are too large to move and must be constructed in the field.

In a welding application, the strength of the bond depends on the shape and preparation of the surfaces. In other words, the weld can only be as good as the surfaces that are being joined. Cleanliness is very important since oxides (rust) and contaminants will reduce the strength of the welded joint.

Two of the oldest and most important welding processes are *gas welding* and *arc welding*. They provide the foundation for many other welding processes that have evolved. Both of these processes are still used extensively.

Gas Welding

Gas welding, often called *oxyacetylene welding*, is accomplished by mixing two gases, oxygen and acetylene, in a welding torch. The heat that is generated is sufficient to melt and weld steel. Consequently, this type of welding is called *fusing*. Often, when two pieces of metal are joined together, it is nearly impossible to fill all of the gaps or to create a strong weld. In these situations a *filler metal* is needed. The use of a filler metal is common practice and worth a closer look.

The heart of the welding process is the gas welding torch. The torch blends the oxygen and acetylene to produce a flame that is able to melt steel. A properly adjusted flame is also *inert*, which means there is no excess oxygen or carbon present to contaminate the weld. More importantly, the inert flame acts as a shield to protect the joint from oxygen in the air while it is in a molten state. The torch is held in one hand and the filler rod in the other. The pieces to be welded are clamped together, and the torch is directed toward the joint. When welding steel, the torch flame heats the metal at the joint to a bright yellow-orange color and a molten pool is created at the point of the flame.

The filler rod is placed into this molten pool, and the filler melts into the pool. The torch and filler rod are moved along the joint, continually creating a molten pool and adding filler. As the heat is moved away from the weld, the joint cools to create a bead of solid metal. This process may be applied to the entire joint or to spaced areas along the joint. Applying welds in small beads that are spaced is called *tack welding*. Gas welding is a slow but relatively inexpensive process.

Gas welding is popular for fabricating sheet metal parts and structures in the field. It is particularly useful for repair and maintenance work. The only equipment that must be moved to the site is the oxygen and acetylene tanks, the torch, hoses to carry the gases to the torch, and regulators to control the gas flow. This equipment is usually mounted on a wheeled cart for mobility.

Torch cutting is a process that uses an oxyacetylene torch to heat metal until a gap is created and the metal is separated. Oxygen and a fuel gas (usually acetylene) are mixed in the torch. Cuts can be made in materials as thick as 14″ (35.6 cm), with a kerf as wide as 3/8″ (9.5 mm). The torch initially heats the steel at the cut line until it is yellow orange in color. Then the operator triggers the torch, sending a jet of oxygen to strike the heated cut line. The oxygen oxidizes the steel, burning a hole at the cut line. The torch then is moved slowly, advancing the cut. Since the process sends out a shower of sparks, it is important to ensure that there are no flammable materials in the area and operators are wearing appropriate protective clothing.

In production applications, the cutting torch or torches are part of cutting machines that can automatically repeat cuts for production runs. The torches are normally guided along paths (cut lines) by programmable controllers or computer numerical control systems.

Arc Welding

Arc welding dates back to the mid-1800s. Over time, the term *arc welding* has come to describe several fusion processes used for joining metals. In each instance, the welding takes place because of the intense heat produced from an electric arc. The arc is formed between the metal workpieces and an electrode (a metal tip, stick, or wire) that is manually or mechanically guided just above and along the joint to be welded. In some applications, the electrode is a metal tip that only conducts the current that creates the electrical arc to the workpiece. But the electrode may also be a metal rod or wire that not only conducts the current but also melts and supplies filler metal to the joint. In almost all cases, some provision is made to protect the molten metal from oxidation and contaminants with either an inert shielding gas or a flux. Today, this simple and versatile process remains popular.

One of the most widely used welding processes is *shielded metal arc welding (SMAW)*. This process is also called *stick welding*, due to the use of a stick-shaped electrode. The electrodes are approximately 0.062″ (1.59 mm) in diameter or larger. This type of welding is reasonably fast, versatile, and relatively inexpensive.

The arc produced in SMAW is created when electricity travels from the consumable electrode to the workpiece. The uncoated metal end of the electrode is clamped in the gripping jaws of an electrode holder. The holder in turn is attached by a heavy conducting cable to an alternating current (AC) or direct current (DC) power source. The workpiece is connected to a second cable coming from the power source. An arc is created when the tip of the electrode strikes the workpiece. Depending on the thickness of the parts to be welded, anywhere from 50 to 300 amps of current are supplied to the holder. The arc is established by touching the tip of a coated electrode to the workpiece, then withdrawing it quickly to the distance required to maintain a suitable arc. The basic principle behind arc welding is the same as with most welding processes. Heat from the arc melts the surface of the workpiece, and the electrode material introduces filler metal into the molten weld pool. The electrode coating, or *flux*, is vaporized by the heat of the arc to create the shielding gas.

The flux also provides scavengers, deoxidizers, and alloying elements for the weld metal. *Scavengers* combine with other impurities that are present in or on the workpiece to prevent the impurities from contaminating the weld. *Deoxidizers*, as the name implies, are there specifically to combine with oxygen before it can combine with the metal in the weld pool, which would weaken the weld. *Alloying elements* combine different metals to increase strength and toughness.

As the arc moves away from the weld pool, a coating of slag forms to cover the weld. This coating continues to protect the weld from the atmosphere as the weld cools. If the weld is to be built up with additional passes, the hard brittle slag coating must be chipped off the weld bead.

SMAW is a fairly labor-intensive process; nevertheless, the process is used extensively in maintenance, shipbuilding, petrochemical pipeline, and construction fields. See **Figure 24-1**. As in the case of all welding and cutting processes, eye protection with shaded lenses is required. Other protective clothing and equipment, such as welding gloves, should be worn as well. Care also needs to be taken to keep flammables out of the area where the welding is being done.

Gas Metal Arc Welding

Gas metal arc welding (GMAW) uses a spool of consumable wire electrode and a tank-supplied inert shielding gas. This inert shielding gas is supplied through a nozzle to protect the weld from contamination and oxidation. GMAW is also referred to in the welding field as "MIG" welding. *MIG* stands for *metal inert gas*. This process was first developed in the 1950s and has gained tremendous popularity. Since it is capable of productivity two to three times greater than SMAW, GMAW has become one of the major welding processes used in industry today.

There are several different shielding gases used with GMAW, but argon, helium, and carbon dioxide are most common. Welding methods vary somewhat, according to the type of gas that is used and the material that is to be welded. The shielding gas is generated to reduce contamination and oxidation of the weld area.

A consumable bare wire electrode is fed automatically through the welding nozzle, which also serves as the shielding gas dispenser. See **Figure 24-2**. The wire carries an electrical charge, while the workpiece is grounded. As the wire approaches the workpiece, an arc is created and the weld begins. The wire is fed by pressing a trigger on the electrode holder or nozzle. The operator moves the nozzle in much the same manner he or she would the electrode holder when using shielded metal arc welding.

One of the major advantages GMAW has over SMAW is that the gas effectively protects the area of the weld from contamination. The almost total absence of oxidation results in welds that have very little slag. Consequently, chipping and removal of slag between welds is not necessary. However, the metal that is being welded must be clean and free of contaminants.

Gas metal arc welding is a simple process to use, so little time is necessary to train operators. It is extensively used in the metal fabrication industry for

Figure 24-1. SMAW is used extensively in pipeline manufacturing. (Westfield Engineering & Services, Houston, Texas)

Figure 24-2. The GMAW nozzle serves a dual purpose: it feeds the consumable wire electrode to the weld pool while dispensing a shielding gas that prevents contamination of the weld.

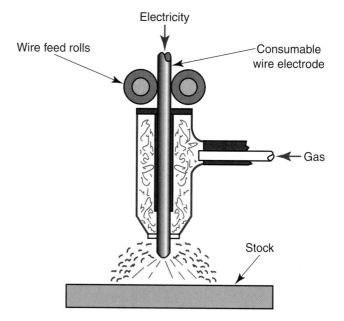

joining both ferrous and nonferrous metals. GMAW is a process that readily lends itself to automated operations.

Gas Tungsten Arc Welding

Gas tungsten arc welding (GTAW) uses a nonconsumable tungsten wire electrode and a tank-supplied inert shielding gas. A separate welding rod is sometimes used to provide filler metal. GTAW is sometimes called **TIG** or *tungsten inert gas* welding. Other than the electrode, the processes of GTAW and GMAW (or TIG and MIG) are basically the same. This is why the terms MIG and TIG are sometimes confused. GMAW is much more widely used in industry than GTAW.

Plasma Arc Welding

The **plasma arc welding (PAW)** process uses a non-consumable electrode recessed in a nozzle that supplies a jet of shielding gas. Plasma arc welding is a further development of the TIG welding process. PAW uses a tungsten or tungsten alloy electrode, similar to TIG, but the electrode is recessed in a nozzle that serves to constrict the arc. Another advantage of recessing the electrode in the nozzle is that electrode contamination is minimized. An electrode can usually last for an entire production shift without needing to be reground.

The plasma is generated by sending a jet of gas through an electric arc. The plasma gas is ionized at very high temperatures in the constricting nozzle and exits the nozzle at high speed. The plasma gas alone is not adequate to shield the molten weld pool from the atmosphere, so shielding gas is supplied around the plasma column, as with GTAW. The flow rate for the plasma gas is much lower than that of the shielding gas to minimize turbulence. PAW produces a very focused heat source and is able to join two plates of differing thicknesses—a daunting challenge for most welding systems. Another unique feature of PAW is how the arc is initiated. High-frequency (HF) current typically is used to establish a pilot arc between the electrode and the copper nozzle. HF is turned off after the pilot arc is started. The pilot arc current usually is fixed at one level or can be set at one of two levels, typically somewhere between 2 and 15 amps. Because of the speed and need to ensure the arc length is controlled adequately, plasma welding is done only on automatic machines. Automated welding ensures consistent spot size and energy density, which produces dependable high-quality welds.

A popular application of PAW is in manufacturing stainless steel vessels or containers and high-quality pressure piping. The process is particularly suitable for applications where deep, narrow welds are needed. The process also creates a minimum of thermal distortion because of its high welding speed.

Resistance Welding

The simplest form of resistance welding is what is usually referred to as "spot" welding. This is a popular process for permanently joining thin-gage metal parts. See **Figure 24-3**. Spot welding is often used in place of riveting.

Resistance welding is accomplished by placing the materials to be joined between two opposing electrodes. The electrodes are pressed firmly against the metal layers. When current is applied, the electrical resistance that exists at the interface between the two pieces of metal to be joined creates heat. This heat is sufficient to melt and bond the two pieces at that spot. The result is a weld "nugget," typically 1/4″ to 3/8″ (6.4 mm to 9.5 mm) in diameter. Other types of resistance welding produce continuous or interrupted weld lines where the metal pieces are joined. The strength of the bond that is created depends on the temperature of the joint, the materials to be joined, and how well the materials were cleaned before welding.

Figure 24-3. Spot welding is the most common form of resistance welding. The two pieces of metal to be welded are pressed tightly together by the electrodes and an electrical current applied. Heat from electrical resistance melts the metal and fuses it together to form the weld nugget or "spot weld."

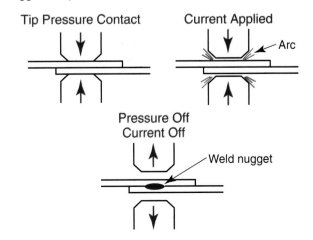

One of the most extensive uses of resistance welding is *seam welding* of beverage cans. Such cans must be leakproof, and often must be able to withstand high pressures. The automatic seam welder shown in **Figure 24-4** uses a pair of *wheel electrodes*. To make the longitudinal seam, one electrode rolls inside the can; the other, outside. Each wheel has a groove cut into its circumference. A thin copper wire can be seen just above and to the left of the center of the picture. The wire wraps around the electrode wheel. When the wheels roll over the weld seam, the copper wire transmits current from the wheels to the work. The hot wire melts the surface of the can and then picks up molten tin from the weld seam. The purpose of the wire is to keep the wheels clean and prolong the life of the wheel. The wheels on the welder shown in the photo are able to weld cans running through the machine at the rate of 197′ per minute.

Butt Welding

Butt welding, also called *flash butt welding*, is similar to resistance welding. Heat is generated from an arc produced when the ends of two parts to be welded make contact. It is an ideal process for joining two rods end-to-end. A variation of this process, *stud welding*, is used to join threaded fasteners to plates.

Butt welding is an arc welding process. Refer to **Figure 24-5**. When the parts to be joined reach the proper temperature, the materials are brought together. This upsetting action forces impurities from

Figure 24-4. Automatic seam welding machines use wheel electrodes to resistance-weld a continuous longitudinal seam for cans. The printed cylinder form entering the machine from right is a tube that will be cut into individual can lengths after seaming. Note wheel electrode, upper center of photo. (Soudronic, Ltd.)

Figure 24-5. In butt welding, an electric arc formed when the two pieces are brought together creating the heat needed to melt and fuse the metal parts.

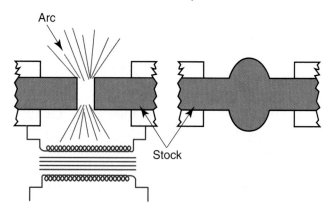

the joint and forge-welds the two materials together. A bond takes place through plastic deformation at the weldment. Some flash is created at the joint.

Butt welding machines are usually automated. Electrical currents required for steel range from 2000–5000 amps/in². The actual forging pressure can be as great at 25,000 psi (170,000 kPa).

Friction Welding

Friction welding was originally developed as a high-volume mass production process for joining metal. The principle of friction welding is simple: the elements to be welded are held in axial alignment while one element is rotated and the other is pushed against it under pressure. The friction developed at the point of contact creates heat. Welding occurs when the heat causes the materials to become molten and fuse together.

The origin of friction welding dates back to 1891, when a patent was issued to cover joining the ends of wire cable. Since that time, major patents covering friction welding processes have been issued in the United States, Germany, the United Kingdom, and other nations.

The process of friction welding has undergone several minor variations over the years to improve its efficiency. This has created some confusion about terminology. The process may be called any of a number of names, including inertia welding, stored energy friction welding, flywheel friction welding, continuous drive friction welding, or spin welding.

Two variations of the basic friction welding process are most frequently used today. One is based on the stored energy approach to friction welding pioneered by Caterpillar Tractor Company and the AMF Corporation. This system relies on energy stored in a flywheel to rotate the workpiece and produce friction. The second variation uses direct power supplied from a continuous drive source. Both of these systems are used extensively in the automotive, aircraft, equipment manufacturing, and aerospace industries.

In *inertia friction welding,* one of the workpieces is clamped in a spindle chuck attached to a flywheel, and the other is held in a stationary holding device. The chuck is accelerated to a prescribed rotational speed. When the desired speed is reached, power is cut and the part held in the chuck is forced against the stationary piece. Friction between the parts causes the flywheel to decelerate, causing frictional heat. The parts are softened at the point of interface, but they are not melted. Just before the flywheel stops, the parts are forced together, stopping the flywheel. This forges the parts together while removing voids and refining the grain structure at the joint.

In 1985, an Indiana company called Manufacturing Technology, Inc. acquired all patent rights to inertia welders and flywheel friction welders in the United States. The company manufactures inertia friction welders ranging in size from micro machines for workpieces smaller than 1 mm (0.039″) in diameter to giant models that apply a weld force of 750 tons to large workpieces. An example of the largest machines is shown in **Figure 24-6**. This huge inertia welder is used to weld aircraft engine components.

There are many advantages of inertia friction welding over conventional forming and forging processes. The power control drive shaft shown in **Figure 24-7** was originally manufactured using upset forging, which required five straightening operations. When forging was replaced by inertia welding, the straightening steps could be eliminated, resulting in time and cost savings.

Figure 24-6. Aircraft engine components are joined by inertia welding on this huge machine. For scale, note operator at left. (Manufacturing Technology, Inc.)

Figure 24-7. The drive shaft blank shown at top was made by upset forging, and required straightening before it could be processed further. The two center views show the shaft and gear blank before and after being joined by inertia friction welding. Note the lack of distortion in the joined parts. At bottom is the finished shaft after machining. (Manufacturing Technology, Inc.)

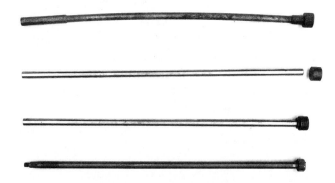

Inertia welding is a machine-controlled process that eliminates operator-caused defects. The process can easily be automated for high-volume production, or it can be adapted for use in flexible machining cells. There is no need to carefully prepare surfaces for welding other than being clean. Joints can be machined, saw-cut, cast, forged, or sheared. Most metals—cast iron is one of the few exceptions—and many nonmetallic materials perform well with friction welding.

Any solid forging in which a bar is upset at one end is a likely application for inertia friction welding. However, parts do not have to be solid. **Figure 24-8**

Figure 24-8. Inertia friction welding can be used with tubular parts, as well as solid components. This hollow shaft was joined to the stamped cover by the inertia friction welding process. (Manufacturing Technology, Inc.)

Figure 24-9. Installing the main injector ports in the space shuttle's engines is done efficiently with the inertia friction welding process. More than 600 welds must be made for each engine. (Manufacturing Technology, Inc.)

shows an application in which a tubular shaft has been welded to a stamped torque converter cover.

Many manufacturing firms are taking advantage of the benefits offered by inertia friction welding. **Figure 24-9** illustrates the type of application that is perfect for inertia friction welding. In this operation, main injector ports for the engine of the space shuttle are welded. There are 600 welds per engine, and each shuttle has three engines.

Friction welders can be fully automated. The CNC welder shown in **Figure 24-10** receives input data on the outer and inner diameters of the workpiece, heating pressure, upset pressure, and amount of heating. All calculations and commands are then performed by the internal computer. This machine features a continuous drive (a motor, coupled to a clutch and brake) to rotate the moving piece.

Electron Beam Welding

High-quality deep and narrow welds can be produced on almost any type of metal with the *electron beam welding* process. The process generates heat by bombarding the weld site with a narrowly-concentrated beam of high-velocity electrons.

Electron beam welding can be used to weld almost any type and hardness of metal. Thicknesses range from thin foils of no more than 0.001″ (0.025 mm) to a plate as thick as 6″ (15.2 cm). Welds can be produced at rates greater than 40′ per minute.

Energy is produced by a gun that accelerates a stream of electrons, giving them tremendous pen-

etrating power. Electron guns have power ratings that can range from 6 kilowatts (kw) to 100 kilowatts. The moving electrons produce kinetic energy, which is converted to heat as it strikes the workpiece. The concentrated beam of electrons is directed toward the workpiece in a vacuum.

The process generates harmful rays, so careful monitoring, maintenance, and systems control is necessary. Electron beams can be projected for a distance of several meters, so the process should be isolated to ensure safe operating procedures. Electron beam welding is used to weld a wide range of metals. A limitation of electron beam welding is its narrow field. It is not appropriate in applications where a gap between materials exceeds 0.005″ (0.127 mm).

Laser Welding

A laser emits a very directional concentrated beam of light that can be focused to a very high intensity. The heat energy created by this light beam can be delivered to a spot as small as a few thousandths

Figure 24-10. This continuous drive friction welding machine operates under computer numerical control, permitting fully automated production of welded parts. (Rocon Equipment Company)

of an inch. The welding process begins by heating a spot until the metal is vaporized, forming a hole in the workpiece. This hole, known as a keyhole, is filled with the vaporized metal (ionized metallic gas), which soaks up all of the laser energy. The extreme heat created within the keyhole radiates out, forming a region of molten metal surrounding the vapor. As the laser beam moves along the workpiece, the molten metal fuses behind the keyhole and solidifies to form the weld. This technique permits welding speeds of thirty to forty inches per minute.

Lasers are gaining popularity for many welding applications, particularly where narrow and deep joints are required. Since it does not have to be done in a vacuum, laser welding has an advantage over electron beam welding. However, in cases when conventional welding processes can be used, they are normally the most cost-effective choice. Laser welding is a process that should be saved for applications where the materials are difficult to weld, or where the parts are heat-sensitive. Laser welding can be performed without filler material.

Laser welding however requires more power than is needed for laser cutting. Laser welding and heat treating require outputs of 2000 watts or higher. Both CO_2 lasers and Nd:YAG (normally called YAG) lasers are used for welding. Both produce pulsed beams. This makes them particularly useful for welding, because they can bring the metal to its melting temperature very quickly. This is an advantage when it is critical to minimize heat distortion or weld heat-sensitive parts such as electronic devices.

When lasers with higher power intensities (6 kW to 25 kW) are used, the efficiency of the system is greatly improved. These lasers provide deep penetration and dense welds.

For more than 20 years, the Chrysler Motors automatic transmission plant in Kokomo, Indiana used traditional electron beam production systems to weld gear assembly components. In 1985, they replaced these systems with multi-kilowatt laser welding systems. Today, this DaimlerChrysler facility has one of the largest high-power laser welding installations in the world.

Laser and electron beam welding are similar, in that both are noncontact fusion welding processes. In laser welding, the heat for fusion is generated by directing a focused beam of photons (light) onto the workpiece. Laser welding does not require the use of a vacuum chamber, but it does depend on shielding the weld pool with inert gas to prevent contamination of the weld.

Laser welding systems can be made fully automatic and do not require operator intervention or close monitoring. The components to be welded can be fed into the welding station and positioned by a robotic pick-and-place system. Once the weld is completed, the assembly is placed on an output conveyor by the weld station pick-and-place robot.

Ultrasonic Welding

Ultrasonic welding works on the principle of changing sound energy to mechanical movement to generate heat for joining metals or plastics. The sound energy consists of frequencies well above the range of human hearing (20 Hz to 14,000 Hz). A mechanism called a *transducer* is then needed to change the sound energy into mechanical vibrations.

For ultrasonic welding, a horn is attached to the transducer to carry these mechanical vibrations to the workpiece. The horn is tuned to vibrate at 20,000 Hz. The horn and tip press together the workpieces to be welded, much like the electrodes in resistance welding. The vibratory energy first cleans away oxides on the surfaces of the metal. Welding occurs when the metals come into contact with each other. The welding tip is usually formed in the same shape as the final workpiece. **Figure 24-11** shows a small to medium-duty ultrasonic metal welding system.

Machines are available for many different types of welding operations. The machine shown

Figure 24-11. Ultrasonic welders, like the system shown here, convert sound energy into vibrations (mechanical energy) to heat and weld parts together. (Stapla Ultrasonics Corporation)

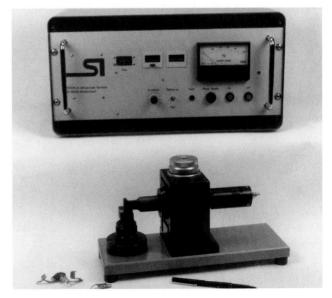

Figure 24-12. This automated ultrasonic welding unit is used for wire splicing. (Stapla Ultrasonics Corporation)

in **Figure 24-12** is one of the most automatic wire splicing units available on the market. This machine has universal tooling that is automatically set up by inputs from an electronic controller.

Brazing and Soldering

Brazing and soldering are both old processes, dating to as early as 3000 BC. Today, both are frequently used for fabrication and joining applications in manufacturing. Both processes incorporate the use of filler metal to span the gap between materials. In soldering, the filler material is a lead-tin alloy called solder, while in brazing, it is normally an alloy of copper—brass or bronze. The major difference in the two is the temperature required to melt the filler. The temperature point that differentiates soldering from brazing is generally considered to be 840°F (450°C). Solder melts well below that point; brazing alloys melt above that temperature.

Brazing

Brazing is a joining method that requires the use of filler metal in the interface area. It joins the base metals by adhesion of the melted (and later cooled and solidified) filler metal—the base metals do not melt. Brazing rods are available in many different alloy compositions for use with different base metals. Brazing is a popular process for maintenance and repair of ferrous castings. It is also useful for joining together some dissimilar metals, such as steel and brass.

There are several different methods used for brazing. Parts can be dipped to apply filler material. However, the most common method involves heating the base metal with a torch using acetylene and oxygen. The filler metal becomes molten when it touches the heated base metal and flows into the gap at the joint.

Normally, brazing metal is in the form of a wire or rod. It can also be purchased in sheets, rings, or powder form. *Flux* is a material added to the braze joint to prevent oxidation, remove oxides that have formed, and reduce fumes. When the filler metal is melted, it flows between the surfaces to be joined to form a bond. When the flame is removed and the joint cools, the materials are permanently joined.

When two tubes are to be joined, another type of brazing called *exotherm brazing* may be useful. See **Figure 24-13**. In this method, the heat is provided by an AC electrical coil situated close to the joint. High-frequency current creates resistance and heats the joint area.

Figure 24-13. Exotherm brazing is a system in which electrical resistance to a high-frequency current creates the necessary heat for joining the parts.

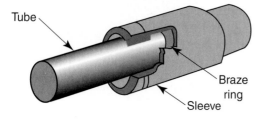

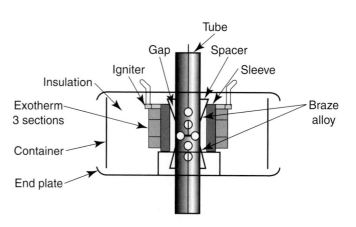

Soldering

As noted earlier, soldering and brazing are similar processes that have been in use for many centuries. Both are *adhesion processes* that form a bond between parts by filling joints with a material that melts at a temperature below the melting point of the base metal.

Until the early part of this century, the filler metal known as *solder* was typically an alloy of tin and lead, although other metals are used in some cases. There were, however, many applications (such as plumbing and food handling) that did require lead-free solders. In those instances, antimony has been used to replace lead. Solder was classified according to the percentage of the metals in the alloy: a 60-40 tin/lead solder would be 60 percent tin and 40 percent lead.

In the summer of 2006, the European Union restricted the use of hazardous substances in electrical and electronic equipment, essentially eliminating the use of lead as a component in solder. The United States has followed a similar policy. The removal of lead from solder has become almost universal, regardless of application. This change to lead-free solders has required manufacturers to review and modify their processes to accommodate the higher costs, different melting temperatures, and characteristics of the lead-free solders.

Today, the common alloys for solder are: tin-zinc, tin-silver, tin-silver-copper, antimony-silver-tin, and several other alloys formulated to meet specific applications. The lead-free solders are available in solid wire, acid or rosin core wire, bars, and paste (a mixture of powdered solder and flux). These solders can be applied using traditional methods such as a torch, induction heating, infrared, or dipping techniques.

The critical factor in creating a successful solder joint is to properly clean the metals to be joined. This can be done by scuffing the surface mechanically or by washing the component. A flux is then applied to clean and protect the work surface. The type of flux used depends on the metal being soldered. Zinc ammonium chloride is common in nonelectrical applications.

Once the materials are cleaned, the metal to be joined is heated at the joint area using a soldering gun or propane torch. Solder is then applied to the work surface, at a point close to the heat source, so that it is drawn into the joint by capillary action. The flux also helps to draw the solder into the joint. The heat is removed and the solder joint solidifies.

For industrial joining applications, specialized methods may be used to apply solder. One of the most common processes is *wave soldering* integrated circuit boards. The process begins by placing the integrated circuit board over a soldering bath with the connections to be soldered just above the surface of the molten solder. Once the circuit board is in place, a wave is created that moves across the surface of the bath. The top of the wave briefly touches the connections. That brief instant when the wave touches the wire ends and the exposed foils is all that is required to make the dozens or hundreds of solder joints on the circuit board.

Soldering can even be done ultrasonically. A common method for coating wires on components or leads is dip soldering: the wire or component is simply immersed in a container of molten solder. The dip pot or solder bath is agitated using ultrasonics. In many cases, the ultrasonic action eliminates the need for flux to complete the soldering process. It also provides a means to bond solder to a number of materials including aluminum and glass that are difficult or impossible to solder using conventional soldering techniques.

Lasers have also been adapted to create soldering systems that are able to provide the heat needed to melt lead-free solder without overheating the surrounding materials and components. These soldering systems frequently incorporate automated dispensing systems that can rapidly deposit very precise amounts of solder paste. The cost of the laser system is partially offset by the savings from the increased speed of the soldering operation, the ability to minimize the amount of solder paste required, and the soldering consistency that can be achieved.

Sometimes, joints are prepared in such a way that the solder does not show. This is called *sweat soldering*. This type of soldering is often done by applying a thin coating of solder to each of the surfaces to be joined. The materials are then clamped together and heat is applied to the joint. Heat is continued until all of the solder is melted and drawn into the joint.

Important Terms

adhesion processes
alloying elements
arc welding
brazing
butt welding
deoxidizers
electron beam welding
exotherm brazing
filler metal
flux
fusing
gas metal arc welding (GMAW)
gas tungsten arc welding (GTAW)
inert
inertia friction welding
MIG
oxyacetylene welding
plasma arc welding (PAW)
resistance welding
scavengers
seam welding

shielded metal arc welding (SMAW)
solder
stud welding
sweat soldering
tack welding
TIG
torch cutting
ultrasonic welding
wave soldering
welding
wheel electrodes

Questions for Review and Discussion

1. What are the disadvantages involved in designing a product that uses mechanical joining of parts, rather than adhesive or cohesive joining methods?

2. Describe the differences between TIG welding and MIG welding.

3. List the welding methods that use a shielding gas.

4. Why is shielding gas necessary in welding?

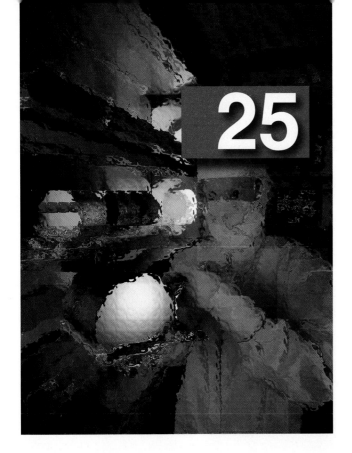

Processes Used to Fabricate Plastic Materials

25

Chapter Highlights

- Cohesive joining is used to permanently assemble plastic parts.
- Embedded wire, used with dielectric heating, is one means of welding plastics.
- High-frequency vibrations are used to generate frictional heat as a means of joining plastics.
- Molecular movement caused by radio-frequency waves creates the heat needed to seal products made of thermoplastic polyurethane or polyvinyl chloride.
- Eight different types of materials are used to bond plastics. Some of these involve cohesive bonding while others involve adhesive bonding.

Introduction

There are many different methods used to fabricate products from plastic materials. Each of these methods can be classified as either mechanical or cohesive fastening.

There are literally thousands of different types of mechanical fasteners that are used to join materials. These range from rivets and Velcro® strips to quick-release thongs and machine screws so small they can be seen only under a magnifying glass. While the use of mechanical fasteners is an important aspect of fabrication, fasteners constitute a significant field in and of themselves, and are beyond the scope of this textbook. In this chapter, our emphasis will be on the manufacturing processes used to fabricate plastic products. Most of these processes involve *cohesive joining* that results in permanent assembly. When referring to cohesive joining of materials, **cohesion** is the intermolecular attraction by which the elements of the body are held together.

Cohesion Processes

There are five major types of cohesive fabrication processes used with plastics: welding, sealing, cementing and bonding, winding, and 3-D physical concept modeling.

Welding is a process that is widely used to fabricate thermoplastics. Recall from Chapter 15 that thermoplastics melt when subjected to heat. This is important to remember when thinking about welding plastics.

Welding of plastics is different from sealing. While both welding and sealing are cohesion processes, *welding* is the joining of plastic components of relatively heavy thickness. *Sealing* is a process used to join thin plastic or film.

Hot Gas Welding

Most welding is done on polyvinyl chloride (PVC) and polyethylene (PE) materials, but all thermoplastics can be welded. Many of the techniques used with gas welding of metal materials also apply to plastics. There are some major differences, however.

Plastics welding uses a plastic filler rod, rather than the metal electrode used to weld metal. The welding temperatures for plastics are much lower than those in conventional gas welding. Such high temperatures would melt and burn plastics.

Plastics welding is accomplished using a welding gun with an electric heating element or a coil heated by a gas flame. Electric guns are preferred because of their lighter weight. Nitrogen or air is blown past the heat and applied to the plastic. A filler rod of the same material as that being welded is melted into the joint. The rod is held at an angle to the joint and heat is applied to the rod and the joint. Several passes are usually made, at a speed of about 1″ to 2 1/2″ (25 mm to 64 mm) per minute. Welding of a plastic container is shown in **Figure 25-1**.

Hot Wire Welding

Hot wire welding is different from plastic welding. You may recall reading in Chapter 15 the separating process known as hot wire cutting. *Hot wire welding*, or *dielectric heating*, works on the same principle. Hot wire welding is a simple process in which a wire with high electrical resistance is placed between two thermoplastic surfaces that are to be joined. With this process, the wire is left in the joint after it is welded together.

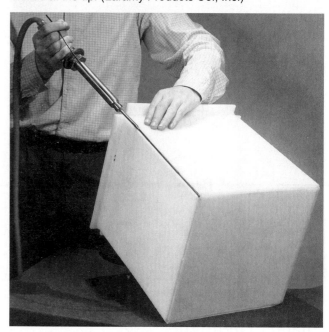

Figure 25-1. A welding gun that heats a stream of gas is used, along with a plastic filler rod to weld plastics. This worker is welding a seam on a plastic container. Note how the filler rod is fed through a channel in the gun so it can be melted at the tip. (Laramy Products Co., Inc.)

The process is used to join plastics, films, fabrics, or foams. Materials typically joined by this method include cellulose acetate, ABS, polyvinyl chloride, epoxy, polyester, polyamide, and polyurethane. Plastics such as polystyrene, polyethylene, and fluoroplastics have very low heat dissipation factors and cannot be joined with this method.

Here is how the process works: the wire that is to be sandwiched between the two sheets of plastic is bent into a zigzag configuration or multiple loops to provide additional strength in the joint. Often, a slot or indentation is made in the plastic to hold the wire. A high-frequency voltage is then applied to the wire. Fusion occurs when the high frequency causes the molecules to rapidly realign themselves. This molecular movement causes frictional heat and the plastic is melted in the area surrounding the wire.

Vibration Welding

We will be discussing two different types of vibration welding. These processes vibrate the workpieces to cause friction and produce heat. The first of these processes is called *vibration welding*. We will also discuss *linear friction welding*; this also uses friction between parts to generate heat. With both types of vibration welding, the heat is high enough to cause thermoplastic parts to melt and weld together. Normally, parts can be vibration-welded in seconds. Vibration welding is practical for parts that are too large for ultrasonic welding. An important requirement for vibration welding is that the parts are designed without obstructions, allowing the parts to contact each other. The process of vibration welding is particularly effective in applications where leak-proof pressure or vacuum joints are desired.

In vibration welding, the two thermoplastic parts are rubbed together under pressure. At the same time, an electrical current at a suitable frequency and amplitude is applied until sufficient heat is generated to melt and mix the polymer. The electrical current is transmitted via wires to the joint area. The high frequency produced causes the plastic molecules to vibrate, creating sufficient frictional heat to melt the thermoplastic.

When the vibration is stopped, the parts are aligned and the molten polymer is allowed to solidify, creating the weld. The process is similar to spin welding except the force between the pieces to be joined is linear rather than rotational.

Vibration welding can be performed using machines that produce linear or angular vibratory action. Low-frequency vibratory machines are classified as those operating from 120 Hz to 130 Hz. High-frequency machines operate at from 180 Hz to 260 Hz.

Linear friction welding has some different features. In linear friction welding, the heat needed to melt the thermoplastic is generated by pressing one of the parts against the other and rapidly vibrating it. The heat generated by the resulting friction melts the material at the joint in 2–3 seconds. The vibration is then stopped, and the parts are aligned and held together under pressure until a solid bond is formed. Such bonds are permanent and have a strength approaching that of the parent material.

Induction Welding

Two major types of *induction welding* are used in industry today: *magnetic induction bonding* and *Hellerbond induction bonding*. Both processes can be used with either thermoplastics or thermoset materials. Heating cycles required for induction welding are short, normally only a few seconds.

As is the case with dielectric welding, a radio-frequency magnetic field is responsible for creating the heat at the joint. The difference is that no lead wires are required to transmit current to the weld area.

In magnetic induction bonding, an adhesive or bonding agent is placed in the joint between the two surfaces to be bonded. The powdered adhesive or bonding agent contains ferromagnetic particles dispersed throughout a polymer. The powder reacts (heats up) in response to the magnetic field generated by an induction coil.

The Hellerbond™ process is similar to magnetic induction bonding, except that the adhesive or bonding agent is replaced with a ceramic powder. The powder normally consists of finely ground magnetic iron oxide. As in magnetic induction bonding, the powder responds to an RF magnetic field. The advantage of this process is that the ceramic powder can be incorporated directly into parts, such as filaments, sealing gaskets, or liquid adhesives.

Many types of thermoset and thermoplastics— ABS, nylon, polyester, polyethylene, polypropylene, and polycarbonates—are induction-welded. Induction welding is also effective for joining materials such as glass, paperboard, textile, paper, and leather.

An important advantage of induction welding is that it can be used with heat-sensitive materials and with dissimilar materials. One drawback is that it can be used only with nonmetals.

The equipment required for induction welding is an RF power supply that can operate in the frequency range of 1.8 MHz to 4 MHz. The tooling required performs two functions. It holds the parts together and also applies the RF magnetic field to the bonding agents through an induction coil embedded in the tooling.

There are many applications for induction welding, since products can be of virtually any size. Handles on kitchen knives and appliances are joined with this process. Thousands of induction-welded products are produced daily for the automotive, packaging, construction, and office equipment markets.

Spin Welding (Friction Welding)

Spin welding, or *orbital friction welding*, is typically used when butt welding cylindrical or tubular parts end to end. One of the parts is held stationary in a nesting fixture, while the mating part is spun rapidly against it. Friction created at the point of contact generates heat and brings the two pieces to their melting point.

Immediately after the material reaches the melting point, the spinning stops and the parts are held together for about half a second for cooling.

After curing, the weld joint created is often much stronger than the substrate itself. Normally, the best weld results from welding times of no more than a second or two.

Almost any thermoplastic material can be spin welded, including filled, structural foamed, crystalline, and solid materials. It is sometimes possible to spin weld dissimilar materials such as thermoplastics and alloys.

There are several points of caution that must be observed when spin welding. When welding a solid bar, the heat generated is localized around the outer circumference of the stock, rather than the center of the bar. This leaves a relatively weak point of adhesion in the center. Also, the process develops internal stresses in the parts. Sometimes, these stresses must be removed through subsequent annealing.

Spin welding has been effectively used for applications as diverse as the production of polyethylene floats, transmission shafts, aerosol bottles, and PVC fittings. One advantage of this process is that welds can be made below the surface of a liquid that might be present in a tank or container.

Ultrasonic Welding

Ultrasonic vibrations move parts at a minimum of 20,000 hertz. The friction caused by these vibrations is enough to weld and melt metal and plastic parts, spot weld, transform adhesives to a molten state, and melt films and fabrics together. Ultrasonic vibrations are used for many different types of thermoplastic sealing applications. Ultrasonic vibrations are also used for *staking*, a process that involves melting thermoplastic studs to mechanically lock dissimilar materials in place. Staking is illustrated in **Figure 25-2**. Another application of ultrasonics is *inserting*, embedding a metal component into a preformed hole in a plastic part. Ultrasonic welders are also used for *slitting*, the process of cutting plastic sheets to a desired width and length.

Ultrasonic assembly is a fast and efficient method for assembling rigid thermoplastic parts or films and synthetic fabrics. Plastic can be joined to plastic, as well as to metal or other nonplastic materials. Ultrasonic welding eliminates the need for mechanical fasteners and adhesives.

Ultrasonic welding of thermoplastics such as ABS, acetal, polycarbonate, and polystyrene can be accomplished using high-frequency vibrations in the 20–40 kHz (thousands of cycles-per-second) range.

Figure 25-2. Staking is a variation of electronic welding. High-frequency vibrations cause plastic studs to heat up and soften so that they can be used to lock dissimilar materials together, as shown.

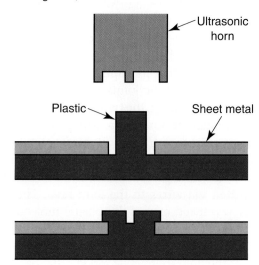

These mechanical vibrations, applied to two mating parts that are pressed together, create rapid movement in the molecules at the joint interface. This generates heat and results in fusion of the parts. The weld that is created often approaches the strength of the material in the original plastic part.

There are six major components in most ultrasonic welding systems. Power is generated through a power supply, which transmits 60 Hz energy to a converter. The converter changes the electrical energy to high-frequency (20–40 kHz) mechanical vibrations. A booster is attached to the converter. The purpose of the booster is to increase or decrease the amplitude of the vibration to the point yielding the strongest weld with the least amount of flash. The *horn* is a key element. The *horn* is a metal wedge-shaped bar that travels down on a vertical spindle to contact the workpiece. The horn transmits the mechanical vibrations to the workpiece. The horn is made of aluminum, titanium, and hardened steel, and is sometimes highly polished and chrome-plated.

Attached to the base of the machine is a *nesting fixture* that securely holds the part. The part to be welded normally consists of a two-piece shell that is snapped together and then ultrasonically joined. The fixture (often made of aluminum) is carefully machined to conform to one-half the shell that is to be welded.

The final component of the system is the actuator. The purpose of the *actuator* is to house the converter, booster, horn, and pneumatic controls. The actuator brings the horn into contact with the workpiece, applies the desired force for the correct length of time, and then retracts the horn after welding has been completed.

A microprocessor-controlled ultrasonic welder is shown in **Figure 25-3**. The power supply, converter, and booster are located in the control console at the top of the machine. The highly polished horn is visible at the bottom of the photograph.

A typical application for welding with the ultrasonic system would be a two-piece cover or shell housing electrical parts. The operator begins by placing half of the shell into the nesting fixture. Normally, this half includes wiring and components. The other half of the housing is snapped on the first shell half. The operator would next place both hands on buttons on either side of the actuator and depress them, lowering the horn onto the two parts. The two buttons are built into the machines as a safety feature. The operator has to have both hands on the buttons when the horn is pressed into the workpiece.

Figure 25-3. The microprocessor control of this ultrasonic welder allows precise temperature, time, and pressure regulation when welding joints in plastics. (Branson Ultrasonics Corporation)

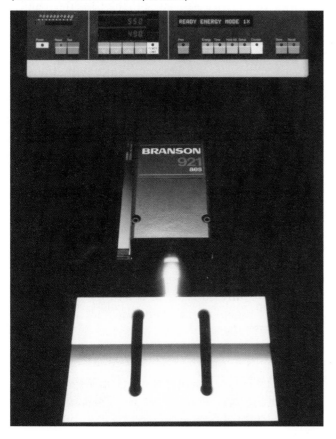

The horn holds the parts firmly together as ultrasonic vibrations are emitted, causing heat to be produced where the two parts fit together. After welding, the horn is retracted and the welded housing removed.

Joint design is critical in ultrasonic welding. The joint should have a small, uniform contact area, and include a method for aligning the parts properly. This is normally accomplished with alignment pins designed into the part.

Ultrasonic assembly techniques are used with a wide range of thermoplastic products. Products that are often sealed with ultrasonic welding include headlamp assemblies, steering wheels, food and beverage containers, blister packaging, small appliances, electronic assemblies, toys, videocassettes, ballpoint pens, and medical components.

Radio-Frequency Heat Sealing

Radio-frequency (RF) heat sealing is often used to join thermoplastic parts that cannot be merged with ultrasonic welding, spin welding, or induction welding. Radio-frequency heat sealing is most frequently used to join rigid plastics to flexible plastics, such as PVC and thermoplastic polyurethanes. Radio-frequency heat sealing has been used for many years, but is still one of the least-known methods for joining plastics.

Radio-frequency heat sealing works much like ultrasonic welding—heat is generated in the area to be joined by molecular movement. Radio-frequency heat sealing uses a frequency of approximately 27 MHz. In radio-frequency heat-sealing, the high-frequency radio waves, combined with the energy and pressure exerted by pressing the parts together, cause the molecules to oscillate and create heat.

Equipment required for radio-frequency heat sealing consists of a generating system, sealing press, and die. The generating system typically includes a generator (normally up to 25 kW in size), a controller, a power supply, an oscillator, and safety protection devices. The power supply converts AC to high-voltage DC. The oscillator converts high-voltage DC to high-frequency AC voltage. An arc detector device protects the system from overload, and will automatically shut down the equipment when preset limits have been reached.

The sealing press consists of a pneumatic ram that provides the pressure required during heating. The press uses a heated bottom platen with controls to regulate pressure during up and down strokes of the ram. The heated platen helps to keep the temperature constant during welding.

A brass die is constructed to provide the desired seal shape. The die is attached to an aluminum plate. A buffer material with high dielectric properties, such as electronic fishpaper or polyester film, is placed between the plate and the machine's bed. The buffer reduces heat loss from the workpiece to the bed.

This process is often used to seal blister packs, medical bags, blood pressure cuffs, tarpaulins, pool liners, inflatable toys, life jackets, and eyeglass cases. It can also be used to bond together some nonplastic materials such as cotton, paper, and woven fabrics.

Dielectric Sealing

Dielectric sealing, using RF vibrations, can be used to bond many thermoplastics, films, fabrics, and foams. Materials with high dielectric loss characteristics (such as cellulose acetate, ABS, polyvinyl chloride, epoxy, polyether, polyester, and polyurethane) are often joined with dielectric sealing. Polyethylene, polypropylene, polystyrene, and the polycarbonates do not have the proper electrical characteristics to be bonded by this process.

The thermoplastic to be sealed is placed as a dielectric between two flat electrodes or sealing bars. These electrodes normally consist of brass strips backed with steel or aluminum. The sealing bars are attached to a high-frequency power source, which is capable of emitting frequencies between 20 and 40 MHz. The FCC has assigned a frequency of 27.18 MHz for dielectric sealing. Most machines use this frequency.

The actual heating occurs when the molecules try to realign themselves with the RF oscillations. This movement of molecules causes frictional heat, causing the point of interface between the two materials to become molten.

Dielectric sealing supplies heat at the surface, causing more intensive heating at the interior of the thermoplastic. Thick material can be heat sealed more rapidly than thin material, because the heat created in thin material is more quickly carried away by the electrodes.

Thermal Heat Sealing

Heat sealing can also be accomplished without the use of high-frequency energy. *Thermal heat sealing* is done with a heated tool or die. The sealing bar or tool is held at a constant temperature,

permitting the heat to flow through one of the layers of the film to the interface joint. This method is practical for thin-film sealing. Some of the common types of seals that are used by the packaging industry are lap, fold-over, step, tear, perforated, and peelable. The lap seal is used to join two wide sheets of film when high strength is required. **Figure 25-4** illustrates some of the many applications for thermal heat sealing.

Thermal impulse heat sealing is a variation on this process in which the dies are heated intermittently. The temperature is increased during the heating cycle and is then decreased to permit the joint to cool under pressure. This intermittent heating produces improved appearance in the joint and is best suited for film with a thickness of less than 0.010" (0.25 mm).

Heat sealing is used for many applications that require sealing plastic to other packaging materials, such as printed card stock. **Figure 25-5** illustrates many of the different applications of thermal heat-sealing for blister packaging.

Cementing and Bonding

In addition to welding and sealing, cohesive bonding of plastics can be achieved by the use of chemicals called *solvents* or *solvent cements*. In order to produce cohesive bonding in plastics, the surfaces must be heated to cause them to flow and allow the molecules to mix together. On thermoplastic polymers, this is normally accomplished by using solvents. The solvents create heat, soften the material, and permit a strong molecular bond.

Figure 25-4. Some common types of seals that are made using thermal heat sealing.

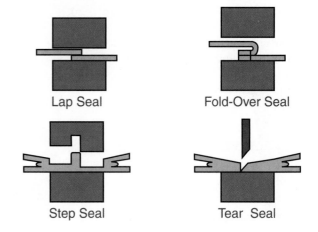

Lap Seal   Fold-Over Seal

Step Seal   Tear Seal

Figure 25-5. Various applications of thermal heat-sealing in blister packaging. This method allows plastics and printed card stock to be bonded together. A—Blister heat-sealed to card. B—Blister clipped and heat sealed to card. C—Blister sealed between double cards. D—Blister sealed to fold-over card.

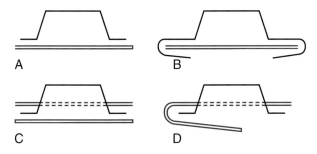

Bonding acrylics, such as clear Plexiglas®, is normally accomplished by soaking, or capillary action. The *soaking method* requires one of the pieces of acrylic to be immersed in the solvent or cement. This soaking produces a softened cushion of acrylic material that will harden when the two pieces are pressed together and cured. *Capillary action* involves squeezing cement or solvent along the joint between the two acrylic surfaces. The cement is drawn into the joint by capillary action so that the softening and bonding can take place. In some cases, the workpiece is not soaked. It can be dipped into the solvent and immediately pressed against the piece to be joined until the two are bonded.

Thermosets and some thermoplastics do not soften easily when exposed to solvents, so they are typically joined by the process of *adhesion* or adhesive bonding. An *adhesive* is a bonding agent that forms a film between the materials to be joined. By "sticking" to both pieces, the adhesive holds them together. Some adhesives remain flexible; others form a rigid joint. Also, some types are best for bonding materials of the same composition, while others are best for joining dissimilar materials.

Choosing the correct bonding agent for adhesive bonding of plastics can be difficult. Adhesives that withstand the stresses of the environment must be selected. There are eight basic types of adhesives used for bonding plastics:

- Silicones.
- Cyanoacrylates.
- Reactive acrylate-based adhesives.
- Polysulfide sealants/adhesives.
- Epoxy resin-based adhesives.
- Polyurethanes and isocyanate-based adhesives.
- Rubber-based adhesives.
- Thermoplastic hot-melts.

Of all of the types of plastic adhesives listed, the cyanoacrylates have superior overall adhesive properties. They require no mixing, cure rapidly, can withstand tensile strength tests of 3000 psi (20,685 kPa), and are suitable for use in temperatures up to 180°F (82°C).

Hot Melts

Hot melts, also known as *reactive melts* or *reactives*, are thermoplastics that become fluid when heated, then solidify upon cooling. Hot melts react with moisture in a substrate or in the air, and then cross-link. Polymerization forms an adhesive bond that resists heat, steam, chemicals, and harsh environmental conditions. When the reaction is completed, the melt will not resoften if heated.

The main polymers that are used in hot melts are polyesters, polyolefins, polyamides, and ethylene vinyl acetate copolymer. Hot melts are available in beads, granules, powders, ribbons, films, pellets, blocks, cards, pillows, slats, slabs, rods, cakes, and liquids.

Hot melts can replace screws, nut-and-bolt assemblies, nails, and other mechanical fasteners. They can also reduce product weight and improve appearance by eliminating exposed fasteners. Many large companies in the automotive, aerospace, and appliance industries are using hot melts.

Hot melt coatings can also be used to protect metal parts from corrosion or damage during handling, shipping, and storage. Parts are dipped into a hot dip-tank containing a solution typically made of a hot-melt resin mixed with oils, plasticizers, and other materials. When the hot melt is removed and cooled, the resilient, airtight, strippable coating protects the parts until they are needed.

One of the major advantages of hot melts is that they are nonvolatile. They do not pollute the air or water, and do not create problems with solvent emissions and hazardous waste disposal.

Hot melts are the preferred method of assembly in the appliance industry for applications ranging from sealing cabinets to bonding of plastics and metals. Top appliance manufacturers such as Amana, Maytag, General Electric, WCI, and Whirlpool use robotic equipment to apply hot melts on their products. Hot melts can even be produced as foams, and used for applications such as foam-in-place gaskets and seals.

Hot melts are applied with applicators that melt the adhesive and pump it to a handgun or spray head.

In many applications on assembly lines, hot melts are applied automatically at high speeds as sensors signal the guns to squirt adhesive. Hot melts can also be applied by hand using applicator guns.

Typical applications for hot melts include positioning switches and wires in automotive products, assembling windshield washer bottles, securing refrigerator insulation, bonding aluminum jacketing to fiberglass pipe insulation, and foam-in-place gaskets for automotive air conditioner housings.

Filament Winding

Filament winding is a process that involves winding continuous fibers saturated with polyester or epoxy resins around a cylindrical mandrel. Filament-wound products have improved glass fiber strength, permitting thin walls and resulting in lighter weight. Many different types of fibers can be used for filament winding. However, the most popular are high-modulus organic fibers such as Kevlar®, which belongs to the family of fibers called aramids. Aramid fibers are much stronger than conventional organic fibers and have a higher modulus of elasticity. These qualities make them particularly suitable for applications requiring tension and compression.

The process of filament winding lends itself to a broad range of product types. Simple cylinders and tanks, spherical tanks, and even complex windmill blades and helicopter rotors are often fabricated using this process. Filament winding is also used to make cylindrical pressure vessels and tanks, missile bodies, pipe, pressure bottles, high-strength tubing, and shotgun barrels. The process is versatile. It is portable enough that on-site fabrication of large tanks is practical. Tanks with surface areas of up to 1500 ft² and cylinders 40′ in diameter used as power plant stack liners are wrapped on single mandrels.

Let us take a closer look at how the process is conducted: The filament is drawn through a resin bath, and then mechanically wound around a rotating mandrel. Mandrels can be removed after the resin cures or they can be left inside the windings as a permanent support liner. When a hollow vessel is desired, hard salt is sometimes used as the mandrel. The salt is then washed out with water. Low-melting-temperature materials also can be used to construct mandrels. Sometimes, balloons are inflated to serve as the mandrel. The balloon is collapsed after the resin and fiber combination cures.

Wax, wood, plaster, rubber, and fiber-reinforced plastics (FRP) are also used to make mandrels.

Filament winding is effective for producing strong parts. Containers made using this process usually have a higher strength-to-weight ratio than those produced with other processes. Another advantage of this process is that it allows reinforcement materials to be placed in areas subjected to the greatest stress. Filament winding provides a high degree of control over uniformity and orientation of the fibers. Filament-wound products can be accurately machined.

3-D Physical Concept Modeling

We will use the term *3-D concept modeling* in this chapter to identify and classify processes used for both rapid prototyping and fabrication of physical models made from plastics for use as prototypes and testing. Terms such as rapid prototyping, desktop manufacturing, solid freeform fabrication, digital light processing, and layered manufacturing are used to describe these processes today.

A *prototype* is a physical model that can be looked at to imagine what the final product will look like. Sometimes rugged prototypes are constructed that can also be tested. But prototypes are normally far from being final products.

Improvement in the efficiency and speed of processes for making physical 3-D models has grown with leaps and bounds over the years. This has resulted in a shift away from the use of the term *rapid prototyping*, because it is now felt by many to be inaccurate and misleading. Many manufacturers are now using these processes to make parts that are used for functional pilot-production testing, or for production of tooling (molds) to manufacture final products. The chrome wheel in **Figure 25-6** shows the degree of detail and realism that can be produced with 3-D models made from CAD drawings.

The creation of **prototypes**, physical representations of the final product for visualization and testing, can be accomplished with processes that are **subtractive**, **additive**, or **compressive**. Often the conventional process of CNC milling is used to produce prototypes. This is an example of a *subtractive* process. Material is removed to create the shape of the final product. *Compressive* processes, on the other hand, are those that force a semi-solid or liquid material into the desired shape, and then subject it to other processes

Figure 25-6. The chrome wheel is an example of a 3-D model produced by Z Corporation. 3-D models are often used for pilot testing and mold making. (Z Corporation, Inc.)

that cause it to harden or solidify. Typical casting and molding processes fall into this category.

The major limitation with both subtractive and compressive processes for producing prototypes for evaluation and/or testing is that machining and casting methods are difficult to use on parts with very small internal cavities or complex geometries. That is where *additive* processes excel.

Three of the major processes used to fabricate 3-D physical concept models through the addition of material will be covered in this chapter. These are:

- Stereolithography (SLA)
- Three-dimensional printing or fused deposition modeling (FDM)
- Selective laser sintering (SLS or SL)

Each of these processes involves adding layers of material on top of each other, much like a laminated sandwich used to make plywood. The difference here is that images or surfaces are bonded or hardened using high-energy laser beams.

Stereolithography

Today, stereolithography is being used by thousands of manufacturers to produce three-dimensional plastic design prototypes, models for testing, molds, dies, and patterns. **Stereolithography** builds a physical object layer by layer by exposing the liquid photopolymer to ultraviolet laser light. **Photopolymer** is a liquid that cures and hardens where the light strikes it. See **Figure 25-7**. Stereolithography is

gaining popularity in the aerospace, computer, medical, consumer, and electronic component industries.

Since the advent of stereolithography, nearly forty different manufacturers have developed and introduced processes that are used to build parts by adding polymeric material one layer at a time on top of subsequent layers of material.

Stereolithography relies on the integration of three major systems:

- Computer-aided design (CAD) equipment and software.
- A photopolymeric elevator.
- Laser exposure and imaging system.

A proprietary computer software program is used to slice the CAD drawing of the part into thin cross-sectional layers, typically from 0.005″ to 0.020″ (0.127 mm to 0.508 mm) thick. The data is then sent from the computer to a laser, which draws each layer on the surface of the vat of liquid photopolymer. Where it is exposed to light from the laser, the polymer hardens. Another laser is used to ensure that the surface of the liquid is in the correct location.

After the thin layer is exposed and hardened, an elevator on the machine lowers the prototype slightly. This allows liquid polymer to flow over the part. The next layer is then exposed and hardened. The process continues until the part has been completed. The elevator platform raises the part out of the resin so that it can be removed from the vat and drained. Liquid that is still trapped in the interior of the workpiece is usually cured in a special oven.

Figure 25-7. Prototype parts developed by the stereolithography process provide outstanding detail and are dimensionally accurate. (3D Systems, Inc.)

Resin that is unexposed remains liquid and is kept in the vat for later use. The part is exposed to ultraviolet light again to harden any liquid that has been trapped in crevices in the part. The final design prototype is normally created in a next-to-finished form. Sometimes it is final finished by sanding, polishing, or painting.

The photopolymeric resin consists of two basic components. One is a *photoinitiator*, which absorbs laser energy and initiates polymerization. The other is an acrylic resin. The acrylic monomers and *oligomers* (polymers made up of two, three, or four monomer units) form solid polymers when exposed to light. The light, working in conjunction with the photoinitiator, causes the polymerizing reaction. This reaction stops abruptly with the removal of the light source.

Three-Dimensional Printing

Three-dimensional printing is a process that requires successive layers of powder and adhesive to build a functional part. Three-dimensional printing has already been used for medical applications ranging from bone reconstruction parts and scaffolds for tissue engineering to devices for delivering timed-release medications. The process is particularly useful for manufacturing tools and parts with greater complexity inside than out, such as intricate pathways or cooling channels that would be difficult to drill.

This process is one of the fastest of all of the rapid prototyping processes. Color can also be added. A disadvantage of the process is that part strength is limited, and final finishing is often required.

Like stereolithography, 3DP creates functional parts and tooling directly from CAD software. However, unlike stereolithography, it can create these parts from virtually any material including ceramics, metals, polymers, and composites. The 3DP process allows rapid fabrication of physical parts with any geometry.

The 3DP process works by spreading a thin layer of 25–200 microns of powder and then selectively depositing binder material from a printhead onto other layers of material. When all of the layers are printed and stacked on top of each other, the loose powder is removed and the final product is cured in an oven. Distribution of the powder is accomplished when a piston moving upward in the supply chamber moves powder to the distribution roller. The roller then distributes the powder at the top of the fabrication chamber.

Next, a jetting head deposits liquid adhesive onto the layer of the powder. The powder bonds in the areas where the adhesive is deposited to form a layer of the object. Once the layer is produced, the fabrication piston moves down as far as the thickness of the layer, and the cycle is repeated until the part is formed. When the final part is formed, it is raised in the chamber and the excess powder is brushed away leaving a *green,* or fragile, object. A 3-D printer is shown in **Figure 25-8**.

Let us take a closer look at how 3-D printer is used to create a large-scale 3-D prototype:

1. Powder housed in the printer's feed box covers the surface of the build piston.
2. The printer applies liquid binder onto the loose powder, forming the first cross section of the part. Where the binder is applied, the particles of powder are glued to each other.
3. After completion of the cross section of the part, the build piston is lowered and a new layer of powder is applied. This process is continued until the build is completed.
4. Loose powder is vacuumed away from the completed part.
5. The completed green part can be used as is, or infiltrated with wax or resin to improve its strength and durability.

Figure 25-8. Z810 3-D printer manufactured by Z Corporation. Physical models can be created in plaster or starch-based materials. (Z Corporation)

Figure 25-9. Companies and their applications of 3-D printing. (Z Corporation)

Company	Product	Application
Athletic Shoe Company	Footwear	Reduction of Product Development Time
Design Continuum	Industrial Design	Increase Use of Models for Design Analysis
Gio' Style	Housewares	Presentation of Design Concepts at Trade Shows
Eagle Precision Casting	Casting	Patterns for Investment Casting/Reduction of Production Time by 50%
Javelin	Animation	Acceleration of Sculpting Process
Medical Modeling Corporation	Medical	Parts to Surgeons for use in Surgical Planning
Orion	Instruments and Electronics	Working Prototypes
Wescast	Transportation	Presentation of Design Concepts to Customers

There are many interesting applications for 3-D printing. The table in **Figure 25-9** lists some current applications. The full-scale (18″ × 19″ × 13″) engine block shown in **Figure 25-10** was printed on a 3-D printer.

Selective Laser Sintering

Firms such as BMS, Boeing, NASA, Reebok, and many service bureaus worldwide are using *selective laser sintering (SLS)* to produce three-dimensional prototypes, parts, molds, tools, and patterns. The process is similar in concept to stereolithography, except that a powder is involved rather than a liquid polymer.

Application of the powder occurs in a fashion similar to the method used with 3-D printing. With SLS, the thermoplastic powder is applied and then tightly compacted. A piston moves upward incrementally to present the powder to the roller and apply the powder. Then, a CO_2 laser beam is projected onto the powder and traces the desired image. The powder is melted, layer by layer, by a computer-directed heat laser. Additional powder is deposited on top of each solidified layer and again sintered. The process is repeated until the part is created in the size desired.

The temperature in the fabrication chamber is kept just below the melting point of the powder so that only a small amount of heat from the laser is needed to cause sintering. The entire cycle is repeated until the object is fabricated. After the object is fully formed, the piston is raised to elevate the product out of the powder. Excess powder is brushed away and final manual finishing may be carried out.

A major advantage of SLS is that functional parts can be made from the material desired for the final product. Many different materials such as nylon, glass filled nylon, and polystyrene are used.

SLS allows for the most diversity in material selection. Materials are available that can mimic the qualities of rigid plastic or flexible rubber. Also available is a new polyamide nylon material called Duraform and copper Duraform for direct tooling. The SLS process provides the most functional rapid prototype available.

The disadvantages of the process is that it is more mechanically complex than most other rapid prototyping processes. Another disadvantage is that surface finishes and accuracy are not quite as good as with stereolithography. SLS has also been modified to provide direct fabrication of metal and ceramic workpieces.

Figure 25-10. Full-scale engine block (18″ × 19″ × 13″) printed on a 3-D printer. (Z Corporation).

Figure 25-11. Firms producing systems for rapid prototyping and creation of 3-D physical models.

Company	Location	Process
3D Systems	Valencia, CA	Stereolithography
Aaroflex, Inc.	Fairfax, Virginia.	Solid Image Stereolithography
Cubital	Raanana, Israel	Solid Image Stereolithography
Stratasys, Inc.	Eden Prairie, Minnesota	Fused Deposition Modeling
Solidscape Inc.	Merrimack, New Hampshire	Three-Dimensional Plotting
Soligen Inc.	Northridge, California	Direct Shell Production Casting
EOS GmbH	Munich, Germany	Laser Sintering
Solidimension, Ltd.	Rosh Ha-Ayin, Israel	3D Printers
Schroff Development Corporation	Mission, Kansas	Desktop Prototyping
Objet Geometries Ltd.	Rehovot, Israel	Solid Object Printer (Objet Quadra)
Z Corporation	Somerville, Massachusetts	3D Printer

There are many innovative firms producing equipment supporting processes to create physical models for prototypes, testing, and mold making. Some of these are listed in **Figure 25-11**.

Important Terms

actuator
additive
adhesion
adhesive
cohesion
compressive
filament winding
Hellerbond induction bonding
horn
hot melts
hot wire welding
induction welding
inserting
linear friction welding
magnetic induction bonding
nesting fixture
oligomers
orbital friction welding
photoinitiator
photopolymer
plastics welding
prototypes
radio-frequency (RF) heat sealing
reactive melts
reactives

sealing
selective laser sintering (SLS)
slitting
soaking method
solvents
solvent cements
spin welding
staking
stereolithography
subtractive
thermal heat sealing
thermal impulse heat sealing
three-dimensional printing
ultrasonic vibrations
vibration welding
welding

Questions for Review and Discussion

1. Why can thick plastics be heat sealed with the dielectric method more rapidly than thin plastics?

2. Describe what happens when plastics are welded.

3. Why can't the dielectric sealing process be used with polyethylene and polystyrene?

4. Describe the differences between ultrasonic welding and radio-frequency heat sealing.

5. How are vibration welding and friction welding accomplished?

26 Processes Used to Fabricate Wood Materials

Chapter Highlights

- Fabrication of wood materials may involve the use of interference fits, mechanical fasteners, or adhesives.
- Nails are classified primarily by length; screws and bolts by head shape, length, diameter, and in some cases, type of thread.
- Glued joints can fail through weakness of the adhesive itself or weakness of the substrate.
- Adhesives may be thermoplastic or thermoset in nature.
- New wood-joining technologies include the use of high-frequency gluing and very high-bond (VHB) tape.

Introduction

The process of fabricating or joining wood parts is accomplished by using one of three basic methods: *interference fits*, *mechanical fasteners*, and *adhesives*. Interference fits, using joints that lock together, are not widely used in manufactured wood products.

Interference fits are created when two mating patterns are constructed in the wood surfaces so that when joined, a joint is created that locks them together. Examples of interlocking wood joints are mortise and tenon, dovetail, T-slots, and tongue and groove. It is interesting to study some of the joints that were made by hand throughout history. The basic mortise and tenon joint, held together with pegs, was used in furniture and construction as far back as the 17th century. See **Figure 26-1**. More recently, interlocking joints were made famous in plastic toys by LEGO™ and are now used by many other toy manufacturers.

Figure 26-1. The mortise and tenon joint secured with pegs was used in the 17th century. T-slot, dovetail, and tongue and groove joints were used up until the early 1900s. Tongue and groove joints running the length of lumber are still popular for securing decking and other construction materials together today.

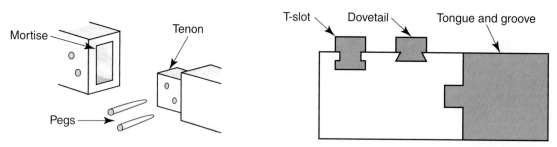

However, most fabrication today is not done using interference fits and interlocking joints. Wood parts are joined using mechanical fasteners or adhesives, which lend themselves more readily to volume production. In production, they are less labor intensive to apply and result in cost savings for assembly of the product.

There are many different types of fasteners used in industry and by the consumer. Some of these, such as screws or bolts and nuts, have threads. Other types use various mechanical methods to permit assembly (hook-and-loop materials, nails, blind rivets, hinges, and slides).

All fasteners must join materials together and carry a load. Some types of fasteners (such as bolts and nuts) can be assembled and then later disassembled, if desired. Other types of fasteners, such as blind rivets, cannot be disassembled after they are joined without damaging either the material being joined or the fastener itself.

Because of the enormous number of different types and applications of fasteners, many volumes would be necessary to describe all of them. For this reason, only the major types of fasteners and adhesives used with wood materials will be covered here.

Mechanical Fasteners

Screws and nails are the most common types of fasteners used to join wood parts. Selection of the correct fastener depends on the type of material being joined and the desired strength of the joint. Screws provide better holding power than nails and are easier to remove if disassembly is necessary. However, when screws are applied using hand tools, they take more time to apply than nails. Pneumatic drivers can also be used to drive both screws and nails.

One of the first places where design engineers seem to focus their attention is on the type of threaded fasteners that are to be used to assemble a product. Threaded fasteners are preferred in cases where interchangeability and large clamping force in a small area is necessary. Disassembly is a major advantage of using threaded fasteners—many products have to be made so that they can be disassembled easily.

Fasteners add to the cost of the product. Also, the time needed to assemble parts in a completed product must be calculated in the cost of the final product. Consequently, one way to save money is to reduce the number of fasteners in the product. Sometimes, using composite materials where several different parts can be produced as one facilitates this. In other cases, the product may be redesigned to reduce the need for fasteners. The designer must not sacrifice reliability when making these cost-saving changes.

Wood Screws

Selection of a particular type of wood screw as a fastener sometimes depends on the type of tool that is available to drive the screw. Screws are classified according to the type of opening in their head, **Figure 26-2**.

This opening (slot, hex, Phillips, square, etc.) naturally dictates the type of device that can be used to drive the screw. A *recessed hex* opening in a screw head has more contact with the driving tool and is desired in applications where there is heavy loading. The design of the head (flat, round, pan, hex, etc.) must also be matched to the job. **Figure 26-3** shows some of the most popular types of screw head designs.

Screws may be installed with a hand or powered screwdriver. Tools are also available which feed screws automatically. Attachments are available to

Figure 26-2. The design of the screw head selected may be influenced by the equipment available to install the fastener.

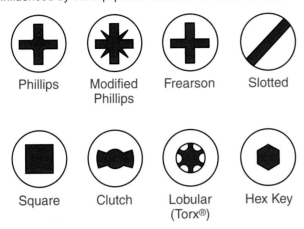

Phillips   Modified Phillips   Frearson   Slotted

Square   Clutch   Lobular (Torx®)   Hex Key

convert a conventional drill into a self-feeding power driver for inserting screws in a production environment. See **Figure 26-4**.

Another popular screw is the self-drilling Phillips-head screw designed for power driving in wood and other materials. Self-drilling screws can be turned into wood surfaces without the use of a pilot hole. They are frequently used in construction to apply drywall, decking, framing, and particleboard. See **Figure 26-5**. Note the different types of threads. The screw on the left has fine threads; all the others have coarse threads. Those intended for softer material (drywall or gypsum board) have threads the entire length of the shank; the screw for use in harder material (wood) has a shank that is only threaded for half its length. The design of the screw and screw threads influences the holding power. Fine threads, for example, provide greater holding power, because there are more threads per inch of the screw length.

Screws are also classified by their length, diameter, surface finish, and composition. Selection of the proper screw for the job starts with determining the length needed. The first consideration is the thickness of the materials to be joined. The screw has to be shorter than this dimension. If two pieces were to be joined, a good rule of thumb would be for the screw to go through the first piece and project at least half way into the second or base piece. When using partially threaded screws, the entire threaded length of the screw should enter the second piece of wood. This will ensure maximum holding strength.

Wood screws are typically purchased by the box. When the desired length is known, the next consideration is the *gage number*, or diameter, of the screw. Wood screws can be purchased in lengths from 1/4″ (6.3 mm) to 6″ (15.2 cm) and in gages from 0 to 24. The diameter of the screw increases with larger gage numbers.

After you know the length needed and the gage number, it is time to choose the head design and material desired. Let us take a moment to study a typical application, and see if we can apply this information effectively.

Suppose an inside building job requires flat-head screws to screw decking on a plywood subfloor. The length of the screws has to be 1 1/2″ (38 mm) for a secure installation. The order could be: "(5) boxes of 1 1/2 × No. 8 flat head steel screws," since this is an indoor project. If the installation was outside, you would most likely select brass, aluminum, stainless steel, or galvanized screws so that they would not rust.

Figure 26-4. This automatic-feed screw driving attachment can be used with a conventional portable electric drill to speed production. (Duo-Fast Corporation)

Figure 26-3. There are a limited number of major screw head styles. Selection of the style to use is normally determined by the application.

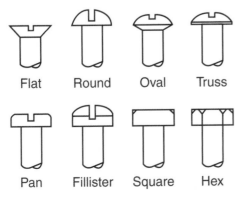

Flat   Round   Oval   Truss

Pan   Fillister   Square   Hex

Figure 26-5. A selection of Phillips-head screws designed to be driven by power equipment. (Duo-Fast Corporation)

Screw Specifications*

Stock No.	S-102	S-105	S-109G	S-108	S-112	S-110
Length	1¼"	1¼"	1¼"	1⅝"	1⅝"	2"
Thread Type	Fine	Coarse	Coarse	Coarse	Coarse	Coarse
Size No.	#6	#6	#6	#6	#6	#6
Head Type	Bugle	Bugle	Bugle	Bugle	Bugle	Bugle
Recess	Phillips	Phillips	Phillips	Phillips	Phillips	Phillips
Point	Pierc.	Pierc.	Pierc.	Pierc.	Pierc.	Pierc.
Materials	Steel	Steel	Steel	Steel	Steel	Steel
Finish	Black	Black	Zinc	Black	Black	Black
Per Box	6,300	6,300	6,300	5,400	5,400	4,500

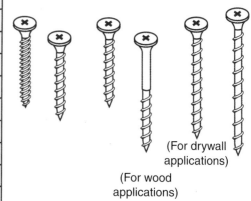

(For drywall applications)

(For wood applications)

*The standard screws listed above comply with ASTM Specification C1002.

When fastening hard wood with conventional screws, it is sometimes advisable to drill two holes. The first is called the *pilot hole*, or anchor hole. It should be slightly smaller in diameter than the largest diameter of the threads. The pilot hole is drilled through the first piece of stock and about halfway into the second. Next, the pilot hole through the first piece of stock is enlarged with a drill of the same diameter as the screw shank (the section above the threads). This hole is called the *shank hole*. The two holes will enable easy insertion of the screw through the first piece of stock and secure threading into the second.

When fastening soft woods together, the pilot holes can be a bit narrower than for hard woods. However, care must be taken to not over-drive (excessively tighten) the screw into the hole. This could destroy the wood in the thread area or split the stock. Proper drilling is even more critical with hard woods, which are more likely to split.

Nails

Nailed joints are often used in industry where the appearance of nails in the joint is not a detriment to the quality of the product. Typically, nails are used for products such as packing boxes, crates, pallets, and frames. Nails with small heads (finish nails) are used as fasteners in many fine furniture applications. These nails are recessed below the surface with a *nail set*, a type of punch. The holes are hidden by filling them with a material stained to match the surrounding wood.

Nails are normally driven by hand using a hammer, or with a powered automatic nailing machine. The air powered (pneumatic) nailer shown in **Figure 26-6** is an example of the type of nailer used on many construction sites today.

There are five major classes of nails: common, box, casing, finish, and brad. See **Figure 26-7**. Nails are typically available in lengths from 1" to 6" (2.5 cm to 15.2 cm), but longer ones are sometimes used. Nails are most often made from mild steel or, sometimes, aluminum. Some nails have a galvanized coating for exterior use. Others are coated with zinc, cement, or resin for special applications. Hardened nails are available for driving into concrete, cement block, and

Figure 26-6. Power nailers and staplers apply the correct amount of force to drive the fastener with minimal effort by the operator. Such devices make fastening work faster and more efficient.

Figure 26-7. These are the major types of nails used to fasten wood. Note the differences in head shape—the brad, casing, and finish nails have small heads, designed to be driven below the wood surface for a more finished appearance.

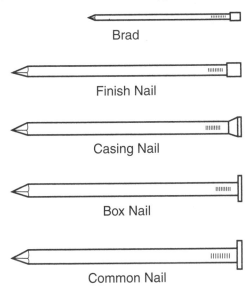

Brad

Finish Nail

Casing Nail

Box Nail

Common Nail

brick. Nails are even made with threaded shanks for special holding applications, such as flooring underlayment or drywall.

The basic unit of classification for nails is called the *penny*, (abbreviated *d*). A 6 penny (6d) nail is 2″ (5.8 cm) long. A 2 penny (2d) nail is 1″ (2.5 cm) long. All types of nails are sold by the pound.

Box nails smaller than 1″ (2.5 cm) in length, and all brads, are sized by their length and gage number rather than a penny designation. These small nails are available in gages from 12 to 20, with the smallest number being the largest diameter.

A rule of thumb for selecting nails to join wood is that the nail should extend into the second sheet for a distance equal to twice the thickness of the first sheet. Larger heads should be chosen for applications in softwoods or drywall.

Common nails are driven with a claw hammer or power nailer. Brads are normally driven using a smaller tack hammer. The power nailer feeds nails in clips, and is often used for construction tasks such as nailing framing members. A similar power tool is used to drive roofing nails into shingles.

Blind Rivets

Blind riveting is a unique process that can be used to fasten just about any type of materials. The *blind rivet* is also known as a *pin rivet*. It is often used

in areas where there is not sufficient room to make use of a conventional rivet that must be compressed by pounding or hammering.

The blind rivet consists of a *mandrel* (shaft) and rivet head assembled as one unit. The rivet is placed in the hole drilled through the materials to be joined. The rivet is clinched by using a rivet tool to pull the mandrel, forming a head on the backside of the material. The mandrel can then be snapped off for a finished appearance. See **Figure 26-8**.

Other Mechanical Fasteners

In addition to nails, screws, and blind rivets, other mechanical fasteners such as wood-biting tee nuts and standard nuts and bolts are popular for fabricating wood materials. Special fasteners with threads on both ends are also used to join wood parts. This type of fastener has machine threads for accepting a nut on one end and wood screw threads on the other so that it can be screwed into the part. Threaded inserts are also available. These require press-fitting a plug with internal threads into a hole in the workpiece. Machine screws are screwed into these threads.

Adhesives

Sometimes, the application is such that the pieces to be joined need to be joined permanently without any exposed hardware. For these applications, fabrication is not done using mechanical fasteners. With the appropriate adhesive and proper surface preparation,

Figure 26-8. Blind rivets are useful in many fastening situations. After being inserted through the materials to be fastened, left, the mandrel is pulled outward by the rivet tool. This forms the second head on the rivet, right. The mandrel can then be snapped off. (Malco Products, Inc.)

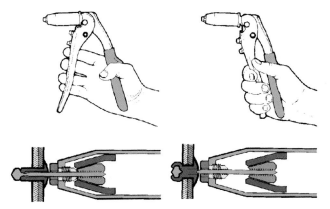

it is actually possible to create a stronger bond than would be possible using a mechanical fastener.

An adhesive is a material that adheres, or sticks to, another material. To better understand what can be accomplished with adhesives, think for a moment about the basic structure of wood. With the aid of a powerful microscope, the wood surface appears much like a sponge cut in crosssection.

When you use a mechanical fastener, the raised surfaces of the wood provide secure contact when held against each other. However, the open pores provide little support to the interface area between the fabricated materials. Since adhesives are normally used in liquid form, they not only coat the raised surfaces but also flow into the depressions. This enables adhesives to provide a more complete bonding surface. However, there is much more to the process of fabrication than just bonding materials in the area where they come into contact with each other.

The linkage between surfaces in the interface area is only one type of bonding that occurs when materials are fabricated. This is actually the poorest type of bonding—mechanical adhesion through the film of glue accounts for only about 10 to 20 percent of the holding force of most glued materials. Most of the bonding strength occurs as a result of chemical reactions between the adhesive and the wood.

Gluing failure in the interface area is referred to as an *adhesive failure*. If the failure occurs in the adhesive itself, it is called a *cohesive failure*. The bond could also fail because of substrate failure. *Substrate failure* happens when both the cohesive strength and the strength of the adhesive bond between the adhesive and substrate exceeds the strength of the substrate. Substrate failure results in the fracture, or destruction, of the substrate.

There are many ways to classify the available types of adhesives. One of the most useful ways is to view them in terms of the type of reaction they exhibit during the bonding process. This is what makes certain types of adhesives stick to some materials and not to others. There are three types of chemical actions through which an adhesive *sets* (becomes hard): through the cooling of a molten liquid, by releasing a solvent, or by polymerization.

Hot Melt Adhesives

An adhesive that must be heated to make it liquid and return to a solid state as it cools is called a *hot melt adhesive*. This adhesive, often referred to as a "hot melt," is a thermoplastic solid that can be purchased in stick or pellet form. Hot melts are applied with a special applicator known as a hot melt glue gun.

Hot melt applications are completed by applying the melted glue on one surface and placing that surface in contact with the other surface to be bonded. An example of this type of application is the gluing of one flap of a box to another flap.

Hot melt bonding normally takes several seconds to complete. In some cases, the material is held in a fixture to provide proper contact during cooling.

Hot melts are popular for edge-banding laminates in furniture construction and in all sorts of packaging applications. Hot melt adhesives are clean, easy to use, and can be used to fabricate almost any material.

The major disadvantage of hot melts is that they are not as strong as many other adhesives. However, for packaging and other sealing applications, this type of adhesive is normally the preferred choice.

Solvent-Releasing Adhesives

Solvent-releasing adhesives give off a liquid as they cure. The liquid may be water or a solvent that is compatible with solid ingredients in the adhesive.

Solvent-releasing adhesives include polyvinyl, hide, and casein glues. Solvent-releasing adhesives are used with porous woods, which will absorb the solvent. Polyvinyl (white glue), aliphatic thermoplastic resin (yellow glue), and alpha cyanoacrylate (instant adhesive) adhesives are all solvent-releasing thermoplastic adhesives.

Other adhesives are designed to withstand a great deal of stretching. These are referred to as elastomers. *Elastomers* are solvent-releasing adhesives that are used in applications where low creep resistance is required, such as in the bonding of drywall to other materials, or in attaching flooring and tile. They are often referred to as *construction adhesives*. See **Figure 26-9**.

The major component in an elastomer is natural or synthetic rubber. Rubber-based adhesives do not cure in the same way as the ureas or phenolics. With elastomers, the adhesive thickens with the loss of solvent, until it reaches a plastic state. Elastomers never become completely hard. Consequently, an elastomeric bond is not rigid and will shift under stress.

Thermoplastic adhesives can be softened by heat and are not recommended for use in high-temperature

Figure 26-9. Construction adhesives are elastomer-based solvent-releasing adhesives that are widely used for fastening wood to various substrates. (Macco Adhesives)

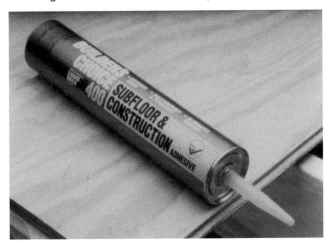

applications. All thermoplastic adhesives acquire tack, or stickiness, when they begin to dry.

Of the three popular types of thermoplastic adhesives, aliphatic thermoplastic glue is stronger and more resistant to heat than polyvinyl glue. Alpha cyanoacrylate is stronger than either polyvinyl or aliphatic glues, and can be used to bond both porous and nonporous materials. Cyanoacrylate glue dries hard in about 24 hours, but it can be reduced to a plastic consistency through the application of acetone.

Thermoset Adhesives

One of the disadvantages of thermoplastic solvent-releasing adhesives is that they often break down in hot or wet conditions. For this reason, thermoset adhesives are often selected when the product application requires exposure to moisture or heat. These adhesives will not soften through exposure to heat. Curing takes place through polymerization. Once the glue has hardened, the joint is strong but brittle.

Thermoset adhesives set when a chemical reaction (polymerization) takes place between the two parts of the adhesive mixture. The ingredients vary, depending on the type of thermoset adhesive. Common types include epoxy, resorcinol formaldehyde, phenol formaldehyde, melamine formaldehyde, and urea formaldehyde.

High-Frequency Electric Gluing

Thermosetting resin glues can be cured through direct contact with a heat source, or by exposing the glued material to a high-frequency electric field. *Electronic gluing*, or electric welding as it is frequently called, makes it possible to create a permanent bond in seconds. Here is how it works:

When materials to be joined are placed in an electric field, a portion of the energy that is generated is changed into heat, thereby increasing the temperature of the exposed material. Increases in electrical frequency result in increased temperature in the material. Frequencies similar to that required for shortwave radio—between 10 and 40 million cycles per second—are used for high-frequency gluing. Household electricity (alternating current) has a frequency of 60 cycles per second.

In high-frequency gluing, most of the heat is generated in the glue area; the wood itself is not heated. Electrodes are placed parallel to the glue line for plywood and perpendicular to the glue line for edge gluing. With high-frequency gluing, the curing time can be much shorter than would be required with contact bonding in a conventional heat press.

Many different types of fabricating can be accomplished with high-frequency gluing. A portable electronic welder is shown in **Figure 26-10**. Welders such as this are being used daily for applications such as veneer paneling; applying laminates and banding to cabinets, furniture, sash, and doors; patternmaking; patching veneers; and bonding plastics to countertops.

A handheld electrode gun is shown in the foreground of Figure 26-10. This gun weighs only a little more than two pounds. All that needs to be done to produce a bond in a glue joint is to place the electrodes on the joint and pull the trigger of the gun. It produces high-frequency energy that penetrates up to 2″ (5 cm) of wood.

Portable machines such as the one shown are popular for small applications. In production, high-frequency batch presses are used for edge gluing. Production machines feed stock into a conveyorized glue spreader. After glue is applied and the pieces are sandwiched together, they are moved into the high-frequency welding machine, which produces the heat to set the glue permanently.

One of the newest additions to the fastening field is *very high bond (VHB)* tape. The new VHB adhesives permit bonding to woods, metals, fiberglass, sealed wood, glass, stone, and plastics. VHB materials can be purchased in either conventional tape roll form or die-cut shapes for specialized applications.

Figure 26-10. High-frequency electronic welders generate waves that can set glues in a matter of seconds. (Workrite, Inc.)

Some VHB material can withstand temperatures up to 500°F (260°C) for limited periods of time, or 300°F (149°C) for sustained periods. VHB joining systems use either adhesive transfer tape with a release liner or an acrylic foam carrier with adhesive on both sides. Both tape and foam use an acrylic polymer adhesive that flows across the surface to which it is attached.

With VHB material, no curing period is necessary. Generally, the tape reaches maximum holding strength in about 72 hours. Residential window manufacturers are finding the VHB tape useful in producing single-unit windows with attached grilles simulating multipane units. The tape attaches the grilles permanently to the glass pane.

With VHB assembly, no drilling, welding or refinishing is required. Although fasteners using VHB tape are not yet as permanent as mechanical fasteners, the technology is rapidly improving. A manufacturing test of various fasteners by a truck manufacturing firm found that VHB fasteners could be more permanent than mechanical fasteners when subjected to certain stressful conditions.

Removable Adhesives

Research is being conducted on the design and application of removable adhesives. Researchers have recognized that in the past, the only requirement for adhesives was that they held parts together. However, as products age and our philosophy toward waste and disposable products changes, there is often a need to disassemble products and replace worn parts.

The conventional way to remove adhesive has been to heat the adhesive to a point just above its melting point, and then pry the materials apart. Sometimes this resulted in damage to the part. Researchers have invented a new approach relying on the use of a reversible chemistry that breaks apart at elevated temperature. When the epoxy is heated, it becomes a liquid, losing its bonding capacity. Then when the temperature is lowered, it regains its ability to bond the materials together.

Important Terms

adhesive failure	penny
blind rivet	pilot hole
cohesive failure	recessed hex
elastomers	shank hole
electronic gluing	solvent-releasing
gage number	adhesives
hot melt adhesive	substrate failure
interference fits	very high bond
mandrel	(VHB) tape
nail set	

Questions for Review and Discussion

1. Discuss the advantages and disadvantages of each major type of mechanical fastener.

2. Why is a pilot hole recommended when drilling hardwood?

3. Which is larger, a 6d nail or a 3d nail?

4. What are hot-melt adhesives made from? How are hot-melts applied?

Processes Used to Fabricate Ceramic Materials

27

Chapter Highlights

- Ceramic materials often must be joined to other materials with joints that are not only mechanically strong, but airtight, watertight, or both.
- Metallizing techniques are used to provide electrically conductive pathways on ceramic substrates.
- Glass-to-glass and metal-to-glass bonds usually involve heat application to form fusion seals.
- Metal coatings are applied to ceramics for decorative, electrical, and sealing purposes.
- Refractory ceramic materials, in forms ranging from bricks to spray coatings, are used to line industrial furnaces and for other high-temperature applications.

Introduction

Fabricating ceramic materials often requires joining the ceramic part to a part made from another material, especially metal. When ceramics are effectively used in conjunction with other materials, they can provide additional characteristics that expand the usefulness and performance of a product. However, the joint and point of interface between the materials must be properly prepared and appropriate processes must be used in order for the joint to be not only mechanically strong, but airtight and watertight. The connection between ceramic parts is normally microscopic (occupying little space compared to conventional weldments on fabricated metal parts). Welding and brazing techniques used with metals are normally appropriate for joining advanced ceramic materials, such as alumina and silicon nitride, to metals. These are the types of materials used in electronic devices, aerospace components, and cutting tools. They are also used to make nuclear fuel rods, oil-less bearings, body armor, abrasives, and thermal liners for high temperature ovens.

Most ceramics are recognized for their high hardness, high compressive strength, and low thermal and electrical conductivity. However, some specific ceramics, such as diamonds, beryllia, and silicon carbide, have a higher thermal conductivity than copper or aluminum. This is why they are good candidates for superconductors.

The processes being used to join advanced ceramics to other materials are still in an early stage of development compared to the much more expansive array of processes that have been used for centuries to join metals and other materials. Ceramic fabrication is an area where much potential for growth lies in the decades ahead.

There are two major types of ceramic fabrication processes: *direct bonding* and *mechanical attachment*. Adhesives, brazing, glass sealing, and electrostatic bonding are all examples of direct bonding fabricating processes. Other processes, such as shrink fit, and the use of bolts, nuts, and screws, are examples of processes used to join materials through mechanical attachment.

Selection of the appropriate type, either direct bonding or mechanical attachment, and the correct process, is dependent on several fitness-for-purpose criteria:

- Temperature limits
- Protection of the product from abuse (important factors may include strength, stress on joint, electrical insulation, resistance to wear, and impact damage)
- Type of materials to be fabricated
- Required level of moisture and corrosion protection for joint (*hermeticity*)
- Design of the part to be fabricated
- Life cycle for part
- Cost

From a process selection standpoint, three of the criteria listed above are most important. These are *cost of the product, type of materials to be fabricated,* and the *temperature of the environment* where the product will be used. There is no need to manufacture a product made of materials that cannot be purchased or a product that cannot be sold. Each combination of materials provides its own set of benefits and limitations related to the performance of the final product, and the selection of processes to make it. But neither of these factors is as important as the temperature extremes that the product must survive as it performs through its life cycle. Each material has its own limits.

Direct Bonding Processes

In some cases, direct bonding processes are used to fasten different types of materials together, such as a metal to a ceramic. Sometimes, a direct bond is applied to protect or seal the seam where the materials are joined to protect the part from contamination, or keep out air or moisture. Often, the purpose for bonding is to provide a method for joining decorative material to a different type of material. Direct bonding is preferred when the parts will never have to be taken apart. Let us take a closer look at some of these processes.

In addition to creating a decorative effect, such as gold or platinum designs on dinnerware, metal coatings are often applied to the surface of ceramic products to perform special technical functions. For example, *metallizing* is often used to provide a transmission path on ceramics for electrical current on electrical devices and components. In electronic applications, the metals silver, molybdenum, and palladium are often preferred because of their superior conductivity.

Glass surfaces can also be metallized to form a base for joining different materials together. Joining is done by coating the ceramic substrate with metal and then soldering or brazing the desired metal to the metallized surface. This is a process that is often used to construct specialized glassware for medical and technical applications.

When pieces of glass are to be joined to each other, one of two techniques is commonly used. Sometimes, *glass seals* are created by melting and fusing the two pieces of glass together. In other applications, powdered glass solder is used to form the joint. The use of glass solder is described in more detail later in this chapter.

Spray metallizing

In addition to metallizing through the use of solder, ceramic parts can also be coated by using such methods as plasma arc spraying or oxyacetylene spraying. Metallizing can be done on cold or hot ceramic and glass surfaces.

When spraying is the desired mode of application, the metal that is to be applied is fed in wire form through a special torch or spray gun. The metal wire has a low melting point. As it is fed through the oxyacetylene flame, it is melted by the flame and atomized by compressed air. The fine droplets of molten metal are blown out the nozzle of the spray gun onto the substrate.

One of the most sophisticated methods, *plasma arc spraying*, uses temperatures as high as 30,000°F (16,650°C). This permits the spraying of materials with high melting points.

In addition to metals, other types of materials (such as cermets, carbides, ceramics, and plastic-based composite powders) are sprayed onto various substrates. Ceramics, for example, can be sprayed at high temperatures, using high-velocity oxyacetylene flames. This process is sometimes used to apply porous protective ceramic coatings to metals.

Other metallizing processes

Coatings are also applied through other processes, such as dipping or powder coating. When the application involves dipping or coating, the material is normally applied to a heated substrate. For permanence, it is then fired on the glass or ceramic part at high temperatures. *Hermetic seals* are painted on by hand or sprayed on to prevent movement of gases or liquids into or out of a product. Many medical products require hermetic seals.

The search for processes and materials to produce moisture-proof seals grew out of needs from the military to guard against system failures resulting from moisture in both electronic and mechanical systems. Moisture can result in corrosion, dielectric breakdown, and loss of insulation resistance between conductors.

In the medical industry, many different types of joining processes are used to produce hermetic seals for a wide range of applications. Seals on implantable devices often involve materials such as nitinol, platinum, or stainless steels. Ceramics and plastics such as polyurethanes, silicones, and perfluorinated polymers are also used. Fabrication processes for implanted devices vary from the use of adhesives and encapsulating (coating) materials to laser-beam welding, reflow soldering, or diffusion bonding.

Metallizing can also be done by mixing metal powders, organic solvents, and binders, then spraying or painting the material onto the ceramic part. Some manufacturers then fire the coating in a kiln containing a *cracked ammonia atmosphere* (hydrogen plus nitrogen). This improves the quality of the coating.

The process of metallizing is used on a variety of different products. Electrical meters are sprayed with brass. The brass is then soldered to a metal bracket to produce a moisture-proof seal. Medical encapsulating tubes, glass-to-metal seals, and instrument windows are also produced with fired-on metallized coatings.

The techniques for producing *glass-to-ceramic seals* requires heating the glass and ceramic together until they are hot enough to melt the glass solder applied to the area to be sealed. Glass-to-ceramic seals are used to bond unlike materials in products such as spark plugs and electronic components.

Fusion Sealing

Many electronic products require pressure-tight seals between metal and ceramic materials. The bonding of metal to some crystalline ceramics involves the use of a molybdenum-manganese layer. When fired under partially oxidizing conditions, the molybdenum-manganese layer creates an oxide that reacts with the ceramic material to produce an adhesive bonding layer. When metals are to be soldered to the sealed area, the part is plated with nickel, silver, or gold before soldering.

When a metallic layer is applied to a ceramic material, there are two different levels of thermal expansion operating against each other. Normally, metals have a higher coefficient of thermal expansion than ceramics. This means that, to avoid problems caused by differential expansion, thin applications of metal with small surface areas are desirable. Otherwise, excessive stresses could be created in the ceramic material.

The *fusion sealing method* uses special glass compositions to make direct glass-to-metal seals in an oxidizing environment. The molten glass combines with a previously formed oxide on the metal to produce a fairly strong bond. The critical sealing temperature required must take into consideration the expansion and contraction requirements of both materials. If this is not done, stresses may break the seal.

An alternative to this process is the *compression seal*, where parts are pressed into their position in the completed assembly. A requirement for creating a compression seal is that the coefficient of thermal expansion (CFE) of the outer member of the part is greater than that of the inner member providing the contact or termination. The outer member is mechanically compressed against the inner member, which is often an insulator, and the contact. Typical product applications for compression seals are electrical components and devices where small leads must join one point on the part to another, such as power surface mounts, and microelectronic packages. The problem with this seal is that such leads are sensitive to vibration.

Robocasting

Robocasting requires no molds or machining, and relies on robotics for computer-controlled deposition of ceramic slurry through a syringe. The slurry, consisting of ceramic powder, water, and small amounts of chemical modifiers, is deposited in thin layers onto a heated base.

Robocasting permits construction of a dense ceramic part in less than twenty-four hours. This makes it a prime candidate for rapid prototyping. More conventional ceramic fabrication processes, such as dry pressing and gel casting, can take weeks to move from the design phase to finished part. However, robocasting is still in the early stages of development. More research is needed to validate its usefulness in the marketplace. See **Figure 27-1**.

Encapsulation

In an earlier chapter, the process of encapsulation as used with polymeric/plastic materials was discussed. *Encapsulation* is also an important process in the manufacture of ceramic capacitors.

Figure 27-1. Robocasting is a new ceramic fabrication process that squeezes slurry from a robot onto a heated substrate. (Sandia National Laboratories)

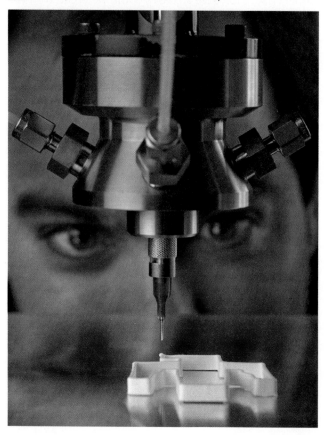

Such capacitors deteriorate when they are exposed to moisture, so encapsulation provides a functional method for sealing them. Small capacitors that form a part of a hybrid circuit can be sealed into ceramic containers, along with other components of the circuit assembly. The leads are typically coated with a polymer. Encapsulation is done by dipping in liquid polymer, by injection molding, or by immersing the assembly in a fluidized bed of molten polymer. Epoxy and thermosetting resins are used for dipping.

Laminating

Laminating is a process that is often used to sandwich polymeric and composite materials together. Laminating is a process that is also important in the ceramic/glass industry. One of the most common glass products produced using lamination is the automobile windshield. Lamination can be viewed as both a forming and a fabricating process. Here is how it works when used to join glass:

First, two pieces of curved glass for the windshield are formed by heating two sheets of glass in a stainless steel mold. After the glass melts and droops into the mold, it is annealed to relieve stresses and cooled. A polyvinyl butryl plastic interlayer is placed between the two pieces, and the sandwich is pressed together in an autoclave furnace to bond under heat and pressure. The vinyl interlayer serves two essential functions: it holds the glass together, and it keeps the glass from shattering on impact. See **Figure 27-2**.

Figure 27-2. A sandwich of two layers of glass with a filling of polyvinyl butryl plastic is being made by these workers in a plant that manufactures automobile windshields. Heat and pressure will be used to permanently bond the layers together. (PPG Industries)

Theoretically, there is no limit to the number of layers of glass that can be fused together. In practice, however, laminating normally involves two sheets of glass. Laminated glass is permanently bonded.

Laminating is also used to join two or more layers of glass with metals or other types of materials. In **Figure 27-3**, an associate at an architectural glass assembly plant inspects sealant being applied to spacers that will be sandwiched between two pieces of glass. The sandwich will form an energy efficient window for a commercial building.

Glass Soldering

In some applications, it is desirable to seal glass together in such a way that the parts can be later separated and repaired. One such application is in the manufacture of a television picture tube. *Glass solder* is used to join the faceplate of the tube to the rear funnel section. The solder enables the installation of new parts without destroying the tube. Several different types of glass solder—vitreous, devitreous, and conductive—are available. *Vitreous glass solder* is a low-melting-point glass that can be applied to glass with a higher melting point to create a seal. *Devitreous glass solder* is much like a thermosetting plastic. It can be reheated without becoming highly liquid and flowing freely. *Conductive glass solder* is used to provide both a seal and an electrical path.

Figure 27-3. An energy efficient window for a commercial building will result from sandwiching these aluminum spacers between sheets of glass. The worker shown is inspecting sealant application on one of the spacers. (PPG Industries)

Refractory Materials

Refractory ceramics are nonmetallic ceramic-like materials that remain solid and intact when they are subjected to high temperatures. Refractory products also are capable of withstanding corrosive conditions in high-temperature environments. Typical applications for refractories are industrial furnace liners, crucibles, and similar high-temperature areas.

Refractories are made from many different types of ceramic materials including high alumina fire clays, alumina, magnesia, and silica. There are many other nonceramic materials that are used in the manufacture of refractories. These include chrome and chrome-magnesite, carbon, and pure oxide refractories. Even a mixture of sawdust, coke, and silica sand has been heated in an electric furnace to produce refractory materials. The silica becomes silicon and combines with the carbon to produce *silicon carbide*. The sawdust burns out, leaving pores in the material for the escape of gases. Because of its hardness, silicon carbide is a popular abrasive used in grinding wheels.

The most common material for making refractory products is clay. Pure fired *kaolin clay*, which is about 46 percent alumina and 54 percent silica, has a melting point of over 3000°F (1700°C).

Other materials, such as high-melting-point alumina, are used in applications requiring even higher temperatures. Alumina has a melting point of 3722°F (2050°C). For many years, the basic refractory materials used by the steel industry have been magnesium oxide (magnesia), chromium oxide, and lime.

Refractory products are installed as linings for industrial furnaces or in similar high-temperature environments. The metal casting industry is the largest user of refractories. Refractories are also used for many special applications such as rocket launching pads, nuclear reactors, and in the technologies involved in processing fertilizers, refining oil, and manufacturing petrochemicals.

Most furnaces use refractory bricks that are produced by dry pressing. *Fire clay* brick is the most commonly used, comprising about 75 percent of all refractories produced in the United States.

Refractory Spraying

Refractory ceramic materials, in the form of a liquid slurry, can be sprayed onto the inside of an open-hearth furnace or boiler firebox. *Refractory slurry* can also be used as mortar in joints between bricks in a furnace lining and to join other types of materials that must withstand exposure to high-temperature environments.

Refractory ceramic coatings can be applied to almost any base material by using the process of metallizing. As noted earlier, coatings are applied by dispersing them through a flame and blowing the atomized material at high speed onto the receiving surface. Ceramic coatings are popular for application to metal components that must withstand extreme environmental conditions such as acids, corrosion, abrasion, erosion, and oxidation.

Adhesive Joining

Adhesive joining materials are used for low-temperature applications where a vacuum or pressure is not required to compress the materials together. Adhesion creates a low-stress joint. Adhesive joining is a critical assembly technique for many products ranging from electrical circuits and miniature components to multimillion pound aircraft. Effective bonding, particularly when it is applied in high volume automated production, requires a good understanding of the materials and processes involved.

There are hundreds of different types of adhesive joining materials, products, and processes. Examples include pressure sensitive labels, electron beam curable resin, electrically conductive adhesive materials, decals, hot melt adhesives, and instant glue.

Low Temperature Bonding

An interesting active soldering process was developed to join metals, ceramics, and composites in air and without flux at temperatures less than 750°F (400°C). This low temperature bonding process adds reactive elements, such as titanium and other active elements, to conventional solder-based alloy bases. The active elements travel to the joint interface and react with the joint surface compounds. The alloys are then heated to a molten state. At this point the alloys become active (that is, the titanium is

allowed to react with the joint metal surfaces and scour its protective oxide layer). This permits the active elements to diffuse into the surface of the two opposing joint materials to form a metallurgical bond.

Mechanical Attachment Processes

Mechanical attachment of one or more materials is often used when the product being manufactured may eventually need to be disassembled. An example is on a musical instrument with a vacuum unit or blower that may need periodic servicing. Many parts on automobiles have seals enabling removal of one part connected to another. In the field of ceramics, seals are frequently used to package fragile components. It is important when working with ceramics not to apply too much pressure when joining parts together. However, the business of fastening ceramics is a lot different than just using bolts, nuts, and screws.

Mechanical Fasteners

The use of mechanical fasteners such as pins, bolts and nuts, and rivets to join ceramic materials to rigid surfaces should be avoided. Localized fractures can be created at the point of contact where the fastener touches the material. When mechanical fastening is required, it is recommended to use soft gaskets made of solid materials (soft nylon, plastic, or metal).

Shrink Fit

Mechanical fabrication of ceramic and glass workpieces is feasible using shrink fit processes due to the high compressive strength and low thermal expansion of ceramics relative to metals. A *shrink fit* is an interference fit created when the outside of one part to be joined to another is heated and the inside part is kept cool for ease of assembly. One example is where the diameter of a hole is smaller than the outer diameter of a metal rod. The ceramic part is heated and the cool rod is inserted. As the ceramic part cools, the hole shrinks over the rod for a tight fit.

Important Terms

compression seal
conductive glass solder
cracked ammonia atmosphere
devitreous glass solder
direct bonding
encapsulation
fire clay
fusion sealing method
glass seals
glass solder
glass-to-ceramic seals
hermetic seals
kaolin clay
laminating
mechanical attachment
metallizing
plasma arc spraying
refractory
refractory slurry
robocasting
shrink fit
silicon carbide
vitreous glass solder

Questions for Review and Discussion

1. Explain why and how sawdust is used in the manufacture of refractory materials.

2. Research and describe several methods used to apply metallic coatings to ceramic workpieces.

3. What is the difference between fusion sealing and the conventional lamination process used to seal vehicle windshield glass?

4. Prepare a briefing explaining different ways that refractory materials could be applied to houses in remote areas where forest fires are a major concern.

Ford and Oxford University are developing a system to create moulds for manufacturing tools and dies by spraying molten metal over a ceramic cast. (Ford)

28

Processes Used to Fabricate Composite Materials

Chapter Highlights

- Cohesive fabrication is the method used for forming a composite in an open mold.
- Secondary bonding is joining together two or more previously cured parts.
- The adhesion process is frequently used in place of cohesive methods to join two parts.

Introduction

In many cases, it is difficult to separate the creation of a composite from its fabrication. People who form parts from metals seldom think about the process that made the material they are stamping or machining. The composite fabricator, however, is involved in creating the composite and forming it at the same time. The successful fabrication of composite components relies largely on the materials selected and the chemistry that creates the matrix binding these materials. In fact, the distinction between the activities of creating, forming, and fabricating composites is easily blurred. To get a clearer understanding of how these processes relate to each other, you may want to review the chapters on *Composites* and *Forming Composites Materials* before you read this chapter.

Fabrication of Composites

The most basic process in the fabrication of composites is *cohesive bonding*. This is a method that permanently joins materials through chemical action. Cohesive bonding relies on the capacity of a substance to adhere to itself. A further property of cohesive bonding is the force that holds the substance together. This *cohesive force* is determined by the strength of attraction between like molecules, called *intermolecular attraction*. Obviously, the magnitude of the strength of attraction will be a determinate on the material selected to form a cohesive bond. As an example, damp sand (a composite of silica and water) at the beach can be formed into an elaborate shape, and one might say that it is the result of cohesive bonding. In fact, the water molecules on the sand granules do provide sufficient cohesion to allow the sand to be molded. However, the cohesive force of the water molecules is very weak and the sand structure will fail easily.

It is also important to recognize the difference between filler materials and the matrix that forms the bond. The sand is a filler material and the water is the matrix in this example of a composite. It is the matrix material that creates the cohesive bonds in a composite. The chapter on composites explains that the matrix, which is frequently a resin, performs two functions: bind the composite together and transfer the load to the reinforcement material or filler. Most popular matrix materials in use today are plastics, which fall into one of two groups or classifications: thermosets and thermoplastics. These plastics are able to create a strong cohesive bond and adhere to the reinforcement and filler material.

The ability of a matrix resin to adhere to a reinforcement and filler materials is essential for a composite to perform satisfactorily. In most cases, the matrix not only adheres to the reinforcement or filler material, but also envelops it. In a few instances, the formation of a composite completes the construction of the product. Simple shapes, such as fiberglass bathtubs and storage containers, are examples of composites that can be completed through the formation process. Boat hulls are also fabricated using a cohesive forming process. See **Figure 28-1**. And, as long as the laminate has not cured, additional items can be bonded to it using the same resin that is used to create the composite matrix. However, there are many instances when large composite structures, such as large boats or truck bodies, will have cured but other composite components must still be attached. You may recall that a part is cured when the matrix resin has completed its chemical reaction and is in its hardened state.

The *adhesion process* is one method used to join cured composites together. This is a distinctly different process from cohesive bonding. In composite fabrication, it is also called *secondary bonding*. The procedure can be thought of as gluing parts together. Adhesive materials may be glues, cements, or the same resin used to create the composite matrix. Most fabricators find that creating satisfactory secondary bonds are one of the most difficult aspects of composite fabrication.

The start of a secondary bonding or adhesion process begins by preparing the surfaces to ensure that they are clean, dry, and free of dust and wax. Surprisingly, wax is an ingredient in some matrix resins. This wax floats to the surface to act as a barrier to promote curing or to suppress styrene emissions. It is essential to remove the wax from the surfaces to be joined. The final adhesion or bond strength will depend as much on the surface preparation as the bonding material.

Immediately after cleaning, the adhesive can be applied and the components can be placed together. The area of attachment on each component is called the substrate. There are many conflicting theories to explain the source and strength of adhesive bonds, but all explanations agree that the adhesive and the substrate must be in intimate contact and that the adhesive must be able to wet the substrates. At the time of application, the adhesive will be a fluid. Once the two surfaces being joined are placed together, they must be held in place and supported if necessary. This is to prevent any stress on the substrates and adhesive until the adhesive has hardened.

The strength of the joint will depend on whether the hardening requirements for the adhesive have been met, the cleanliness of the surfaces, the size of the area of attachment, and the configuration of the joint. These factors will determine if the joint strength matches or exceeds the cohesive strength of the composite components.

Some composite parts are joined with friction fits, while others are joined with *mechanical fasteners*. Mechanical fasteners are quite common in joining components to composite structures. In general, many smaller fasteners will perform better than a few larger fasteners. The same concerns about equalizing stress along the joint surface apply to mechanical fasteners.

Figure 28-1. Open molding is one of the most common methods for forming a composite. Recreational boats are created in this manner.

If mechanical fasteners are used, often a *sealant* must be placed along the joint to make the assembly dust, water, or gas tight. Sealants are often confused with adhesives because they share some of the same characteristics. Many of the sealants used with composites cure to a tough, flexible, rubbery consistency. These sealants are normally a paste rather than a liquid, and often need to be worked onto the substrate so that they are able to wet the surface in order to adhere to it. Unlike resins or glues, sealants should remain flexible so they can withstand some movement of the substrate without cracking or losing adhesion. This is the one major difference between sealants and adhesives; allowing limited movement.

One of the advantages of using composites, however, is that mechanical fasteners can often be designed out of the product. Usually the principle benefit for using mechanical fasteners with or without a sealant is being able to take the joint apart.

However, if the composite structure does not need to come apart, then it would be more economical to eliminate the fasteners and permanently adhere the two components. Generally, designing out the fasteners by adhering the components will improve the strength while reducing the cost of the assembly.

Attaching Components to Composite Matrices

The technology for joining wood or metal workpieces is fairly well-established. It may include using fasteners such as bolts, screws, or rivets. Other methods, such as gluing or soldering, may also be possibilities. In many applications, metal parts must be fastened to a composite structure. Furthermore, watertight seals must be maintained in many of these applications. One of the problems that occurs is caused by the difference in expansion and contraction of metal and fiberglass-reinforced composites. The metal part can be securely bolted to the composite, but it will require a flexible bedding compound to ensure a watertight seal. The bedding compound also acts as an adhesive that aids in distributing the stress to the composite.

Another problem can occur when attaching parts to a composite. Most composites can tolerate shock loads if the composite is able to flex. An example is demonstrated when the hood of a composite truck body can be struck without causing a dent. However, if the hood is not able to flex, it may crack. This is particularly true if there is a hard spot that limits the movement of the composite. Designers of composite structures have to take precautions to limit hard spots or reinforce these spots to distribute the load.

Areas where latches, locks, handles, and tie-downs are located on composite structures need to be reinforced so that the loads can be distributed away from the mounting points. See **Figure 28-2**.

Another concern in fabricating is attaching components to composites that have foam cores. The overall strength of cored composites can be tremendous, but the *crush strength* for *spot loads* (loads placed in a relatively small area) may be very low. Visualize a 4′ × 8′ piece of foam insulation 2″ thick placed on a concrete floor. It is possible to place a layer of concrete blocks uniformly across the entire surface of the foam without damaging it. However, if the same weight of concrete is concentrated on just one square inch, the weight of the concrete will crush the foam. The weight that can just be supported on 1 in² without crushing the foam defines its crush strength. A more relevant example might be an aircraft wing that can support

an engine, but the same wing will be damaged if a mechanic walks on it. Another example is a cleat that is through-bolted on a foam-cored fiberglass boat deck. During assembly, the cleat can crush the foam core if the bolts are overtightened, creating a high spot load. In the case of the cleat, the designer will have to replace the foam in the area where the cleat is to be mounted with a material that will spread the load to the composite in the area around the cleat attachment point. In the case of the wing, the design may not be able to be changed and other provisions for human access will have to be developed.

Important Terms

adhesion process
cohesive bonding
cohesive force
crush strength
intermolecular attraction
mechanical fasteners
sealant
secondary bonding
spot loads

Questions for Review and Discussion

1. When can cohesion be used to attach a component to a composite?

2. When is adhesion used to fabricate a composite structure?

3. Describe some of the precautions that should be taken when using mechanical fasteners to fabricate a composite structure.

Figure 28-2. Point or concentrated loads need to be distributed in composite structures.

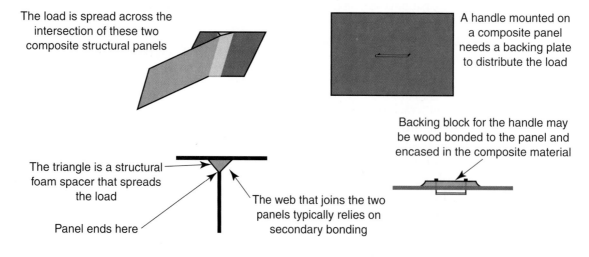

The load is spread across the intersection of these two composite structural panels

A handle mounted on a composite panel needs a backing plate to distribute the load

The triangle is a structural foam spacer that spreads the load

Panel ends here

The web that joins the two panels typically relies on secondary bonding

Backing block for the handle may be wood bonded to the panel and encased in the composite material

29

Processes Used to Condition Metallic Materials

Chapter Highlights

- Manufacturers have the ability to change the mechanical properties of metallic materials.
- Selecting the appropriate alloy is essential to obtain the desired mechanical properties.
- Heat-treating is the most widely used process to alter the mechanical properties of metals.
- Hardness, strength, and workability are some of the mechanical properties that can be changed by heat-treating.

Introduction

One of the advantages that metal fabricators have is the ability to change the mechanical properties of metallic materials. This process is called *conditioning*. The condition of a metallic material that makes it suitable for fabrication may not be appropriate for its intended use. Conditioning processes alter the internal structure of a material to give the part the desired strength, hardness, and toughness. Ferrous materials such as steel are particularly well suited for conditioning.

The Crystalline Structure of Steel

One of the most widely used conditioning processes is heat-treating. The process is fairly simple to do once the basics are understood, however, it takes a great deal of skill and knowledge to do it well consistently. To begin, it is necessary to have at least a basic understanding of the phases, or changes in molecular structure, that take place when metal is heated to change it from a solid to a liquid.

Pure metals (those containing only one kind of atom, such as iron or copper) have a clearly defined temperature at which they melt, or change from a solid to a liquid. They also solidify (change from a liquid to a solid) at specific temperatures. *Alloys* solidify over a range of temperatures. When the temperature of a molten alloy drops below the *liquidus line*, it starts to change to a solid. At this point, the material becomes a mushy solid. If the temperature is lowered even more, the alloy will eventually harden as it reaches the *solidus line*. See **Figure 29-1**.

There are thousands of possible alloys and types of metals now available, so it might seem nearly impossible to accurately predict the effect that a heat-treating process would have on a particular metal. However, thanks to the development of the *phase diagram* or *equilibrium diagram*, it is easy to determine the structure of the material when it is at a particular temperature. See **Figure 29-2**. As steel is heated or cooled, it goes through phase changes. For each change or phase, there is a term that identifies the resulting structure of the material.

One of the most useful aids for heat-treating is the iron-carbon equilibrium diagram for steel and iron. See **Figure 29-3**. The X-axis of the diagram indicates the percentage of carbon. When you look

Figure 29-2. Phase changes take place as metal is heated or cooled. When steel is heated sufficiently, it changes to its austenitic form. Depending on the speed of cooling and other factors, it may be transformed to one of the other structures shown.

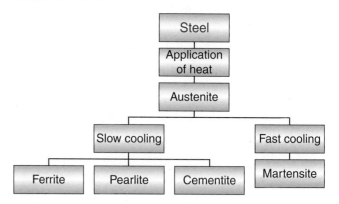

at the diagram notice that the range between 0.1 and 2.0 percent carbon represents steel while cast iron begins at 2.0 and continues to 4.0 percent carbon. In addition to the carbon content, temperature has a major influence on the material's properties.

Using this diagram, let us take a moment to examine how temperature and carbon content influence the process of heat-treating. If you pull a piece of low-carbon steel, such as 0.4 percent carbon, from a storage rack at room temperature, it will be in solid form and will contain a small amount of carbon spread throughout the steel. At this point, the steel is referred to as *ferrite*. Ferrite is the softest and most ductile form of low-carbon steel found on the equilibrium diagram. It is magnetic from room temperature

Figure 29-1. A typical graph representing phase changes in an alloy. Note that the liquidus and solidus lines reflect differing ranges of temperatures as the composition of the alloy changes.

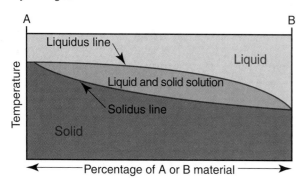

Figure 29-3. This simplified iron-carbon equilibrium diagram shows the phase changes of iron at different temperature and carbon-content levels. The lower transformation temperature and upper transformation temperature lines show where phase changes begin and are completed for iron with different carbon percentages.

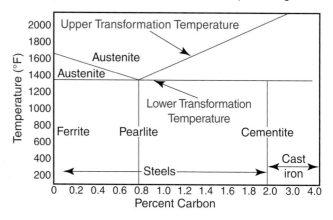

up to about 1400°F (761°C). Ferritic iron may contain up to 0.8 percent carbon.

When steel contains 0.8 percent carbon, it begins to change to another state, called *pearlite*. This is a mechanical mixture of bands, or layers, of ferrite (the metallurgical name for iron) and cementite (iron carbide, Fe_3C) in the proportion of 88 percent ferrite and 12 percent cementite. Under the microscope, this material has the appearance of mother-of-pearl, thus the name. Pearlite is soft and relatively ductile.

Steel containing 1.0 percent carbon (low-carbon steel) has both pearlite and *cementite* structures. When steel reaches 2.0 percent, it is truly cementite. The cementite phase is reached when steel is slowly cooled after being heated to above 1341°F (727°C), the point where it becomes *austenite*. Austenite is the densest form of low-carbon steel. Another term to consider is *martensite*. A martensitic structure is created when steel is rapidly *quenched* or cooled, usually by plunging it into a liquid bath.

Referring again to Figure 29-3, check the steel section of the diagram and locate the regions containing ferrite, pearlite, and some cementite. While the diagram shows definitive points where pearlite and cementite occur, it should be remembered that steel changes gradually from one state to another. There is an overlap between ferrite/pearlite and pearlite/cementite.

We have discussed the influence that the percentage of carbon has on phase changes. Now let us examine more closely the impact of temperature. Two additional terms are needed. The *lower transformation temperature* and *upper transformation temperature* represent limits for changes between phases, depending on temperature and percentage of carbon.

For example, when ferritic iron is heated, it changes to austenite. The lower transformation temperature is the temperature where the material begins to change to austenite. The upper transition temperature is the temperature at which the change is complete—no ferrite exists and the iron is completely austenite. All of the area on the diagram above the upper transformation temperature line represents austenite.

The point where the two transformation temperature lines come together indicates that pearlite, with 0.8 percent carbon content, has a *eutectoid* composition. See the intersection labeled with a circled 2 in **Figure 29-4**. The term *eutectic* is usually applied to an alloy that immediately changes from solid to liquid

Figure 29-4. The relationship of temperature and percentage of carbon on material structure. The point where the upper and lower transformation temperature lines intersect represents the immediate transformation of solid pearlite to solid austenite, without any liquid phase.

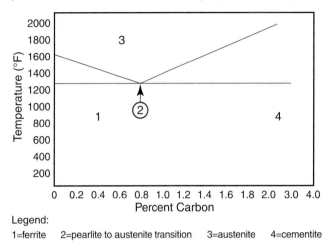

Legend:
1=ferrite   2=pearlite to austenite transition   3=austenite   4=cementite

at a specific temperature. In the case of pearlite, however, the change that takes place immediately when the transformation temperature is reached is that it changes from solid pearlite to solid austenite.

Equilibrium diagrams can be obtained from steel manufacturing firms and metal suppliers. In many cases, these diagrams may seem quite complex; however, only three things really need to be known about a sample or part made from an alloy in order to use the diagram:

- The sample's temperature.
- The percent of carbon in the steel.
- Information on any previous heat treatment that the sample has received.

Figure 29-4 shows a simplified version of an equilibrium diagram. In the illustration, the horizontal scale represents the percentage of carbon in the sample. Temperature is shown on the vertical scale. Take a moment to locate the various regions shown on the diagram.

Processes Used for Conditioning

Nearly all conditioning processes used with metallic materials involve heat and are described generally as *heat-treating*. The Society of Manufacturing Engineers defines **heat-treating** as "any process whereby metals are better adapted to desired conditions or properties in predictably varying degrees by means of controlled heating and cooling in their solid state without alteration of their chemical composition." Both mechanical and physical properties can be changed by heat treatment.

The controlled application or removal of heat causes the atoms of a metal workpiece to move, thereby changing the internal structure of the material. Another important aspect of heat-treating is the rate and method used to cool the metal after it is heated. These also influence the internal structure of the material.

There are many different types of heat treatments, including case hardening, annealing, normalizing, and tempering. Each of these processes will be covered in this section. It is easy to get confused when thinking about heat-treating processes, because there are so many different types of alloys and methods for heat-treating. One way to simplify all of this information is to classify heat-treating processes by type. Generally, there are four major types of heat-treating processes:

- Full hardening of a workpiece
- Partial hardening of a workpiece
- Softening of a workpiece
- Relieving of stresses in a workpiece

Figure 29-5 lists specific processes classified under each of these types of heat-treating. Heat-treating is done to improve the toughness, wear resistance, machinability, tensile strength, ductility, bending quality, corrosion resistance, and magnetic properties of metal. Steel is the most

frequently heat-treated metal, but iron, copper, aluminum, and many other metals are conditioned by heat-treating.

Heat-treating of metals can be done using a conventional heat-treating furnace or any of a variety of other heat sources. Heat treatments can be conducted over a wide range of temperatures—the temperature required depends on the type of metal involved and the condition desired. Some conditioning can be conducted at room temperature, but this is unusual. Heat-treating temperatures required for steel usually begin around 600°F (316°C).

Heat-treating furnaces are designed for continuous or batch production. Parts are fed into continuous furnaces on a moving conveyor. *Continuous furnaces* are long heating chambers, used for high volume production, that receive material at one end and heat the material as it is transported to the exit end. *Batch furnaces*, used for heat-treating small quantities of parts, are loaded with material, closed, and then opened so the material can be removed. *Fluidized bed furnaces*, which use hot gases to suspend heated aluminum oxide particles in a chamber, are also being utilized.

An oxyacetylene torch can also be used to heat-treat individual parts. Using torches or other focused heat sources are part of a conditioning method called *flame hardening*, which is discussed in more detail later in this section. Heat-treating can be accomplished with CO_2 lasers that have power outputs above 500 watts. More powerful and larger diameter beams are better for heat-treating of metals. Smaller beams are better used for such processes as cutting and drilling.

Figure 29-5. The major heat-treating processes can be classified into four types: full hardening, partial hardening, softening, and stress-relieving.

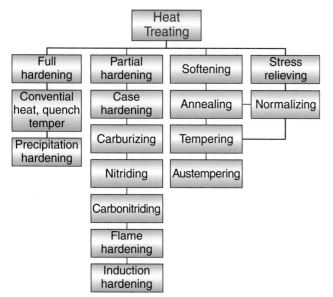

Full Hardening

The heat/quench/temper hardening process is the most common way to fully harden steel. It is not practical to try to full-harden soft steel with less than 0.35 percent carbon (sometimes referred to as *35 points of carbon*). Look again at Figure 29-3 and you will see that no matter how high the temperature, it is impossible to bring low-carbon steel out of the austenite region. Hardening requires transformation to cementite.

Ferrite, or steel with less than 0.8 percent carbon, has very little ability to dissolve carbon. Carbon is what gives steel its ability to become strong and hard. In order to harden low-carbon steel, it is necessary to increase the amount of carbon on its surface by forming a thin outer layer or *case*. This outer case can then be hardened by heat-treating the steel to a point slightly above its upper transformation temperature, then quenching it. This creates a part with a hard outer shell and a ductile center.

The process of hardening steel with the heat/quench/temper process begins by heating the workpiece above its upper transformation temperature (its critical temperature) and then rapidly cooling or **quenching** it in water, warm oil, or air. Other quenching media, which are less frequently used, are brine, molten salt, and sand.

Water is the most frequently used quenching medium, especially with low-carbon steels. Water provides a sudden quench that creates good hardness and strength. The problem with this type of quench is that the very rapid temperature change sometimes causes internal stresses and distortion in the part.

Oil is a less drastic quench than water, and is preferred for use on thin or delicate workpieces. Typical parts that are oil-quenched are thin and fragile cutting blades, such as razor blades. While oil is gentler than water, it does not create the same degree of hardness that is possible with a water quench.

The slowest type of quench is achieved by letting the part slowly cool in air. The advantage of this quench is that it produces fewer stresses in the workpiece; the disadvantage is that it does not produce the hardness and strength possible with other quenches. High-alloy steels such as molybdenum or chromium have greater internal stress than milder alloys, and are more likely to crack or distort when quenched. At the same time, they do not require such a drastic quench to improve their hardness. Air is the preferred quenching medium for high-alloy materials.

After the part is hardened and quenched, it is usually tempered. In tempering, some of the hardness is sacrificed in order to improve tensile strength and toughness. A part is **tempered** by reheating it to just below the lower transformation point, then cooling it—any rate of cooling can be used.

Precipitation Hardening

To heat-treat nonferrous metals and some types of stainless steel, different methods must be used. These metals do not undergo phase transformations in the same way as iron and steel. Some stainless steels and **nonferrous metals** such as aluminum alloy and copper alloy are hardened by adding **precipitates** (small particles of a different phase) into the original

matrix. The process used to accomplish this is called **precipitation hardening**. Precipitation hardening is also called *age-hardening* or *aging*.

There are three steps involved in precipitation hardening: solution treating, quenching, and aging. **Solution treating** involves heating the alloy, which exists as a two-phase solid solution at room temperature, to a temperature that causes it to become a single-phase solid solution. When an excessive number of particles from one phase are dispersed throughout the matrix of a different phase, many of these particles cannot be dissolved completely into the original matrix. This results in precipitates being formed.

The second step involves quenching this single-phase solution to create a supersaturated, but unstable, solid solution. Quenching is usually done in water. At this point, the material is softer than it might be in the annealed condition, and therefore, is in ideal condition for forming or machining.

The third and last step illustrates the real value of this process. An alloy that has undergone the first two steps in precipitation hardening can be supplied to the manufacturer in a "solution treated" condition. This material can be readily formed or machined. After the machining or fabrication is completed, the part is then heated to a low temperature by the manufacturer to increase the strength of the alloy. This step in the process is carried out at low temperature to prevent distortion. This last step is known as aging or age-hardening.

Some aluminum alloys harden and become stronger when aged over a period of time at room temperature. This type of aging is called **natural aging**. Usually, this is accomplished by first heating, quenching, and forming the part, then letting it age at room temperature for an extended length of time.

When the process is conducted at heat levels above room temperature, it is called **artificial aging**. This method of age-hardening must be undertaken with care. Exposure to too much heat will result in **overaging** the part, making it weaker. One advantage of overaging, however, is that dimensional stability is improved.

Case Hardening

Often it is practical to use low-carbon or mild steel to make gears or other types of parts that must survive high-load and wear conditions. For a part such as a gear, it is undesirable to harden the entire workpiece. Such a fully hardened part is excessively hard and brittle, and would lack the toughness necessary to withstand heavy loads. A gear with hardened teeth and a softer, more ductile and tougher center is preferred. In order to accomplish this, carbon must be added to the outside surface of the part. This is referred to as *case hardening*. Case hardening produces a hard shell or case on the outer surface of the workpiece. The material becomes softer as one moves into the part from the outer shell. **Figure 29-6** illustrates the effect of case hardening on various types of metal workpieces.

There are a number of different methods used for case hardening to produce various degrees of hardness. All involve heating the workpiece to a high temperature, typically about 1700°F or 927°C, exposing it to a particular gas, liquid, or powder, then quickly cooling it. Some of the more popular methods include carburizing, cyaniding, nitriding, carbonitriding, flame hardening, and induction hardening.

Of the various methods of case hardening, the oldest is called *carburizing*. This method involves covering the heated workpiece by dipping or dusting it with a carbon-hardening powder. The powder will melt on the hot surface, forming a thin coating. If a thick coating is desired, the workpiece is packed in a container of carbon-hardening powder called *Kasenit*, and kept in a furnace for an extended length of time. The longer the steel is kept hot, the thicker the coating that will be produced.

Carburizing treatments can also be accomplished with methane gas, liquids, or charcoal packs. With each method, the heated metal is placed in a carburizing environment for the required length of time.

At this point, case hardening is still not complete. The part now must be reheated to a bright red/orange color and held at this temperature for several minutes. The part can then be quickly quenched

by immersing it in cold water, oil, or saltwater. The liquid used for the quench depends on the nature of the conditioning that is desired and the type of steel.

Cyaniding

Cyaniding is another surface-hardening method that has been widely used. An iron-based alloy is heated while in contact with cyanide salt so that the surface absorbs carbon and nitrogen. The process of cyaniding is followed by quenching and tempering to produce an outer shell that is tough and hard.

This process is also called the *liquid-salt method*, since the heated steel is quenched in a bath of liquid salt, heated to approximately 1600°F (871°C). After immersion, the part is then quenched in water. This is a popular case-hardening method when a thin case layer is desired. A typical layer of about 0.010″ (0.254 mm) is produced by leaving the part in the salt bath for about one hour before quenching it. This process is dangerous to use, however, because of the toxic fumes that are given off by the potassium cyanide bath. If the cyaniding method is used, the area must be properly ventilated and all personnel must wear proper eye protection and protective clothing.

Nitriding

One of the hardest surface coatings is accomplished using a gas in a process called *nitriding*. The nitriding process begins by placing the workpiece in a sealed container in a heat-treating oven. The container is then heated to between 900°F and 1150°F (482°C and 621°C). High-temperature ammonia gas is then introduced into the container. The ammonia breaks down into nitrogen and hydrogen. The nitrogen attacks the surface of the steel and combines with iron-forming nitrides. Nitrides are compounds that give the steel the desired hardness.

The major limitation associated with nitriding is that it is a time-consuming process. It would normally take about 50 hours to produce a case-hardened coating about 0.015″ deep. Other methods to surface-harden steel can be accomplished in just a few hours. There are some forms of carburizing that can be completed in one hour or less.

One of the advantages of the nitriding process is its ability to create a thicker hard shell than can be normally achieved with cyaniding. Nitriding is particularly useful in situations where warpage,

Figure 29-6. Case hardening results in a hard shell surrounding a core of softer material. The illustration shows the effect of case hardening on round and irregularly shaped workpieces.

distortion, and cracking of the part may result from exposure to extreme temperatures. Nitriding can be used with very little heat, only 900°F to 1000°F (482°C to 538°C). This is the only case-hardening process that permits the use of such low temperatures to produce a hardened case.

Carbonitriding

In the process of nitriding, a nitrogenous gas, usually ammonia, is united with the ferrous metal's surface to form iron nitride. With **carbonitriding**, low-carbon steel parts are hardened by exposing them to dry gases that are rich in carbon and nitrogen. Carbonitrided parts are harder and have greater wearability than carburized materials.

The process is conducted as follows: First, the parts are placed in an atmosphere of carburizing gas and ammonia at temperatures between 1400°F and 1700°F (760°C and 927°C). The ammonia gas creates nitrogen; both carbon and nitrogen are absorbed into the parts. Once the parts are saturated, they are quenched in oil to reduce internal stresses and improve ductility. Carbonitriding is a slower process than nitriding. A coating of 0.010" (0.254 mm) would normally require about 90 minutes.

Flame Hardening

Metallic workpieces are often **flame-hardened** with coatings up to 0.250" (6.35 mm) in thickness, using nothing more than oxyacetylene torches to generate heat. The surface of the iron-based alloy in the area to be hardened is heated to the upper transformation temperature. This changes the composition from ferrite to austenite. The metal is then quenched to create martensite. Both heating and quenching must be done quickly.

The advantage of flame hardening is that it can be used to create localized hardening. With this process, it is not necessary to heat the entire part. Heat needs to be applied only to the localized area that is to be hardened, thereby reducing distortion and overall changes to the steel.

Induction Hardening

This is another process than can be used to create localized case hardening. **Induction hardening** creates heat by passing a high frequency electrical current through a coil of wire. The part to be hardened is placed inside the magnetic field created by the coil. Eddy currents of from 3000 to 1,000,000 hertz attempt to pass through the metal, generating a resistance that causes heat. The heated part is then quenched in water or oil.

One of the major advantages of induction hardening is speed—hardening can be accomplished in seconds, instead of the minutes needed for other case-hardening processes. This makes it practical for use in many high-volume production situations. Another advantage is that it is well-suited to hardening such irregularly shaped parts such as gear teeth, crankshaft bearing surfaces, and roller bearings.

Softening Processes

Thus far, this chapter has covered processes that are used for full hardening and case hardening of ferrous and nonferrous metals. Now, we can examine the methods that are used to soften ferrous and nonferrous metals.

Annealing

The first process that usually comes to mind when thinking about softening metal is **annealing**. However, annealing is a process that accomplishes much more than this. Annealing is also used to refine grain structure, to restore ductility after hardening, to remove internal stresses caused by forming processes, to improve electrical and magnetic properties, and to improve machinability.

Annealing involves holding the material for an extended length of time at a temperature high enough to achieve an austenitic structure, then slowly cooling it. The temperature that is required varies depending on the material and type of annealing. Some of the more common types of annealing processes are full annealing, isothermal annealing, intermediate annealing or process annealing, and soft annealing.

Full annealing is accomplished by heating to a temperature just above the upper transformation temperature. The workpiece is then cooled slowly in a furnace to a point just below the lower transformation temperature, followed by further cooling in still air to room temperature. Metal can also be annealed by submerging the red-hot workpiece in lime, ashes, or other noncombustible material until it is cooled. A general rule for determining the rate of cooling is to allow approximately one hour per inch of material thickness at the largest section of the part.

Isothermal annealing is accomplished by heating the ferrous alloy part near the critical temperature (upper transformation temperature) to produce a structure that is partly or wholly austenitic. The part is then cooled and held to an intermediate temperature, for example 500°F to 800°F (260°C to 425°C). This causes a transformation of the austenite to a relatively soft ferrite-carbide aggregate.

Intermediate annealing begins by heating the part to a temperature close to or just below the lower limit of the critical temperature. After heating, the part is allowed to cool. The annealing softens and stress relieves the part so it can undergo further cold working. Since this is done between operations during manufacture, this process is referred to as intermediate annealing.

Soft annealing involves heating the metal to a temperature above its critical range and then cooling it appropriately. The objective is to achieve the greatest softness or ductility possible.

When the exterior of a ferrous part is exposed to the atmosphere during annealing, a bluish oxide will form on its surface. Because of this color, the process is sometimes referred to as *blue annealing*. However, if the part is annealed in a controlled atmosphere furnace or in a vacuum, oxidation will be nearly eliminated and the surface will remain relatively bright. This is called *bright annealing*.

Brass and copper can be annealed after they are cold-worked by heating them to approximately 1100°F (593°C) and then cooling the workpiece. The rate of cooling is not important.

Normalizing

Steels are normally annealed in still air to avoid excessive softening. This application of annealing is called *normalizing*. Normalizing requires heating the part to a temperature about 100°F (56°C) above the upper transformation temperature, holding it at this temperature until the steel is heated through, then removing it from the furnace to cool to room temperature in still air.

The purpose of normalizing is to produce a more uniform and fine-grained structure in the metal. It also provides greater strength (retaining some hardness) and fewer internal stresses than does annealing.

Stress Relieving

Complex parts that have been welded, cast, or machined often develop internal stresses that could cause distortion of their shape or affect their strength or serviceability. To relieve these stresses, parts made from carbon steel are heated to between 1000°F and 1200°F (538°C and 649°C), held at that temperature for at least an hour, and allowed to cool in air. Alloy steels usually must be heated to somewhat higher temperatures for effective stress relief.

Tempering

Tempering of steel is done to remove some of the hardness from metal so that it is less brittle. This makes the metal easier to form or machine and also improves its *toughness* (resistance to breaking). Tempering is done by reheating the metal, after hardening, to a temperature well below the transformation temperature. The metal is then allowed to cool through any of the various methods mentioned. When the steel is tempered, its microstructure changes from martensite to a softer ferrite or pearlite. Recall that martensite is formed by rapid cooling of a low-carbon steel part that has been heated, changing it from ferrite to austenite.

Another method of tempering is called *austempering*. This is accomplished by heating the part to a temperature below that required for case hardening. After heating, the workpiece is cooled in a salt bath kept at a temperature of about 800°F (427°C).

The advantage of austempering is that it increases ductility, toughness, and hardness. Austempering also reduces the possibility of distortion and cracks, because the method of quenching does not shock on the material.

If the metal has already been hardened and a martensitic structure has been created, then it is not possible to use austempering. In this case, the part can be tempered by heating it to a point just below the lower critical temperature, which may be between 300°F to 1050°F (149°C to 566°C), depending on the type of steel involved.

Normally, if the part must be resistant to wear, it will be tempered below 400°F (204°C). If the part must be tougher, tempering will be done above 800°F (427°C). Quenching of tempered metal is often done in a heavy-oil bath. The oil is heated to between 500°F and 600°F (260°C and 316°C). The part is heated, immersed in oil, and reheated to the tempering point. It is then dipped in a bath of caustic soda, followed by a quench in hot water.

Quenching can also be accomplished in baths of molten lead or molten salt. With these quenches, the part is heated to the tempering point and then immersed in the bath. It is left there until it is cooled to the bath temperature.

In manufacturing, it is sometimes necessary for tool-and-die makers, machinists, or technicians to fabricate parts for repair or maintenance. They may have to heat-treat the replacement part with a torch. To determine the proper temperature for various heat-treating processes, they may use an interesting method that involves watching color change in the metal.

When heat is transferred to a polished part by a torch, the polished surface will change color. At 430°F (243°C) tool steel will turn yellow; at 530°F (277°C) purple, and at 610°F (321°C), pale blue. When the desired temperature is achieved, the metal is quenched. Skilled tool-and-die makers know that a part will lose some hardness but it will be less brittle if it is heated to the point where it is turning from a yellow to a purple color and then quenched.

In production heat-treating, time and temperature controls must be more precise. To achieve this control, often the heat-treating cycles are controlled by programmable logic controllers. It is also important to ensure that the parts are evenly heated. A good deal of skill and experience is needed to manage a successful heat-treating operation.

Cryogenic Conditioning

There are also cold and cryogenic conditioning processes that have come into use. The difference between cold and cryogenic conditioning is the temperature required for processing. The optimum cold treatment temperature is –120°F (–84°C). By comparison, *cryogenic conditioning* occurs at a temperature of around –310°F (–190°C) and is able to improve certain properties beyond the capability of a cold treatment.

A typical cryogenic treatment consists of a slow cool down at a rate of 4.5°F per minute (2.5°C per minute) from room temperature to the temperature of liquid nitrogen. Liquid nitrogen is the medium used to chill the parts. When the material reaches the temperature of liquid nitrogen, the material is soaked for a duration up to 24 hours. After the soak, the part is removed from the liquid nitrogen and allowed to warm to room temperature. It is important to note that cryogenic processing is not a substitute for heat-treating. The desired properties of hardness must still be accomplished through heat treating. A part that is poorly heat-treated will result in unsatisfactory parts, even with cryogenics.

Cryogenic processing is able to alter the crystal structure of metals to enhance fatigue resistance and reduce abrasive wear. Processors using cryogenic conditioning indicate that carbide precipitates form within the martensitic structure of the material during the processing. It is the formation of martensite that is responsible for the improvement in the part's wear resistance. A cryogenic processing is not limited to iron and steel. Nonferrous alloys including aluminum, magnesium, titanium, copper, nickel alloys, and even plastics such as nylon may be treated as well.

With cryogenics, everything must be handled with precision. Times are extremely critical, and temperature changes must be controlled with the utmost precision. Careful handling is vital—if a gear or tool is accidentally dropped after it has been removed from liquid nitrogen, it could shatter.

While cryogenic processing seems to offer many advantages in terms of the strength of the part and wear life, it does have one drawback: inconsistent results. Efforts are now underway to use automation to remove human factors, the primary source of variability in the process.

Important Terms

annealing
artificial aging
austempering
austenite
batch furnaces
blue annealing
bright annealing
carbonitriding
carburizing
case hardening
cementite
continuous furnaces
cryogenic conditioning
cyaniding
equilibrium diagram
eutectoid
ferrite
flame-hardened
fluidized bed furnaces
full annealing
heat-treating
induction hardening
intermediate annealing

isothermal annealing
Kasenit
liquidus line
lower transformation temperature
martensite
natural aging
nitriding
nonferrous metals
normalizing
overaging
pearlite
phase diagram
precipitates
precipitation hardening
pure metals
quenching
soft annealing
solidus line
solution treating
tempered
toughness
upper transformation temperature

Questions for Review and Discussion

1. Explain the difference between hardness and toughness.

2. What is the difference between tempering and annealing?

3. What is the difference between annealing and normalizing?

4. Describe the process to full harden low-carbon steel.

30 Processes Used to Condition Plastic Materials

Chapter Highlights

- Chemical blowing agents are used to produce lightweight cellular plastics.
- To relieve stresses caused by processing activities, plastics are often annealed.
- The rate of polymerization for certain plastics can be increased by exposing them to radiation.
- Polyanisidine (PANIS) is a conductive polymer that can be used as a temporary replacement for solder in some electronics applications.
- Polyphenylene sulfide plastic, providing up to 50 times more thermal conductivity than conventional plastic, is used by manufacturers of stepper motors to carry heat away from the motor and into the air.

Introduction

Processes used to condition plastics are responsible for changing the internal molecular structure of the material. This may involve heating to promote cross-linking of polymer chains or to relieve stress, exposure to ionizing or nonionizing radiation to improve such qualities as heat resistance, or the incorporation of reinforcements, fillers, or chemical additives to provide desired characteristics.

Since plastics are engineered materials that are designed and constructed by people, there are many ways to alter their internal characteristics. Many of the innovative processes related to conditioning plastics deal with the use of the plastic as a support matrix carrying one or more different types of fillers or reinforcements. Some of the processes will be covered in this chapter, but most will be discussed in detail in the chapters on composite materials.

As a material, plastic has its own set of behavioral characteristics. Many of the surface-hardening processes used with metals, for example, have no impact on plastics.

Conditioning of thermosets and thermoplastics is also done through heat, which is usually generated or applied during the forming process.

It has been established that heat has a major impact on the internal molecular characteristics of polymeric materials. If the material is a thermoset, heat speeds the rate of polymerization. With thermoplastics, heat changes the form of the material from solid to liquid, allowing it to be formed.

In order to reduce part breakage and relieve internal stresses, it is sometimes necessary to alter the structure of the plastic material. In such cases, annealing, a process that is commonly used with metals, can be useful with plastics as well.

Annealing

Like many other materials, plastics develop internal stresses when their external shape is changed. Some manufacturing processes bend, flex, or reshape the material in such a way that drastic realignment of the internal molecular structure takes place. At first glance, it might appear that only those processes that result in changing the external shape of the material would create internal stresses. This is not true—even some finishing processes and chemical treatments can create internal stress in a part.

Annealing is used with plastics to relieve the internal stress in the material and to improve its workability. Plastic materials are annealed by subjecting them to prolonged heating at a temperature slightly above their molding temperature, then slowly cooling them under controlled conditions. The slow cooling lets the molecules gradually settle back into alignment. The length of time necessary to anneal a given plastic material varies according to the size of the part and the type of plastic involved. A typical heating/cooling cycle for 0.025″ (0.635 mm) Plexiglas® is heating for 2.5 hours at 230°F (110°C) and then cooling for 1.5 hours to reach a removal temperature of 158° F (70° C). See **Figure 30-1**.

Annealing is recommended when the plastic part or product needs to have superior visual appearance, have good dimensional stability, or exhibit high resistance to solvents. Machining of acrylic plastics causes *crazing*, or fracturing in areas of localized stress. Reducing the internal stresses will reduce crazing.

Figure 30-1. Charts used to determine proper heating and cooling times, in hours, when annealing Plexiglas. Note that annealing at higher temperatures may cause parts to show deformation to an objectionable degree. (Rohm and Haas Co.)

Heating Times for Annealing of Plexiglas									
Time in a Forced-Circulation Oven at the Indicated Temperature									
Thickness (mm)	Plexiglas G, II, and 55					Plexiglas I-A			
	110 °C*	100 °C*	90 °C*	80 °C	70 °C**	90 °C*	80 °C*	70 °C*	60 °C	50 °C
1.5 to 3.8	2	3	5	10	24	2	3	5	10	24
4.8 to 9.5	2½	3½	5½	10½	24	2½	3½	5½	10½	24
12.7 to 19	3	4	6	11	24	3	4	6	11	24
22.2 to 28.5	3½	4½	6½	11½	24	3½	4½	6½	11½	24
31.8 to 38	4	5	7	12	24	4	5	7	12	24

Note: Times include period required to bring part up to annealing temperature, but not cooling time.
*Formed parts may show objectionable deformation when annealing at these temperatures.
**For Plexiglas G and Plexiglas II only. Minimum annealing temperature for Plexiglas 55 is 80 °C.

Cooling Times for Annealing of Plexiglas									
Time to Cool from Annealing Temperature to Maximum Removal Temperature									
Thickness (mm)	Rate (°C)/h	Plexiglas G, II, and 55				Plexiglas I-A			
		230 (110 °C)	212 (100 °C)	184 (90 °C)	170 (80 °C)	194 (90 °C)	176 (80 °C)	158 (70 °C)	140 (60 °C)
1.5 to 3.8	122 (50)	¾	½	½	¼	¾	½	½	¼
4.8 to 9.5	50 10	1½	1¼	¾	½	1½	1¼	¾	½
12.7 to 19	22 - 5	3¼	2¼	1½	¾	3	2¼	1½	¾
22.2 to 28.5	18 - 8	4¼	3	2	1	4	3	2¼	1
31.8 to 38	14 - 10	5¾	4½	3	1½	5¾	4½	3	1½

Radiation Processing

As you will recall from an earlier chapter, the process of polymerization of thermosets typically involves the interaction of a resin and catalyst. Polymerization can also be created through radiation. This type of polymerization is accomplished with a process called *radiation processing*. Radiation processing produces internal change through exposing the stock to high-energy ionizing radiation. There are two major types of radiation processing—ionizing and nonionizing.

Ionized Radiation Processing

Ionizing radiation processes produce high-level radiation, either gamma rays generated from radioactive materials such as Cobalt 60, or high-energy electron rays created by electron accelerators. Gamma rays are able to penetrate thick objects, but usually provide very slow processing speeds.

High-energy electrons do not penetrate as deeply as gamma rays, but they are very powerful, often up to several hundred kilowatts (kW) of power. Industrial electron acceleration machines are presently available with voltages up to five million volts. Because of the versatility of these machines, electron acceleration has become the most popular radiation processing method. Electron beam acceleration is desired over other processes that expose the workpiece to radioactive material because it enables the operator to direct the source of radiation to the workpiece and switch it on and off as desired.

Electron beam machines have been used for many years by the plastics industry to improve thermal, impact, wear, and chemical resistance of commodity thermoplastics. Electron beam processing has also been used with thermosets, and has improved their dimensional stability, ability to withstand higher temperatures in use, resistance to chemicals and water, and abrasion resistance.

Today, some of the most popular uses of radiation processing are the cross-linking of insulation material for wire and cable, the cold sterilization of surgical supplies and medical disposables, and the grafting of polymers onto textile webs to improve soil resistance and the ability to be dyed. Other applications include cross-linking of foams, depolymerization of cellulose pulp, curing of coatings, and cold pasteurization of food.

Nonionizing Radiation Processing

Induction heating is a common **nonionizing radiation process**. Induction heating provides no contact, and energy-efficient heat in a short length of time. Normally, induction heating is used with metals and other conductive materials, but it can be used with plastics as well. When used with nonconductors, a conductive metal attachment is used to transfer the heat to the workpiece.

With induction heating processes, typical RF power supplies used to heat or cure workpieces range from 1 to 5 kW, depending on the type and shape of the parts and the nature of the thermal application. Induction bonding is effective for adhesive bonding because it cures the material from the inside out.

There are several advantages for using induction heating over high temperature processes for heat treating. Heat is provided only to the area where the heat is needed. This reduces the amount of warp and distortion in the part. One other advantage of this process is because the workpiece is placed inside a coil, no contact is made with the workpiece. This reduces the numbers of rejects because the part is never touched.

The process is relatively simple. An ac power supply and induction coil is all that is needed. The power to the coil is turned on and the workpiece or part to be heated is placed in the coil. The power supply transmits an alternating current to the coil, creating a magnetic field. An eddy current is generated inside of the workpiece, causing heat from the inside out.

Nonionizing radiation can also be generated by exposing the workpiece to microwave, infrared, or ultraviolet energy sources. Nonionizing processes are normally used when low-level radiation is needed to cure or dry adhesives or coatings.

Exposure to either ionizing or nonionizing forms of radiation, if carefully monitored and controlled, can improve the physical characteristics of a plastic material. Thermoplastics that have been irradiated are able to withstand exposure to higher temperatures and caustic environments. Surface exposure to radiation can also be used to improve the material's resistance to static electricity.

Exposing formed parts to controlled irradiation can be used to achieve polymerization, cross-linking, and grafting. *Grafting* occurs when two or more elastomers attach to a molecular chain without the use of a catalyst.

If the radiation source is not carefully controlled, however, it can cause damage. Radiation damage can break molecular bonds, destroying the alignment of atoms. This can lower the molecular weight of the material. It can also cause cross-linking, polymerization, branching, or oxidation.

Some plastics are more sensitive to radiation than others. The polystyrenes, polycarbonates, and polyesters show good stability when exposed to radiation. The polypropylenes and acetals have poor stability. Plastics with poor stability can be damaged with far less radiation exposure than those with good stability. Epoxy has a particularly good resistance to radiation, requiring as much as 10,000 mrads (millirads) of exposure to damage the material. In contrast, acrylic plastic can be damaged with as little as 5 to 10 mrads.

Polymerization and cross-linking caused by radiation processing can be used to shrink wire and cable insulation, increase the bonding strength of adhesives, and improve the resistance of plastic to cracking. It is also used to cure plastic coatings and film layers.

Conductive Plastics

Several years ago, it was unknown how to make *conductive plastics*, those that would be good electrical and thermal conductors. In the past ten years, however, developments and applications with new types of composite materials in medicine, space, and electronics have led to exciting new opportunities.

Research and development on applications for conductive plastics has resulted in new applications for plastics used as electronic devices, batteries, and chemically modified electrodes and sensors. A new field of research was born called *synthetic metal research*. The Department of Defense, the Office of Naval Research, the National Science Foundation, the Air Force Materials Laboratory, and firms including Allied Chemical, IBM, Xerox, Bell Labs, Eastman Kodak, Exxon, General Electric, and Westinghouse are all investing heavily in synthetic metal research.

Today, many methods for making conductive plastic coatings are used in Electromagnetic Interference (EMI) shielding. Plastics are also being designed to perform as efficient thermal insulators able to spread or dissipate heat. The ability of plastics to serve as thermal conductors was an important

accomplishment that led to significant improvements in the life cycle of many plastic products. Prior to the invention of thermal plastics, plastic parts exposed to heat from appliance operation systems often failed prematurely.

Plastics as Electrical Conductors

Plastic is made conductive by adding metal to the plastic base material, resulting in a material referred to as *synthetic metal*. Metal fibers have been used as conductive elements in antistatic fabrics and yarns. Some of these applications have involved compounding stainless steel fibers with the polymer during molding.

While conductive thermoplastics are frequently used in electronic, business-machine, computer, and industrial applications, the medical community has capitalized on the improved characteristics of these materials for products ranging from instruments, and trays, to communication devices and laboratory equipment.

Approaches for creating conductive materials involve the use of specialty fillers, reinforcements, and modifiers to protect against static accumulation, electrostatic discharge (ESD), and electromagnetic and radio-frequency interference (EMI/RFI). Other attempts to produce conductive plastics have involved the use of conductive webs.

Most plastics can be modified with conductive fillers. **Figure 30-2** shows some of the common polymers and their uses for medical and healthcare applications. Some of the advantages of conductive thermoplastics are that the final parts are normally lighter in weight. They are also more durable and have greater resistance to scratching and denting.

Polyaniline films and powders are also being used for synthetic metal applications, not only because they are conductive, but also because they are relatively stable in air. A new type of conductive polymer, polyanisidine (PANIS), has proven its ability for use as either conductor or nonconductor. A conductive coating is produced when polyanisidine is blended with the nonconducting polymer polyacrylonitrile. The blend provides a tough coating that is not susceptible to oxidation. PANIS can be produced in various forms, ranging from particles and sheets to a conductive gel that can be shaped much like putty. Research is even being conducted into using the conductive gel as a replacement for conventional solder in some industrial electronic

Figure 30-2. These are conductive plastics commonly used in medical and healthcare applications.

Plastic	Method of Sterilization	Performance Advantages	Product Applications
Polyether ether ketone (PEEK)	Autoclave, Ethylene Oxide (EtO) gas, or high-energy radiation	Good chemical resistance	Catheters, disposable surgical instruments, and sterilization trays
Polycarbonate	All sterilization methods	Good toughness and impact resistance	Reservoirs and equipment covers
Polyurethane	Dry heat, radiation and EtO gas	High-clarity	Catheters, tubing, connectors and fittings, pacemaker leads, wound dressings, and transdermal drug-delivery patches
Liquid-Crystal Polymer	All common sterilization methods	High strength and stiffness	Surgical instruments, dental tools, and sterilizable trays
Polysulfone (PSO)	Autoclave, EtO gas, or radiation	Excellent thermal stability and toughness; resistant to many chemicals	Handles and holders, microfiltration devices, reusable syringe injectors, respirators, nebulizers, sterilizer trays, and dental tools

applications. Wires can be inserted into the conductive gel and then easily removed as desired.

EMI/RFI Shielding

For a number of reasons, plastic has become the dominant choice for the construction of enclosures for computers, appliances, microphones, business machines, medical instrumentation, and telecommunication devices. See **Figure 30-3**. One important reason for this popularity is the ability to add conductive materials to the polymer as a means of controlling the emission of EMI/RFI. The Federal Communications Commission has set limitations on the amount of EMI/RFI interference, or *noise*, a product can release.

EMI-shielded products easily dissipate electrostatic charges and provide improved reception of electromagnetic signals through antennas and other receiving systems. These products can be made using standard injection molding and blow molding equipment. Adding metal modifiers makes the plastics conductive. The stock that is used to modify the plastic is available in various forms: flakes, fibers, powders, or metal-coated substrates. Aluminum flakes, nickel-coated fibers, and stainless steel fibers are commonly used.

There are other methods that are also used to improve the EMI shielding of products. The most popular alternative to mixing conductive plastics is to use a paint that is heavily pigmented with a metal, such as silver and nickel. Vacuum metallizing

(described in the chapter on finishing) is also used to reduce EMI noise. One process that produces a very pure coating is called *electroless plating*. Let us take a closer look at how this process works.

Electroless Plating

Electroless plating is a process that produces a continuous and uniform coating of metal on plastic to achieve EMI shielding of plastic cases used on electronic products. The coating is applied in several

Figure 30-3. Conductive materials are added to the plastics used to fabricate enclosures for electrical devices to provide shielding against excessive EMI/RFI radiation. The enclosure shown houses a high-performance drive for an ac (alternating current) industrial motor. (MagneTek)

thin layers using an *autocatalytic*, chemical plating process. A chemical reaction is autocatalytic if the reaction itself is a catalyst, increasing the speed of that reaction without itself being changed as a result of the chemical reaction.

The process of electroless plating starts by etching the plastic part in a chromic acid solution. The purpose of this step is to improve adhesion. After etching, the acid is neutralized. At this point, the plastic part is placed in a solution containing tin and palladium salts. Palladium metal nuclei are deposited on the plastics in this stage of the process. This stimulates nickel or copper growth.

There are several different types of conductive paints and coatings used today for EMI and RFI shielding. Most of these are made by adding metal flakes of nickel, silver plated wire, or pure silver coatings. Paint that contains flakes of nickel is cheaper, but also provides a lower level of shielding than paints with particles of silver-plated wire or pure silver coatings. Since the advent of silver coatings in the late 1980s and early 1990s, this technology has expanded and prices have continued to drop. Today, copper is used less frequently. Silver-loaded paints are normally used in cell phones, computer monitors, and modems.

The technology used for electroless plating does not involve any current to aid in deposition of the material on the surface of the plastic. Instead, a chemical process is used to make the surface of the part autocatylitic. This causes the surface to join with the catalyst on the surface. In some cases, the catalyst is painted onto the surface of the substrate.

Plastics as Thermal Conductors

Normally when you think of a conductor, you probably think of something that conducts electricity. Thermal conductors are objects that transfer heat quickly. Conventional plastics have a thermal conductivity of approximately 0.2 W/mK (Watts/meter Kelvin). Most thermally conductive plastics have from 1–10 W/mK, about 10 to 50 times the thermal conductivity of conventional plastics.

Thermally conductive plastics have other advantages over conventional materials. They weigh about 40 percent less than aluminum. This makes these plastics an ideal candidate for use in parts such as sensors, gaskets, and coils used in appliances. Lightweight material such as thermally conductive polyphenylene sulfide plastic is used by manufacturers of stepper motors to carry heat away from the motor and into the air. Computer manufacturers are using thermally conductive plastics to dissipate heat away from critical parts in confined spaces.

Plastic Additives

Various additives may be used to provide plastics with the characteristics desired for a finished product. Generally, an additive will make up no more than one-tenth of the finished product's weight; many additives represent a much smaller proportion of product weight. Review the table of commonly used additives and the benefits each provides.

Commonly Used Plastic Additives	
Plastic Additive	**Purpose**
Ultraviolet light absorbers	Retard damage from exposure to sunlight
Flame retardants	Slow or stop any burning of the plastic
Plasticizers	Increase flexibility
Lubricants	Provide a slippery or nonstick finish
Antistatic agents	Dissipate static electricity charges
Preservatives	Protect against attack by microbes
Reinforcements	Add strength
Fillers or foaming agents	Change the density of the material

Preservatives

Most plastics are relatively impervious to destruction by microorganisms, rodents, or insects. However, elastomers, polyvinyl chlorides, and other plastics that are heavily plasticized, or softened to make them moldable, are susceptible to attack, especially when used in continuously warm and wet environments. Such products as shower curtains, boat covers, pool liners, and convertible tops are subject to attack by mildew, fungus, mold, and bacteria. Deterioration of the base material may also encourage attack by rodents and insects.

The best way to discourage such attacks is by adding chemical preservatives like mildewicides, fungicides, and rodenticides to the plastic. Use of these chemicals is carefully regulated by federal agencies to ensure safety of both production workers and end users.

Franklin Fiber

Franklin fiber is a microscopic crystalline fiber derived from gypsum. It can be used as a reinforcing filler in plastics. However, it also improves the tensile strength of plastic and, in most cases, improves its modulus of elasticity. This makes the plastic less brittle and increases its impact resistance.

Franklin fiber is a crystalline form of calcium sulfate, with crystals so small that they must be observed with a microscope. The fibers are white in color and feel soft and silky. When they are compressed, the Franklin fibers have a tendency to mat.

The material is produced by agitating gypsum in suspension for several minutes in a steam reactor. The thick slurry that results is then forced through a filter press to remove most of the water. The fibrous cake is then shredded, dried, and calcined at high temperature to remove the remaining water.

Franklin fiber has excellent temperature resistance (even better than asbestos) and provides increased strength for the plastic to which it is added. The fibers are nonabrasive and are easy to form using conventional plastic molding equipment. The fiber can be used with thermoplastics such as PVC, polypropylene, and nylon. It is also used with sheet molding compounds and other thermosets.

While it provides many advantages, the greatest value of using Franklin fiber with thermoplastic and thermosetting resins is in reducing costs and conserving petrochemical resources. The basic raw materials (petrochemicals) used for making plastics are much more expensive than gypsum, the raw material for Franklin fiber. Gypsum is also readily available in the United States. When the fiber filler is added to plastic, less plastic is needed. This reduces the cost of manufacturing, while making better use of limited petrochemical resources.

Microspheres

Hollow glass bubbles called ***microspheres*** are mixed with many different types of plastics to reduce assembly costs, decrease density, reduce weight, and generally improve the physical properties of the part. Microspheres are chemically stable, water resistant, and nonporous. The spheres are normally made of glass, ureaformaldehyde, silica, or phenolics, and are embedded in a matrix made of polyester or epoxy.

These thin-walled, hollow glass spheres are typically around 75 microns (75 millionths of an inch) in diameter, but are available in sizes as small as 20 microns and as large as 200 microns. Wall thicknesses range from 0.5 to 2 microns.

The spheres can be used by mixing them with polyester sheet molding compounds. They can also be used with processes such as hand layup, sprayup, and casting. The addition of these spheres to products usually improves sanding and machining rates. Products made with microspheres also have improved dimensional stability at high temperatures and improved water resistance.

Using synthetic hollow glass spheres as fillers also offers an additional advantage—the spheres are as much as eighteen times lighter than fillers like calcium carbonate. In addition to their light weight and high strength, they have the added advantage of being compatible with many types of resins. This makes it possible to design composite materials that are more able to survive extreme weather conditions.

Industry is using this technology for many applications, including marine flotation devices, automotive parts, compression molding compounds, high performance aerospace composites, adhesives, furniture parts, coatings, sporting goods materials, cultured marble, and carpet backing.

Glass Flakes

Glass flakes are used, in a manner similar to that described for microspheres, to create moisture barriers and corrosion-resistant coatings. One major manufacturer produces a glass flake that is chemically resistant and another that is electrically conductive.

The U.S. Air Force has been experimenting with glass flakes since the early 1960s to develop high-strength composites for rocket cases, fins, and missile radomes, as well as moisture barriers and corrosion-resistant coatings. It was found that glass flakes mixed with polymer increase the flexural modulus and reduce the coefficient of thermal expansion of the polymer.

Foam-Frothing

The ***foam-frothing*** process is also known as *cellular*, *blown*, or *bubble expansion*. Cellular plastics have the advantage of being light in weight, with a very high strength-to-weight ratio.

The basic process of foam frothing is a variation of the urethane foam pour-in-place method. What is unique about the frothing method is that a urethane chemical mixture is dispensed in a partially expanded state, rather than as a viscous liquid. Frothing is achieved by adding chemical blowing agents to the mix. When the blowing agents decompose at a particular temperature, gases are given off, creating a cellular structure in the polymer.

A special machine is required to create the desired frothing action. The ingredients are partially expanded, then mixed under a pressure of about 100 psi (690 kPa). The blowing agents create frothing, which facilitates filling of narrow cavities in plastic-injection molds.

The foam-frothing process produces lower density foams with thin-walled sections. The major disadvantage of frothing is the cleanup problems presented by the process—a solvent such as methylene chloride must be used to flush the mixer at the end of the pour.

Frothing can also be achieved by forcing nitrogen gas into the resin or melt, just before it expands in the mold. This variation of the process is called *physical frothing*, or *physical foaming*. Frothing can also be accomplished by aggressively whipping air into the resin, and then rapidly cooling the polymer.

Important Terms

autocatalytic
conductive plastics
crazing
electroless plating
foam-frothing
Franklin fiber
grafting
ionizing radiation processes
microspheres
nonionizing radiation process
physical frothing
radiation processing
synthetic metal
synthetic metal research

Questions for Review and Discussion

1. Describe the procedure used for annealing plastics.

2. How can conductive plastics be used to reduce EMI/RFI interference?

3. What is the purpose of Franklin fiber?

4. What do Franklin fiber and microspheres have in common in terms of their use as fillers? What is similar and different about these two processes?

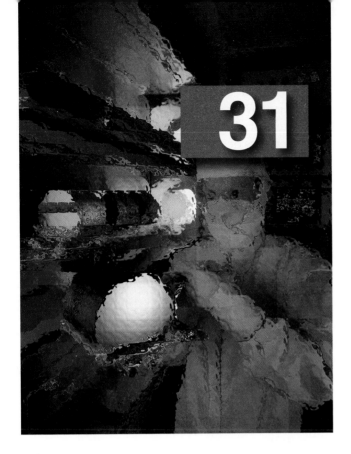

31 Processes Used to Condition Wood Materials

Chapter Highlights

- Conditioning processes are used to make wood products resistant to moisture and temperature extremes.
- Microwave drying (RF heating) improves the strength, impact resistance, and finished quality of wood.
- Various processes can be used to make wood more supple and easier to form.
- Acetylation provides wood with increased toughness and impact resistance, while allowing it to be cut and shaped with conventional woodworking tools and techniques.
- Injecting chemicals into wood under pressure is a method used to make wood more resistant to moisture and other conditions encountered in outdoor use.

Introduction

Wood is an *anisotropic* material. This means that drastic structural changes occur as the wood's moisture content changes. The capacity to modify the internal structure of wood, through adding or removing moisture from its cells, creates an entirely new set of design requirements for products made from this material. These factors are particularly important if the product is to be used in wet environments where it also will be exposed to hot or frigid temperatures. Often, wood products can be conditioned to resist moisture.

Conditioning processes used on wood materials influence the *porosity* (moisture-absorbing capacity) of the internal structure of the material, making it more dimensionally stable. Sometimes, this is accomplished with specialized drying methods; in other cases, this is accomplished by modifying the structure of the wood using chemical treatment or processes such as high-pressure impregnation. Both conditioning and finishing processes are effective in helping to reduce the shrinkage and swelling that would occur even after conventional kiln drying.

Radio-Frequency Dielectric Heating

A process that is gaining increasing use as a means of reducing the moisture content of wood is *radio-frequency (RF) dielectric heating*. This is also referred to as *microwave drying*.

In RF heating, moisture is vaporized through induced heat. One of the reasons for using RF heating is to speed the drying time of stacks of wood. Another is that the process improves the impact resistance, shear strength, and overall finished quality of the wood. Radio-frequency heating also prevents defects, such as end checking and distortion, that can occur with conventional drying processes. The RF process is most effective when it is used with hardwoods, such as oak, walnut, or ash.

The RF heating process uses high-frequency current (microwaves) to achieve uniform distribution of heat throughout the wood. Besides being used for drying and curing wood, RF heating can be used to speed the drying of coatings, paints, and inks on wood. If excessive energy levels are used, however, RF heating can cause damage and burning of the material.

Researchers have discovered that microwave energy can heat and dry pine and spruce 20 to 30 times faster than traditional methods, with no degradation. Studies conducted by Oak Ridge National Laboratory (ORNL) on microwave pretreatment of hardwoods found that RF drying pushes water out of wood and reduces the time required for drying. This can lead to significant energy savings through the use of microwave pretreatment prior to conventional drying.

Wood Plastic Composition (WPC)

The process referred to as *wood plastic composition (WPC)* uses radioactive isotopes to transform wood into a material that is much like plastic. The material that is created is called *irradiated wood*.

To make irradiated wood, the material is placed in a vacuum chamber, and all of the air is removed. A plastic monomer, normally methacrylate, is then introduced and is drawn by vacuum into the cells of the wood.

After the wood cells are filled with plastic, the workpiece is bombarded with gamma rays from radioactive isotopes. This exposure to radiation generates heat in the material, and hastens polymerization. When the workpiece cures, the wood has been transformed into a material with many of the characteristics of a plastic. See **Figure 31-1**.

The major advantage of WPC is that it creates workpieces that are much more resistant to moisture than conventional wood. Another important reason for selecting WPC material over conventional wood is that it has improved strength and dimensional stability.

Much of the early work on WPC was done by Dr. James Kent at West Virginia University. This research, begun in 1969, was sponsored by the Atomic Energy Commission. Researchers found that the wood specimens had greatly improved mechanical properties after being irradiated with gamma rays for twenty hours. The treated material was as much as eleven times harder than natural wood, seven times more abrasion-resistant, and up to six times more able to withstand compression perpendicular to the grain.

Today, there are many other types of "plastic wood." Some of the most popular variations are impreg, compreg, and acetylated wood. These variations are discussed in the following sections.

There are also other methods used to chemically modify wood to make it more resistant to moisture and chemical attack. For years, materials such as beeswax, tar, paraffin, or asphalt have been used to

Figure 31-1. Wood is transformed into a material with many of the properties of a plastic in the wood plastic composition (WPC) process. A—A vacuum is drawn in the chamber. B—The plastic monomer is introduced and drawn into the cells of the wood. C—Gamma rays bombard the plastic-impregnated wood. D—Heat generated by the action of the gamma rays speeds up the curing process.

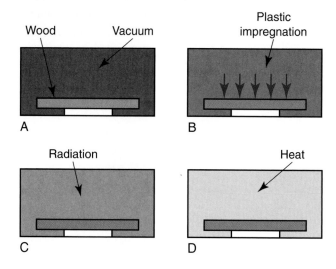

fill the pores of wood and make it more resistant to moisture and chemical attack. Even such materials as molten sulfur and inorganic salts have been used as impregnating agents. Sulfur improves the wood's resistance to chemical attack; inorganic salts reduce shrinkage as the wood dries out after being subjected to moisture.

Impreg

Another process that involves saturating the cells of wood with plastic is called *impreg*. With impreg, a fiber-penetrating thermoset resin is used. The plastic or plasticized wood is then cured without compression.

The most successful use of thermosetting resins for making impreg has involved water-soluble phenol-formaldehyde resins with initially low molecular weights. Thus far, no thermoplastic resins have been found that can be used to saturate the cells of the wood and provide dimensional stability.

The thermoset resin penetrates the cell wall, keeping the wood in a swollen condition. It is then polymerized by applying heat to produce a water-insoluble resin incorporated in the cell walls of the wood stock.

Impreg is usually made using *green veneers* to facilitate resin pickup. Veneers that are green are those that have not undergone much drying and aging. The green veneer is either soaked in the resin or impregnated with the resin under pressure. When pressure-induced impregnation is used, the workpiece is kept under pressure until the resin content equals a 25 to 35 percent gain in weight (when compared to the dry weight of the wood). If the stock is impregnated to the 35 percent level, shrinkage and swelling when exposed to moisture changes may be one-third less than wood dried by conventional methods.

When impregnation is accomplished by soaking, solution temperatures of 70°F to 100°F (21°C to 38°C) are used to facilitate resin penetration. Treatment is usually limited to stock less than 1/3" (8.3 mm) thick, because treating time increases rapidly with increased thicknesses. Also, thicker wood may be subject to checking and cracking when drying. Often, veneers are laminated together to make products such as handles for knives and other kitchen utensils.

Once the wood is impregnated with resin, it is dried at 175°F to 200°F (79°C to 93°C) for approximately 30 minutes. This produces moisture content of about 10 percent. The drying process is conducted using this low temperature to keep from forcing the uncured resin to the surface of the wood.

After the treated veneer is initially dried, it is then cured by heating at high temperature, usually 310°F (154°C) for 30 minutes. When impreg has been dried and cured, it is normally reddish-brown in color. If another color is desired, the material can be finished with most paints and varnishes.

Most adhesives can be used with impreg. However, those containing a high percentage of solvent should be allowed to evaporate somewhat before they are applied.

Compared to unmodified wood and plywood, impreg has improved properties. Swelling in water is reduced to 35 percent of that exhibited by untreated wood. This, in turn, creates a reduction in grain raising and surface checking. Impreg also exhibits improved compressive strength.

Impreg is useful when resistance to decay, rot, termites, and marine borers is required for an application. This property is probably due to the fact that the treated cell walls cannot take up sufficient moisture to support biological attack.

Impreg also has a high resistance to acid and can even be used to make storage containers for batteries. It has also been used for electrical control equipment because it offers electrical resistance higher than that of untreated wood.

Despite impreg's many advantages, the product does have some limitations. While the addition of the resin increases the weight of the material, it does little to improve the wood's overall structural strength, and actually *reduces* its impact bending strength. For these reasons, it is not recommended for structural applications. Impreg is most often used in the construction of pattern and die models, and in the manufacture of plywood paneling. A model for an automobile roof master die, made from mahogany impreg, is shown in **Figure 31-2**.

Stabilizing Wood

Spinoffs from variations of the process of impreg led to the discovery of several new processes for conditioning wood using *stabilization* processes. Craftsman making pens and knives found that softwoods often had fantastic grain patterns, but they were not waterproof, or resistant to abuse through daily use. The use of stabilizing chemicals meant that they could use these softwoods and make them much

Figure 31-2. Impreg is often used to make master die models, such as this one for an automobile roof panel. (Forest Products Laboratory, USDA Forest Service)

more durable. The process involves injecting dyes into the material while it is being pressure treated led to new promise and potential. The advantage of this process is that the resulting product is still wood and still has the look and feel of wood. It does not look like an imitation—plastic designed to simulate wood. There are many unique grain patterns and textures that can be created.

Methods used today for stabilizing wood vary from nontoxic brush-on water solubles to highly toxic chemicals that require protective masks, protective safety clothing, and special ventilation. Some of these chemicals can cause severe burning. When these chemicals are dried, they produce toxic dusts that are so corrosive they must be removed through special air-filtration systems. Some of the processes used today to stabilize wood are shown in **Figure 31-3**.

Compreg

Another type of plastic wood is made with a process known as *compreg*. Compreg is similar to impreg, but the wood is compressed prior to curing. Impreg involves no compression.

While impreg is limited to use on veneers and thin stock, compreg is used to make larger solids such as water-lubricated bearings, gears, and aircraft parts. When large solids are desired, it is often popular to sandwich treated veneers into panels and compress them without the use of bonding resins.

During the mid-1990s, *Diamondwood* and *Pakkawood*™ became popular. Both are types of plasticized hard woods. Pakkawood and Diamondwood are made from very thin layers of wood that is dyed and then pressure-treated with resin.

With the introduction of this type of plasticized wood, many beautiful colors and textures became available. Hobbyists making pens and knives found these materials perfect for the precision work that they did. Wood turners also liked these woods because there were no knots or imperfections for the cutting tools to catch on.

Water-soluble phenol-formaldehyde resin is often used for making compreg. The treated wood is compressed while the resin is curing in its cells. The percentage of resin used will vary, depending on the product application. When maximum impact strength is required, only 10 to 20 percent resin is used. If the product requires high dimensional stability, as much as 30 percent resin might be used.

When compreg is made from veneers, each strip of veneer is treated with water-soluble phenol-formaldehyde to a level of 25 to 30 percent resin. The veneers are then dried at a temperature of no more than 175°F (79°C).

Figure 31-3. This table describes popular wood stabilization methods.

Stabilization Process	Approach	Application
Surface coating	Surface treating rotten wood to make it accept fillers for reuse.	Used by antique restorers to permit use of original wood and strengthen areas where defects occur.
Pressure impregnation	Forcing strong mixture of chemical into the wood.	Used with softwoods to produce hard, plasticized stock that will not splinter or crack. Similar appearance to wood.
Selective chemical impregnation	Applying hardening chemicals to end checks, cracks, and voids in hardwoods and softwoods.	Used on areas that did not absorb materials through pressure/injection.

The next step is to compact the plies to bond the sheets together, forming the product. The pressure used depends on the resin and type of wood, but it is typically 1000 psi to 1200 psi (6895 kPa to 8314 kPa). After compression, the product is heated to produce the final cure. This is normally done at a temperature of approximately 300°F (149°C).

The compression process results in improved moldability and appearance, when compared to impreg. Compreg is used to make products such as cutlery and tool handles, model airplane propellers, and antenna masts. It is also a popular choice for constructing dies and jigs, water-lubricated bearings, pulleys, and even electrical insulators.

Case hardening, a process often used with metallic materials to produce a tough outer shell on the workpiece, can also be done with compreg. It is used to produce a tough cladding on the exterior surface of the wood by adhering compreg panels to a core of ordinary wood. An example would be compreg used as the outer ply on a plywood panel where abrasion resistance is needed. It is also used for handles of knives and other utensils.

Neither compreg nor impreg will swell and crumble like particle board or other untreated wood when used in moist environments, but both are more brittle than untreated wood.

Both compreg and impreg are hard materials because of the cured resin they contain. Consequently, conventional wood tools will dull quickly when used to cut or shape them. Metal cutting tools are recommended when machining is necessary.

Acetylated Wood

Acetylated wood was developed while searching for a modified wood that would provide the strength of compreg and impreg, but without their brittleness. Acetylation is a chemical process that is conducted in an airtight chamber.

Many different chemicals have been used to impregnate conventional wood in the acetylation process. These include the anhydrides, epoxides, isocyanates, acid chlorides, carboxylic acids, lactones, alkyl chlorides, and nitriles. The most popular chemical for making acetylated wood is acetic anhydride.

Acetylation requires the use of dry wood with less than 5 percent moisture content. The wood is placed in a pressure cylinder, with the temperature elevated to between 230°F and 250°F (110°C and 121°C). Acetic anhydride is introduced to the chamber to acetylate the wood. A pyridine catalyst is used to speed up the reaction. The acetylated wood process is popular for treating thin stock and veneers, and can be done on all types of hardwoods and softwoods. It reduces wood shrinkage and swelling by as much as 30 percent, compared to wood processed by conventional kiln drying.

The process also makes the wood resistant to attack by fungi, termites, and marine organisms, and causes little change in color. Once a wood has been acetylated, it is more stable in sunlight. It also will have better acoustical properties. This makes acetylated wood particularly useful for making speakers and sound systems.

Like compreg and impreg, acetylated wood creates a product with increased toughness and impact resistance. A major advantage of acetylated wood is that it is more flexible than either compreg or impreg and is much easier to machine. Unlike compreg and impreg, it can be cut and shaped with conventional woodcutting tools. Acetylated wood can be finished with the same processes used on conventional wood.

Plasticized Wood

Wood can be made supple, or *plasticized*, by exposing it to ammonia. This allows it to be more easily shaped or formed. The process is still in the early stages of commercialization, but appears to have considerable potential for the wood material processing industry.

After the material is exposed to the ammonia, it can be formed and then dried. It will keep its original properties in the new form. There is also the advantage of more rapid processing of the workpiece. The common wood-forming processes, such as laminating, soaking, or steaming, require holding the stock in a form for an extended period of time until it is dried and set. The anhydrous ammonia process requires keeping the wood in a form for only a brief period of time, and springback is minimal.

Research has demonstrated that wood can be plasticized by using anhydrous ammonia in either the liquid or gaseous state. With the liquid ammonia process, the wood stock is soaked in a tank for a period of time (normally from one to three hours for 3/4″ stock). The process makes the wood so flexible that small strips of it can even be tied in knots. There are some limitations to the process, however. The anhydrous ammonia that is used must be kept below –31°F (–33°C), so that it remains in a liquid state. The wood that is to be plasticized must be precooled before treatment.

The gaseous ammonia process also begins with cooling the workpiece. The workpiece is then placed in an airtight vacuum chamber. The chamber is filled with anhydrous ammonia gas. Once the sample has been exposed long enough for it to be plasticized, the gas is removed from the chamber. The stock is then placed in a form for a brief period of time to set in the desired shape. See **Figure 31-4**.

Staypak

Since compreg and impreg are both more brittle than untreated wood, the U.S. Forest Products Laboratory conducted research to find a suitable alternative. They developed another type of modified wood, called *staypak*, which does not use any resin. Instead, wood preconditioned to between 30 percent and 65 percent relative humidity is compressed and heated. The pressure (1,400 psi to 1,600 psi, or 9,653 kPa to 11,032 kPa) and heat (350°F or 177°C) cause the natural lignin between the wood fibers to liquefy, and the entire mass to realign itself. This realignment process reduces internal stresses and creates a new type of wood with improved tensile and flexural properties.

Both veneers and solid wood can be used to make staypack. If solid wood is used, it must be free from knots. Pieces of the resulting material can be glued together and finished like untreated wood.

Figure 31-4. Anhydrous ammonia gas is used to plasticize wood. In the gaseous ammonia process, the wood is cooled in a soaking bath. A vacuum is then drawn in the treatment chamber, causing liquid ammonia to vaporize into a gaseous form that will penetrate the wood.

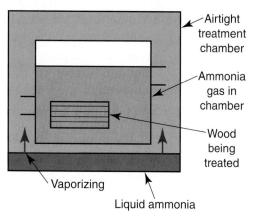

Like other modified wood products, staypak has increased impact strength. Its strength increases almost proportionally to the amount of compression. The greatest increase is in impact bending strength. Staypak is not as resistant to water or biological attack as compreg. However, it does absorb water at a slower rate than uncompressed wood, and is more dimensionally stable.

Typical product applications for staypak are tool handles, mallet heads, and tooling jigs and dies. Materials similar to staypak are produced in Germany (under the tradenames *Lignostone* and *Lignofol*). In England, *Jicwood* and *Jablo* are two types of compressed plywood that are similar to staypak.

Pressure-treated Wood

Both government and private agencies have conducted a great deal of research and experimentation on chemical treatment for the preservation of wood. All industries in the United States must comply with regulations that implement various statutes at the federal and state level. Those regulations have direct impact on the availability, manufacture, and use of treated products such as utility poles, railroad cross ties, posts, and construction lumber that must be treated to withstand exposure to insects, weather, and moisture.

All of the major wood preservatives used today face the same regulatory programs. Primary concerns address pesticide regulation, waste management, human and environmental exposure, and availability of forest resources. The regulations concerning these issues continually evolve.

The processes for applying the wood preservative to the properly conditioned wood are covered by national standards published by the American Wood-Preservers' Association (AWPA). **Figure 31-5** shows some of the regulatory acts and agencies controlling manufacture, use, and disposal of pressure-treated wood products.

Hundreds of volumes have been written on methods for using wood preservatives, the impact of processes used to manufacture these products on the environment, disposal of pressure-treated products, and the impact of these products on people, animals, and the environment. As a result, the industry has made major strides forward to comply, and to manufacture products which are safe when properly used and discarded.

Figure 31-5. Legislation regulating use and disposal of pressure-treated wood

Legislation	Function or Type of Regulation	Impact
RCRA (Resource Conservation and Recovery ACT)	Hazardous waste regulations	Wastes from wood preserving plants, including inorganic arsenical preservatives, creosote, and pentachlorophenol (penta). Also addresses groundwater and site contamination.
FIFRA Fungicide and Rodenticide Act)	Regulation and use of pesticides such as wood preservatives	Use of three major types of pesticides: creosote, pentachlorophenol, and inorganic arsenicals. To minimize general public exposure to these preservatives, retail sales were eliminated by classifying them as restricted use pesticides. This required that they be used only by, or under the direction of, a trained and certified pesticide applicator.
OSHA (Occupational Safety and Health Act)	Remissible Exposure Limits (PEL) for wood preservatives; Hazard Communication Regulations (1984) requiring Material Safety Data Sheets (MSDS) be prepared for all wood products (untreated or treated) that might be sawn or machined. This regulation was introduced by the International Agency for Research on Cancer (IARC), which classified wood dust as a potential carcinogen.	All industries
Clean Water and Air Act	Any discharge of process wastewater or storm water must be covered by appropriate permits	All industries
SARA (Superfund Amendments and Reauthorization Act)	Inventory of chemicals and releases and transfers of hazardous chemicals	All industries

It is misleading to talk about safe and unsafe chemicals. There are no chemicals that are safe, only safe ways to use, store, and dispose them. Pressure-treated products are safe if they are used properly. Treated wood should not be used to store drinking water, or to make bowls for eating, counter tops, and kitchen utensils. Treated wood needs to be disposed of through trash collection methods or by burial, not by burning.

The Southern Pine Association classifies chemical wood preservatives into three types: (1) Creosote, (2) Oil-borne preservatives, and (3) Water-borne preservatives.

Preservation treatments are available that can be used to make wood resistant to almost any type of harsh environment. *Preservatives* can be fire-resistant,

moisture-repellent, colorless, and odorless. Some are even capable of receiving a surface finish. Most treatments impregnate the cells of the wood under pressure. Pressure-treating processes are used to produce a uniform deep penetration of the wood.

Typically, the processes involve holding the workpiece in a closed horizontal pressure cylinder, called a *retort*, while the preservatives are introduced into the wood. The retort usually is 6' to 8' in diameter and 100' to 150' in length with a door on both ends. The stock to be treated is carried on a *tram* (train car) that runs into the retort.

There are two basic types of pressure treating systems used today: *empty-cell impregnation*, used to apply creosote, and *full-cell impregnation*, used to apply fire-retardant and moisture-protectant chemicals.

Empty cell processes are used to apply oil-borne preservatives; full cell treatments are used for water-borne preservatives. With the empty cell process, air pressure is applied to the wood prior to filling the retort with preservative. Full cell treatments require the application of vacuum prior to filling with preservatives.

The empty-cell process starts by placing the wood to be impregnated in a stream of high-pressure air. The preservative is then introduced and the pressure increased to force it into the wood cells. After a given length of time, the air pressure is returned to normal. This allows the air trapped inside the wood cells to push any excess preservative out of the cells.

The advantage of the empty-cell method is that it produces more uniform penetration than the full-cell method. The empty-cell process is used to treat telephone poles, posts, and lumber for piers and bulkheads.

Pressure-treated products have been used in the United States for telephone utility poles since the early 1930s. Many tests have proven the effectiveness of treated lumber to resist failure from decay or attacks by insects.

The preservative treatment of wood used for utility poles requires replacing some of the natural moisture in the wood with a liquid preservative solution. However, before the preservative can be placed into the wood cells, sufficient moisture must be removed. This is accomplished through natural air seasoning, or with accelerated seasoning processes such as kiln drying, steam conditioning, or heating in the preservative under vacuum (**Boulton drying**).

Pressure treating wood does not make it waterproof, but it does protect the wood by making it useless as a food source for wood destroying insects and organisms. What takes place is a chemical reaction where the original water soluble solution of CCA (chromated copper arsenate) is reduced by wood sugars to form highly insoluble and leach-resistant precipitates. These precipitates are fixed in the wood, are nonvolatile, and will not vaporize or evaporate.

In the full-cell process, the wood is placed in the treatment chamber and a slight vacuum is applied. In this method, the preservative slowly penetrates the cells, rather than being forced in aggressively as it is with the empty-cell process. A typical 40' utility pole might contain as much as ten pounds of preservative.

The full-cell method is used when products must be able to withstand continued exposure to harsh conditions. Products treated by the full-cell method include those that must be *fire-retardant* or must withstand continued exposure to water.

Other pressure-treating processes are gaining popularity. Some make it possible to treat only the end of a timber or post that will be exposed to weather or other harsh conditions. Others involve drilling holes in the product and treating only the areas around the drilled holes. Other processes use liquefied petroleum gas (LPG) to distribute the preservative.

Cautions When Handling Pressure-treated Wood

Wood that has been pressure-treated using CCA (chromated copper arsenate) should be used only for applications requiring protection from insects and decay. The chemical penetrates deeply into the wood. However, over time some chemical residue may work its way into the soil. If improperly handled, particles may come into contact with the skin.

Precautions should be taken when using treated products. Treated wood should not be used where it will contact food, humans, or animal feed. In addition, it should never be used as mulch, cutting boards, counter tops, animal bedding, or structures and containers for storing human or animal food and drinking water.

Disposal of treated wood should be made through trash collection methods or landfill. Do not burn treated wood, unless it is done in a commercial or industrial incinerator. The toxic chemicals produced with the smoke and ashes must be contained.

Avoid prolonged inhalation of sawdust from treated wood. Wear a dust mask when machining treated stock. When possible, cut treated wood outside to minimize accumulation of airborne sawdust. Use safety glasses or goggles and gloves when handling the wood.

Polyethylene Glycol

One chemical treatment that results in improved dimensional stability is conducted by soaking the wood in **polyethylene glycol (PEG)**. PEG is a white, waxy chemical that looks similar to *paraffin*. Polyethylene glycol melts at 104°F (40°C), dissolves readily in warm water, and is noncorrosive, odorless, and colorless. PEG is used widely in pharmaceuticals, cosmetics, agricultural products, and many industrial applications.

Polyethylene glycol is normally used to fill the cells and cell walls of green wood. No pressure is required. The process involves immersion of the wood product in a solution of PEG dissolved in warm water. Treatment is usually done at temperatures ranging from 70°F to 140°F (21°C to 60°C). Diffusion of the material into the wood is greatly accelerated by increasing the temperature and keeping the concentration of the solution at about 30 percent. The PEG treating solution can be used many times.

PEG treatment is slow. How long it takes depends on the wood thickness and density. In many cases, the treatment takes weeks. After parts are treated with PEG, they are *stickered* (stacked using spacers between boards) and dried in a well-ventilated heated room. Time necessary for air-drying depends on thickness, temperature, and relative humidity.

Treatment with PEG is not permanent. No curing takes place, so the chemical remains water-soluble. If the treated material is exposed to water, the polyethylene glycol will leach out. For this reason, PEG-treated wood will become sticky in high humidities (above 60 percent). Usually, treated wood is finished with two coats of a polyurethane varnish to seal in the glycol.

Polyethylene glycol is used to help in eliminating checking and cracking of wood during drying. It has even been successfully used to preserve antiques and to prevent rare wood artifacts discovered in the ocean from cracking during drying. The material can be used in any application where the wood is wet and needs to be dried gradually without cracking.

Important Terms

acetylated wood
anisotropic
Boulton drying
compreg
Diamondwood
empty-cell impregnation
fire-retardant
full-cell impregnation
green veneers
impreg
irradiated wood
Pakkawood™
paraffin
plasticized
polyethylene glycol (PEG)
porosity
preservatives
radio-frequency (RF) dielectric heating
retort
stabilization
staypak
stickered
wood plastic composition (WPC)

Questions for Review and Discussion

1. What is the advantage of using the polyethylene glycol (PEG) process in preference to other processes? What is the major advantage of the PEG process?

2. What is the greatest problem with irradiated wood?

3. What process would be best for making wood impervious to chemical attack? Why do you feel that this is true?

4. What are some of the methods you might select for impregnating wood to make it resistant to moisture and swelling?

The process of pressure treating wood involves moisture monitoring, drying, and computer-controlled infusion of preservatives in large pressure cylinders. This company redries the wood after treating to reduce warpage and shrinkage. (Cox Industries, Inc., Orangeburg, SC)

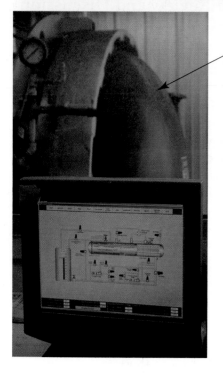

32 Processes Used to Condition Ceramic Materials

Chapter Highlights

- Sintering is the most common method of increasing the density of a ceramic workpiece.
- Calculation of the bulk density of a ceramic workpiece is important in identifying its porosity and potential for absorbing moisture.
- Most ceramic products must be dehydrated before they can be fired.
- There are four major types of sintering used in producing ceramic products.
- Reaction sintering, used to produce ultradense parts, results in a near-net-shape product.
- Pyrometric cones, developed in the 18th century, are still widely used to identify the proper firing range for ceramics.

Introduction

The behavior of any solid, liquid, or gaseous material is often described in terms of *density*, or *weight of the material per unit volume.* The behavior of a material depends on such factors as grain size, distribution of the atoms, and surface area. These factors are used to describe how powder will pack, flow, react, and yield to pressure.

Density is normally described in terms of pounds per cubic foot (lb/ft^3) or grams per cubic centimeter (gm/cm^3). Density is one of the most frequently recorded measures of the morphological characteristics of ceramics.

Primary Processing

Most ceramics have traditionally come from oxide minerals or other minerals that can produce oxides, like phosphates, sulphides, hydroxides, and carbonates. These minerals incorporate most of the elements that occur on Earth's surface. About half of all known elements occur as oxides, normally as complex oxides such as silicates. It is hypothesized that weathering of silicates, followed by sedimentation, is what leads to formation of clay minerals.

It is interesting to know that the important refractory MgO (magnesium oxide) and several other ceramics are recovered from seawater. However, most of the minerals important in ceramics come from igneous rocks, such as granite or basalt, that crystallized from magma.

Most traditional and advanced ceramic processes require the use of high purity powders that are *beneficiated* (milled and refined) to produce crude minerals, industrial minerals, or industrial chemicals. See **Figure 32-1**.

Primary processing operations (crushing and grinding) are used to separate the desired minerals from undesired components (impurities and organics). This is accomplished by settling, magnetic separation, and collecting the fine-particle clay powder using flotation processes. The clay powder that results varies widely in composition and particle size, depending on the site from which it originated and the process methods. For example, high quality Georgia kaolin is known for its low iron content (high purity), which is helpful for making white wares. Clay with high iron content causes discoloration. Ball clays contain more organic material and are more plastic.

Figure 32-1. Beneficiated clay materials.

Beneficiated Material	Types of Material
Crude minerals	Ball clay, shale, tile, stoneware, and others
Industrial minerals	Purified ball clay, kaoline, talc, feldspar, glass sand, patter's flint, bauxite, zircon, and others
Industrial chemicals	Calcined magnesia (from seawater), fused alumina and magnesia, silicon carbide, soda ash, titania, iron oxide, stabilized zirconia, and others

Densification

The process of increasing the density of a ceramic part is called *densification*, and it is most often accomplished by sintering. As you may recall from a previous chapter, sintering transforms the loosely bonded green compact formed by dry pressing into a dense ceramic product.

Sintering is simply diffusion on an atomic scale. Moisture and organic materials are burned out of the green body during the initial stages of firing. Then, the temperature of the furnace is raised to the point where diffusion occurs. At this point, matter is channeled from the particles into the openings between the particles. This results in densification and shrinkage of the workpiece.

Density is a physical property of matter that describes the compactness, how closely packed atoms of an element or molecules of a compound really are. The more closely packed the atoms, the denser the material is. Different materials have different degrees of density, so this is often a useful way to determine the type of material.

The type of density that is often measured on ceramic parts is referred to as *bulk density* (P_b). This measurement is based on the bulk volume of the part. Ceramic products are generally porous and cells are generally open or closed. A ceramic workpiece with open pores will allow a liquid such as water to penetrate it. Closed porosity refers to those pores that have become sealed within the body or mass of the ceramic. Sealing of the pores can be accomplished with binders and through compression. In either case, the pores can affect the strength of ceramics by creating stress concentrations. Since ceramic workpieces lack the property of plastic-deformation, once the stress on the part reaches the critical level, a crack will form and spread unchecked. This may lead to fatigue or breakage.

This is an important principle to understand, because reducing or eliminating pores in ceramics is essential if the full strength of the ceramic is to be realized. Aside from concentrating stress, pores reduce the strength of ceramics because they also reduce the cross-sectional area over which a load can be applied. This lowers the stress that these materials are able to withstand.

One method to determine density of porous materials is defined by ASTM C693, the *Standard Test Method for Density of Glass by Buoyancy*, developed by the American National Standards Institute. This is based on *Archimedes' Principle*, which states that a body, or mass of material immersed in a fluid is buoyed up by a force equal to the weight of the displaced fluid. The procedure involves submerging the part in water. Other liquids can be used for materials with densities less than 1.0 grams/cm³ (a ceramic that would float in water). This method compares the weight of the part in air to the weight of the displaced water (or other liquid of known density) of the submerged part. Depending on the nature of the specimen (open or closed cells), the resultant value may deviate from the true mass.

Here is how the measurement is obtained. A clean specimen is weighed accurately in air using a laboratory balance. The same specimen is weighed while suspended in water or other liquid of such density that the specimen will sink. Next, the volume of the specimen is calculated by noting how much liquid was displaced by the submerged part (Archimedean principle).

It is important to understand the relationship of density on the strength of a ceramic product. We know that density relates directly to the porosity of the ceramic workpiece. The greater the porosity of the part, the more likely it will be to absorb liquids and vapors into the material, and the less dense that it will be. During curing of a freshly molded part, the individual grains are drawn together until they contact each other, resulting in shrinkage. Usually, absorption of moisture into the porous compact will result in excessive shrinkage and structural damage to the material during drying and firing. The higher the moisture content, the greater the shrinkage that will result.

Depending on the product's end use, pores may or may not be desirable. The density of the grains and the number and sizes of the pores in the ceramic product must be carefully controlled. For example, unglazed tableware would be undesirable, because it would become permanently stained and would be unsanitary. On the other hand, for applications such as residential construction, a product such as a highly porous, lightweight brick may be desirable.

The major factor influencing the conversion of powder particles into a solid body is the reduction of surface energy. Thus, the alignment of ceramic particles in the green compact is critical to the thermodynamics of densification. This is not always a simple matter to achieve.

One of the most difficult problems encountered in preparing fine ceramic powder is achieving homogeneity in the powder. The individual grains are so small that even after thorough mixing, inconsistencies arise. Several approaches are being used to obtain powder consistency and improve the capacity for densification of the final product.

Chemical processes are sometimes used with materials such as uranium, zirconium, thorium, and several other metal oxides. All of the desired metal ions are prepared in solution, and then precipitated as hydroxides or oxalates. Chemical precipitation of materials is performed during intensive mixing. One of the problems with this process is the low concentration of solids. Often, a mixture of 10 percent solids and 90 percent water will form a gel that is difficult to dehydrate by filter pressing. Normally, the solid must be dried, then compacted or pelletized before it can be further used.

The fluid energy milling process discussed in a previous chapter is also used to produce the finely ground particles needed for densification. The mill utilizes a superheated stream of air—1500°F to 1700°F (800°C to 900°C)—to grind the particles. This enables the grinding operation to be performed at temperatures that are above the decomposition temperatures of many nitrates, sulfates, hydroxides, and oxalates. The high temperature helps break down the particles much faster and produces finer grains in the mixture that is milled.

Materials that are ground in the fluid energy mill can be mixed slurries, salt solutions, molten salts, or even solid compounds. The oxide particles that are produced are so fine that it is difficult to separate them from the heated gas in which they are suspended. The particles that are generated from fluid energy milling are like thin-walled bubbles or hollow spheres, many of which are so fragile that they will be destroyed before they can be collected. To make these tiny crystals useful for ceramic processing, a process called *calcination* is used to grow the particles to the desired size.

Calcination

Calcination is a high-temperature heat-treating process for ceramics. This process is used to achieve uniformity in particle sizes and to prepare ceramic powders for later processing. Calcining improves the interaction between the constituents in a ceramic body, prior to sintering, and results in improved densification. A calcined part is more dimensionally stable and more desirable for many electrical applications.

The temperatures used and approaches taken depend on the material characteristics of the powder being calcined. Calcining can be used to decompose a salt or a hydrate to an oxide.

Calcining can also be used to dehydrate materials, as is done when producing plaster from gypsum. Heat treatment dehydrates the gypsum and results in plaster of paris. If water is added, the material once again forms gypsum. If too high a temperature is used to dehydrate the gypsum, all of the water will be removed, and an anhydrous powder will be created. This powder cannot be easily rehydrated.

In other instances, calcining is used to coarsen the powder, increasing the size of the ceramic particles so that the powder can be more easily compressed. Coarsening is usually accomplished by loosening particles to create aggregates that function like larger particles.

Freeze-Drying

The conventional method of drying ceramic materials before firing is to use either ambient air or heated air from a kiln. Drying is sometimes done in a vertical oven that moves ceramic pieces up and down on a conveyor through heated air.

A less conventional method is *freeze-drying*, also known as *cryochemical drying*. The process rapidly freezes the water in the clay particles, encasing drops of salt solution that hold the desired metal ions. The cryogenic process produces spheres (frozen droplets) that are 0.004″ to 0.20″ (0.101 mm to 5.08 mm) in diameter.

To obtain usable oxides from these frozen spheres, the spheres must be carefully dried by heating them in a vacuum. The moisture is removed through *sublimation* (a process in which a substance passes directly from the solid state to the gaseous state, without becoming liquid); the sphere remains intact. The resulting particles are anhydrous sulfate or salt. The salt spheres are then calcined to remove the sulfate and grow the oxide particles to the desired size.

Firing

Ceramic products that have been dried and finished are called *greenware*. At this point, the products are fragile and vulnerable to cracking and damage through handling. *Firing* consists of heating the part to an elevated temperature in a carefully controlled environment. It is normally initiated in a kiln or furnace to fuse the materials together, providing strength and permanence to the product.

Firing is necessary for all types of ceramics and glasses. It improves the densification of the materials. The strength or hardness that is developed in the ceramic product is due to the glassy bond between the oxide particles in the ceramic material. When heat removes moisture from the part, the pores close, resulting in shrinkage by as much as 15 to 20 percent.

It is common to fire some ceramic parts *twice*. The first firing is done without a glaze (coating) applied to the product, creating **bisque ware**. This firing, called the **bisque firing**, solidifies and fuses the body so it can be handled without problems in glazing. The second firing is referred to as the **glost firing**. In this firing, the clay body is sintered and the glaze develops.

Generally, there are three stages in firing:

- Presintering (bisque firing).
- Sintering (glost firing).
- Cooling.

Presintering

The actual process of sintering usually does not occur until the temperature in the product exceeds one-half to two-thirds the melting temperature of the material. The **presintering** heating process causes drying, decomposition of organic binders, vaporization of water from the surface of particles, changes in the states of some ions, and decomposition of additives. During presintering, stresses from the pressure of gas produced or from thermal expansion must not cause cracks in the fragile greenware.

Sintering

Sintering can be defined as the process of densifying a ceramic compact with the application of heat. Sintering eliminates the pores between the particles of powder, and therefore results in shrinkage of the part. The first melting phase in sintering normally takes place between 700°F and 750°F (371°C and 399°C). The major types of sintering are shown in **Figure 32-2**.

Figure 32-2. Processes used to sinter industrial ceramics.

Sintering Industrial Ceramics
Vapor-phase
Solid-phase
Liquid-phase
Reactive-phase
Viscous

Vapor-Phase Sintering

In *vapor-phase sintering*, or *pressure-assisted sintering*, pressure causes ceramic powder to be shifted from the surface of the particles to the point where one particle contacts another. Vapor-phase sintering reduces both the size of the particles and the pore space between them. Sintering is accomplished by evaporation and condensation or by vaporization pressure. The strength of the product is increased by reducing the pore space.

When hot pressing is used to compact the material, the density of the compact is increased by one order of magnitude at 20–40 MPa pressure. Hot isostatic pressing increases the density of the compact by two orders of magnitude at 200–300 MPa pressure. One order of magnitude means that the quantity is ten times as great. A pressure of 20 MPa increased by one magnitude would be 20 × 10 or 200 MPa pressure. A pressure of 200 MPa increased by two orders of magnitude would be 100 times as great, 200 × 100 or 20,000 MPa pressure. Pressure-assisted techniques are effective for sintering materials that are difficult to sinter, such as composites or covalent solids.

Solid-Phase Sintering

Solid-phase sintering involves movement of surface material by diffusion. Diffusion can occur as movement of surface atoms or may take place throughout the material. This process may or may not result in shrinkage, depending on the sintering time and temperature. Solid-phase sintering results in material being fused where the particles are in contact with each other. See **Figure 32-3**.

Liquid-Phase Sintering

Liquid-phase sintering involves wetting the ceramic particles with a viscous liquid before heating the part to the sintering temperature. Liquid-phase sintering is accomplished by mixing powders of different chemical compositions and melting points.

Liquid held between the grains of the material creates capillary pressure. The capillary pressures generated by silicate liquids can exceed 1000 psi. This enhances densification by reorienting the particles to improve compacting. It also increases the contact between the particles, resulting in plastic deformation.

Liquid-phase sintering can be used in cases where grain-boundary impurities, such as frozen solids, do not interfere with the performance (electrical, mechanical, or thermal) of the sintered component. This is frequently the case with technical silicates used at room temperature. The liquid additive remains in the grain boundaries during the early stages of sintering. In some cases, the liquid dissolves in the solid or is chemically or structurally changed through sintering. This is what occurs when processing high-temperature ceramics such as silicon nitride (Si_3N_4).

In most cases, liquid-phase sintering is much faster than solid-state sintering, because of liquid capillary action. This causes the equivalent of great external pressure and helps to rearrange particles early in the sintering process.

Reactive-Phase Sintering

A variation of liquid-phase sintering is called *reactive-phase sintering*. This is known by many names including *reaction bonding* and *transient sintering*. As in liquid-phase sintering, a liquid is present during sintering. However, in reactive-phase sintering, the liquid breaks down or disappears. Because of the effect of the liquid on the sintering process, parts processed by this method sometimes can even be used at temperatures exceeding the initial sintering temperature.

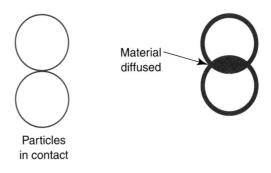

Figure 32-3. In solid-phase sintering, material diffuses from one particle to another through physical contact.

Material diffused

Particles in contact

Reactive-phase sintering is a process that yields a part with less shrinkage than conventional sintering processes. Reaction-sintered silicon nitride can be produced from silicon powder. Pressing, injection molding, or slip casting is used to produce the green compact. Once the compact is formed, it is placed in an oven with a nitrogen, nitrogen/hydrogen, or nitrogen/helium atmosphere. Initially it is heated to a temperature of approximately 700°F (1200°C).

The heat forces the nitrogen into the pores of the compact, and the reaction begins. As the temperature is increased, the reaction rate increases. Reaction-sintered *silicon nitride* is a material with tremendous potential for manufacturing. Silicon nitride has a relatively low elastic modulus and coefficient of thermal expansion, and relatively high thermal conductivity. This makes it a material with good *thermal shock resistance*. Possible product applications include turbine engine and space station components.

Silicon carbide is another material that has been produced using reaction sintering. A mixture of powdered silicon carbide and a thermosetting phenolic resin is formed using plastic processes such as casting, pressing, extrusion, or injection molding. After the molding process is completed, the plasticizer is removed by charring. This produces the carbon that is used later in the process.

The preform is then exposed at high temperatures to molten silicon. The silicon reacts with the carbon and bonds the original silicon carbide particles together. The material that is produced is nonporous silicon/silicon carbide composite with a high modulus of elasticity. This makes it particularly useful for forming complex shapes. The process is good for near-net shapes, as well—final products normally have less than one percent shrinkage during densification.

Viscous Sintering

Viscous sintering is used when the powder contains a high amount of silicates. The silicate, when used together with alkaline oxide additives, is what creates the "viscous" or syrupy behavior of the mixture. Viscous sintering occurs in the production of porcelain.

Sintering porcelain is more complicated than many advanced ceramics. This is due to the large number of impurities, ingredients, and resulting chemical reactions that can take place when producing and sintering the final product. A viscous liquid reaction results when low-melting point alkaline-alumino silicates integrate with other ingredients such as kaolin (clay) and quartz (silica sand).

Vitrification

Selecting the correct sintering temperature is particularly important in product applications such as brickmaking. When brick is fired at a temperature of approximately 1800°F (1000°C), a glassy bond begins to form and pores begin to close. If the temperature is increased, more pores will close. Consequently, the brick is made denser and stronger.

The process is known as *vitrification*. Finding the proper balance between density and strength is important. If the firing temperature is too low, the bricks will be too porous and fragile; if the temperature is too high, they will be too dense and brittle. The span of temperatures between the point of sufficient vitrification and too much vitrification is called the *firing range*.

The proper firing range depends on the composition of the material being fired and cannot be left to chance. To determine the proper firing range, a system using what are referred to as *pyrometric cones* was developed in the 18th century. The system is still used today.

Pyrometric cones are wedge-shaped pieces of clay, each with a specific melting point. Cones are purchased in a set or series. The series is arranged so that a cone will slump over at a temperature about 70°F (20°C) below that of the next one in the series. Thus, the higher the number of a cone in the series, the greater the temperature that is required to get it to slump over. See **Figure 32-4**.

When the cones are prepared for use, they are first seated in a base plaque of clay and placed in the kiln at a point where they can be observed through a viewing port in the kiln door. The cones are carefully selected by the operator so that the highest number in the series will slump at a temperature close to the melting point of the material being fired. The successive slumping of cones as the temperature rises is a visual indicator to determine when the product is approaching the melting point.

Transformation Toughening

Transformation toughening is a process that is used to increase the strength and toughness of

Figure 32-4. Pyrometric cones are purchased in a set or series. The series is arranged so that each cone will slump over at a temperature about 20°C below that of the next cone. The kiln operator uses the successive slumping of cones as a visual indicator that the product being fired is approaching the melting point. The use of cones enables firing different types of materials in the same batch.

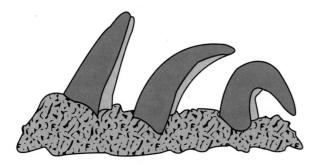

ceramics in products that must exhibit high wear resistance. Today, this process is used most frequently to toughen zirconia, the principal ceramic material in zirconium oxide. Transformation toughening is accomplished by carefully controlling the heat treatment cycle, composition of the powders, and particle size.

Zirconia goes through a phase transformation from its tetragonal crystal form to a monoclinic crystal form while cooling through a temperature of approximately 2100°F (1150°C). The transformation results in an increase in the volume of the material of about 3 percent. By controlling the conditions of compacting and processing, zirconia can be densified at higher temperatures and cooled with the tetragonal phase being carried all the way down to room temperature. What this means is that if a fault appears in the part, the crack will start to propagate. Immediately, the high stresses in the tip of the crack will halt the transformation of the adjacent tetragonal crystalline grains into monoclinic form. This will cause them to expand by approximately 3 percent, compressing the crack tip and preventing the crack from expanding. Transformation toughening produces a ceramic that is tough and strong.

Vapor Deposition Coating Processes

Various vapor deposition processes can be used to deposit nonporous ceramic coatings on a substrate. Chemical vapor deposition (CVD) and sputtering are two of the most widely used chemical deposition processes.

Chemical Vapor Deposition

Chemical vapor deposition (CVD) is a *thermochemical* process that is typically accomplished by heating the part to be coated in a vacuum chamber while flowing a controlled gas mixture over it. CVD is a process used to transform vapors (gaseous molecules called precursors) into solids, and is often used to grow thin film and powder materials. See **Figure 32-5**.

Many different variations of chemical vapor deposition are used today:

- Low-pressure chemical vapor deposition (LPCVD)
- Atmospheric pressure chemical vapor deposition (APCVD)
- Plasma-assisted chemical vapor deposition (PACVD)
- Photochemical vapor deposition (PCVD)
- Laser chemical vapor deposition (LCVD)
- Metal-organic chemical vapor deposition (MOCVD)
- Chemical beam epitaxy (CBE)
- Chemical vapor infiltration (CVI)

In each of these different approaches, the chemical vapor reacts when it contacts the heated part. This results in a very-fine-grained coating that is harder than those achieved with most other ceramic fabrication processes. The ceramic coating is applied at a rate of about 0.010" (0.254 mm) per hour. Care must be taken to obtain a uniform coating.

Figure 32-5. In chemical vapor deposition, gas reacts when it contacts the heated parts inside the vacuum chamber, producing a hard, very fine-grained coating.

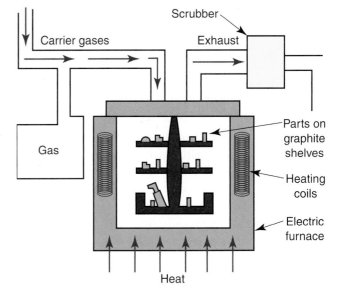

Sputtering

Sputtering is done by placing the part to be coated next to a flat plate of coating material inside an evacuated chamber. The plate, called a *target*, is bombarded by a beam of electrons. The electrons break atoms off the target so that the coating material can be deposited on the part. Only the area of the part facing the target receives the coating. Thin coatings are applied consistently, but the coating rate is very slow.

In the sputtering process, the part itself is not heated during coating. This makes it particularly useful for coating electrical substrates where insulation and protection are desired.

Ion Implantation

Ionic bonds occur when an atom of one material releases an electron that bonds to an atom of a different material. Ion implantation is a process that introduces ions into the surface of the part to improve hardness, wearability, and corrosion resistance.

In the semiconductor industry, crystals are grown using a form of ion implantation called doping. Doping involves adding *dopants* (alloying elements) to liquid metal to produce unique properties. The most commonly used semiconductor materials are silicon and germanium. The addition of impurities (dopants) is necessary to give these materials their desired electrical properties.

Doping Methods

Two major processes are popular for growing crystals. The first of these is called *crystal pulling*. With the crystal pulling method, a small, single-crystal seed is brought into contact with molten semiconductor material. The seed crystal, attached to a rotating pulling bar, is then slowly pulled away. Where the semiconductor touches the seed, it influences the growth of the seed's crystalline structure. Single crystals of materials such as titanium and zirconium are made with this process.

Semiconductors are materials somewhere between an insulator and a conductor in terms of conductivity. There are two different types of semiconductors—*intrinsic semiconductors* and *extrinsic semiconductors*. Intrinsic semiconductors are made from pure elements such as silicon. The flow of electrons in intrinsic semiconductors increases with a rise in temperature. Extrinsic semiconductors are those materials with impurities or dopants.

Crystal pulling is a process that is effective in refining and doping silicon and germanium crystals with carefully controlled orientation and purity. Most crystal pulling is done using high frequency induction heating processes. Radio frequency power supplies from 1 to 10 kW are normally used, depending on the material that is involved.

The process begins with *zone refining*, a technique developed by Bell Telephone Laboratories as a method for preparing high purity materials used for making transistors from geranium crystals. Here is how it works. When the semiconductor materials are melted, the molten material travels along a rod or ingot of the material. This causes impurities to travel along the rod and be deposited on one end of the rod. Impurity levels are dependent on the number of passes made along the rod by the molten zone.

At this point, the semiconductor material is melted in a crucible made of carbon graphite. The melt is held at a specific temperature, and a single crystal with the desired orientation is inserted into the molten mass on the end of a rod. The crystal is then quickly withdrawn. The molten material adheres to the crystal and solidifies with the same orientation as the crystal.

Another process that is used to produce single crystals is called the *floating zone method*. This process is used with materials that have high melting points, such as silicon. See **Figure 32-6**. In the floating

Figure 32-6. In the floating zone method, energy from a radio frequency (RF) coil is used to create a molten zone around the interface between a single crystal seed and a rod of polycrystalline silicon. The coil (and thus the molten zone) is slowly moved upward along the rod. The material that solidifies below the coil has the same crystalline structure as the seed.

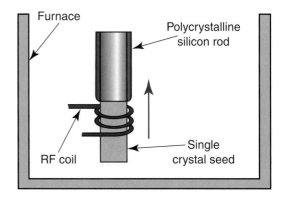

zone process, a rod of polycrystalline silicon is brought into contact with a single crystal seed. Both are suspended in a furnace. Energy from a radio frequency (RF) coil is used to create a molten zone around the interface between the seed and rod. By slowly moving the RF coil upward, the molten area moves along the originally polycrystalline rod. The material that solidifies below the coil will have the same crystalline structure as the seed. The long crystal rod is then sliced with a diamond saw into wafers about 0.020″ (0.508 mm) thick. The wafers can then be cleaned and further cut into tiny chips for electronic applications.

Crystal growth experiments have flown on shuttle missions since 1985. The ISS Expedition 2 was the first flight that tested CPCG-H (Commercial Protein Crystal Growth—High Density) hardware. The crystals that are grown in space are larger and more organized than those grown on Earth. As part of one space flight for NASA, the space shuttle *Endeavour* grew single crystalline material using the floating zone technique. Let us take a closer look at why these studies are important, and what was learned.

Studies of the structure of microgravity-grown crystals have provided important information for the development of new drugs. Crystals grown on shuttle flights have been used in the design of inhibitors that may serve as broad-spectrum antibiotics.

One of the most popular methods for studying protein structures is crystallography. Proteins are the primary building blocks of life, and the way they react to other biochemicals in living organisms provides important information on the health of a creature or organism. Each protein has a particular chemical structure or favored shape. When researchers are able to determine how the protein works, they can develop new treatments for diseases in humans, plants, and animals.

According to the NASA Marshall Space Flight Center, preliminary analysis from the crystal experiments in space indicated that at least 65 percent of the macromolecules tested in the CPCG-H experiments produced diffraction-sized crystals. X-ray diffraction studies of crystals were conducted and the data was used to determine and refine the three-dimensional structures of these macromolecules. Three important proteins, ML-I, Thermus flavus 5S RNA, BARS and Mb-YQR were used in the experiments to validate the performance of the hardware. ML-I (Mistletoe lectin) is an enzyme that has the ability to inactivate ribosomes and inhibit cell replication. It is a target for new cancer treatments. Data was also collected on human recombinant insulin crystals. It is hoped that this data will help to model the structure of this type of insulin, and that pharmaceutical companies will be able to use this information to improve insulin treatments used to control diabetes.

Another crystal-growing process used in industry is the *Czochralski method*. With this process, the seed on the end of a rod is dipped into a crucible containing melted silicon and the temperature is reduced slightly until the liquid starts to crystallize onto the seed. The seed crystal is then withdrawn, slowly at first and then more quickly until a long, thin crystal has formed on the end of the seed. This long crystal is called the *neck*. When the crystal has grown at a sufficient rate, the temperature is reduced and the growth rate slows down. This enables the crystal to become larger in diameter. This is known as the *crown*. When the diameter of the crystal is about the size of the diameter of the rod, the growth rate is increased and temperature adjusted to create a smooth transition between crown and body. This area of the crystal is called the *shoulder*. The final stage of crystal formation occurs when the melt is depleted and the crystal diameter is reduced in a cone shaped taper known as the *tail*. The Czochralski crystal growth process is used to produce the majority of commercially available single crystals. One of the leading manufacturers of silicon crystals and crystal growing furnaces is Kayex Corporation, Rochester, New York. See **Figure 32-7**.

Figure 32-7. These Kayex KX150 furnaces are used to grow silicon crystals. (SUMCO USA and Kayex, Photo by Kelly James)

Annealing

Annealing is a process that relieves the strains in ceramic or glass products through reheating and gradual cooling, just as it does in metals. Annealing in ceramics could be likened to the glost firing.

Annealing is a much more *critical* process with glass. After shaping, all commercial glass products must be reheated to a temperature at which frozen-in stresses can be relaxed by internal flow of the glass. The temperature for annealing is below that at which the products would seriously deform under their own weight. Annealing is accomplished in an annealing lehr, and is followed by slow cooling.

Tempering

Tempering is a heat-treating process used with glass products. It gives these products three to five times their annealed strength.

Glass is tempered after forming by heating it to a temperature close to the softening point. It is removed from the heat and quickly chilled from about 1200°F (650°C) to room temperature by rapid air-jet cooling. This places the outside surface under high compression stresses. For example, tempered borosilicate glass products are produced with working stresses of 4000 psi (27,580 kPa). This is about the same strength as ordinary gray cast iron.

Glass breaks only when it is subjected to tension, not compression. When tempered glass shatters, it breaks into small harmless pieces that hold together. Glass cannot be cut or drilled after tempering, since the cutting or drilling action would cause it to shatter.

Important Terms

bisque firing
bisque ware
bulk density (Pb)
calcination
chemical vapor deposition (CVD)
cryochemical drying
crystal pulling
Czochralski method
densification
density
dopants
extrinsic semiconductor
firing
firing range
floating zone method
freeze-drying
glost firing
greenware
intrinsic semiconductor
liquid-phase sintering
presintering
pressure-assisted sintering
pyrometric cones
reaction bonding
reactive-phase sintering
silicon carbide
silicon nitride
sintering
solid-phase sintering
sputtering
sublimation
target
tempering
thermal shock resistance
thermochemical
transformation toughening
transient sintering
vapor-phase sintering
vitrification
zirconia

Questions for Review and Discussion

1. What is the difference between reaction bonding and transformation toughening?
2. Which process is easier to accomplish, chemical vapor deposition or sputtering?
3. Select one of the methods used in industry for growing crystals and explain how it works.
4. What is the purpose of glost firing?

33 Processes Used to Condition Composite Materials

Chapter Highlights

- There are a variety of methods that can change the behavioral characteristics of a composite.
- Several processes can make a composite surface more conducive to resin adhesion.
- The thermal conductivity and specific gravity of a composite can be changed by introducing additives to the resin.

Introduction

Composites are a combination of different materials that are joined together to create a material that has its own unique properties. The physical properties of a composite are generally established during the forming and fabrication of the composite component. However, there are a variety of methods that can change the behavioral characteristics of a composite. Unfortunately, these methods are not as extensive as the conditioning methods available for metallic materials.

Irradiation

Exposure of a material to energy from a radioactive source is called *irradiation*. Both ionizing and nonionizing methods of radiation are used to heat, dry, and cure adhesives that are used to join different materials that make composite parts. Ionized radiation is produced with electron beam accelerators, using radioactive materials such as Cobalt 60. Ionizing atoms and molecules causes them to become highly reactive.

Prolonged exposure of polymeric materials to radiation can destroy their structure and composition. However, controlled irradiation can be used to improve the ability of plastic workpieces to withstand continued exposure to extreme temperatures. This process can also be used to improve cross-linking, and can enhance the electrical resistance characteristics of wire insulation. When used with elastomers, radiation can improve vulcanization (increasing strength and elasticity) and reduce stress cracking.

Another advantage of controlled irradiation of thermoplastics is the improvement in weatherability, impact strength, and resistance to static electricity. Controlled radiation is also used on composites to improve their resistance to abrasion and harsh chemicals.

Induction heating and dielectric heating are examples of nonionizing radiation processes. Microwave, ultraviolet, and infrared wavelength exposures are also used to generate nonionizing radiation. Processes such as these do not make the composite workpiece reactive, but they do have the ability to destroy molecular bonds in the materials. Normally, nonionizing radiation is used to dry or cure adhesives or UV initiated matrix resins.

Radiation is not always beneficial to polymers. It may cause damage to plastics, resulting in a reduction in their molecular weight. It can also cause cross-linking and polymerization when it is not wanted.

Improving Bonding of Composites

A common step of the fabrication of composite parts is bonding materials together. This is not always an easy task because of the specific properties of the surfaces to be joined. Being able to print directly on a composite or to get paint to adhere to the surface may also be a problem. In each instance, the adhesive, ink, or paint must be able to wet the surface of the composite material. To achieve good adhesion, the composite surface must be prepared to be wettable. *Wettability* is dependent on the *surface energy* (surface tension) of the material, which affects how well the liquid wets the surface.

Figure 33-1 shows two liquid droplets (ink, adhesive, or paint), each set onto a smooth solid horizontal surface. One droplet has spread out over the substrate, the other has not. Wettability is described by the angle between a tangent line at the contact point and the horizontal line of the surface of the solid. Look again at the drawing showing "poor wettability" in Figure 33-1. Angle CBD is the angle between the tangent line on the droplet and the point of contact with the surface being wetted. If only partial wetting takes place, the contact angle will range from nearly 0° (indicating almost complete wetting) to 180° (indicating very poor wetting). The higher the surface energy of the solid material in relation to the surface tension of the ink or adhesive, the better its wettability. A smaller contact angle shows better wettability. Refer to the drawing showing "good wettability" in Figure 33-1. Note that the angle GFH is much smaller than the angle CBD.

In order to determine the wettability for a particular material, it is necessary to know the surface energy or tension of the solid and liquids involved. Surface energy is measured in units of dyne-centimeters (dyne-cm). Some of the values for common solid and liquid materials are shown in **Figure 33-2**.

Figure 33-1. The wettability of a surface is indicated by the angle of a tangent line drawn from the point of liquid contact with the surface.

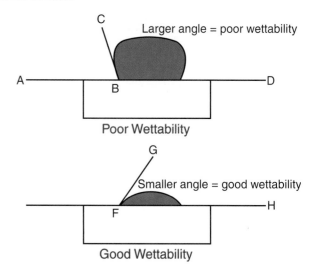

Figure 33-2. Some dyne-cm values for typical solids and liquids. To be wettable by a given liquid, a solid must have a value at least 10 dyne-cm higher.

Surface Energy (Dyne-Cm)		
Dyne-CM	**Material**	**Form**
100+	Adhesives	Highly viscous liquids
100+	Metals	Solid
100+	Glass & ceramics	Solid
50–80	Most finishes	Liquid
46	Nylon	Solid
43	Polyester	Solid
33	Polystyrene	Solid
31	Polyethylene	Solid
29	Polypropylene	Solid

Thick paints and adhesives have greater surface energy than high-viscosity liquids; consequently, they have a greater contact angle. Wetting a substrate with liquid adhesive or paint is also influenced by the cleanliness of the substrate. Air bubbles, oil, dust, or moisture buildup will naturally interfere with the ability of the coating to wet the substrate. Good wetting requires strong surface tension in the liquid or droplet that must stick to the substrate. This surface tension must be less than the adhesive force between the liquid and the substrate. This is an important concept, worth a closer look. A liquid with a high surface energy, such as an epoxy adhesive, cannot be used to wet a low-surface-energy polyethylene solid; this epoxy does not stick well to the polyethylene. This reaction occurs because the surface tension in the epoxy material is greater than the adhesive force between the epoxy and the polyethylene. On the other hand, a low-surface-energy polyethylene liquid adheres well to an epoxy solid, since the liquid would have a lower surface tension.

If a good bond is to occur between the solid and liquid, the substrate's surface energy should exceed the liquid's surface tension by 10 dyne-cm. The surface energy of some plastics, such as polyethylene and polypropylene, is insufficient to permit adequate bonding of printed images or adhesives. Often, these plastics are desired because of their useful properties: chemical inertness, low coefficient of friction, high wear resistance, puncture resistance, and tear resistance.

Electrical Surface Treating

The process called *electrical surface treating*, or *corona discharge treating* is a way to increase the surface tension of a material. Increased surface tension means that other materials such as ink or an adhesive will adhere better to the treated material. This is how the process works. The air near the material surface is exposed to a high-voltage, high-frequency discharge between the two electrodes (a corona). This environment causes the oxygen molecules in the discharge area to divide into their atomic form. These oxygen atoms are then available to bond with the molecules on the surface of the material being treated. This changes the molecular structure of the surface so that inks, coatings, and various adhesives are able to wet it. Since most plastic films and sheet materials have a smooth, slippery surface (low surface tension), the corona treatment essentially roughens the surface chemically (raising the surface tension), which improves the adhesion of inks, paints, and adhesives.

Figure 33-3 shows an electrical surface treating system designed for cabling. The system consists of a high-frequency signal generator, high-voltage transformer, treating electrodes, and ozone filter. The high-voltage transformer intensifies the output

Figure 33-3. This equipment is capable of generating a high-frequency, high-voltage signal needed to create a corona discharge for electrical surface treatment of insulated cabling. (Tantec, Inc.)

signal from the generator to the level necessary to create a discharge of the desired intensity. The cable is pulled between the electrodes at a rate up to 600 meters/minute.

Materials that are electrically surface-treated should be finished soon after conditioning, since there is a shelf life for the surface treatment. The temperature at which the material is stored also influences shelf life. After a coating has been applied to the substrate, the bond becomes permanent.

Coupling Agents

There are ways to improve the interfacial bond between matrices (resin), reinforcement materials, fillers, and laminates. This can be done by adding *coupling agents* (adhesion promoters). *Silanes* are one of the most popular and widely used coupling agents for increasing adhesion between two different materials. Typically, resins modified with silanes show improved adhesion to inorganic materials and enhanced resistance to weathering, acid, heat, and solvents. Silane can also be used to treat the surface of the inorganic materials before mixing with the organic resin, or it can be added directly to the organic resin. There are other coupling agents, too. When the composite contains a particulate or fiber-reinforcement, titanate can be added to improve strength.

Protection from Weathering

Composites exposed to the weather need to be conditioned so that the composite is not degraded by exposure to ultraviolet rays. Ultraviolet stabilizers and carbon black are sometimes used with polyesters, polystyrenes, and other plastics to reduce damage from ultraviolet rays. *Carbon black* is a soot-like material produced by the incomplete combustion of petroleum products. These are added to the matrix resin or the gel coat to block the ultraviolet rays, which could damage the composite matrix. Other additives, called *antiozonants*, are used to help prevent the composite from being damaged by the ozone in the atmosphere.

Changing Composite Characteristics

For some applications, additives and surface conditioning are not sufficient to provide the needed characteristics for the composite to perform its function. The thermal conductivity and specific gravity of a composite can be changed through the introduction of a chemical agent. One additive is a *foaming agent* that causes the resin to expand into a foam. The material cures quickly into a rigid cellular structure. Cellular foams are sometimes desired in applications where improved thermal insulation, light weight, and impact resistance are necessary. One approach for producing cellular foams is adding chemical blowing agents or volatile liquids to the resin or melt. *Chemical blowing agents* are materials that decompose when the resin they are added to is heated to a specified temperature during curing. This causes gassing, which creates a cellular structure. When a volatile liquid is used, the curing temperature causes it to change to a gas. As the gas expands, it creates voids or cavities within the softened polymer and fills the entire mold. Blowing agents are useful when forming a composite structure in a closed mold and are frequently used to ensure that the mold is completely filled.

Another method for making cellular composites is to create *syntactic polymer* by adding microspheres or microballoons to resin. In some cases, the small balloons actually expand when the resin is heated during polymerization.

Temperature Degradation of Composites

The degradation of composite materials through exposure to high temperatures is quite different from what might be expected with other materials, such as glass or steel. With glass, internal changes occur through phases of melting and crystallization. Steel is routinely heat treated and can be hardened, annealed, and re-hardened.

However, most polymeric-composite materials exposed to intense heat will char and decompose—they do not melt and then reform. This decomposition occurs because the resins used are predominately thermoset resins. The cross-linking that occurs when a thermoset resin cures is not reversible.

Even if a thermoplastic resin was used, the composite structure would not stay intact when the resin melts. The fibers, particles, or other additives that are held together by the resin matrix will behave independently in terms of how well they will survive exposure to extreme conditions. Therefore, each component must be evaluated for its heat resistance to determine if the composite will function properly.

Graphite, ceramic, or boron fibers are typically resistant to high temperatures. However, these fibers are normally used in conjunction with a matrix that may be much more vulnerable. For example, graphite fiber composites are as strong as metals, and they have the advantage of being lighter. However, if the matrix holding the graphite fibers is an epoxy, it will burn and the fibers will no longer be locked in place.

There are applications for polymeric-matrix composite that require them to be flame resistant. One method for improving their ability to meet this requirement is by adding boron or graphite powders to the matrix. The additives help to retard the char that begins to occur in thermal oxidation. Flame-resistant additives can also be added to the matrices.

In other end-use environments, a composite may have to withstand exposure to fuels or chemical solutions. Exposure to a particular chemical can sometimes cause the matrix to swell or dissolve.

Examples of product applications where this would be critical include an underground fuel tank or an automobile engine component exposed to hot oil. Other products, such as kitchen counter or laboratory countertops, must also be constructed so that the matrix can withstand chemical attack. Sometimes, multilayered barrier films are used to separate the chemicals from the matrix resin.

Important Terms

antiozonants
carbon black
chemical blowing agents
corona discharge treating
coupling agents
electrical surface treating
foaming agent
irradiation
silanes
surface energy
syntactic polymer
wettability

Questions for Review and Discussion

1. How does the corona discharge method of conditioning improve wettability?

2. What is the relationship of viscosity to wettability?

3. What is the purpose of a coupling agent?

4. Explain why a blowing agent or microspheres are added to a resin.

A lightweight cellular plastic provides dimensional stability, firmness, and thermal insulation when used for liner panels in such products as refrigerators. This cutaway view shows the cellular material sandwiched between two layers of a styrene alloy specially formulated to resist blistering from the blowing agent used to form the cellular material. (Dow Plastics)

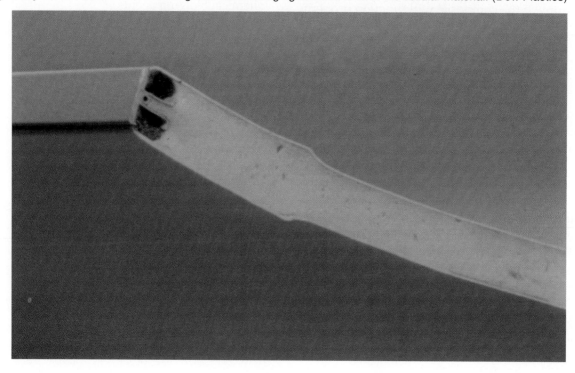

34 Processes Used to Finish Metallic Materials

Chapter Highlights

- Detergent washing is a common process used to prepare metal parts prior to applying a decorative or protective finish.
- Burrs are an unwanted result from many forming and fabricating operations and must be removed prior to finishing.
- Electroplating applies a decorative and protective metallic coating to metal.
- Painting is a popular method for finishing metal parts and can be accomplished using a variety of processes.

Introduction

Metal products are usually finished to protect the metal from corrosion, improve its appearance, or to enable it to perform its intended function. Functional finishes are frequently overlooked because they have nothing to do with cosmetic appearance or corrosion resistance, but nonetheless are critical for specific applications. An example of a functional finish is the insulating varnish that is applied to the copper wire that is used in the windings of an electrical motor. The failure of that finish would cause the motor to short out and cease functioning. In the manufacture of automotive parts that go into engines and transmissions, the surface finish on many of these parts is critical to the part's function. The elimination of burrs, tool marks, and surface imperfections are considered finishing operations. Cleaning may be the most universal finishing operation in manufacturing. Even the simplest parts need to be free of dirt, oil film, and other debris that collects on a part during manufacturing. Therefore, surface preparation is generally the beginning for most finishing operations.

Surface Preparation

Before any surface finish can be applied, it is necessary to prepare the surface for finishing. This must be done to remove unwanted scratches, burrs, and defects. Surface preparation is also necessary to clean the part so that the finishing material can properly adhere to the surface.

Washing Systems

Cleaning is frequently carried out by washing the part in a detergent and, in some cases, a solvent degreaser. The detergent washing is very similar to a dishwashing process. However, most detergent washers are progressive systems, meaning that they have several separate stages. The parts are carried in a basket or hung on a chain and conveyed through the various stages. There are many factors that must be considered when designing a washing system.

As a starting point, it is important to identify the size, shape, and configuration of all the parts that will need to be cleaned. Parts with large surface areas, acute angles, closed-off sections, and blind holes will require special attention during the design of the washing system. Heavy parts also mean that the conveying system will have to be stronger and the heat load on the system will be higher. The type of contaminant(s) to be removed will also influence the type of chemicals to be used as well as the size and number of stages required for the system to ensure proper cleaning of all parts.

Detergent cleaning is a washing process using mild chemicals that rely on the performance and flexibility of the washing system. Therefore, the equipment must provide enough mechanical agitation to supplement detergent action to achieve good cleaning.

There are several ways to achieve the mechanical agitation needed for a good washing system. One of the most common methods is *platform oscillation*. This method is generally used in batch parts washing systems. Platform oscillation is the simplest way for reaching all the surfaces of an irregular part. It is particularly effective in flushing dirt off the interior surfaces of a part. In operation, the part is immersed and oscillated with an up-and-down motion that swirls solution over the part. With each oscillation, the mechanical force of the moving solution continues to wash off dirt and oil from the surfaces of the part. This allows fresh chemical cleaning agents to attack the newly exposed layers of dirt, thereby speeding the cleaning action.

Solution turbulation (spray under immersion) moves solution over, under, around, and through a part. This method uses a pump to spray the solution. The turbulent flow caused by the spray is directed at the part from a manifold(s) containing *jets* or *eductors*. These jets increase the speed of the water by forcing it through small holes. Specific problem areas can be addressed by repositioning or adding jets to the manifold. This method produces higher solution impingement on the surface than straight platform oscillation. In difficult cleaning situations, solution turbulation can be used with platform oscillation to produce constantly changing flow patterns.

Cavitation cleaning (also called *ultrasonic cleaning*) is a special method used mainly for precision cleaning. It is most effective when it is used in conjunction with other methods, such as part oscillation and solution turbulation. At the start, the parts are completely immersed in the cleaning solution. Next, an ultrasonic transducer in the tank produces high frequency sound waves that cause an intense microscopic scrubbing action at the surface of the part. The energy released by the transducer also heightens the chemical activity of the cleaning solution, which helps to strip away contaminants.

Once the parts have been cleaned, they are rinsed thoroughly and dried. Ferrous materials require rust inhibitors to prevent the newly cleaned parts from tarnishing. Washing is usually done at an elevated temperature to speed the drying process. Another important feature in cleaning systems is filtering and recirculation of cleaning and rinsing solutions. Companies are responsible for trapping and properly disposing of contaminates that are removed from the parts.

There are many times when a part or a surface on a part needs to be cleaned but not washed. As an example, a part that has been machined may have rough edges that have to be removed before further processing can take place. This can be done using a mechanical method such as abrasive cleaning, tumbling, blasting, or brushing.

Deburring

Burrs are sharp edges that are produced when metal is deformed by shearing, trimming, stamping, or machining. Burrs are created, for example, when a cutting tool on a mill runs off the edge of a workpiece. See **Figure 34-1**. Typically, burrs are no more than .001″ to .005″ high (0.025 mm to 0.127 mm) and .003″ thick (0.08 mm). In most cases, burrs must be removed before the part or product can be used.

Burrs are sharp and can cause injury to persons handling the product. Burrs can interfere with proper assembly of parts and can cause premature fatigue. In electrical fields, burrs can cause arcing and short circuits.

Deburring is the process of removing burrs through finishing processes. Burrs can be removed by filing, wire brushing, polishing, vibratory finishing, shot blasting, abrasive machining, ultrasonic machining, or waterjet machining.

In cases where there are only a few parts to be deburred, the most cost-effective method of burr removal is often filing, brushing, or sanding. In high-volume production, deburring is often done by *vibratory finishing*. This is accomplished in a closed container that holds the parts and an abrasive finishing media. Simple systems involve placing the parts and medium in a drum, then rotating the drum so that the parts and medium roll against each other long enough to remove the burrs. Other production deburring machines are more complex, and some are fast enough to be used in line with a conveyor. Many incorporate automated feeding and unloading operations controlled by programmable logic controllers or computers. See **Figure 34-2**.

Sometimes the part has scale or flux that a detergent washing machine cannot remove. Chemical processes such as *pickling* (acid cleaning) or steam cleaning may be used to remove the unwanted material. Normally, parts are conveyed to such chemical processes and cleaned automatically.

Figure 34-1. Burrs are produced when parts are stamped or a cutting tool slices across a flat workpiece.

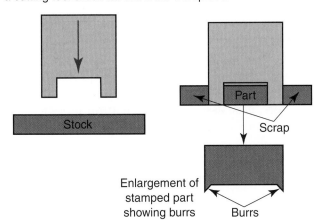

Figure 34-2. This centrifugal-disk finishing machine uses treated walnut shells and corncob meal as the grinding media to deburr hard metals, such as cermet (ceramic/metallic) magnets. (Roto Finish Company, Inc.)

One interesting electrochemical finishing process is called *electropolishing*. With electropolishing, the workpiece is placed in a hot alkaline solution. The solution serves as the electrolyte, while the part acts as the anode. A cathode is placed in the solution to complete the electrical circuit. The process initially de-plates the workpiece and then goes on to produce a highly polished, lustrous, mirrorlike surface.

Here is how the process works: When electricity passes between the anode and cathode, the current dissolves material from the part. Oxygen bubbles form on the surface of the workpiece. When these bubbles burst, a scrubbing action is created that cleans dirt and scale from the part. In electropolishing, there is no mechanical contact between the polishing tool, which is the cathode, and the workpiece. This makes the process particularly useful when polishing irregular shapes. Electropolishing is seldom used as a surface preparation process, because the work to be polished must be quite smooth. Electropolishing is not economical for removing more than 0.001″ (0.025 mm) of material. Electropolishing is widely used on stainless steel hardware used on yachts and commercial trucks.

Abrasives

Abrasive materials have been discussed in conjunction with grinding processes. However, a great deal of final finishing is accomplished using coated abrasives. Coated abrasives perform much like a grinding wheel, but the abrasive materials are attached to cloth or paper backing with a glue or resin binder. Coated abrasives are available in the form of sheets, discs, belts, rolls, and cones. Some abrasives are used for mechanical hand polishing while others are used with hand grinders to clean up welds and surfaces on castings.

Emery paper and *crocus cloth* are two coated abrasive products that are used most frequently to polish and buff metals. Emery is made from aluminum oxide and iron oxide. Crocus is made from purple iron oxide. Both products are available with different types of backing, and with different degrees of coarseness.

Abrasives are both natural and synthetic. Natural abrasives include emery, sandstone, garnet, corundum, and diamond. The most common synthetic abrasives used in industry are silicon carbide, aluminum oxide, and bauxite.

Finishing Processes

Once the surface preparation and final finishing processes have been completed, the final finish can be applied. Finishing processes protect and beautify the surface of the workpiece. Finishing includes all of the surface-protecting and surface-decorating processes. Finishes can be applied through coating processes: brushing, rolling, spraying, or dipping. They can be applied by electrochemical means: plating or anodizing. Some of the finishing methods in general used for metals are described in the following sections.

Electroplating

Electroplating is a process that can be used to apply a metallic coating of one type of metal up to 0.002″ thick (0.05 mm) on any other type of metal. Common coating materials used in industry are chromium, copper, nickel, zinc, and tin. Careful cleaning and surface preparation of the workpiece is critical to successful plating.

Parts to be plated are placed in racks or a barrel and are lowered into a water-based electrolytic plating bath. In electroplating, the workpiece serves as the cathode. The process is the opposite of electropolishing. The material that will plate the workpiece makes up the anode. In the plating process, electrical energy causes ions to be given off by the anode. These ions then combine with ions in the electrolyte and are attracted to the cathode (workpiece), where they are deposited as a metallic coating.

One of the major problems with electroplating is the difficulty of achieving a uniform plating (layer of metal) on workpieces that are irregularly shaped. Plating takes place on the raised areas of the surface before it is attracted to the cracks, interior corners, and depressions.

Electroless Plating

Plating can also be accomplished without electricity. The process known as *electroless plating* is done entirely by chemical reactions. It is most often used to plate nickel, but it can be used on any metal. It can also be used with nonconductive materials, such as plastics and ceramics.

In electroless plating, the workpiece serves as a catalyst. When this process is used to plate nickel, sodium hypophosphite acts as a reducing agent, converting nickel salts to nickel metal that is suspended in a solid solution of phosphorous nickel. Since this is entirely a chemical process, the coating that is created is relatively uniform in thickness.

Anodizing

Anodizing is an electrochemical process that turns the surface of the workpiece into a hard and porous oxide layer that is resistant to corrosion. The process is also called *anodic oxidation*. In addition to creating a hard coating, anodizing also produces an attractive finish for many product applications. Aluminum and magnesium are the most frequently anodized metals. The aircraft industry is a major user of anodized parts because anodizing provides a tough, hard corrosion resistant finish. Another important property of anodizing is its anti-galling properties. Galling can occur when very-close-tolerance parts, such as threaded fasteners, are being fitted together during assembly. As the parts slide against each other, the friction can cause a surface to stick and become even rougher, creating more friction. However, the lubricity imparted by the anodized surface will minimize this friction and, therefore, ease assembly. This natural lubricity also allows anodized aluminum sheet stock to be formed and drawn into shapes without marring the anodized surface or destroying its protective qualities.

When anodizing, the workpiece is immersed in an acid bath and acts as the anode. Current flows from the cathode to the anode, causing a chemical reaction. Oxygen is absorbed from the bath, causing the reaction to create a thin layer of aluminum oxide on the surface of the metal. Anodizing can be thought of as *reverse electroplating*. Instead of the coating being added to the surface, a reaction is created inward, resulting in a thin protective layer of aluminum oxide developing on the aluminum. The layer that is created is normally from 0.0005″ to 0.001″ thick (0.0127 mm to 0.025 mm).

Sometimes, dyes are added to the bath to penetrate the pores of the metal and provide a colored finish. Anodizing may also be used to create a clear oxide layer that does not change the color of the metal. Typical anodized products include aluminum utensils, furniture, automobile trim, and keys. An anodized surface also serves as a good foundation for paint application, which is useful when finishing hard-to-paint materials like aluminum.

Ion Implantation

Ion implantation is a process that can be used to harden the surface of metal, improving a part's ability to withstand wear, friction, and corrosion. The process, which is conducted in a vacuum, implants ions into the surface of the workpiece by accelerating them to the point where they are implanted into the substrate layer of the metal. Ion implantation is a rapidly growing process for producing semiconductor devices. When it is used in this way it is called *doping*, which means alloying with small amounts of different metals.

Metallizing

Metallizing is the process of spraying or vacuum impregnating metallic coatings on various metals and nonmetals. Because the base material to be coated is not heated, metallizing can be used on paper, wood, and plastics, as well as on metals.

There are three major types of metallizing:
- plasma arc spraying
- vacuum metallizing
- wire metallizing or flame spraying

The **plasma arc spraying** process creates an arc between two electrodes to heat argon or another inert gas. The heated gas is accelerated to supersonic speed. A powdered metal coating material is fed into the stream of gas, where it is melted and blown onto the part.

Plasma arc spraying is often used to coat parts for use in high-temperature environments.

This technique involves greater temperatures than other metallizing methods (in excess of 30,000°F or 16,650°C), therefore only materials with higher melting points can be sprayed. Coatings of tungsten carbide, ceramics, nickel chromium alloys, and exotic alloys are applied in this manner. In addition to metals, materials such as cermets, oxides, and carbides can also be sprayed.

Vacuum metallizing is usually the preferred process when a coating of aluminum metal is desired. Plastic as well as metal can be coated in this way. The process begins by placing the parts to be coated in a vacuum chamber that contains a small crucible holding aluminum pellets. With the parts suspended in place, the chamber is evacuated and the crucible is then heated until the aluminum vaporizes. The aluminum vapor then condenses on all the parts, covering them with a thin coat of aluminum.

Once the aluminum vaporizes, the vacuum metallizing process takes only moments to create the desired coating. Product applications include coating jewelry, automotive trim, and toys. The process is also popular for placing thin conductive coatings on plastic film used in electrical/electronic capacitors.

Coatings applied by **flame spraying** (also called **wire metallizing**) are useful in corrosion protection. They are also used to rebuild worn wear surfaces on motor shafts. In wire metallizing, a wire of the desired metal is fed into a metallizing spray gun, where it is melted by electric arc or flame. The molten metal is broken up into fine droplets by a stream of compressed air and blown on to the workpiece.

The spray gun used in this method is somewhat similar to a chopper gun used for creating fiberglass. The gun may be handheld or mounted on the tool post of a lathe for automatic spraying. The stand-off distance between the workpiece and the spray gun is normally from 6″ to 10″ (15.2 cm to 25.4 cm).

Before the base material can be coated, it is necessary to roughen and clean the spray surface. This produces a better bond. Coatings as thick as 1/2″ (13 mm) can be applied with flame spraying, but most applications do not require a coating more than 1/16″ or 1/8″ (1.6 mm or 3.2 mm) thick.

Powder Coating

Metal parts are often finished by coating them with powdered organic resins. The powdered coating is then fused with heat and cured. Common systems used for powder coating include fluidized bed coating, electrostatic fluidized bed coating, electrostatic powder spraying, and heat transfer coating. While the method of applying the powder varies from process to process, the principle is the same. Normally, the part is heated, the powder is applied, and additional heat is supplied to melt the powder, fusing it into a seamless coating on the surface of the material.

Spray Painting

Spray painting has been used to coat surfaces for over one hundred years. Joe Binks, a maintenance supervisor for a Chicago department store, devised a paint spray machine to whitewash the basement under the store. In 1893, painters were unable to keep up with the construction of buildings and exhibit halls being erected for the Columbian Exposition, an international fair that was to be held in Chicago. Binks' whitewash spray painting machines were brought in to do the painting. Using these machines, the painters were able to easily complete all the painting on time.

After the turn of the century, spray painting evolved to handle solvent-based paints as well as water-based finishes. In the 1920s, the rapidly expanding automotive industry employed spray painting as the premier method for applying paint. Spray painting's initial advantage was speed, but it soon became apparent that the resulting film of paint applied with a spray gun was more uniform and smoother. Glossy paints applied in this manner created stunning cosmetic finishes. These two characteristics, speed and finish quality, made spray painting one of the most widely used processes in manufacturing.

Electrostatic paint systems

The first electrostatic painting process was developed and patented by Harold Ransburg in the late 1940s. *Electrostatic paint spraying* was designed to minimize *overspray*, the paint that flies past the part and ends up as waste. This improvement in transfer efficiency was accomplished by designing a spray gun that charged atomized paint particles so that they are attracted to the grounded workpiece. Because of the electrostatic attraction, paint that would normally be sprayed past the part

Figure 34-3. This diagram shows the components of an electrostatic spray gun and how the "wrap-around" effect is achieved to reduce overspray.

Paint particles that would miss the part (overspray) are attracted to the positively charged workpiece and change direction. This is the wrap-around effect of electrostatic painting.

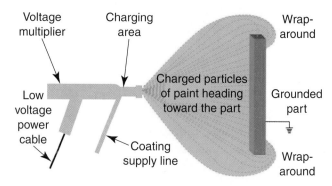

actually bends toward the charged part. This process coats the back and sides of the workpiece to produce "wrap-around" coverage. See **Figure 34-3**. Today, electrostatic spray guns can apply liquid and powder coatings with good efficiency.

Here is how the process works: When the paint comes out of the gun, it is charged as it passes through a high-voltage electrode corona discharge area. The paint can be charged with either a negative or positive charge, depending on how the gun is set up and what charge is placed on the part to be coated. As the paint becomes charged, it is sprayed out of the gun toward the part. Further atomization is achieved as the charged particles repel each other to form a fine cloud that is drawn to the part surface. Once on the part, the liquid particles wet the substrate and are left to dry. If the coating being applied is a powder, the charged particles cling to the surface until the part is placed into an oven for curing.

After the part is coated with powder, handling must be done with care because a bump or blast of air can easily dislodge the uncured powder. As the part passes through the oven, the powder coating material melts and fuses together. Most powder coatings are thermoset materials, requiring heat to cure. During the curing process, the melting powder wets the surface of the part adhering to it and then cures, fusing into a single conforming sheet covering the part.

HVLP paint systems

High-volume, low-pressure (HVLP) paint guns have become very popular for paint finishing operations. The HVLP guns produce a soft, easy-to-control spray that floats toward the part. Conventional spray guns atomize paint at pressures of anywhere from 35 psi to 90 psi. The result is a finely atomized spray that is carried at high velocity toward the part. The speed of the spray and turbulence at the part surface creates bounce-back, causing the paint to miss the part and adding to the overspray. Consequently, the transfer efficiency for conventional spray guns is typically 25 to 35 percent, two-thirds to three-quarters of the paint becomes overspray.

The design of an HVLP gun differs significantly from the conventional spray gun. A major difference is in the larger air line going to the gun. Instead of using an air compressor to supply the air, an HVLP gun uses a turbine to deliver a large volume of low-pressure, low-velocity air, which moves the atomized particles slowly toward the part. There is little turbulence and virtually no bounce-back at the part surface. A well-designed HVLP spray system delivers paint at 10 psi or less with a transfer efficiency of 65 to 85 percent. HVLP paint systems are well suited for most industrial painting in a wide range of applications.

E-Coating

E-coating is used mostly in the automotive industry for body, suspension, and brake components that need to be protected from corrosion but do not require a cosmetic finish. The "E" in E-coating refers to *electro* since the process, which is very similar to electroplating, uses an electrical current to deposit a paint film on a part. The prime advantage of E-coating is its ability to apply a uniform thickness of paint on all surfaces that are in contact with the coating solution.

The process begins by washing and pretreating ferrous parts with a phosphate conversion coating prior to E-coating. Cleaning and phosphating are important steps in creating a corrosion resistant finish. After washing, the racked parts receive an electrical charge and are lowered into a vat. The solution in the vat consists of 80 to 90 percent deionized water and 10 to 20 percent paint solids. The deionized water acts as the carrier for the paint solids, which are under constant agitation. When the charged parts are immersed, the paint comes out of solution and begins coating the surface of the part. As the coating thickens, the electrical conduction diminishes which is how the coating thickness is controlled. The parts are then removed from the electro-coating vat and rinsed to reclaim any unattached paint solids. In the final stages, the parts are dried and baked to cure the coating. **Figure 34-4** shows the stages in a typical E-coating system.

Figure 34-4. E-coating systems are large and can efficiently handle high-volume painting. This diagram shows the stages in the E-coating process.

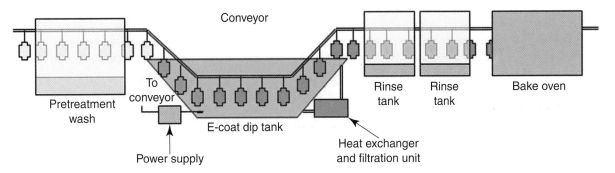

Electrocoating (E-coat) is an immersion painting process. When the part enters the dip tank, charged paint particles are attracted to an oppositely charged part surface. Paint continues to build up on the part until the coating thickness is sufficient to insulate the part from further paint attraction.

Important Terms

anodic oxidation
anodizing
burrs
cavitation cleaning
crocus cloth
deburring
doping
E-coating
eductors
electroless plating
electroplating
electropolishing
electrostatic paint spaying
emery paper
flame spraying
High-volume, low-pressure (HVLP) paint guns
ion implantation
jets
metallizing
overspray
pickling
plasma arc spraying
platform oscillation
solution turbulation
ultrasonic cleaning
vacuum metallizing
vibratory finishing
wire metallizing

Questions for Review and Discussion

1. What are the similarities and differences between anodizing and electropolishing?
2. How does an HVLP spray system differ from a conventional spray gun?
3. What is the difference between mechanical finishing operations and washing?
4. How is E-coating similar to electroplating?

35 Processes Used to Finish Plastic Materials

Chapter Highlights

- Phenolic resins can be applied at the worksite by spraying to protect the interior of large chemical storage tanks.
- Plastisols are widely used for dip-coating and dip-casting applications.
- Extremely fine particles of thermoplastic resin can be suspended on air currents and act as a fluid for finish-coating of thermoset materials.
- Metallic coatings can be applied by vacuum metallizing, electroplating, or electroless (chemical) plating.
- Electrostatic powder coating, one of the most popular processes used in industry today, creates a uniformly distributed coating, even on the back side of a round or irregularly-shaped part.

Introduction

Many of the processes used to finish metallic, ceramic, composite, and polymeric/wood materials—sanding, buffing, polishing, filing—are also used with plastics. Other processes, such as *solvent polishing* or *in-mold decorating with foils*, are unique to plastics. Special surface treatments, such as electrostatic powder coating or electroplating, are used for plastics as well as metallic products. A number of the most important finishing processes are discussed in this chapter.

Material-Removal Processes

Many finishing processes involve the physical removal of material through *abrasion* to produce a smooth and lustrous surface. Other methods use heat or chemical action to soften and smooth the surface.

Tumbling

Tumbling, or *barrel finishing*, is one of the most economical methods for quickly finishing plastic molded parts. It is also used to remove flash, smooth rough edges, grind, and polish. This process is often used with metallic and ceramic materials, as well.

The parts to be finished are placed in a drum with finishing materials known as *tumbling compounds*. These compounds may include abrasive particles, waxes, sawdust, wood plugs, and a variety of other materials.

Rotation of the tumbling drum causes the parts and finishing materials to rub against each other. This produces a polishing action. A small amount of material is removed, depending on the type of abrasive used, the length of the tumbling cycle, and the speed of the barrel.

Other methods used with tumbling to finish plastic parts include abrasive spraying and the use of dry ice. Dry ice is sometimes used in the barrel to improve removal of thin flash on the parts. The ice chills the flash, making it brittle, and tumbling separates it from the product.

Smoothing and Polishing

The primary step in finishing rough plastic parts is often machine sanding using open grit sandpaper. The coarse, open grit of the paper will remove the maximum amount of stock without becoming clogged by plastic particles. Number 80 grit silicon-carbide abrasives should be used for rough sanding. When rough sanding is completed, the parts may need further finishing. Final finishing processes can be completed with fine 400 or 600 grit paper. Sanders normally run at speeds ranging from 1750 rpm for disk models to 3600 fpm for those using belts. Dry sanding must be done using light pressure to keep from overheating and melting the plastic.

Machine processes such as polishing or buffing are sometimes used with a soft muslin or a chamois wheel. *Buffing* is done when small surface defects must be removed, or when a polished surface is needed. Normally, buffing is completed before polishing. Buffing is done using muslin disks sewn or collected and tied together to provide either a firm or flexible buffing wheel. Firm wheels are best suited to products with simple shapes. Flexible wheels are softer, and are used for buffing irregularly shaped parts. Both types of wheels are *charged*, or lightly touched, with a very fine abrasive material such as tripoli or red rouge. A buffing sequence often includes ashing, polishing, and wiping.

Ashing involves a wet abrasive, usually number 00 pumice mixed with water. The slurry is applied to a loose muslin wheel. The use of a wet abrasive reduces overheating of the plastic and improves the cutting action.

Polishing is sometimes referred to as *luster buffing*. When polishing plastics using hand or conventional buffing processes, a wax compound that includes a fine abrasive (such as whiting or levigated alumina) is applied to a clean chamois or loose flannel buffing wheel. The wax fills in small defects in the part and protects the highly polished surface. Polishing compound is applied to half the wheel, while the other half is left clean. The part is first held against the charged half, then against the uncharged part to wipe off the compound. Excessive wheel speeds and hard wheels should be avoided.

Wiping is accomplished by polishing the part with an uncharged wheel. Sometimes, cleaning compounds are applied to remove grease and wax.

When polishing or buffing, the part should be pulled toward the operator from beneath the wheel, using rapid even strokes. Care must be taken to avoid applying too much pressure, or the wheel will be distorted toward the edge of the part. This could cause the piece to be jerked from the operator's hands.

Flame Polishing

Flame polishing is sometimes used to smooth or reduce imperfections and to smooth rough surfaces on the edges of acrylic sheet plastic. Flame polishing is accomplished with a rapid back and forth motion of the flame from an oxygen-hydrogen welding torch. The flame is pointed downward to intersect the edge at an angle of approximately 45°. It is an effective method for smoothing hard to reach areas.

It is easy to know when the plastic is hot enough because it should become glossy. When it has reached the proper temperature it should look wet. Too much heat will warp or burn the material.

Solvent Polishing

Cellulosic and acrylic parts can be polished by dipping or spraying them with solvents. This is often effective in removing the matte finish left from sanding, drilling, cutting, or other operations. A small amount of acrylic solvent, such as ethylene dichloride, can be applied to the finish by dipping or spraying. Contact time should be from 15 to 60 seconds, depending on the extent of the surface roughness. Solvents can also be used to polish edges or holes. The solvent is then removed and the surface is allowed to dry. The solvent-polished area will have a fine polished finish, but it will not have the clarity of a fine-sanded and highly polished (buffed) surface. Parts to be solvent-dipped should first be annealed to keep them from *crazing* (developing a network of fine cracks on or under the surface).

Coating Processes

Application of a coating is a common form of finishing with plastics, as well as many other materials. Coatings are applied to plastics for appearance, protection, and electrical conductivity. Some coatings are only thousandths of an inch in thickness, others may be up to one-half inch thick.

Dip Coating

Polyvinyl chloride powder can be dissolved in a large amount of plasticizer to create a viscous solution called a *plastisol*. An important use for plastisols is *dip coating* the gripping surfaces of metal products, such as tool handles. Other typical dip-coated items are laboratory ware, kitchen utensils, and dish draining racks. In addition to dip coating, plastisols can be used to manufacture products using slush casting or rotational molding processes.

Sometimes dip coatings are applied by dipping heated parts into plastisol. Materials that do not require the part to be heated are also available. Once the part is coated with plastisol, it must be cured. Depending on the type of plastisol used, this may be done by air-drying, or by heating the part to 330°F to 450°F (166°C to 232°C). After curing, the material becomes solid and rubbery to the touch.

Plastisol can also be sprayed. Vinyl plastisol spraying produces an impact resistant and chemical resistant finish. It is flexible, and protects the product from corrosion. Plastisol sprays can be applied to aluminum, steel, and some plastics. See **Figure 35-1**.

A related application to dip coating is the *dip molding* or *dip casting* process used to make products such as surgical gloves and spark plug covers. In this application, a heated metal tool, or *mandrel*, is immersed in a dip tank containing plastisol. The mandrel conforms to the internal dimensions of the part that is being molded. You will recall that a plastisol consists of PVC, or another resin, dissolved in a plasticizer. The resin usually is in powder form, and thus—even though a solution ("sol") has been created—fine particles of resin will remain in suspension. After the mandrel is coated with the thick plastisol, it is removed from the tank and placed in an oven. The heat of the oven cures the plastisol and fuses any remaining fine particles of resin together. When curing is complete, the mandrel is cooled and the part is stripped from it. **Figure 35-2** shows typical PVC dip-molded parts.

Figure 35-1. Vinyl plastisol spraying produces a finish that is impact, abrasion, and chemical resistant. It is flexible and protects against corrosion. Available in glossy, matte, smooth or textured finishes, vinyl plastisol spray can be applied to aluminum, steel and some plastics with excellent results. (Wright Coating Company)

Figure 35-2. These insulating covers for battery connectors were manufactured by the dip molding process. (Arbonite Corporation)

Coating Storage Vessels in the Field

Sometimes, because of the size, weight, and location of large storage vessels and holding tanks, it is necessary to coat them in the field. Some manufacturers spray phenolic coatings onto large metal tanks in thicknesses up to 0.5" (13 mm) without removing them from the work site. **Figures 35-3** and **35-4** show large chemical tanks about to be coated internally with a phenolic material. While this coating is not a true plastisol, it does behave in a similar way. Let us take a closer look at how this works. Preparation involves insulating the exterior surface of the metal tank to prevent heat loss over the large surface area.

The phenolic coating is then sprayed on the inside surface. Hot-air generators are used to bake the phenolic material onto the inside surface. The hot air speeds polymerization and crosslinking. After the lining is inspected, the insulation is removed and the tank exterior is blasted and painted to suit the environment in which it is used.

Electrostatic Powder Coating

Electrostatic powder coating is one of the most popular coating processes used in industry today. Particles of plastic powder are given a negative charge as they are sprayed on to a positively charged, grounded part. The charged powder is attracted to the part much like metal filings are attracted to a magnet. The particles do not strike the part in a straight line, but follow the charged field. Consequently, the coating will be evenly distributed, even on the backside of a round or irregularly shaped part.

Once the part has been sprayed, it must be cured in an oven to fuse the plastic to the part. Without curing, the powdered plastic would simply fall off or be easily wiped off of the part when it lost its electrical charge.

Many automotive and heavy equipment manufacturers use electrostatic spraying, since it provides excellent coverage with low material loss. Powder that falls off the part is captured and reused. Powder coatings are hard, tough, and attractive. They do not chip and peel like conventional paint coatings. Coated products can be bent or deformed and the coating will flex with the product.

Electrostatic powder coats can be applied in a wide range of colors and finishes. It is often used in

Figure 35-3. This reactor tower for use in chemical processing has been insulated to retain heat. A phenolic coating sprayed on the inside surfaces will be baked (cured) by the introduction of heated air. After the coating is cured, the insulation will be removed and the tower installed in a vertical position. (Arbonite Corporation)

Figure 35-4. Like the reactor tower shown in Figure 35-3, this large storage tank has been insulated prior to baking its protective interior coating of phenolic resin. The completed tank will be used to hold highly corrosive concentrated sulphuric acid. (Arbonite Corporation)

Figure 35-5. Electrostatic powder coating used on this hospital bed provides durability and protection from chemicals and corrosion. Electrostatic powder coating is a popular process used by manufacturing firms today. (Wright Coating Company)

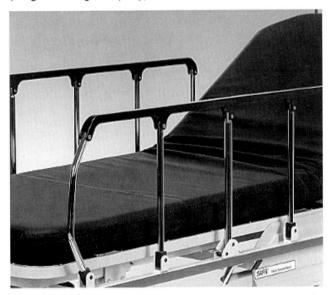

place of conventional paints because of the excellent adhesion, durability, and chemical resistance. Vinyl, epoxy, and polyester materials are available in many different gloss levels. See **Figure 35-5**.

An important advantage of electrostatic powder coating is that, unlike conventional paint spraying, there are no air pollution worries associated with the coating process. The powder that is not attracted to the part is contained and recycled for reuse. The powder contains no solvents, so no solvents are given off into the air during curing.

Fluidized Bed Coating

Fluidized bed coating is a process that is used for powder coating of thermoset plastics. Polyethylene is a popular polymer for use with this process. Fluidized bed coating changes powder to a fluidized state by mixing it with air. This holds the powder in suspension so that it behaves like a fluid. The fluidizer tank has a porous bottom through which air or inert gas is injected at low pressure. The powder is vibrated to make it mix more readily with the air.

The part to be coated is heated to a point above the fusion temperature of the powder and is then dipped for a few seconds into the fluidized bed. The powder attaches itself to the part, then melts and bonds to the heated surface in a layer of uniform thickness. Sometimes, additional heating is used to improve the smoothness and coverage.

The application of fluidized bed coatings usually takes place with a single dipping. Areas which are not to be coated are usually covered with a film of fluoroplastic material (such as polytetrafluoroethylene or polytetrafluoropropylene) to which the powder will not adhere. Coatings applied by the fluidized bed method are flexible and durable and are excellent electrical insulators. The coatings applied by this process are from 0.005″ to 0.080″ (0.127 mm to 2.032 mm) in thickness.

One advantage of fluidized bed coating is that it provides a drip-free, run-free finish with outstanding edge coverage. Vinyl, epoxy, and nylon coatings are available. The process is also helpful in hiding scratches and minor imperfections on the workpiece. See **Figure 35-6**.

The disadvantages of the process are related to the difficulty in applying coats to thin workpieces. Thin materials are difficult to cover because they do not hold enough heat to build up a coat. Sometimes, it is difficult to produce a coating that is the same thickness across the entire surface of the part. Also,

Figure 35-6. Fluidized bed coatings provide drip-free and corrosion-resistant coatings on products such as this grocery cart. (Wright Coating Company)

if the production environment is not temperature-controlled, the thickness of the coating will vary with seasonal changes in temperature.

Flame Spray Coating

Flame spray coating is a process that is often used to apply a plastic coating to tanks, vessels, or other products too large to cure in an oven. In flame spray coating, a fine plastic powder (metallic or nonmetallic) is heated and dispersed through a specially designed spray gun burner nozzle. The powder is quickly melted as it passes through the flame, and bursts of coating are distributed on the workpiece. The flame also heats the part to be coated. Thus, the hot plastic is sprayed onto the heated substrate to form a smooth and uniform layer. When the plastic cools, it forms a hard permanent coating.

Conductive Spray Coating

Zinc-arc spray metallizing is used to apply conductive shielding material for a variety of electrical applications. Conductive zinc coatings deposited by this method are dense, durable, and abrasion-resistant. The major limitations of this process are lack of uniformity in the layering, adhesion problems, and the creation of zinc dust.

In a zinc-arc spray gun, two zinc wires are fed through the gun and brought together so that they are almost touching. An electrical arc between the wires is created by applying dc current. This melts the wires, allowing air pressure from the gun to eject the molten zinc and deposit it on the plastic substrate.

The arc spray process is more efficient than flame spraying. It can produce a film 0.002″ to 0.004″ (0.05 mm to 0.10 mm) in thickness using only one-ninth the melting energy required for flame-spraying a film of the same thickness. Reduced heat results in less part warping and distortion.

Vacuum Metallizing

Vacuum metallizing is a process used to apply very thin metal coatings to a plastic substrate for decorative or functional purposes. Coating of the plastic occurs by vaporizing the metal (usually aluminum, copper, or chromium) in a vacuum. The metallic film is deposited in thicknesses that are less than could be accomplished with electrolytic or electroless plating—only a few millionths of an inch thick. Coating adhesion to the substrate is not as good as that accomplished with electroplating. After the parts are metallized, a final protective coating of lacquer is normally applied.

Vacuum metallizing is used to frost and decorate glass, and fabricate computer chips and web material for packaging. It is a popular process for coating alloys, costume jewelry, bottle caps, and automotive trim.

There are two basic types of vacuum metallizing, evaporative vacuum metallizing and sputtercoating. The hardware for these processes is quite different.

Evaporative vacuum method

In the **evaporative vacuum method**, the metal to be deposited on the plastic is evaporated from disk-shaped targets placed in a vacuum chamber. Usually, the targets are made from chromium or copper. Evaporation is accomplished through resistance heating or electron beam bombardment. The vaporized metal molecules travel through the chamber and condense on the cooler plastic substrate to form the coating. The appearance of the metal coating that forms on the plastic is slightly different from that of the material that was first evaporated, because the original metal has vaporized and reformed on the substrate. Most applications require a basecoat and a number of topcoats of lacquer to improve adhesion and protect the surface.

Sputtercoating

Sputtercoating is a process that is widely used to coat the surfaces of electronic components. Sputtercoating equipment that is magnetically enhanced creates a magnetic field in the vacuum chamber. The electromagnetic field directs chromium atoms from the disk targets to the substrate.

In this process, a high level of kinetic energy is produced in a magnetron sputtering gun. The gun has a water-cooled anode and a cathode that emits electrons. The electrons bombard the argon gas molecules and ionize them to create the sputtering action. When the argon ions are accelerated to the disk target, they become stronger, more dense, and exhibit stronger adhesive bonding. These ions liberate or dislodge (sputter) atoms of the coating material, which are directed to the substrate. The sputtering process is used with high-temperature materials, or for more precise deposition of material on the substrate.

With both vacuum metallizing processes, coating of parts can be accomplished either in batches or continuously. In a batch coater, the parts are placed on racks in lots and are inserted into the processing chamber for coating. In the continuous coater, parts enter the processing chamber through an airlock. This airlock ensures that the processing chamber is always under vacuum.

Electroplating

The two processes generally used to plate plastics are electroplating and vacuum metallizing. Vacuum metallizing, as described in a previous chapter, produces a very thin coating. This coating is not as durable as that produced by electroplating, but it is less costly.

You might think that, since plastics are not conductors of electricity, they cannot be electroplated. This is not true—chromium, silver, nickel, gold, and other shiny metallic coatings can be applied to plastics. However, before electroplating can occur, a conductive coating must be applied to the plastic.

The *electroplating* process is used to deposit metals from solutions onto metallic surfaces using *electrolysis*. The coating that is produced is tightly bonded to the surface of the metal, and provides outstanding protection against rust. It can also serve as an electrical conductor, or to provide decoration.

Approximately 50 percent of all of the plastics electroplated in the United States is used by the automotive industry, and an additional 25 percent by appliance manufacturers.

Electroplating begins with careful cleaning of the part. A chemical bond of thin copper is then created. Once the copper coating is completed, electroplating can be carried out using the same methods employed for plating metals.

Electroless Plating

The term *electroless plating* is used to describe those processes that use chemical rather than electrolytic action to produce the desired metallic coating on a substrate. Two of the fastest-growing areas for electroless plating are the construction of copper-plated printed circuit boards (photofabrication) and production of conductive plastics for electromagnetic interference shielding in the electronics industry. Automotive parts producers and aerospace manufacturing firms are also big users of electroless plated parts.

Electroless plating is accomplished by autocatalytic reduction, using a solution of approximately 3 percent formaldehyde and 10 percent methanol stabilizer. The solution is further modified with caustic soda to adjust the pH (acid/alkaline balance). As the pH increases (the solution becomes more alkaline), the formaldehyde increasingly transforms the copper ions to copper metal for deposit on the substrate.

Electroless plating is generally a more expensive way to produce a copper coating than electrolytic deposition (electroplating). Because of the cost factor, thin coatings are normally produced by electroless plating. The coatings that are applied are used primarily as functional coatings of metals and nonconductors to provide a hard, ductile, wear resistant, and corrosion free surface for use in environments up to 750°F (400°C).

Other Finishing Processes

There are numerous other types of finishing processes used in manufacturing. Many of these are used with materials outside of the major material families (metallics, plastics, ceramics, and composites). Many of these processes are used in the printing or textile manufacturing industries, and are thus outside the scope of this book. Two processes are worth mentioning here, since they are used with a variety of manufacturing materials: in-mold foil decorating and coloring.

In-mold Decorating with Foils

In-mold decorating can be done by applying foils to compression-molded and transfer-molded products made with melamine, urea, or ammonia-free phenolic thermoset and thermoplastic materials. A *foil* is a printed, melamine-impregnated paper that is introduced into the mold.

With thermoset materials, the molding cycle is interrupted at the earliest possible moment after curing, so that the printed foil can be placed on top of the part. The mold is then repressurized for several seconds at 2500 psi to 3000 psi (17 238 kPa to 20 685 kPa) and temperatures of 290°F to 350°F (143°C to 177°C). The pressure and temperature must be maintained long enough to cure the foil and produce the desired gloss.

Foils used may be opaque or transparent with reverse lettering to achieve a color-on-color effect. They can be die-cut pieces or gaskets with multiple decorations for different parts of the product. Foils are capable of producing high-quality pictures and intricate drawings. Any subject that can be commercially printed can be used as a foil.

With thermoplastics, insertion of the foil is accomplished at a different point in the process—the die-cut foil is placed in the mold just before it closes and the plastic is injected. The injected molten plastic bonds to the foil and suspends it inside the molded part.

Various methods are used to hold foils in correct alignment in the mold. They can be held securely by pinching the foil between the two halves of the mold. In some cases, they are held in place electrostatically or by the use of mechanical attachments on the mold.

Almost any molded product made with polystyrene, acrylic, ABS, polycarbonate, and polypropylene materials can be decorated using in-mold foils. Dinner plates molded from melamine resins are often decorated with this process. Highly reflective foils have been made to produce unbreakable optical mirrors.

Coloring

While foils can be used to produce colored designs in many molded products, the most economical way to achieve color entirely through a product is to blend pigments directly into the resin base. Additives such as plasticizers and fillers, and the conditions of the molding process can also affect the color of the final product.

Color is produced by mixing pigments in the form of dry powders, pastes, organic chemicals, or metallic flakes with the resin in a Banbury (batch-type) mixer or a continuous mixer. Mixing disperses the pigments throughout the resin. Water- and chemical-soluble dyes are used by some manufacturers. When such dyes are used, coloring is done by dipping the product into a dye bath and then letting it air-dry.

Vibrant colors are mixed into the resin of plastisols before the handles of tools are dipped into the resin. This produces a handle covering without requiring any other finishing. See **Figure 35-7**.

Important Terms

ashing
barrel finishing
buffing
charged
crazing
dip casting
dip coating
dip molding
electroless plating

Figure 35-7. Plastisols are widely used in the hand tool industry to form dip coatings on handles of tools such as the pair of snips at top. The coating provides greater user comfort and helps to prevent slipping. Other tools use molded slip-on grips, like those at the bottom of the photo.

electroplating
electrostatic powder coating
evaporative vacuum method
flame polishing
flame spray coating
fluidized bed coating
foil
in-mold decorating
in-mold decorating with foils
mandrel
plastisol
polishing
solvent polishing
sputtercoating
tumbling
tumbling compounds
vacuum metallizing
wiping
zinc-arc spray metallizing

Questions for Review and Discussion

1. Identify several different products made today that were probably created using dip molding of plastisols using a heated mandrel.

2. Describe the differences between electroplating and electroless plating.

3. How do dip coating and dip casting differ? When would you use one process over the other?

4. In electrostatic spraying, how is it possible for the coating to be distributed completely around the object, even though the spraying is done only from one side?

36

Processes Used to Finish Wood Materials

Chapter Highlights

- Surface preparation is the most difficult part of finishing a wood product.
- A filler must be used before applying the final finish on open-grained wood such as oak.
- Electrostatic coating and the waterfall method are types of finishing that have achieved widespread acceptance in industry.
- Both synthetic and natural abrasives are available, but synthetic types are most often used in industry.
- Electrophoresis was one of the first manufacturing processes to be utilized in space, as part of the Material Processing in Space (MPS) initiative undertaken by the National Aeronautics and Space Administration (NASA).

Introduction

When you think about finishing wood products, the first thing that may come to mind is brushing on paints, varnishes, or stains. However, there is much more to finishing wood than just applying a coating over an unfinished product. Normally, there is a great deal of surface preparation that must take place before the actual application of finishing materials. Once the product is ready to accept a finish, the material is typically applied using industrial processes such as spraying or curtain coating.

As a living tree in the forest, wood often will survive for hundreds of years, despite exposure to frequently adverse conditions. This is made possible through the amazing process of photosynthesis—the foundation of growth for the tree. Growth not only occurs internally in terms of cells and passageways for nourishment, but also externally in terms of the tree's protective layer of bark.

Once the tree is harvested and processed into lumber, the green wood is exposed directly to the elements without the protection of bark. While the pattern of the grain in unfinished wood may be beautiful, that grain may discolor and develop defects if the wood is left in an unfinished state. Untreated wood will absorb dirt, oil, grease, and moisture. Discoloration from dirt and oils is difficult to remove from unfinished surfaces, especially open end-grain. Excess moisture will cause unfinished wood to swell and deform. If drying is done too rapidly, the wood may warp or develop checks or cracks.

Finishing processes protect wood from exposure to contaminating influences in its environment. If finishes are properly applied, they can preserve and improve the natural beauty of wood. Hasty or careless finish application, however, can destroy the quality product that has been created.

Surface Preparation

To successfully apply any finish, you must know about the structural characteristics of the wood being finished. Not all finishes work equally well on all woods. Also, the surface preparation needed for finishing hardwoods is different from the process used for softwoods.

Finishing of wood surfaces is a labor-intensive process, and often involves more planning than is necessary with most other materials. It is simpler to finish most metals and plastics than it is to finish wood and ceramics. There are many different surface preparation methods that may be used, depending on the type of stock and the desired finish. Most of the time consumed in the final finishing of wood products is devoted to surface preparation. The first step in surface preparation is usually sanding.

Sanding

Sanding can be done by machine or by hand; machine sanding is most common in industry. Machine sanding involves the use of an abrasive-coated sheet, belt, or disk attached to power-driven equipment. There are many different types of natural and synthetic abrasives used to smooth wood. Abrasives are normally fastened to a backing with adhesive. Each individual abrasive particle works like a tiny cutting tool against the surface of the wood. Manufactured abrasive materials are usually referred to in industry as **coated abrasives**, although

they are still called *sandpaper* in common usage. Sand is not actually used for abrasives, although the natural abrasive quartz (silicon dioxide) used to coat **flint paper** is chemically similar. Despite the fact that there is no sand in sandpaper, material removal by use of abrasives is still called sanding.

Types of sanding

There are three basic types of sanding (whether done by machine or by hand): roughing, blending, and fine finishing. **Roughing** is done with a coarse-grit abrasive to remove the maximum amount of stock. If the product has surface scratches or gouges, rough sanding may be a necessary first step to remove the imperfections.

If the product is relatively free of blemishes, it may be possible to start off with blending. **Blending** is done with a medium-grit abrasive. Blending results in some material removal, producing a fairly smooth finish.

Fine finishing involves using a fine-grit abrasive to remove light scratches. After fine finishing, the stock is often polished or burnished with rubbing oils or rubbing compound. Most rubbing oils are made of varnish diluted in oil. They penetrate well and produce an attractive finish that is maintained by using furniture polish, wax, or mineral oil. No further finishing or surface treatment is necessary.

The most common abrasives used to remove stock from products made of wood are aluminum oxide, silicon carbide, and garnet. Garnet is the softest of the three, while silicon carbide is the hardest. Abrasive hardness, like the hardness of other minerals, is measured on Mohs scale. The scale ranges from the hardest, 10 (diamond), down to 1 (talc). Silicon carbide is rated at 9.5, aluminum oxide at 9, and garnet at 7.

Abrasive grain sizes

The **grain size** of an abrasive material is used to grade the coarseness or fineness of the finishing material. A coarse abrasive has large grains, while a fine abrasive has much smaller grains.

The grain numbers used to classify abrasives refer to the number of openings in each inch of mesh through which the abrasive is sifted. As the numbers become larger, the grain size becomes smaller: a particle with a grain size of 12 (extra coarse) would be a virtual boulder next to an extra-fine particle with a grain size of 600. The standard numbering system applies to most types of abrasives, which might be designated as a "No. 80 paper" (a medium grit) or a "No. 120 paper" (a fine grit). Flint abrasives are the exception. They are graded as extra fine, fine, medium, coarse, and extra coarse.

The extra coarse, very coarse, and coarse grain types range from 12 to 50 grains per inch. These highly abrasive materials are used for heavy-duty industrial sanding machines. Medium-grit papers range from 60 to 100, fine from 120 to 180, very fine from 220 to 280, and extra fine up to 600 grains per inch.

The grade of the paper selected will greatly influence the speed of stock removal and the number and depth of scratches generated on the surface of the material. The most efficient process is to start with a coarseness that will remove the imperfections or finish without creating unnecessary scratches. When the entire product is adequately smooth, replace the sandpaper with one that is two grades finer. Continue this process until the product is sufficiently smooth.

Coated abrasives are manufactured by attaching the abrasive particles to a paper or cloth backing with an adhesive. The material is then cut into the shape required to fit the desired sanding application. Sanding materials are available in sheet, disk, belt, or cylinder form.

Sanding equipment

Sanding is normally done with a belt sander, disk sander, or spindle sander. Sanding also can be accomplished using a radial arm saw by removing its blade and installing a sanding disk in its place. Sanding can even be done on a band saw by using a special segmented blade with small abrasive disks attached to each segment.

Sanding must be done in the same direction that the grain of the stock is running. Sanding *across* the grain will produce scratches that are difficult to remove. **Figure 36-1** shows an edge sander being used to edge-sand a long strip with the grain.

The abrasive belt planer is a machine used for stock removal in most furniture-making plants. On this machine, the sanding is done by a large belt held tightly on two drums. The stock is fed into the machine on a conveyor; as much as 1/16″ (1.6 mm) of material is removed from one face of the board with a single pass through the machine. Abrasive planers have motors heavier than those of conventional sanders. The abrasive belt, which moves at speeds up to 7500 surface feet per minute (sfpm), is normally made of tough cloth and can withstand tremendous abuse. There are many variations of

Figure 36-1. Sanding is typically done with the grain of the wood. This machine is being used to edge-sand a thin strip of material. Note the ear protection worn by this worker because of the high noise level in this production area. (Robersonville Products)

the basic abrasive planer. The machine shown in **Figure 36-2** is a *production molder,* a type of abrasive planer used to shape long strips of molding.

Applying Fillers

With open-grained woods such as walnut, oak, mahogany, or ash, a *filler* should be used to close the pores before any finish is applied. Fillers also help to produce a smoother surface and enhance the beauty of the wood by highlighting the grain pattern. Fillers are usually purchased in paste or semi-paste form, but are also available in liquid form.

Paste filler contains powdered quartz, linseed oil, turpentine, and drying agents. Its natural color is tan. It can also be purchased in other colors, or

Figure 36-2. A production molder is a special-purpose abrasive machine used to create the desired crosssection on a length of molding. (Robersonville Products)

colored as desired before use. Varnish and lacquer are often used as clear fillers.

Before wood is filled, you must first clean the pores with a wash coat of shellac or lacquer. The wash coat consists of one part shellac mixed with seven parts of alcohol, or one part sealer and five parts lacquer thinner. The wash coat is applied with a brush and allowed to dry.

The filler is then applied with a stiff brush. The first coat is applied in the direction of the grain. The second coat is applied against the grain to ensure complete coverage. While the filler is still wet, any excess material is removed by wiping with a rag. Wiping should be done with the grain.

Once the filler has been applied and dried, the wood is ready for its final finish of stain, varnish, paint, enamel, or lacquer. Sometimes, further treatment is needed to change or enhance the color of the wood before finishing.

Bleaching Wood

Sometimes, it is necessary to lighten the natural wood coloring of an entire workpiece. Other times, you may need to lighten some areas and leave others alone in order to equalize the color of the wood. This change can be done by *bleaching*, a process that involves applying a caustic liquid to the surface with a sponge or rag. In large volume industrial applications, the bleach is typically applied by spraying it onto the wood surface. Bleaching removes the color by oxidation.

Bleaching is done to achieve honey-toned finishes. Normally, lighter-colored woods are used, since it is difficult to drastically lighten dark woods. Bleaching often is done when the wood is to be stained, or to bring out the desired grain pattern.

Commercial bleaches usually consist of two solutions, hydrogen peroxide and caustic soda, that are sprayed on the surface. The two parts mix to produce the desired bleaching action. When bleach is applied in this manner, no caustic residue remains, so there is no need to neutralize the surface after bleaching.

Homemade bleaches can be made by mixing oxalic acid crystals in hot water. Whenever bleaches are used, rubber gloves and eye protection should always be worn—the solutions are very caustic and require careful handling.

The bleach is applied and allowed to soak into the pores of the bare wood. The bleaching time varies according to how much the wood has to be lightened.

Once the desired degree of bleaching has taken place, lightly sponge the stock with warm water to *neutralize* (stop the action of) the bleach and remove any residue. When the surface has dried, the wood can be lightly sanded to prepare it for final finishing.

A major disadvantage of bleaching is that washing off the residue adds moisture to the wood. This sometimes causes the wood to *cup* (assume a concave shape). To keep this from occurring, water is often applied to both face surfaces of the wood to equalize the swelling.

Shading Wood

Shading or glazing is a process that is sometimes used to improve the color uniformity of the wood surface. Shading differs from bleaching because it is carried out after the wood has been sealed with filler. Most shading is done on low-cost furniture to equalize the color produced by a previous application of stain.

The process involves spraying color-tinted shading lacquer over the area to be shaded. The lacquer adds a lighter color to the area to equalize the color throughout the product.

Shading can also be applied by mixing a color tint with commercial glazing liquid, then brushing the material on the surface. To achieve the desired look, some of the glaze is wiped off with a rag or brush. This serves to highlight certain areas and produce an aged or antique look in the product.

Staining Wood

Manufacturing firms use stain to change the overall wood color or to beautify the grain by either adding emphasis or lightening the grain pattern. *Stain* is made of either finely ground pigments held in suspension or chemicals in solution.

Stain is applied with a brush, sponge, or rag and allowed to dry. The length of drying time depends on the type of stain. Oil stains are wiped off with a soft rag after being left on the wood long enough to achieve the desired degree of darkness. Stains can be applied to filled or unfilled wood, depending on the type of stain and effect desired.

There are four basic types of wood stain: water stain, oil stain, non-grain-raising stain, and spirit stain.

Water stains are used to improve the consistency of color. When a darker tone is desired, more pigment is mixed with water. The major disadvantage of a water stain is that it will raise the grain in the wood. This can be controlled by wetting the wood with a sponge prior to staining. Water stains are often sprayed.

There are two major types of *oil stains*: pigmented oil stains and penetrating oil stains. Both types consist of pigments suspended in oil and can be purchased in either powder or liquid form. Oil stains are applied by brushing, spraying, wiping, or dipping. As noted earlier, they are wiped off with a soft rag to control the depth of color.

Non-grain-raising (NGR) stains are widely used by furniture manufacturers. NGR stains contain dye mixed in a *methanol* (methyl alcohol) solvent, rather than water. Since alcohol evaporates fairly rapidly, these stains have a short drying time (often 15 to 20 minutes). NGR stains are normally applied by spraying or dipping.

Spirit stains have the fastest drying rate of all of the stains. They consist of *aniline dyes* (rather than pigments) suspended in alcohol. Spirit stains are the most difficult type to apply, primarily because of the rapid evaporation rate of alcohol. These stains are used mostly for repair or touch-up work.

Once a stain is applied and allowed to dry thoroughly, it should be *sealed* with shellac or a lacquer-base sanding sealer. The sealer encapsulates the stain and permanently protects the finished surface. Sealers are normally brushed or sprayed.

After the lacquer-base sanding sealer is applied, it should be lightly sanded. The abrasive paper that is used is so fine that it produces a very fine dust and leaves no scratches in the finish.

Finish Coatings

After the wood has been filled, finished, and sealed, it may or may not be necessary to provide another protective coating. If such a topcoat is desired (as is the case with fine furniture), it is usually a clear coating of oil varnish, lacquer, or synthetic-resin varnish. The type of topcoat selected depends on the durability needed in the product, the extent of contamination in the environment where it will be used, and the characteristics of the material. Depending on the type of finish used, topcoats can be applied by spraying, brushing, or wiping.

Varnishes and Lacquers

Varnishes are different from stains or paints. There are synthetic varnishes (polyurethane), oil varnishes, spirit varnishes, and acrylic varnishes. All of these different types of varnishes contain oil, resin, solvent, and driers. A varnish topcoat is applied to protect the surface from water spotting, chipping, and abrasion.

Sometimes the topcoat applied is a *lacquer.* Lacquer is made of nitrocellulose, acetone, varnish resins, a thinner such as benzene, and plasticizers. Lacquer dries quickly and provides a hard finish. The preferred method for applying lacquer is by spraying, but sometimes several coats are applied by brushing. Lacquer is highly volatile, and the fumes are toxic. It should be handled and applied with care, making sure there is adequate ventilation.

Paints and Enamels

Paints and enamels are opaque finishes that hide the grain of the wood. Generally, these finishes are applied as surface protection when it is not important to see the grain and texture of the wood. For this reason, less surface preparation is necessary when paints or enamels are to be used.

There are two basic types of paint; oil-base and latex (water-base). Naturally, the type of base determines how the paint can be thinned or removed when the painting job is completed. *Latex paints* are soluble in water. *Oil-base paints* are thinned with paint thinner, turpentine, or mineral spirits.

The type of base used is important to know when refinishing any painted product. The general rule to remember is that oil-base paints can be applied over a previous coating of latex paint. However, if latex is applied over oil-base paint, then the finish will bubble and peel.

There are many different types of oil-base and water-base paints. A *paint* consists of pigments suspended in a vehicle. *Pigment* gives the paint its opacity and color. In most cases, a greater amount of pigment would result in a duller appearance. High-gloss paints have little pigment, but more vehicle and solvents.

The *vehicle* functions as the carrier for the pigment. When the surface is painted, the vehicle forms a layer of film with the opaque pigment suspended in it. When the vehicle dries, a protective surface layer remains. Paints are normally applied by brushing, spraying, dipping, or other coating processes.

Enamel is different from paint; it is made by adding pigment to varnish. Enamel has a gloss finish and is more durable than latex paint. Since they are not opaque, however, enamel will not cover as well as paint. Often, a base coat, or undercoat, is required to seal pores and provide a stable foundation for subsequent coats.

The first coat of enamel is applied to the base coat. Several coats may be necessary, with 24 hours allowed for drying between them. Light sanding is often done between coats to smooth the surface and aid adhesion. Enamels are thinned with mineral spirits.

Application Methods

Because of the volume of material to be finished, industrial production usually requires the use of automated application methods for finishing materials. These include several spraying methods, electrodeposition, and coating by means of roller, curtain, or dip application.

Atomized spraying

In *atomized spraying*, pressure is used both to break the finishing material into very fine drops (atomize) and to blow it onto the item to be coated. After the product is sprayed, the finishing material is either air-dried or heat-dried. Atomized spraying uses air or hydraulic pressure. Atomized spraying can be done on many different types of materials.

When hydraulic pressure is used to force the finishing material to the spray gun, the technique is referred to as *airless spraying*. Airless spraying is popular for spraying latex and varnishes, but it can also be used with other types of finishes. In addition, airless spraying saves as much as 15 percent of the material that would be lost as overspray with conventional spray systems.

Airless spray systems force the paint through a small opening in the spray nozzle. This process atomizes the paint much like water coming out of a garden hose. Airless systems are able to apply heavier finish coats than conventional spray systems. This results in fewer coats and faster application. With airless spraying, the pump delivers finishing material to the gun under fluid pressures as high as 3000 psi (20 685 kPa).

Airless spraying is a process that is popular in the marine and offshore oil industries. It is also used by many tank and container manufacturers to make pipe and pipeline coatings, coatings for railcars, and coatings for large area surfaces such as silos and cooling towers.

Spraying is a relatively straightforward process. Conventional spraying systems consist of a compressor with air hoses, tank, and pressure regulator, a filtration device or *extractor* to remove oil and water, and a spray gun. Airless systems consist of a compressor, pump and hoses, a controller, and a lubrication/filtration system. Airless systems are sometimes equipped with a heater.

There are two basic types of guns used for atomized spraying: the siphon gun and the pressure-feed gun. The *siphon gun* is fed by suction, drawing finish upward to a fluid cap, then mixing it with air. With the *pressure-feed gun* system, the finishing material is fed to the nozzle, where it is mixed with air. The siphon type is used for smaller jobs, since a limited quantity of finishing material can be stored in the fluid cap. Most production spraying is done with the pressure-feed system. Robotic painters are used in many companies to ensure a consistent paint thickness, reduce spray time, and remove workers from the spraying environment. See **Figure 36-3**.

Sometimes, production spraying (atomized or airless) is done using *heated* finishing materials. Heating finishes during application reduces their viscosity for better spraying without the addition of undesirable solvents. The process uses special guns that heat the finishing material to a temperature of 160°F to 200°F (71°C to 93°C).

Electrostatic spraying

Electrostatic spraying is a process that was developed as a method for coating metallic materials.

Figure 36-3. Atomized spraying is a widely used industrial method of applying finishing materials. This robotic painter is spraying the finish on wood shelf components. (Robersonville Products)

It can also be used to finish woods, but the wood must either be high in moisture content or coated with a conductive material. Electrostatic spraying of wood has been done in Europe, where the moisture content of the test samples is more than 12 percent. Wood in the United States seldom has more than 6 percent moisture, so precoating with conductive material is necessary.

In electrostatic spraying, the finishing material is fed to a specially designed spray gun, where it receives a negative charge. The product to be coated is given a positive charge.

When the negatively charged finishing material is sprayed out of the gun, it is drawn by attraction to the positively charged product. With this coating method, coats of even thickness can be applied with little overspray and wasted material.

There are three different techniques used to apply electrostatic coatings. These are referred to as the gun method, the disk method, and the bell method. The gun operates on conventional air pressure.

The disk method and bell methods both make use of centrifugal forces. Fluid is pumped directly to the disk and the bell. When the devices are rotated, centrifugal forces disperse the fluid to the devices' outer edges. The material then receives a negative charge and is pulled away by the positive charges surrounding the product to be coated.

The great advantage of electrostatic spraying is the cost—only about 40 percent of the finishing material required with conventional spraying is used with electrostatic spray coating. Overspray material (powder that does not stick to the part) is saved and often recycled.

Electrodeposition

Electrodeposition is similar to electroplating and is normally used on metals. It can also be used with coated polymers. Electrodeposition is also called *electrophoresis*. This is essentially a method for separating materials based on the speed of migrating ions in a conductive fluid that is influenced by an electrical field. Electrophoresis can be used to separate organic and inorganic ions, peptides, proteins, enzymes, amino acids, and biological particles and cells. It has been used since the 1930s to separate materials in the medical field.

The process is accomplished by placing the part to be coated in a vat of special paint, and then attaching it to the positive side of a charging circuit.

The part becomes the anode. The paint particles contain a negative charge and are attracted to the positive anode. After the product is coated, it is placed in an oven for drying and curing. Like electrostatic spraying, the process produces a uniform coating.

There are several limitations to the process in a 1G (the gravity of earth) environment. It is limited to use with lightweight materials. Other limitations experienced when using electrophoresis in the earth's atmosphere are convection caused by heating of the electrical field and gravitational settling. These limitations are nonexistent in outer space.

Electrophoresis was one of the first manufacturing processes to be utilized in outer space, as part of the Material Processing in Space (MPS) initiative undertaken by NASA. In the decades to follow, many other experiments were conducted with electrophoresis in space. NASA and McDonnell Douglas encountered unexpected problems and high costs that kept the technique from reaching its full potential. Today, the field of genetic engineering has overtaken this research.

Roller coating

Roller coating is a process that is economical for high-speed coating of flat stock such as wood, metal, plywood, and particleboard. The process consists of squeezing the stock to be coated through rubber rollers carrying a film of finishing material.

In roller coating, the rollers are partially submerged in a vat of finishing material. When the stock travels through the rollers in the same direction that they are turning, the process is referred to as *through coating*. Today, most manufacturers use what is referred to as the *reverse roller process*, in which the rollers turn in one direction and the stock moves in the other. This results in better coverage of material on the workpiece.

Roller coating is often used to apply a wood grain finish to particleboard or hardboard panels. The panels are first roller-coated, then pass through a grain printer. Next, a polyester topcoat is applied by another set of rollers. Panels are normally dried using ultraviolet (UV) light waves, rather than a higher intensity source of heat.

Curtain coating

Stock that is to be coated is conveyed under a coating head that provides a continuous, sheet-like, falling stream of finishing material. The process is called *curtain coating*. Sometimes the curtain is created by flowing the finish over a dam, where it drops by gravity onto the workpiece to coat it. Excess material then runs off the product, is captured in a gutter, and flows to a reservoir for recycling through the system. The process is also referred to as *waterfall coating* or *flow coating*. The advantage of curtain coating over conventional spraying methods is the elimination of overspray.

Many panel manufacturers finish their products using curtain coating. Films with a thickness of 0.001″ to 0.0025″ (0.025 mm to 0.064 mm) can be applied using curtain-coating machines.

Flow coating is a variation of the basic process of curtain coating. With this method, the workpiece is placed beneath a bank of low-pressure nozzles that eject streams of nonatomized finishing materials. This process helps to reduce the amount of uncontrollable airborne contaminants, because less air is used to distribute the finishing material.

Dip coating

In *dip coating*, the finishing material is applied through complete immersion of the product or part. The viscosity of the finishing material must be carefully controlled to prevent the lower surfaces and edges of the product from retaining an excessive amount of film.

Dip coating is fast and economical. The process is practical for coating parts that are difficult to finish by any other method. Auto bodies are often dip-coated. As noted, the critical element in dip coating is the viscosity of the finishing material. If it is too thin, it will produce inadequate coverage; if the finishing material is too thick, sags and runs are likely to occur.

Tumbling

Tumbling can be used to finish wood knobs, buttons, golf tees, beads, and other small parts. The objects to be finished are placed in a drum or barrel with a small amount of lacquer, enamel, or wax. The barrel revolves at about 25 rpm causing the parts to tumble over each other. Additional finishing material is placed in the barrel as needed. When the finish

Figure 36-4. This overhead conveyor system is used to convey parts into an oven for forced drying of the finishing material. (Robersonville Products)

dries, the tumbling action continues and the parts polish against each other, creating a smooth, satiny finish.

Forced drying

Forced drying is typically accomplished by loading parts or assemblies onto conveyors that carry them through heated ovens. **Figure 36-4** shows an overhead conveyor system used to transport products to drying ovens.

Forced drying is often done with ultraviolet (UV) radiation. Radiation drying is preferred for curing flat sheets. The finish coat is cured by light rays, rather than heat, converting the coating to a tough, solid film in a matter of seconds.

Some wet surfaces cannot be exposed to high temperatures immediately after finishing or they will bubble and blister. These finishes are given a *flash-off period* to allow excess thinners and solvents to evaporate. This is normally done by air drying, either at room temperature or in a warm-air enclosure.

Important Terms

airless spraying
aniline dyes
atomized spraying
bleaching
blending
coated abrasives
cup
curtain coating
dip coating
electrodeposition
electrophoresis
electrostatic spraying
enamel
extractor
filler
fine finishing
flash-off period
flint paper
flow coating
forced drying
grain size
lacquer
latex paints
methanol
neutralize
non-grain-raising (NGR) stains
oil-base paints
oil stains
paint

paste filler
pigment
pressure-feed gun
production molder
reverse roller process
roller coating
roughing
shading
siphon gun
spirit stains
stain
through coating
tumbling
varnishes
vehicle
water stains
waterfall coating

Questions for Review and Discussion

1. What is the disadvantage of atomized spraying when compared to electrostatic spraying?

2. What finishing methods do not produce airborne contaminants and are available for use with woods?

3. Why must a filler be used before applying a final finish on oak? Is a filler needed when finishing an item made from pine?

4. What method is used to remove dust from the pores of wood before applying a final finish?

Properly applied finishes, such as seen on these cabinets and prefinished floorboards, enhance the beauty and durability of the wood surfaces.

37

Processes Used to Finish Ceramic Materials

Chapter Highlights

- Polishing and lapping can be used to remove minor imperfections on the surface of a ceramic product.
- Grinding can often be used to remove larger defects in ceramic workpieces.
- Glazes can be applied by spraying, dipping, or brushing, but spraying is the most common.
- Decorations can be applied to ceramics before glazing or after glazing.

Introduction

In many cases, ceramic products are ready for the consumer when they leave the kiln, and no final finishing is necessary. Most of the time, however, some additional processing or postfiring operations are necessary.

There are essentially two different types of finishing that apply to ceramic products. The first of these includes processes that address the final surface preparation of the workpiece after firing. It includes such processes as scribing, etching, grinding, and the application of decals and decorative stencils.

A second type of finishing includes processes that are used to place a decorative or protective coating or design onto the *bisque ware*. Bisque ware refers to a clay product that is dried, then fired in the kiln without a glaze, normally at 1700°F or 1800°F. Bisque firing makes the product stronger and easier to handle.

Processes, such as silkscreen printing and glazing, may require the application of decorative *frit* in glaze to provide the desired graphic details on the ceramic body. The addition of frit lowers the necessary firing temperature of the glaze. Frit is a type of ceramic glass, normally a combination of flux, silica, and powdered minerals that are melted, quenched in water, and then reground into fine powder. This powder is then used in glazes and enamels. Frits can make soluble materials insoluble. For example, oxides such as boron that are found in some minerals are soluble and cannot be used in glazes because they change the fluidity and viscosity of the glaze. Frits make these oxides insoluble and make the glaze more defect free. Frits also help to improve the process safety of toxic metals by driving off undesirable and unsafe compounds.

Grinding

The process of grinding has already been presented as a basic material removal process for use with ceramic products. Grinding as a finishing process is normally required when the design of the ceramic product requires support by refractory kiln furniture to keep it from slumping in the kiln.

Setters, which are solid masses of clay of the same composition as the part, are placed under certain areas of the workpiece for support. These setters cause imperfections that may be large enough to require grinding. In other cases, where the extent of surface marring is minimal, processes such as polishing or lapping may be adequate.

Precision Grinding

Silicon carbide grains provide extreme hardness and uniformity of shape. This makes them an ideal candidate for use in grinding, polishing, lapping, and pressure-blasting applications on hard materials including glass and ceramics.

Flame Polishing

Flame polishing melts the thin surface layer of the product to reduce the size and quantity of surface defects. This process has been used to polish the finished edges on sheet window glass. Today, there is very little polishing necessary, since glass is normally made by the float glass process. In this process, flat glass is formed by flowing the material onto a bed of molten tin. See **Figure 37-1**.

Flame and Plasma Spraying

As you will recall from an earlier chapter, ceramic powder can be changed to molten droplets by passing it through a high-temperature plasma, such as an oxyacetylene flame. The molten particles flow out of the gun through the nozzle and strike the substrate to be coated, at a velocity of approximately 150 feet per second (fps). Most carbides, borides, oxides, nitrides, and silicides that do not decompose can be applied by molten particle techniques.

The first widely used molten particle process was the oxyacetylene powder gun, called the *flame spray gun*. Another approach is the *oxyacetylene rod gun*. With this gun, ceramic powder is not used. Instead, a sintered rod of coating material is fed into the oxyacetylene flame. Air bursts blow the molten ceramic at the tip of the rod onto the substrate at a rate of approximately 600 fps.

A newer version of the molten particle process is accomplished with the *plasma arc gun*. A high-intensity, direct-current arc is produced in a chamber. Helium or argon gas is passed through the chamber, heated by the arc, and forced through a water-cooled nozzle in the form of a high-temperature plasma. Ceramic powders are ejected into the plasma, where they are melted and sprayed onto the workpiece. Velocities as great as 1500 fps have been reported.

The major disadvantage of this process is the high temperature of the gas. This means that metallic parts have to be cooled to keep them from melting, while ceramic parts must be preheated to reduce thermal shock and keep them from cracking.

The process has many present-day applications, ranging from spraying chromium on ship propellers

Figure 37-1. This float glass machine produces a continuous ribbon of flat glass by floating the molten material onto a bed of molten tin. The process virtually eliminates the need for grinding and polishing. (PPG Industries)

Figure 37-2. This laser scribing system for ceramic workpieces is precisely operated by a CNC machine housed in the console at left. (Coherent General)

to providing thermal coatings on superalloys. Molten spray techniques are particularly advantageous in applying refractory linings and coatings that are chemical-resistant or wear-resistant.

Laser Processing

You have already learned about forming processes using lasers with ceramics and other materials. Some of the applications for these products in aerospace, medical, automotive, and engineering application require very close tolerances. This makes it necessary to use final finishing, or *micro-machining* processes. Thermally assisted machining processes such as *laser-* or *plasma-assisted machining* are sometimes used for final finishing.

One reason for using lasers on ceramic materials such as glass or electronic substrates is that energy transmitted from the laser beam occurs in very short, rapid pulses. This reduces thermal shock to the material that is being processed. The most common general-purpose laser for such applications is the CO_2 laser.

The most popular of the processes that use laser technology are scribing, drilling, and cutting. These will be discussed here as final finishing processes, since laser processing is normally done after firing. This enables final tolerances to be obtained in a single step. Lasers are used to perform such processes as *scribing* on a continuous basis, without concern for material thickness. An automated laser scribing system for use with ceramics is shown in **Figure 37-2**.

To successfully perform scribing on ceramic and glass materials, it is necessary to minimize the heat transmitted to the workpiece to avoid thermal shock. This is accomplished by drilling a series of tiny holes into the substrate, at a depth of 20 to 30 percent of the material thickness, to form the scribe line. This is accomplished by using very short, rapid pulses. Laser manufacturers produce equipment capable of varied rates and scribing depths.

Glazing

A *glaze* is a specially formulated glass that melts on the surface of the ceramic workpiece and adheres to the body after cooling. Glazes are often applied to provide an impermeable surface on what would normally be a porous product. This results in the creation of a surface coating that can be easily cleaned.

Glazes are also applied to improve the appearance of the product by giving it a glossy surface. A glaze can also serve as a base for a decoration, or as an overglaze to protect a decoration. Firing melts the glaze and helps it to flow evenly. See **Figure 37-3**.

Glazes are applied by dipping, brushing, or spraying directly on the surface of the ceramic piece to be finished. Both glazes and porcelain enamels contain silica glass, flux, colorants, and opacifiers.

Before 1920, ceramic floor and wall tiles were glazed by hand dipping or on a roller machine.

Figure 37-3. These teacups have had glaze applied by the spraying method. They are shown exiting from a glaze dryer prior to firing. (Lenox China)

Today, most glazing of tile is done automatically by curtain coating or by spraying.

Glaze can be applied to either bisque ware or greenware. It can also be applied to completely vitrified (nonporous) ceramic parts. The most common approach is to apply the glaze to pieces that have been bisque-fired.

Application of Glaze

When glaze is applied by curtain coating, a typical product, (such as a floor tile) moves through a curtain, or waterfall, of continuously flowing glaze. See **Figure 37-4**. This method has gained considerable popularity in recent years because it reduces the amount of airborne contaminants produced. After the finish flows down onto the product, any excess falls into a trough of water; the water is then collected in drums for disposal.

This method contrasts sharply with conventional spraying, which generates a mist of finishing material in the air. Such mists can result in explosions or fires, as well as posing health hazards to workers.

Once the glaze is applied, pieces are normally loaded into *saggars*. These are refractory boxes that protect the workpieces from exposure to flames, gases, and contamination in the firing operation. The pieces are then pre-sintered, fired, and cooled in the kiln. See **Figure 37-5**.

Glaze Characteristics

Glaze is classified according to the percentage (by weight) of the oxide it contains. Glaze compositions vary a great deal, but the major constituent in most is silica. Exact recipes depend on the thermal expan-

Figure 37-4. The waterfall system, shown here being used to apply a glaze to ceramic tile, eliminates harmful and potentially explosive spray mist. A stream of water carries away excess material for proper disposal.

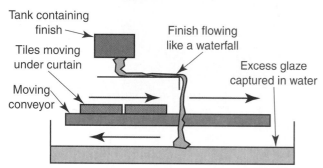

Tank containing finish

Tiles moving under curtain

Finish flowing like a waterfall

Excess glaze captured in water

Moving conveyor

Figure 37-5. This worker is stacking small plates on saggers—fixtures or boxes that are used to protect the workpieces from exposure to flames and gases and contamination in the firing operation. (Lenox China)

sion required of the glaze, whether it is to be colored or not, whether it is to be transparent or opaque, and the temperature that is needed to cause it to adhere to the clay body. The composition of glaze is carefully formulated so that it will expand at a slightly lower temperature than the ware on which it will be applied.

Glaze Additives

Glaze is prepared with additives to color it or make it opaque. The purpose of using a glaze is to provide a protective coating on the workpiece. There are several different types of glazes, to address different types of materials and applications. We also discussed the use of frits in glaze in a previous section.

Some ceramic products must be decorated using a coloring compound. This is usually a metal oxide, which is mixed together with the glaze. Colors can be applied before glazing (an *underglaze* decoration), or after glazing (an *overglaze* decoration). The application of decorative gold *gilt* after glazing on products, such as final china, is normally done by hand. This process is called *gilding* or *banding*. See **Figure 37-6**.

Many manufacturers of fine dinnerware and other ceramic products provide a final finishing touch using gilding, or banding. Both of these processes are often done by hand, requiring great skill and patience. Gilding is done with gold, while banding

Figure 37-6. Gold decoration (gilt) is applied to the rim of a plate by this automated system. (Lenox China)

Figure 37-7. Ceramic dinnerware typically has decoration applied either before glazing (underglaze decoration) or after glazing (overglaze decoration). Finishing processes for ceramic products often provide both surface protection and decorative appearance. (Caleca USA Corp.)

is normally done with silver, or with a color complimenting a featured component of the graphic design on the product. There are several different ways to apply these materials, depending on the base material that the gilt or band is to be attached to. In the case of porcelain or ceramics, the process normally requires burnishing or rubbing the gilt against the product.

Underglazes are protected from wear by the overlying glaze, but do not provide the extensive range of colors available with overglazes. Colorants for underglazes are fine powders that can be easily mixed with feldspar or china clay.

Overglazes consist of powders ground with frits containing soft fluxes such as soda and potash. Overglazes must melt at temperatures lower than the glaze over which they are applied. See **Figure 37-7**.

Overglaze decorations can be applied by hand painting, by silkscreen printing, or by applying decals. The most popular method is to apply printed designs with water-slide decals, often called *paper transfers*. Heat-release transfers are also used. The printed image is prepared on the material and then either released with water or heat.

The printed image for paper transfers consists of varnish applied to a paper backing, usually by lithography. Dry ceramic powdered colors are applied and adhere to the wet varnish. After the varnish sets, the decal can be stored until needed. To apply the decal to a product, it is wetted to release it from the paper backing. The image is slid off the backing and onto the workpiece. When the piece is fired, the colors are permanently affixed. See **Figure 37-8**.

Heat-Blocking Ceramic Finishes

There are materials used today with ceramics to provide heat-resistant protective coatings. Heat barrier paint is used to block the heat from entering your home. A coating of ceramic paint can act like six inches of fiberglass insulation when applied to almost any surface, including roofs. Thermal coatings are also available in mildew proof or lifetime coatings.

Figure 37-8. Paper transfers are applied by hand to fine china to provide the desired overglaze decoration. (Lenox China)

Important Terms

banding
bisque ware
flame polishing
flame spray gun
frit
gilding
gilt
glaze
laser-assisted machining
micro-machining
overglaze
oxyacetylene rod gun
plasma arc gun
plasma-assisted machining
saggars
scribing
setters
underglaze

Questions for Review and Discussion

1. How are printing processes used to direct print images on ceramics without breakage?

2. What is the purpose of a *setter*?

3. Why are glazes necessary?

4. How are decals applied? Are they overglaze or underglaze decorations?

38

Processes Used to Finish Composite Materials

Chapter Highlights

- A finish for a composite may be integral to the composite or applied after the composite has been formed.
- Colorants may be added to the matrix resin to create an integral finish.
- Finished surfaces can be created using a laminate to impart unique characteristics.

Introduction

Composites are used in military, commercial, and consumer products. Many of these products have no finish other than that formed by the matrix resin. However, some consumer products, such as yachts and bathtubs, require finishes that meet cosmetic requirements while providing a tough protective surface. Finishes of this type can be considered as being integral to the composite. There are also some finishes that are not part of the composite structure. These surface finishes may be a paint that is applied after the composite has been formed. Although they are not integral to the composite, these finishes are also very important.

All too often, the quality of a part or product is based on its appearance. Generally, composite parts can achieve a cosmetic finish that equals those found on metals. Many of the processes used to create the composite produce an exterior surface that is readily suitable for surface finishing. However, as with most materials, composites require cleaning and some may need surface imperfections filled before a surface finish can be applied.

Surface Finishes

When a finish is applied after the composite has been formed, its purpose may be primarily cosmetic or to act as a barrier to protect the composite from environmental stress and degradation. In either case, some of the more commonly used finishing processes include spray painting, calendar coating, electroplating, and powder coating. These processes are described in the chapter covering finishes for metallic materials.

Transfer coating

One type of coating process that is particularly useful with composites is called *transfer coating*. With this process, a surface skin coat is rolled on a paper that is coated with a release agent. Once the coated material sets up or gels, a second coat called a *tie coat* is rolled over this skin coat. At this point, the coating is transferred to the composite surface and allowed to dry. When the coating has set, the release paper is removed.

With the transfer coating process, it is possible to join many different types of materials using a film or fabric coating. Transfer coating is also used to place printed images onto composite materials. The image is printed on transfer paper or film and is then heat-set or ironed onto the actual workpiece. The carrier paper or film is then carefully peeled off.

Printing

Many of the finishing processes are specifically designed to print images on a composite's substrate. Transfer printing or coating is one of these processes. Describing all of the printing processes available are beyond the scope of this textbook, however, several are worth mentioning here. The most commonly used processes for imprinting images on composites in a continuous production environment are pad printing, hot stamping, and embossing. Fully automated production machines can be purchased for pad printing and hot stamping.

Pad printing is accomplished using heat-set inks. The raised printed design contacts an inked pad and then is pressed lightly against the workpiece. The wet ink that has transferred onto the product is dried by the application of heat.

Hot marking is a stamping process that can be used to impress or "brand" images onto the workpiece. It is also used with foils to transfer color to the product. An image, created in relief, is heated and then pressed into the stock. When foils are used to color an image, they are inserted between the relief form and the workpiece. This form is not effective when the surface is a thermoset resin.

Embossing is a process that is the opposite of branding, since it produces a raised image rather than one that is sunken into the product's surface. In embossing, the design or type is constructed in sunken or female form. Often, the embossing die is made of wood and is included as part of the die used to transfer-mold, or vacuum-form, heated polymeric sheets. The embossing die presses down onto the sheet as pressure from below pushes the stock up into the recesses of the die.

Integral Finishes

Fiberglass boats and bathtubs have a finish that is an integral part of the composite. The *gel coats* used in these applications are usually a polyester resin that is sprayed or flow coated onto the mold surface. The gel coat's primary function is to protect the composite from the effects of the environment and to provide a pleasing cosmetic finish. The gloss and fairness of the gel coat is dependent on the finish of the mold surface and not on the method of application. As an example, one builder of small boats routinely pours gel coat out of a mixing container directly into the center of the open mold and moves the excess gel around with a 4" paint brush to cover the remainder of the mold surface. When the boat hull is completed and pulled from the mold, the gel coat gleams and has the same mirror finish as the mold surface. The builder claims that this method provides a uniform coverage without the mess of spray application. For large composites, spray or flow coating is a more efficient way to apply gel coat. See **Figure 38-1**.

Gel coats contain colorants and fillers to provide the desired environmental and cosmetic properties. However, gel coating (particularly for truck bodies, aircraft components, and other commercial applications) may not provide the colors and designs that the end user wants. Therefore, the gel coat may be the base surface for the final finish. These finishes are frequently two part polyurethane enamels. These enamels can be applied with spray equipment to provide a high quality cosmetic finish. These urethane enamels will maintain their gloss for years and provide good UV protection.

Figure 38-1. Gel coat is applied using flow coaters or spray guns. This boat hull is being coated with a dark blue gel coat. Mold surfaces are usually finished in a bright contrasting color to help the operator see areas where the gel coat is too thin. (Bauteck Marine Corporation)

This open mold has been cleaned and waxed prior to gel coating. The mold is a bright orange color.

Gel coat is sprayed on the mold.

The mold is now completely covered with a dark blue gel coat.

Colorants

When a particular color is needed, it is possible to add a colorant to the base resin. *Colorants* may be dyes or pigments. Although their primary purpose is to impart color, they often serve other functions. Colorants can improve a composite's strength, and electrical properties, or help it resist degradation from heat or ultraviolet light.

Solid colors

Cast composites, which have become popular for sinks and countertops, have colorants mixed with the matrix resin. This provides a uniform color throughout the composite. When these items are fitted at a job site, any raw edges, mars, or scratches can often be removed by buffing and polishing. Besides colors, cast composites can contain particles that make the composite look like stone or some other form of architectural material.

Laminates

Some fiberglass bathtubs and spas have a surface that is made from a clear acrylic sheet of plastic. This surface is created using a thermoforming process. The first step is to heat and, with the aid of a vacuum, draw the acrylic sheet into a female mold so that it conforms to the shape of the finished part. See

Figure 38-2. Once the sheet has cooled, it is removed from the mold and taken to the fiberglass lay-up area. The first coat applied to the formed sheet is the gel coat in the desired color. This is sprayed on the back of the sheet. The front is the finish side. After the gel coat is applied, the fiberglass and reinforcement materials are applied as needed.

The use of a clear sheet of acrylic plastic provides several advantages both for manufacturing and for the customer. First, thermoforming is much faster than laying up fiberglass. Therefore, a manufacturer

Figure 38-2. This is a view of a thermoforming mold used to form an acrylic plastic sheet for a fiberglass bathtub. (Seawolf Design, LLC)

can make more parts per hour from a single mold. This means one mold can supply several fiberglass lay-up stations, which lowers mold-tooling costs and allows for a higher level of production. For the consumer, the clear sheet of plastic provides a clear coat finish that can withstand abrasion and minor scratches without harming the color coat. The acrylic can also be buffed to restore some of the luster that is lost over time.

Important Terms

colorants
embossing
gel coats
hot marking
pad printing
tie coat
transfer coating

Questions for Review and Discussion

1. What is the difference between integral and surface finish?

2. When a gel coat is applied to an open mold, what determines the quality of finish?

3. Explain the difference between transfer coating and gel coating.

4. Why would a composite manufacturer choose to use thermoforming as a way to provide a finish for a composite?

A plastics paint inspector ensures a perfect finish on a plastic rear bumper. (Ford)

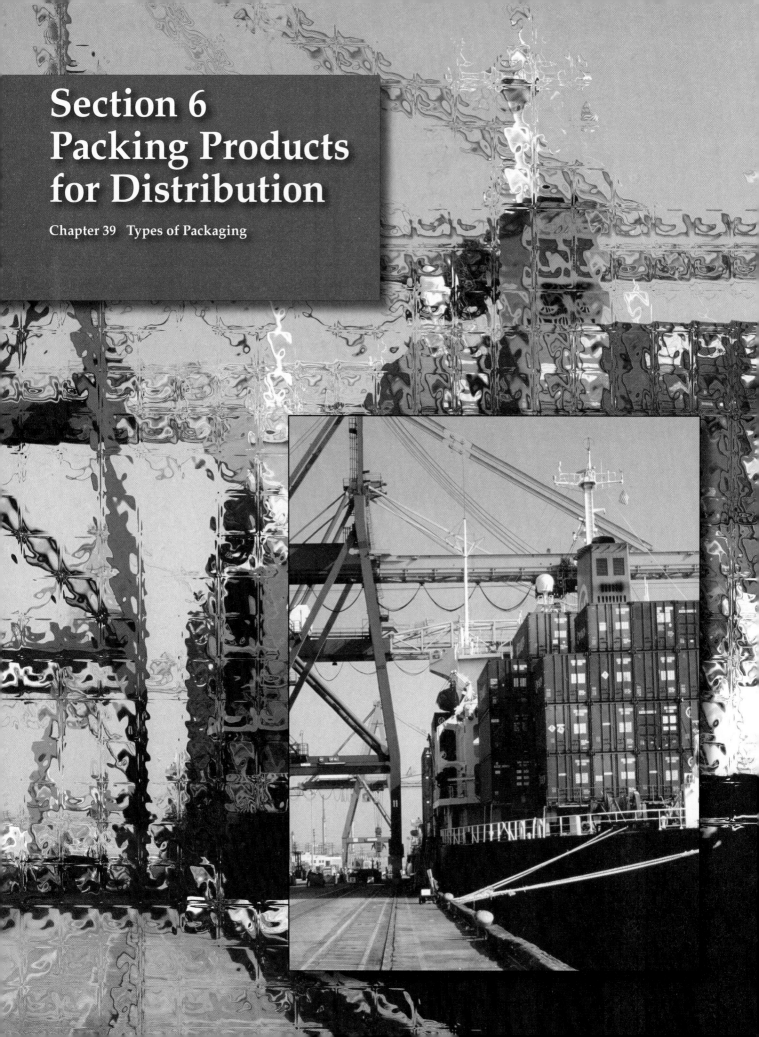

Section 6
Packing Products
for Distribution

Chapter 39 Types of Packaging

39 Types of Packaging

Chapter Highlights

- Packaging is normally referred to as "packing" in manufacturing.
- The major reasons for packing products are to provide protection, make them easier to transport and use, improve their salability, assist the user in assembly, and reduce theft while products are displayed on store shelves.
- Biodegradable packaging includes starch-based products made from corn, biodegradable plastic mulch, and even plates made from potatoes.
- Common types of packaging used today include corrugated, aseptic, food and drug, tamper proof/tamper resistant, automated, stretch and shrink wrap, and skin packaging.
- Aseptic packaging eliminates the need to use preservatives and is used to store products such as milk and juices without refrigeration.

Introduction

When most people think about packaging, cardboard cartons or (corrugated) boxes, grocery bags, food containers, shrink-wrap, and die-cut inserts holding consumer products safely in their boxes come to mind. These are some of the different types of packages used to protect and store materials. But the job of packaging, often referred to as *packing* in manufacturing, covers much more. See **Figure 39-1**.

We will not try to cover all of the different types of materials, applications, equipment, and methods for handling and transporting products here. Packaging is an entire field in and of itself. There are many schools that provide packaging programs, and many hundreds of books written on this subject alone.

Manufacturing and design engineers need to have a basic understanding of the different types of packaging that can be used to pack and ship manufactured products. It is also important for the marketing team to take full advantage of these containers to promote the product, establish brand loyalty, and convey information about the company responsible for making the product. The packaging solution is just as important as selecting the proper materials and producing a well-designed and high-quality product.

Figure 39-1. Packaging covers a lot more territory than just boxes and containers.

Categories	Examples of Product Application
Adhesives and Applicators (Hot Melts)	Chemicals, Baked Goods, Meat and Poultry
Cartons and Cartoning	Apparel, Beverages, Dairy, Pet Care, Pharmaceuticals
Casing and Corrugated Containers	Appliances, Foodservice, Hardware, Household Products
Labels and Decorations	Stationery, Agricultural, Apparel, Container Manufacturing, Sporting Goods
Marking and Imprinting	Pharmaceuticals, Snack Food, Dairy Products, Beverages
Package Manufacturing	Bottle Making, Flexible Packaging, Industrial Packaging
Instrumentation and Control	Inspection Components, Measuring and Weighing Devices
Filling, Capping, and Form/Fill Sealing	Toiletries, Beverages, Personal Care Products
Palletizing, Conveying, and Material Handling	Material Handling Systems, Robotic Packaging
Resins and Coatings	Food Services, Electronics, Laundry Products
Rigid and Semi-Rigid Containers, Caps, and Closures	Bottle Caps, Childproof Enclosures
Bundling Equipment	Wrapping, Strapping, Protecting Consumer and Industrial Products

Packaging is used for many reasons, depending on product design and materials, the form of transportation required to get the product to market, and the environmental conditions that the package will have to face before the product is unpacked by the consumer. The major reasons for packaging products follow:

- To protect the product from damage during transport
- To preserve the product until it is opened for use by the consumer
- To contain small parts so they will not be lost before assembly
- To help sell or market the product with name, brand, and product information
- To help the user assemble the product
- To aid in ease of handling and safe transport
- To reduce theft by packing small products in larger containers

Many products have protective finishes that can be damaged and need to be protected during shipping. Other products are fragile and need to be protected during assembly. Some products, such as liquids or foods, must be contained (held securely) while they are being poured and cut. Bags can be used to group small parts together, providing an organized method of assembly for the consumer. The graphics on containers and shipping cartons are designed to inspire confidence, trust, enjoyment, and any other emotion that will help to sell the product.

Package designers also have to worry about factors such as how the product will be transported, shipped, and distributed. This means that a system may need to be used to contain the components of the product, as well as another system (shipping container) to hold the box and provide additional protection. The most important concerns dictating selection of the packing solution are the size, weight, type of material, and shape of the final product. The same type of package used to protect electronic components from static electricity is not appropriate to store eggs or preserve frozen food.

Important considerations that have an impact on the environment include the material, manufacturing process, and disposal method of the packaging. The packaging design process starts with determining how long the package and packing material have to last, and then searching for a material that will ideally be recyclable. There is no need to over-design the package and make it last for a hundred years if the product has a short shelf life.

The packaging industry recognizes three different levels of packaging: primary, secondary, and tertiary packaging. *Primary packaging* is the outer container that displays the product to the consumer.

The primary packaging provides information about the product. Examples of primary packages include cartons, drums, tubes, bottles, cans, and cartridges. The carton that holds eggs in the store is an example of primary packaging.

Secondary packaging is sometimes referred to as *unit packaging*. This is used mainly to transport, display, and store several primary boxes or packages. Secondary packages are often the first packaging seen by the consumer at the point of sale. These packages are often placed in aisles to display quantities of products. A common example of secondary packaging is the cardboard, or corrugated paperboard, box. Most of the time secondary packaging boxes have very little printing. The purpose of secondary packaging is to hold other boxes and convey information through a UPC (*universal product code*) bar code label. Boxes used for secondary packaging must be durable and stackable.

The third level of packaging is called *tertiary packaging*. This type of packaging is used for protection during transportation, shipping, and distribution. The package must be designed for safe handling and must protect the primary and secondary packaging and their materials from damage. Normally, these packages are used to transport several of the same type of products and contain little graphic information. The most common type of tertiary packaging is the cardboard box. Pallets of tertiary packaging are a common sight outside of many stores and warehouses.

Our goal in this chapter is to introduce you to the major classes of packaging used with manufactured products. Types and examples of packages will be provided in conjunction with each of these classifications. See **Figure 39-2**.

Figure 39-2. Major classes of packaging used with manufactured products.

Classes of Packaging
Cartons and Corrugated Packaging
Flexible Packaging
Aseptic Packaging
Food and Drug Packaging
Cello and Shrink Wrap Packaging
Antistatic Packaging
Skin Packaging
Container Packaging

Impact of Packaging on the Environment

A report from the U.S. Environmental Protection Agency (EPA) stated that in 2001 the average person in the U.S. generated more than 4.4 pounds of municipal solid waste each day. 35 percent of this came from paper and paperboard. Evaluating the paper and paperboard component is one way that the EPA determines the level of municipal solid waste. Of the 81.9 million tons of paper and paperboard municipal solid waste generated, 44.9 percent was recovered and recycled.

According to the EPA, the level of municipal solid waste generated in the U.S. declined slightly from the year 2000. However, population growth throughout the world is placing more stress than ever before on the earth's natural resources and environmental ecosystems.

We have a natural tendency to view packaging as the "culprit," as something that is bad because it results in wasted materials, particularly materials from nonrenewable sources. Coupled with this is our understanding that packaging is bad because it requires the use of additional land space for incineration or burial.

Our goal needs to be better informed so we can make wise choices about the design and use of recyclable and biodegradable materials for packaging. Education of consumers and users is important to guard against improper disposal of both products and packages.

Biodegradable Packaging

One measurement of a package's impact on the environment is the package's ability to degrade into an innocuous material, and essentially vanish from the face of the earth. When this degradation occurs by the action of living things, typically microorganisms, the packaging is considered to be *biodegradable*.

The major problem with biodegradable packaging is that the cost of the materials is about twice that of conventional materials. Consumers are excited about the new materials, but are reluctant to pay for them. However, improved processing, education, and possible govenmental legislation will eventually result in the move toward increased use of biodegradable materials. Some researchers feel that biodegradable packaging is already price-competitive when environmental costs such as energy usage in production and greenhouse gas emissions are properly considered.

Several manufacturers are making starched-based dinner plates from potatoes. These plates are completely biodegradable, nontoxic, and great for all fast-food applications. They are also stronger, more rigid, and provide better short-term insulation than conventional disposables. Potato plates can be easily stacked, handled and shipped due to their rugged nature. Potato-based plates can be manufactured in virtually unlimited shapes and designs, and are able to contain liquids better than conventional plates and eating containers.

Types of Packaging

There are many different types of packages that are needed to store, transport, protect, and preserve today's manufactured products. The unique characteristics of the product and conditions it must face serve as design constraints when creating the appropriate package. If the material is a food or pharmaceutical, it may require packaging that will preserve its contents. A different type of package will be needed if the product is a hard good, needing only to be protected from breakage under normal handling conditions. Exposure to unusual environmental factors such as extreme moisture, high heat, high pressure, or ultra cold may also require unique packaging applications. Some products are delicate and can be damaged if unsettled during shipping. Others must withstand exposure to desert heat or pressure and ultra-cold ocean depths. Let us take a closer look at the major forms of packaging that are used today.

Corrugated Packaging

Corrugated packaging is often called *containerboard* by the industry. Containerboard has at least three components: two pieces of flat paper called *liners*, or *linerboard*, and one piece of crimped paper called *corrugated medium*. The liners are the flat facing sheets that are glued to the corrugated medium and provide a smooth surface on the outside of a cardboard box. The liners and corrugated medium provide a packaging system that can support a great deal of weight. The rigid flutes standing on their ends serve as support columns holding up the weight.

The flutes have several other advantages. They act like shock absorbers, providing cushioning for the contents of the box. They also serve as insulation, providing some protection for the product against sudden temperature changes.

There are many different sizes and profiles of flutes, depending on the manufacturer and type of machine used to make them. Flutes are made in five sizes: A (1/4″), B (3/32″), C (3/16″), E (3/64″), and F (1/32″). Profiles of the flute are determined by the number of flutes per foot. The larger the profile, the greater the vertical compression strength and cushioning. Flutes are produced in 40, 42, 50, 94, and 128 flutes per foot. See **Figure 39-3**.

According to The Fibre Box Association, there are more than 1,550 corrugated packaging plants in the United States and Canada. These can be broken down further into types of plants: corrugator plants, sheet plants, and sheet suppliers. Approximately 94 percent of these are corrugator and sheet plants.

Many corrugated boxes contain die cut cartons and inserts, so individual components or subassemblies to be transported will be carefully supported in the shipping container. Corrugated paperboard is made with longer pulp fibers than other types of paper and chipboard. This long pulp fiber, used in conjunction with fluted medium, provides a strong packaging material.

Flexible Packaging

Flexible packaging such as pouches, bags, tubs, and trays has been around for a long time. One of the earliest forms of flexible packaging was paper made

Figure 39-3. Corrugated board is a combination of at least three sheets of paper, collectively called containerboard. The paper layers on the outside are known as the liners, or linerboard, while the fluted or wave-shaped material in the middle layer is called corrugating medium. Tripplewall is also available.

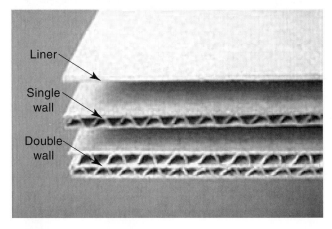

by the Chinese in the first or Second century B.C. They used bark from the mulberry tree.

Products that may need flexible packaging can be in the form of powders, pastes, free-flowing liquids, or thick syrupy liquids. Flexible containers are made in a variety of sizes, shapes, and configurations. Some of these packages are simple and low cost while others are more costly to ensure increased product freshness or better product presentation and marketing appeal. Flexible packaging is used in markets ranging from car parts, bakery goods, cereals and beverages, to fertilizers, seafood, cosmetics, and pharmaceuticals. Just about any product can be protected with flexible packaging.

The largest market for flexible packaging is for food products, accounting for about 50 percent of shipments. Medical and pharmaceutical markets account for an additional 15 percent of revenues. Other markets, such as household goods, pet food, garden supplies, cosmetic samples, and industrial, agricultural, and industrial products, also rely on flexible packaging. See **Figure 39-4**.

Flexible Pouches

Flexible pouches are made from single or double-layer films and can be small or large. One type of flexible, or *retort packaging*, used by the military, NASA Astronauts, and members of the Forrest Service is called *Meal-Ready-to-Eat* (MRE). MRE pouches made of multilayered aluminum foil and plastic sandwiched together, and are used to store food in field environments when refrigeration is not

Figure 39-4. These food packaging trays provide flexible packaging. (Foodservice and Packaging Institute, Inc)

available. The food in the MRE is precooked using aseptic processing (more on this process later). Food in the MRE can be eaten cold, but when possible is usually heated by boiling the pouch in water. SOPAKCO, a company that manufactures and packages military rations, provides data on estimated shelf life. They say that at 70°F (20°C)an MRE should last 66 months, and at 100° F (38°C), eighteen months.

The content to be stored in pouches can be liquids, syrupy liquids, or solids. These products can be dispensed all at once or as needed. Flexible pouches are particularly useful for storing foods, beverages, and some pharmaceuticals. Pouches made of foil film laminate provide stability, stand up on shelves, and in the case of beverage packages are designed to hold a straw that can be retrieved when needed.

Flexible Foams

Flexible foam packaging is ideal for packaging fragile and lightweight materials. Flexible foams can be cut and fabricated to create carton inserts, corner pads, end caps, and other configurations that protect products such as computers, computer accessories, medical tools and devices, delicate analytical equipment, and many other delicate products. Flexible foams have soft, nonabrading surfaces that are helpful in protecting delicate surfaces. The durability of flexible foams also enables reuse of the packaging component.

Air Pouches

Some manufacturers offer air bags that suspend the product in an inflated pouch of air called an *air pouch*. These pouches provide protection against damage by reducing shock and vibration. Sometimes the products are placed in an air bag that is suspended in an inflated pouch. When the products are stored in the inflated bag, they are held snugly in a pocket formed between two inflated sheets of heavy gauge film. When the product reaches its final destination, the pouch is punctured to deflate the bag. Sometimes the pouch is deflated using a straw and is then reused. Air pouches can be used for large as well as small parts.

Flexible Bags

Flexible bags used to package liquids and solids include antistatic bags, antistatic zip lock bags, antistatic shielding bags, woven polypropylene sand bags, white woven puncture-resistance bags, drawstring bags, shopping bags, specimen bags, and many others. Woven and polypropylene bags are used in the chemical, agriculture, construction, and mining industries. Bags are available to package, transport, and store materials in just about any environment. Polypropylene bags containing sand are often able to hold back floodwaters and remain strong for more than 1,200 hours in the sun without disintegrating.

Desiccant bags made of nonwoven crepe material are used to dry the air within a package or container during shipping or storage. These disposable bags or packets are able to protect components and moisture-sensitive products for many months. Drierite desiccant bags are often used to protect items such as electronic equipment, instruments, pharmaceuticals, circuit boards, and photographic equipment.

Desiccant bags are filled with drierite, an anhydrous calcium sulfate that completely dries the air in a package or container during shipment or storage. Drierite desiccant bags are approved by the FDA. The drums of drierite come with a 30 percent relative humidity card that changes color from blue to pink when the base is no longer active. See **Figure 39-5**.

Antistatic bags can be identified by their pink or blue color. ***Antistatic bags*** are used to protect static-sensitive parts. Antistatic bubble bags with zip lock seals or slider zippers are also used to provide heavy-duty protection from shock.

Figure 39-5. Drierite desiccant bags used to remove moisture in packages and containers containing instruments and electronic products. (Drierite, Inc.)

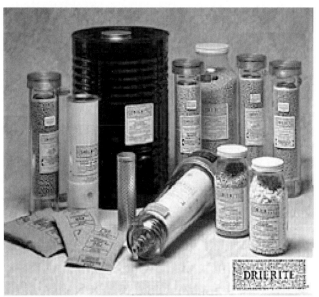

Aseptic Packaging

Did you ever think about how great it would be if you no longer had to worry about refrigerating milk, juice, or dairy products before they are opened, right in their own protective container? Well, this is no longer just a dream in the minds of many consumers with the invention of ***aseptic processing***, and ***aseptic packaging***. The consumer drink box, is a beverage and liquid food containment system that was used in Europe and Asia for several decades, prior to its introduction in the U.S. in the 1980s.

The drink box was referred to as "the most significant food science innovation of the last 50 years" by the Institute of Food Technologists. In 1996, an environmentally friendly aseptic package won the Presidential Award for Sustainable Development in the United States, the only package that has ever won this prestigious environmental prize. See **Figure 39-6**.

Before a product can be stored in an aseptic package the product must undergo aseptic processing. This destroys harmful bacteria and pathogenic microorganisms through thermal processing. Then the packaging film barrier of the aseptic packaging protects the sterile product from contamination. Processing is more effective than traditional canning or hot-fill canning. Both of these processes require products to be heated in a container. The aseptic-processed foods and beverages are sterilized outside the package using a flash-heating and cooling process that heats and then cools the product before filling. This usually takes from 3 to 15 seconds, at a temperature up to 285° (140°C).

Aseptic packaging eliminates the need to use preservatives. The U.S. Food and Drug Administration (FDA) regulates aseptic food processing and packaging.

Here is how the aseptic package works. The box consists of six layers of high-quality paperboard, polyethylene, and aluminum that are laminated together. The paperboard provides substance and stiffness to the package. Polyethylene is used for the innermost layer that comes into contact with the liquid. Polyethylene is also used on the outside of the package to protect the package from moisture. Aluminum is used as a protective barrier to keep the contents away from light and oxygen, and is what eliminates the need for refrigeration. The order of application of the materials from the outside of the package is polyethylene, paperboard, polyethylene, aluminum foil, polyethylene, and polyethylene.

Figure 39-6. The shelf life of food products such as liquid milk and fruit juice can be extended for years using aseptic packaging. They require no refrigeration, but must be refrigerated once they are opened.

Aseptic packaging has been applauded as the package of the future and was recognized as having the lowest environmental impact of any beverage container. Aseptic packaging uses less packaging than other comparable containers. 70 percent of the package is paper, 24 percent polyethylene, and 6 percent aluminum. Aseptic packaging also uses less energy to manufacture and is easily recycled through a process called *hydrapulping*. **Hydrapulping** works like blending and agitates the package for about 40 minutes until the paper, plastic, and foil layers separate from each other. These materials are then reused.

Food and Drug Packaging

Cans, bottles, tubes, and many other types of packaging containers are used to contain solid and liquid foods and pharmaceutical materials. The FDA closely regulates how packaging is used and how well it performs in terms of preservation and storage of materials. The Consumer Product Safety Commission has *Good Manufacturing Practices* (GMP) stated in the form of regulations and rules that must be carefully addressed by manufacturers related to child-resistant packaging, poison protection, tamper-resistant enclosures, protection of food for humans and animals, and more. The U.S. Food and Drug Administration provides the industry with these GMP in section 520 in the Quality System (QS) of the Food, Drug and Cosmetic (FD&C) Act. GMP are the requirements domestic or foreign manufacturers, processors, and packagers must meet in designing, manufacturing, packaging, labeling, storing, installing, and servicing finished medical devices intended for commercial distribution in the United States. The purpose of GMP is to help ensure that products are safe and effective, and that they do what they are intended to do.

The Consumer Product Safety Commission administers the Poison Prevention Packaging Act, requiring special child-resistant packaging and special labeling of all household substances that may cause serious injury to children if handled, used, or swallowed. This requires the packaging to be difficult for children under the age of five to open, while not being difficult for senior adults. The following are some of the materials that require special child-resistant and senior-friendly packaging:

- Acid and alkali drain cleaners
- Alkali oven cleaners
- Aspirin and Ibuprofen
- Dietary supplements

- Ethylene glycol (antifreeze)
- Glue removers
- Over the counter and prescription drugs
- Mouthwash containing alcohol
- Paint solvents and thinners
- Permanent wave neutralizers
- Turpentine
- Methyl alcohol

The types of packaging used today to comply with these requirements are shown in **Figure 39-7.**

Packaging of foods and drugs are protected through laws and regulations established by the FDA and Consumer Protection Agency. The primary application environments for packaging of food and drugs are shown in **Figure 39-8.**

Tamperproof and Tamper-Resistant Packaging

Many of today's products must be packaged in ways to keep others from accessing or changing the contents of the package. This can come to be something that we as consumers have accepted as a necessity for keeping products safe for our use.

In the 1980s, a case involving tampering with over-the-counter medication for pain relief occurred.

Figure 39-8. Application environments for packaging of food and drugs.

Food and Drug Packaging
Beverage Packaging
Dairy Product Packaging
Candy Packaging
Confectioner Packaging
Meat Packaging
Snack Food Packaging
Drug Packaging
Bakery Goods Packaging

Since that time, there have been other cases of tampering with pharmaceutical products, foods, and chemical products. Today, everything that is ingested by humans is either tamperproof or tamper-evident. A good example of a tamperproof package is a metal can. The only way someone could tamper with the contents is to open it.

Tamper-evident packages are those that show when tampering has taken place. This can be accomplished though either visual inspection or sound.

Figure 39-7. Child-resistant, tamper-proof packaging.

Child-Resistant and Tamper-Proof Packaging	Description
Blister or strip packs	The product is housed in an individual compartment, which must be cut open to obtain the product.
Film wrap	The film is wrapped around the product and must be cut off to open the package.
Bubble packs	The product is sealed in clear plastic and mounted on a display card. The plastic must be broken into to remove the product.
Heat shrink bands or wrappers	The bands, applied where the cap meets the container, must be broken in to open the package.
Foil or plastic pouches	The product is enclosed in a protective pouch that must be broken in to obtain the product.
Bottle seals	Paper, foil, or thermal plastic sealed to the mouth of a container beneath the cap. The seal must be torn or broken or removed in order to use the product. Bottle seals applied with a pressure sensitive adhesive are not considered to be an effective tamper-resistant mechanism.
Breakaway caps	The container is sealed by a plastic or metal cap that either breaks away completely when the cap is removed from the container or leaves part of the cap attached to the container.
Sealed tubes	The mouth of a tube is sealed shut, and the seal must be punctured in order to use the product.

Visual systems for indicating that the product has been either accessed or tampered with include a film or paper seal that covers bottle openings of over-the-counter medicines. One can tell that the bottle has been opened if the seal is broken or removed. Rings and seals on jar lids are also used to provide indication that the bottle has been opened.

Audible methods for providing proof that the product has been accessed are less evident. Many jars or bottles are vacuum-sealed with a metal lid. This lid has a convex center that is normally out, but is pulled into the jar from the vacuum pressure. When the lid is removed, the pressure is released and the center of the lid "pops" back to its normal position. If the jar has already been opened, then there will be no audible pop and consumers should be suspicious that someone has opened the package.

Stretch and Shrink Wrap Packaging

Flexographic printing is a process used to print on rolls of plastic and other materials. Flexible inks are applied to a substrate using rubber printing plates. These plates are attached to a high-speed printing press running a web (roller) of cellophane wrap. The finished product can be used as overwraps for bakery goods, candies, and snacks.

Stretch wrap

Stretch wrap film is desired when loads need to be held together. Stretch wrapping is often the most efficient method for holding multiple packages and products together on or off pallets. Loads must often withstand severe vibration during transportation both in the plant and during shipping. Stretch film has the elastic property needed to withstand these forces and hold parts and packages together.

Stretch wrapping is also good for keeping loads clean and dry and to help protect against vandalism and pilferage. On a cost-per-load basis, stretch packaging is one of the most cost effective ways to wrap heavy loads. See **Figure 39-9**.

Shrink wrap

Shrink wrap is a flexible sheet of plastic that is sold on rolls and used to wrap pallets, boxes, bottles or cans, and for securing products to trays or flat boards to aid in shipping. Shrink wrap packaging is a simple, two-step process. First the product is wrapped with a layer of shrink wrap. Then the shrink wrap is heated with a heat-sealing machine

Figure 39-9. This stretch wrapping system is popular for securing heavy loads and is designed for use with a pallet jack. (ABS Packaging)

or a heat gun. Often products are conveyed through a heat tunnel where they are exposed to heat. The tunnel or heat gun shrinks the wrap about 25 percent of its surface dimension.

Shrink wrap is also used on construction sites to contain sandblast or chemical abatement hazardous materials, such as lead, asbestos, or heavy metals. It is also used to provide protection from weather for irregular shaped objects, road freight, and palletized truckloads of equipment and materials, and for temporary housing and greenhouses. In the marine industry, shrink wrap is used to provide a protective coating and protection from weather during shipment.

Skin Packaging

Skin packaging is the method of choice for packaging products such as drills and screwdrivers on chipboard for display on store racks. Skin packaging is accomplished using thermoforming. First, the card stock, chipboard, or corrugated medium is positioned on the bottom platen of the vacuum forming press. Then, a sheet of tough plastic film is heated by air from the top platen. The top platen is then lowered down and pushes the plastic in contact with the part. The film is normally carried on a roller that feeds and advances film as needed in a location between the two platens. When the product is in the desired location on the platen, the film is advanced, and a vacuum is pulled through holes in the lower platen. This vacuum pulls the sheet down tightly against the card. The platens are then opened and the card is moved on to final processing where the excess plastic is cut away from the card.

Important Terms

antistatic bags
aseptic packaging
aseptic processing
biodegradable
containerboard
corrugated medium
desiccant bags
hydrapulping
linerboard
liners
primary packaging
retort packaging
secondary packaging
shrink wrap
skin packaging
stretch wrap
tertiary packaging
unit packaging

Questions for Review and Discussion

1. What are the differences between primary, secondary, and tertiary packaging?
2. What are the major types of biodegradable packaging? What causes them to decompose in the environment?
3. What are the primary reasons for packaging a product?
4. Explain the construction and benefits of aseptic packaging.
5. What manufactured product ends up as litter more than any other?

Glossary

A

abrasive belt planer. A high-production wood-planing machine that uses continuous abrasive-surfaced belts.

abrasive jet. Stream of pressurized water mixed with small amounts of abrasive particles, such as garnet.

abrasive jet machining. A grinding process that suspends tiny particles of abrasive material in a low-pressure stream of gas (dry air, carbon dioxide, or nitrogen) sprayed through a sapphire nozzle.

abrasive shot method. A glass-frosting method in which a mixture of abrasives and lead shot is placed inside a frame with a sheet of glass, then agitated.

ABS. See *acrylonitrile-butadiene-styrene*.

absolute sensitivity. The ratio of the change in output of the sensor per unit change in the parameter being measured.

accuracy. The largest expected error between actual and ideal (expected) output signal, often expressed as the percentage of the full range output.

acetylated wood. A modified wood that provides the strength of resin-impregnated woods, without the brittleness associated with them.

acrylonitrile-butadiene-styrene (ABS). A polymer made up of acrylonitrile, butadiene, and styrene monomers. ABS is widely used for plumbing pipe and fittings.

actuator. A device in an open-loop control system that responds to the signal from the controller and performs an action.

adaptive control systems. Systems that provide control in response to changing conditions.

addition polymerization. The linking together of monomer molecules to form long chains. The process is mechanical; there is no chemical change in the material.

additives. Materials that are added to other materials to change the original material's characteristics. Fillers, pigments, and thixotropic agents are examples of additives used in plastic gel coat.

adhesion processes. Metal-joining methods (soldering and brazing) that form a bond between parts by filling joints with a material with a melting point lower than that of the base metal. Also, any method of producing a product that involves joining materials with a glue or other bonding agent.

adhesion. The process of joining materials by using an agent that sticks to both pieces.

adhesive. A bonding agent that forms a film between materials to be joined.

adhesive failure. Failure of a glued joint that takes place in the interface area between the adhesive and the material being joined.

advanced ceramics. Sector of the ceramics industry that manufactures products from engineered materials.

aerogel. A rigid material that is extremely light and remarkably strong, developed by Dr. Steven Jones and NASA's Jet Propulsion Laboratory.

age-hardening. See *precipitation hardening.*

air drying. Method of drying wood that reduces the moisture content to around 15 percent.

air pouch. An inflated and sealed plastic bag used in packing to provide protection for a product by reducing shock and vibration.

airless spraying. A spray method that uses hydraulic pressure, rather than compressed air, to force the finishing material into the spray gun.

alloying elements. These combinations of different metals are used in welding to increase strength and toughness.

alloys. Mixtures of several different metals, sometimes with the addition of other nonmetallic elements.

amorphous. Polymer molecular arrangement in which the molecules are arranged randomly and are intertwined.

analog device. Produces a signal that is proportional to the characteristic being measured.

aniline dyes. A class of dyes used in spirit stains.

anisotropic. The property of wood that results in drastic structural changes as it absorbs moisture or gives up moisture (dries).

annealing. A process for softening metal and relieving stresses by holding it at high temperature for an extended length of time, then slowly cooling it. Also used to relieve stresses in ceramic or glass products.

annealing lehr. An oven through which molten glass is pulled to relieve internal stresses.

annular ring. The layers of cells appearing in concentric circles and representing each year's new growth of a tree.

anode. A positive electrode.

anodic oxidation. See *anodizing.*

anodizing. An electrochemical process, often used on aluminum, that converts the surface of the metal into a hard oxide layer that is resistant to corrosion. Also called *anodic oxidation.*

anti-kickback device. A safety mechanism attached to power saws to eliminate the problem of kickback.

antiozonants. Additives used to help prevent breakdown of plastics by ozone gas in the atmosphere.

antistatic bags. Plastic bags that are used to protect electronic components from static electricity.

arbor. In a machine tool, the rotating spindle on which a cutting tool is mounted.

arc welding. Several fusion processes used for joining metals. In each instance, the welding takes place because of the intense heat produced from an electric arc.

artificial aging. A hardening process for aluminum alloys conducted at heat levels above room temperature.

artificial intelligence (AI). A control system concept in which inputs from sensors, coupled with predefined rules and relationships stored in the computer's memory, allow the control system to provide expert control in well-defined situations.

aseptic packaging. Packaging that is used for storing milk and other food products without refrigeration for several years. Once the package is opened, the product shelf life is reduced to approximately two weeks.

aseptic processing. Processing used by firms to prepare materials for storage in aseptic packaging.

ashing. The use of a very fine wet abrasive to reduce overheating of plastic and improve the cutting action.

atomized spraying. Pressure is used both to break the finishing material into very fine drops (atomize) and to blow it onto the item to be coated.

attrition milling. A batch operation in which the materials to be ground are placed in a stationary chamber, along with the grinding media. A rotating shaft provides the agitation needed for grinding action.

austempering. A form of tempering in which metal is cooled in a salt bath to increase ductility and toughness.

austenite. The most dense form of low carbon steel.

autoadhesion. A strong bond that some surfaces have when they come into contact with each other.

autocatalytic. A chemical reaction in which the reaction itself is a catalyst, increasing the speed of that reaction without itself being changed as a result of the chemical reaction.

automated control. A control system that simultaneously monitors the quality of the product and the functioning of the system; it must be able to start, stop, and sequence production.

automated manufacturing system. Production system that results from linking together several processes or operations that require no human intervention.

automation. The process of applying automatic control devices to production equipment.

B

bagasse. A cellulosic by-product of the pressing of sugar cane. It provides the fiber for manufacture of insulation board.

ball mill. A large steel cylinder that is partially filled with spheres of heavy steel and tumbled or rotated to grind materials. Also, a small porcelain jar mill used to mix clay and additives.

band casting. A continuous production process used to produce thin strips of ceramics for use as electrical substrates and heat exchanger devices. Also called *tape casting*.

band. The continuous loop blade used in a bandsaw.

banding. Also known as gilding, applying gold gilt on the edges of ceramic china cups and plates.

bandsaw. A saw with a continuous loop blade, used primarily for cutting curved shapes.

balsa wood. A popular core material used in marine hulls and decks to create a lightweight but rigid composite.

barrel finishing. See *tumbling*.

batch. A number of essentially identical parts or units manufactured at the same time. Also called a *job lot*.

batch furnace. A type of furnace used to heat-treat small quantities of parts carried on a rack or flat car that is loaded into the furnace.

bed width. The width of stock that can be milled in a planer or similar machine.

bed-type machine. A type of milling machine in which a worktable is mounted directly on the machine bed.

bending. A process used for producing curved shapes in either solid or laminated wood materials.

beneficiated. The practice of processing or refining raw material at plants located near the mine in order to reduce transportation costs.

beveling. The process of changing the sharp (90°) angle where a vertical and horizontal surface join into a less-sharp angle, usually 45°.

binary device. A device that only has two states or values.

binders. Phenolic resins used to bond together wood chips or flakes when producing various types of composition board. Also, chemicals that are added to dry ceramic powders to hold the mixture together when it is pressed.

bioceramics. Ceramic devices used in the medical field.

biodegradable. Able to degrade into an innocuous material by the action of living things, typically microorganisms.

bisque firing. The first firing of a ceramic material, done without glaze; bisque firing solidifies and fuses the body so it can be handled.

bisque ware. The ceramic product that is fired without a glaze (coating).

bits. Tools used in a drill press or boring machine to produce round holes.

black plate. Light-gage cold-rolled steel without any metallic coating.

blank. The stock on which a flat die is used to roll threads. The term is often used for any metal piece being shaped by forging or stamping processes.

blanking. A process used to punch a flat blank from sheet metal. Blanking produces very little waste.

blanks. Slices cut from a mass of highly viscous ceramic stock and used in the jiggering and jollying processes of production. Also called *pugs*.

bleaching. A process that involves applying a caustic liquid to the surface of a workpiece to lighten the natural wood coloring.

blending. A type of sanding that results in some material removal, producing a fairly smooth finish; it requires a medium-grit abrasive.

blind holes. Holes that do not run through the workpiece.

blind rivet. A fastener, consisting of a mandrel (shaft) and rivet head assembled as one unit, for riveting when only one side of the workpiece is available.

blow and blow process. A bottle-forming process using a blast of air to blow the molten glass into a desired shape.

blown film. Film that is produced by extruding a tube of plastic material which is then expanded in diameter and reduced in thickness by internal air pressure, and finally slit to form a flat sheet.

blue annealing. Annealing process in which the exterior of a ferrous part is exposed to the atmosphere, forming a bluish oxide on its surface.

board foot. The basic unit of measurement for dimension lumber, 1 inch thick by 12″ by 12″.

bonding. A forming process that involves the use of heat and pressure to compact particles or chips of wood into wood composition board.

Border Industrialization Program. Program under which the first maquiladoras were built in 1966 in Baja, California and Ciudad Juarez, Mexico.

boulton drying. Process used to pressure treat wood under vacuum.

bow warp. Warpage that occurs with the length of the board.

brazing. A joining method that holds base metals together by adhesion of a melted filler metal, without melting the base metal.

breakdown maintenance. The process of responding to unplanned interruptions of work.

breaking point. The point in the necking process at which increasing strain on the part will result in it breaking.

bright annealing. Annealing taking place in a controlled atmosphere furnace or in a vacuum, nearly eliminating oxidation and leaving the surface relatively bright.

brittleness. The tendency of a material to break with minimal flexibility.

broach. Cutting tool that creates a planing action when pushed or pulled across or through a workpiece.

broaching. A process used for the internal machining of keyways, splines, and irregularly shaped openings.

B-stage. A partly cured preform that is dry, but still slightly tacky.

buffer stock. Excess work in progress generated to cover longer cycle times and bad parts produced.

buffing. Polishing process that uses muslin disks on a wheel to provide a smooth glossy finish.

building blocks of automation. The basic ingredients of automation: a repeatable manufacturing operation or process, a control system, and a material placement system.

buildup. See *lay-up schedule.*

bulging. One type of process used to expand tubular metal shapes.

bulk density. The type of density that is often measured on ceramic parts. It is based on the bulk volume of the part.

bulk molding compounds (BMC). Putty-like materials used as stock in some closed molding processes.

bulk systems. Handling systems designed to convey liquids, gases, and granular solids to a dispensing mechanism.

bullnose forming. See *stretch forming.*

burrs. Sharp edges produced when metal is deformed by shearing, trimming, stamping, or machining.

bus. The pathway that connects all the devices in a control system.

butt welding. A form of arc welding in which heat is generated from an arc produced as the ends of two parts to be welded make contact. Also called *flash butt welding.*

C

calcination. A process in which material is first heated, then milled, to produce particles that are sufficiently fine to be mixed with the clay for ceramic production.

calendering. Pulling and squeezing pliable thermoplastic stock in a one- or two-pass operation between a series of turning rollers to produce the desired film thickness.

cam drive. A type of mechanism that provides an accurate and reliable means for converting rotary motion to and from linear motion.

captive suppliers. Those for whom a manufacturing company is the principal, or sometimes only, customer.

carbon black. A soot-like material produced by the incomplete combustion of petroleum products.

carbon steel. Steel that includes carbon as an alloying element.

carbonitriding. A process similar to nitriding that is used for hardening low-carbon steel parts.

carbonization. The heating step in the production of carbon fiber, during which heat is applied to burn off most of the noncarbon elements of the fiber.

carburizing. A method of casehardening in which the heated workpiece is covered with a carbon-hardening powder that melts and forms a thin coating.

case hardening. Process that is used to produce a hard shell or case on a workpiece.

cast iron. One of the two basic forms of iron. Cast iron contains between 1.7 percent and 4 percent carbon. It is used when strength and durability are critical.

casting. A process in which a liquid resin is poured into a mold and solidifies.

catalyst. An agent that enables a chemical reaction to proceed at a faster rate.

cathode. A negative electrode.

cavitation cleaning. A precision cleaning method. An ultrasonic transducer in a tank produces high frequency sound waves that cause an intense microscopic scrubbing action at the surface of the part. The energy released by the transducer also heightens the chemical activity of the cleaning solution, which helps to strip away contaminants.

cell. Forming chamber made up of plate glass sheets held together with spring clips.

cell casting. A molding process in which acrylic sheets are cast and allowed to cure between sheets of plate glass.

cellulose. One of the two major ingredients in wood, making up about 70 percent of volume.

cellulosics. A group of commercial plastics made from a natural polymer, cotton cellulose.

cemented carbide. A cutting tool tip material (tungsten or titanium carbide, suspended in a cobalt matrix) brazed on a tool blank to cut harder material at higher speeds.

cementite. The phase that is reached when steel is slowly cooled after being heated to above 1341°F (727°C). Cementite has a carbon content of 2.0 percent.

centerless grinding. A special type of cylindrical grinding that can be used for external or internal applications.

centrifugal casting. Process used with many different types of materials. Centrifugal forces created by the rapidly turning mold cause the material to be distributed against the walls of the mold, causing the material to creep upward as the mold continues to turn.

ceramic. Products made from clay, an inorganic, nonmetallic solid material derived from naturally decomposed granite.

ceramic engine. An internal combustion engine with components that are either entirely ceramic or ceramic-coated metal.

ceramic-matrix composites (CMCs). Composite materials in which the matrix is a ceramic.

cermet. A ceramic composite material, made from carbide and sintered oxides, that is used to make disposable tip inserts for cutting tools.

characteristic waste. An approach for determining whether a material is hazardous. The material is considered in terms of its characteristics when humans and animals are exposed to the material. Materials classified as "characteristic waste" are seldom included on EPA lists. Examples would be materials that are readily ignitable, corrosive (has a pH level of 2 or less or 12.5 or higher), reactive (normally unstable and readily undergoes violent change), or toxic.

charged. The addition of a small amount of fine abrasive to the wheel while buffing.

chassis. Serves to locate and orient a product to make assembly easy and logical.

chemical blanking. An etching process that forms a part by etching completely through the workpiece.

chemical blowing agents. Materials that decompose when the resin they are added to is heated to a specified temperature during curing.

chemical etching. The process by which glass is frosted, or made opaque, with the use of an acid.

chemical milling. Process used to etch or form metallic parts. Also called *chemical machining* or *photoforming.*

chemical properties. Attributes used to distinguish one material from another, including the density and classification information, such as corrosion resistance and flammability.

chemical vapor deposition. A conditioning process in which a gas reacts and contacts a heated part, resulting in the production of a very fine-grained coating.

chopped mat. A mat that contains randomly distributed fibers cut to lengths typically ranging from 1.5″ to 2.5″.

clear cuttings. The felling of all trees in a chosen area in one operation.

climb milling. A process in which the cutter is positioned on top of the workpiece, with the work moving in the same direction as the cutter's rotation.

closed molding. A forming process in which stock is compressed between the halves of a two-piece mold.

closed-die forging. A method of forging in which a preheated metal billet is placed between dies that completely encompass the billet and restrict its flow.

closed-grain woods. Hardwoods that have small pores, such as maple.

closed-loop control systems. A basic control system in which the actual output is measured, compared to a preset standard or reference, and adjusted if necessary.

CO_2 laser. A device that produces a narrow beam of coherent (all one wavelength) light that can be used to cut, scribe, solder, weld, or heat-treat many materials.

coated abrasives. Finishing materials consisting of an abrasive adhered to a backing material, such as paper or cloth.

coated preform. A shape made by spraying chopped fiberglass and a binder material onto a form.

coefficient of thermal expansion. The amount of increase in volume of solid, liquid, or gas for a rise of temperature of 1° at constant pressure.

co-fired. Co-fired products are those that contain a sandwich of several materials, such as metals and ceramics, and are fired together. This not only joins the materials together, but also permits the formation of signal connections and buried pathways.

cohesion. The intermolecular attraction by which the elements of the body are held together.

cohesive bonding. A method that permanently joins materials through chemical action, relying on the capacity of a substance to adhere to itself.

cohesive failure. A bonding failure that occurs in the adhesive itself.

cohesive force. In cohesive bonding, the force that holds the substance together.

coining. A precision closed-die forging process that causes metal to move, as the die halves are closing, from thinner to thicker areas.

coinjection molding. A molding process in which two or more materials are injected into the mold to make an individual part.

cold compression molding. A molding process in which the material to be formed is compressed between unheated male and female molds. It is also referred to as cold press molding.

cold flow. Creep that takes place at room temperature.

cold forming. See *upsetting.*

cold heading. See *upsetting.*

cold isostatic pressing. A dry powder pressing process used in ceramics manufacture. Higher pressure is used than required for dry pressing.

cold-rolled steel. Steel that is rolled while cold, and has a shinier surface finish than hot-rolled material.

cold-roll forming. See *roller forming.*

collet. A socket, equipped with threads on one end and split jaws on the other, which is used in a lathe to automatically center stock that is round, square, octagonal, or hexagonal.

collimated. The focused, parallel light rays of a laser beam.

colorants. Dyes or pigments added to plastic to impart color, and often, to improve other properties of the product.

column-and-knee-type machine. A type of milling machine in which a worktable is mounted on a knee that moves up and down the column.

combination blades. Circular saw blades with teeth that can be used for both crosscut and rip sawing.

commodity resins. Standard-grade resins such as low-density polyethylene (LDPE), polypropylene homopolymer (PP), crystal polystyrene (PS), and rigid polyvinyl chloride (PVC).

common lumber. Wood that is graded by number, ranging from 1 to 4, with number 1 presenting the best grade (no knots or knotholes).

compact. The formed part that results from green machining. Also, the action of squeezing metal molecules more closely together through forging or embossing in a press.

composites. Material family that consists of stock that results when two or more distinctly different but complementary substances are physically combined.

composition board. A sheet product made from wood that has been broken down into particles or fibers, then reconstituted. Types of composition board include insulation board, hardboard, waferboard, particleboard, and oriented strand board.

compreg. Wood treated with water-soluble phenol-formaldehyde resin and compressed while the resin is curing.

compression molding. A closed molding process that can be done with either thermosetting or thermoplastic polymers.

compression seal. Bond in which parts are pressed into position in the completed assembly.

compressive. Process that forces a semisolid or liquid material into the desired shape, and then subject it to other processes that cause it to harden or solidify.

computer numerical control (CNC). A system in which a program is used to position tools and a workpiece and carry out the sequence of operations needed to produce a part.

computer-integrated manufacturing (CIM). A plant-wide or company-wide management philosophy for planning, integration, and implementation of automated production. CIM is an integration of separate computer systems that allows data to be shared, transferred, and modified with ease.

condensation polymerization. A chemical reaction in which polymer chains become cross-linked. Usually, a by-product such as water is formed and given off.

conditioning. Internal change in a material caused by exposure to extreme hot or cold temperatures, or forces such as magnetism and electrical energy.

conductive glass solder. A material that is used to provide both a seal and an electrical path.

conductive plastics. Plastic that is designed to be a thermal or electrical conductor.

conifers. Cone-bearing trees that have needles which remain green all year long. Softwood lumber is produced from conifers.

consumable products. Substances that are "used up" after purchase, such as foods and many types of paper products.

container board. Cardboard; used to make corrugated boxes.

continuous casting. A molding process, used for acrylic sheet, in which a highly viscous syrup is slowly cured between two continuously moving stainless steel belts.

continuous die cutting. A process that allows a material in roll form to be cut as it passes through either a flatbed press or a rotary machine.

continuous extrusion blow molding. A process in which a parison is continuously formed and molded in a series of molds.

continuous fibers. Unbroken, long, glass reinforcing fibers. They are sometimes sprayed with binders to hold them together in a continuous mat.

continuous furnace. A type of reheating furnace in which the charge introduced at one end moves continuously through the furnace and is discharged at the other end.

continuous path control. A type of NC control system in which the tool feed rate is controlled at all times, and cutting tool offsets or compensations are specified for every point on the tool path. Also known as *contouring*.

continuous-strand mat. A mat formed from swirls of unbroken fiber strands.

continuous variable. A variable that could have any value from zero to the maximum force the transducer is able to measure.

contour machining. A process that selectively etches a desired area to some specified depth.

contouring. A method used to remove metal from surfaces of irregularly shaped parts by selectively etching to the desired depth.

contour roll forming. See *roller forming.*

controlled factors. Settings used during an experiment.

controller. That part of an open-loop control system which provides the logic and governs the action of the actuator.

cope. One-half the frame that makes up the flask used in sand casting. See *drag.*

copolymer. A molecule that results when two or more kinds of monomers are combined to form the long-chain molecule.

core materials. Solid material used in plastic laminates.

core. A molded sand shape used to provide a cavity in a part produced by the sand molding process.

corona discharge treating. See *electrical surface treating.*

corrugated medium. Container board component that consists of one piece of crimped paper.

coupling agents. Substances added to the resin component of a composite to improve the strength of the bond between the resin and reinforcement materials or fillers. Also called *promoters.*

cracked ammonia atmosphere. A mixture of hydrogen plus nitrogen used in a kiln to improve the quality of the metallizing coating on a ceramic part.

crazing. A defect in plastics consisting of a network of fine cracks on or just under the surface.

creep. The slow and continuous increase in length, over time, when a material is placed under a steady load and constant temperature.

creep failure. A change in the dimensions of a material due to creep (stretching) caused by stress from a constant load.

critically damped response. Adjustments must closely follow input requirements.

crocus cloth. A coated abrasive product that is made from purple iron oxide and is used to polish and buff metals.

crook warp. Warpage that causes the board to twist in an arc or partial circle.

cross direction. Another term for fill yarn.

crosscut sawing. Saw cuts made across the grain of the wood.

crosslinking. The process of joining polymers into long chains.

crush strength. The ability to support a specific weight without being damaged.

cryochemical drying. See *freeze drying.*

cryogenic. Ultracold temperatures as low as –300° F (–185° C).

cryogenic conditioning. Conditioning process for metals that occurs at a temperature of around –310°F (–190°C) and is able to improve certain properties beyond the capability of a cold treatment.

crystal pulling. A crystal-growing method in which a small, single-crystal seed is brought into contact with molten semiconductor materials which influence growth; the crystal is then attached to a rotating pulling bar and slowly pulled away.

crystalline. Polymer molecular arrangement in which the molecules are arranged closely and in a distinct order.

crystallites. Long-chain cellulose molecules that are arranged in a nearly-parallel orientation and help to form the cell wall in woods.

cup. A concave shape that may result from using too much water to wash the bleach off a wooden workpiece.

cupola. A high-temperature furnace used to produce cast iron.

cup warp. Warpage that occurs across the width of the board.

curing. A chemical reaction in which cross-linking of polymer chains takes place.

curtain coating. A process in which stock to be coated is passed through a continuous, sheet-like falling stream of finishing material.

cutoff saw. A circular saw, mounted on a pivot, that is often used on construction sites to cut angles on framing lumber.

cutting speed. The distance in feet per minute that the cutting edge of the drill actually travels. It may also be called peripheral speed.

cyaniding. A case-hardening technique used to create a high nitrogen-low carbon case.

cycle time. The time it takes to complete a task on one unit.

cylindrical grinding. Grinding process used on workpieces with curved surfaces or cylindrical shapes.

Czochralski method. A crystal growing process in which the seed is dipped into a crucible containing melted silicon, then withdrawn. The molten material freezes to the crystal.

D

Danner process. A method used to make glass tubes by flowing molten glass down over a hollow mandrel. Air is blown through the mandrel to keep the walls of the tubing from collapsing.

dead-end escapement. A dispensing device that aligns and holds a part at a barrier (dead end) until it is picked up by a tool or workholding fixture for use.

de-air. To remove the air from highly viscous ceramic stock, and produce blanks or pugs.

de-airing pug mill. A machine that consists of knives on a rotating shaft that cut and fold clay stock to remove bubbles. The process increases the density and strength of parts produced from the clay.

deburring. Removing burrs through various finishing processes.

deciduous. Trees that produce hardwood lumber. They are broad-leafed species, typically shedding their leaves each fall.

decoupled. Making a task an independent activity.

deep drawing. Process in which a female die is pressed into thin sheet stock, stretching it over a male forming punch.

defiberize. The process of drying out wood particles to be used for hardboard, so that most of the chemicals are removed.

deflocculant. A water-absorbing chemical that is used to help make slip more fluid or less fluid.

degating cutter. Cutter that is used to remove the gates attached to runners that carry molding material to the part after it is cast.

densification. The process used to increase the density of a ceramic part. It is usually done by sintering.

density. Weight of a material per unit volume.

deoxidizers. Deoxidizers prevent oxygen from contaminating a weld pool, which would weaken the weld.

dependent variable. The output from the process that is critical to the performance or quality of a part.

depth of cut. The amount of material being removed from the workpiece in a single pass of the cutting tool.

desiccant bags. Bags made of nonwoven crepe material containing *Drierite* (anahydrous calcium sulfate) used to dry the air within a package or container during shipping or storage.

design for assembly (DFA). A process that simulates the actual assembly of a product, allowing it to be evaluated in terms of its suitability to modern methods of assembly.

devitreous glass solder. A material similar to a thermoset, in that it can be reheated without becoming highly liquid and flowing freely.

dielectric. Handheld moisture meter that does not use pins, and therefore does not create damage to the sample. It relies on surface contact with a flat electrode that does not penetrate the wood.

diamond point tool. A chisel used to produce specialized contours on a lathe.

diamond wire. A high-tensile-strength copper-plated wire impregnated with fine particles of industrial diamond to serve as cutting teeth; used to saw composites.

diamondwood. Plasticized hardwoods made from very thin layers of wood that are dyed and then pressure treated with resin.

die block. A closed die consisting of a hardened cylinder with a hole running through its center, used for cold heading nails, bolts, and rivets.

die casting. The process of forcing molten metal under high pressure into the cavity of a die to form a part.

die cavity. The open area inside a mold that will be filled with the material being formed.

die cutting. A process that is used to cut paper, plastic sheets, and similar materials by forcing sharp-edged cutting rules through the stock.

die set. The combination of the die block assembly and one or more punches.

diffusion. A process used with ceramics to move particles of material into the void spaces between them, improving the density of the part. Diffusion also helps decrease shrinkage during firing.

dimension lumber. Wood that is normally purchased in standard-length boards up to 1″ thick and 12″ wide; it is ready to use, without any additional sanding or surface preparation.

dip casting. See *dip molding.*

dip coating. Applying finishing material through complete immersion of the product or part. Also, a method used to apply a plastisol to the gripping surfaces of metal products, such as tool handles.

dip molding. An application in which a heated metal tool is immersed in a dip tank to acquire a coating of plastisol. Products such as surgical gloves are made using this method.

direct bonding. Adhesives, brazing, glass sealing, and electrostatic bonding are all examples of direct bonding fabricating processes.

direct extrusion. Type of extrusion in which the ram or punch moves down on a heated metal billet. The pressure from the ram forces the stock to extrude (flow) out of the die opening.

discrete device. A switch that is either closed and completing the circuit or open and providing information on a discrete variable.

distributed numerical control (DNC). A control system in which individual NC machine tools are connected through communication lines to a central command computer.

doctor blade casting process. A process that uses a metal squeegee (doctor blade) to regulate the amount of ceramic slurry that is allowed to flow onto a metal belt.

dopants. In semiconductor production, a form of ion implantation in which alloying elements are added to liquid metal to produce unique properties.

doping. An ion implantation process used in producing microelectronic devices. It permits alloying with small amounts of different metals. In ceramics, the process of adding organic binders and lubricants to the clay mix to reduce it to a semiliquid plastic state.

double planers. Knife-blade planers with cutter-heads above and below the stock, so that both sides of the stock can be planed at the same time.

double wall corrugated. Package containing two walls, each separated with corrugated medium, to provide improved strength and durability.

dowels. Wooden connecting pegs often used in the furniture-making industry.

drag. One-half the frame that makes up the flask used in sand casting. See *cope.*

drain casting. See *slip casting.*

drape forming. A forming process in which a thermoplastic sheet is clamped, heated, then drawn down over a mold.

drawing. A metal-forming action involving both stretching and compressing to form a three-dimensional part from flat stock.

drop. The material separated from a part.

dry bag isopressing. A process using a male die punch with a liquid-filled rubber bag in an open-cavity lower die block. The filled bag exerts pressure, forcing the ceramic powder against the punch to form a part.

dry felting. Process used to make hardboard, similar to wet felting, but with dry felting the fibers are blown directly into the dryer where moisture is evaporated.

dry pressing. A process of forming ceramic parts by placing the dry powder, binders, and lubricants in a mold and applying pressure. After pressing, the part will be fired in a kiln.

dry winding. A process that involves winding a pre-preg tape or roving on a warmed mandrel.

ductile cast iron. Cast iron that is heat-treatable and is used for making parts such as crankshafts, camshafts, and connecting rods for both gasoline and diesel engines.

ductility. A material's ability to be formed plastically, without breaking.

dust. The stock used in dust pressing. It is composed of 62 percent talc and 38 percent clay.

dust pressing. A process very similar to dry pressing, but with a higher percentage of water in the mix.

dwell points. Stops in one revolution of an indexing table.

dynamic compaction. A new field of study that has evolved from explosive forming. It permits firing of ceramics at lower temperatures, thus decreasing the amount of contaminants in the part.

dynamic range. The distance between the maximum and minimum measurable values of a sensor.

E

e-coating. Process which uses an electrical current to deposit a paint film on a part.

eductors. See *jets*.

E-glass. Glass used in electrical applications because of its insulating properties.

elastic deformation. This is when a material changes shape, from an external force or from supporting its own weight, and returns to its original shape.

elastomers. A class of highly resilient plastics that are more like rubber than plastic.

electrically erasable programmable read-only memory (EEPROM). Retains its contents even when the power is turned off.

electrical surface treating. A high-voltage discharge process used to change the surface tension and wettability of plastic materials. Also known as *corona discharge treating*.

electrochemical machining (ECM). A method of material removal that shapes a workpiece by removing electrons from its surface atoms. In effect, ECM is the opposite of electroplating.

electrodeposition. A finishing process, similar to electroplating, that is used on metals and coated polymers. Also known as *electrophoresis*.

electrodischarge machining (EDM). A process that uses electrical energy in the form of sparks to erode stock from a metal workpiece.

electroless plating. A process that uses chemical rather than electrolytic action to produce the desired metallic coating on a substrate.

electrolytic tin plate (ETP). Black plate that is electrolytically coated with tin.

electromagnetic forming. A process that forms a workpiece by using intense pulsating magnetic forces. Sometimes called magnetic pulse forming.

electron beam machining (EBM). A thermoelectric process that focuses a high-speed beam of electrons on the workpiece.

electron beam welding. A fastening process that generates heat by bombarding the weld site with a narrowly-concentrated beam of high-velocity electrons.

electronic gluing. Curing of thermosetting resin glues by exposing them to heat generated by a high frequency electric field.

electrophoresis. See *electrodeposition*.

electroplating. A means of depositing a metal or alloy from an anode through an electrolyte solution onto an article to be plated (connected as the cathode).

electropolishing. Smoothing and enhancing the appearance of a metal surface by electrochemically removing an extremely thin layer of material.

electrostatic paint spraying. Process using a spray gun that charges atomized paint particles so that they are attracted to the grounded workpiece.

electrostatic powder coating. A finishing method in which negatively charged particles of plastic powder are attracted to a positively charged part, resulting in an evenly distributed coating.

electrostatic spraying. A process that was developed as a method for coating metallic materials. It uses negative and positive charges to evenly coat the workpiece with finishing material.

embossing. A stamping process that produces a raised image on the product's surface.

emery paper. A coated abrasive product made from aluminum oxide and iron oxide. It is used to polish and buff metals.

empty-cell impregnation. A type of pressure treating that uses high-pressure air to force a preservative (creosote) into the cells of the wood.

enamel. An oil-based paint that is made by adding finely ground pigment to varnish; it provides a hard, glossy, very durable finish.

encapsulating. A protective method similar to potting, except that the mold is removed from the product after casting, rather than becoming part of the product.

encapsulation. Process in the manufacture of ceramic capacitors in which they are sealed by dipping in liquid polymer, by injection molding, or by immersing the assembly in a fluidized bed of molten polymer.

end effector. The tool or device mounted on the end of the arm of an industrial robot.

end grain. The top surface of a stump or end of a log.

engineered materials. Materials that are designed and constructed by people, rather than occurring naturally.

engineering resins. High-performance resins, such as those formulated to be exceptionally resistant to chemical attack, extreme heat, or impact.

equilibrium diagram. See *phase diagram*.

error. The difference between the true value of the quantity being measured and the value indicated by the sensor.

escapement. A device for orienting and dispensing individual components to a workholding fixture or tool.

etchant. The agent that is used to etch (chemically machine) metals, glass, or other materials.

European economic community. An organization established in 1958 by treaty between Belgium, France, Italy, Luxembourg, the Netherlands, and West Germany (now Germany), known informally as the Common Market. In 1960, Britain proposed that the Common Market be expanded into a transatlantic free-trade area and engineered the formation of the *European Free Trade Association* (EFTA). In 1967, the European Community (EC) was formed. Today, most of the EFTA members have transferred their memberships from EFTA to the European Union.

eutectoid. An alloy that changes immediately from solid to liquid at a specific temperature.

evaporative vacuum method. A process in which the metal to be deposited on the plastic is evaporated from disk-shaped targets placed in a vacuum chamber.

excess WIP (work in progress). If the part or assembly is not being worked on, having value added, then it becomes excess WIP.

excimer lasers. A type of laser, still in the developmental stage, that removes stock, molecule by molecule, rather than vaporizing it. Since they do not generate heat, excimer lasers can be used for procedures such as eye surgery.

exotherm brazing. A joining method that uses high-frequency current through an AC electrical coil to heat the joint area.

expanding. A process used to increase the size of tubular parts that have been formed by drawing.

expansion molding. A molding process in which a prepreg is placed inside a silicone rubber mold. In a curing oven, the mold expands to provide the forming pressure necessary to compress the prepreg.

experimental design. The precise way different settings will be tried and adjusted during the experiment.

experimental objective. The problem to be addressed in an experiment.

experimental procedure. The steps carried out in the experiment.

explosive forming. The process that uses the shock wave from an explosion to force sheet stock against a forming die.

extender. A filler used with the sole purpose of reducing cost by extending the volume of the resin with less costly filler.

external cylindrical grinding. A form of grinding that uses a grinding wheel which runs over the outside surface of a workpiece. The work is secured between centers or in a chuck.

extractive industries. Industries that collect raw materials and convert them into refined materials.

extractives. Materials found in small quantities in wood, such as starches, oils, tannins, coloring agents, fats, and waxes.

extractor. A device used to filter oil and water out of the air as it passes through an air-pressure spray gun system.

extrinsic semiconductor. Extrinsic semiconductors are those materials with impurities or dopants.

extruder. A machine that converts thermoplastic powder, pellets, or granules into a continuous melt, then forces the material through a die to produce the desired shape.

extrusion blow molding. A process in which a melted plastic is forced through an extrusion die to form a tubular parison. The parison is then clamped in a mold and expanded by air pressure to fill the cavity.

extrusion. A process in which a hot or cold semisoft solid material is forced through the orifice of a die to produce a continuously formed piece in the desired shape.

extrusion cutter. A machine that uses a high-velocity rotating knife to cut the extrudate.

F

fabricating. Process of joining parts or components together to create another product.

faceplate turning. A turning operation done with the stock mounted on the faceplate of the lathe. Also known as *facing.*

faceplate. A disk fixed with its face at a right angle to the live spindle of a lathe for the attachment of the work.

factory and shop lumber. Wood that is used primarily for remanufacturing purposes in mills that produce fabricated doors, windows, cabinets, moldings, and trim items.

FAS. A hardwood grading term, meaning firsts and seconds. The best grade of hardwood stock, thus the highest quality grade, is firsts; the next best grade, seconds.

fatigue. The tendency of a material to crack and fail due to repeated stress.

Federal Water Pollution Control Act (FWPCA). Federal legislation that provides information on pretreatment requirements for wastewater to be discharged into rivers, streams, lakes, and other water sources.

feed. The distance a cutting tool travels in one minute.

feedback. In a control system, the process of measuring the actual output, then comparing it to a preset standard or reference.

feeder system. Material handling system that contains and conveys discrete components while simultaneously orienting and sorting them.

fence. An adjustable stop running perpendicular to the cutterhead axis on a saw, jointer, or similar woodworking machine.

ferrite. The softest and most ductile form of low-carbon steel found on an equilibrium diagram.

ferrous metals. Metals that are produced from iron ore.

fiber reinforced plastic (FRP). Polymeric matrix composite with fiber reinforcement.

fiber tows. Bundles of fiber material used as prepregs in some composite production processes.

fibrils. Bundles of cellulose molecules that form the wall of wood cells. Also known as *lamella.*

filament winding. A process for fabricating a composite structure by winding a continuous fiber reinforcement over a rotating core or mandrel. The fiber may be impregnated with a matrix material before or during winding.

filament. Flexible fiber that is drawn very fine.

filler. A material that is suspended in a matrix binder to form a composite.

filler metal. In welding, a metal such as brass, solder, or iron, that is used to help fill any voids in a joint.

film blowing. Process in which polymer film is produced by extruding a tube of plastic material which is then expanded in diameter and reduced in thickness by internal air pressure, and finally slit to form a flat sheet.

filter cake. The squares or thin slabs of material that result when a filter press dewaters slip.

filter presses. An iron frame with nylon filters that is closed with hydraulic pressure to press water out of slip.

fill yarn. Yarn oriented at right angles to the fabric length.

fine finishing. A type of sanding in which the stock is often polished or burnished with rubbing oils or rubbing compound; it requires a fine-grit abrasive.

finishing. Actions that improve the outward appearance or protect the exterior surface of a part.

fire clay. A material used to make refractory brick that is produced by dry pressing.

fire-retardant. A material that inhibits the spread of fire.

firing. Heating greenware to an elevated temperature in a kiln or furnace to fuse the materials together, providing strength and permanence to the product.

firing range. The span of temperatures between the point of sufficient vitrification and too much vitrification.

fitness for use. Product quality is based on a robust design that ensures the product is fit for its intended use in the hands of the customer.

fixture. A special workholding device.

flame hardened. Metallic workpieces that have been heated and quenched. Flame hardening can be used for zone or spot hardening (treating only part of a workpiece).

flame polishing. A process used to reduce the size and quantity of surface defects in small-diameter rods and filaments by rotating the material in a helium/oxygen flame.

flame spray coating. A process in which a fine plastic powder is dispersed through a specially designed spray gun burner nozzle, so it is melted quickly to form a smooth, uniform, permanent layer.

flame spray gun. A finishing process in which molten particles of ceramic material are sprayed onto a substrate.

flame-spraying. The process of melting metal in a special gun, then spraying it as a thin coating on a substrate. Also called *wire metallizing*.

flash. The excess material that is forced out between the halves of forging dies. It must be removed as a finishing step.

flash-off period. Time allowed after finishing to allow excess thinners and solvents to evaporate before the finish is subjected to forced drying.

flask. The rectangular box used to hold the mold in sand casting.

flat-platen-pressed. Designation of a type of particleboard that is produced when pressure is applied in a direction perpendicular to the faces.

flint paper. A type of coated abrasive consisting of grains of quartz (silicon dioxide) adhered to a paper backing.

float glass process. A process used to produce sheet glass by floating it on top of a bath of molten tin.

floating zone method. A crystal growing process in which a rod of polycrystalline silicon is brought into contact with a single crystal seed. A molten zone around the interface between the seed and rod will solidify with the same crystalline structure as the seed.

flow coater. A device used in open molding that looks like a spray gun except the nozzle has a series of holes. It sends out a fan of droplets to apply resin to the reinforcement material.

flow coating. See *waterfall coating*.

fluidized bed coating. A finishing process in which coating powder is mixed with jets of air. The air holds the powder in suspension so that it behaves like a fluid.

fluidized bed furnace. A type of furnace that suspends heated aluminum oxide particles in a chamber with hot gases.

flutes. Channels on a drill bit through which chips are carried out of the hole being drilled.

flux. A material used to prevent the formation of oxides or other surface impurities in a weld, or to dissolve them as they form.

fluxing. In the calendering process, the mixing of dry ingredients prior to adding liquids.

foamed blow molding. Blow molding process in which foaming of the resin is accomplished through the use of a nucleated foaming agent, based on reactions of baking soda and citric acid. This combination results in a release of carbon dioxide and creation of tiny voids in the part, resulting in less material needed for the completed part.

foam-frothing. Process in which a urethane chemical mixture in a partially expanded state is mixed with chemical blowing agents that give off gases to create a cellular structure.

foaming agents. Chemicals added to plastics to create a lighter, more cellular structure.

foil. A printed, melamine-impregnated paper that is introduced into a mold in the in-mold decorating process.

follower. A freewheeling spindle attached to the tailstock of a lathe in the spinning process. It helps hold the stock in place.

forced drying. A process usually accomplished by loading parts or assemblies on conveyors that carry them through heated ovens. Forced drying is often done with ultraviolet (UV) radiation.

forming block. The mandrel or mold used to shape stock in the spinning process.

forming tool. The device used to apply pressure against the face of the workpiece in a spinning operation.

fracture strength. See *breaking point*.

franklin fiber. A microscopic crystalline fiber, derived from gypsum, that can be used as a reinforcing filler in plastics.

free blowing. A process in which a heated plastic sheet is clamped between an upper forming ring and lower metal platen and air is blown through a hole in the platen to force the plastic upward into a bubble.

freeze casting. Thick slip is poured into a smooth-walled, rigid rubber mold. The mold is then frozen to force the clay to expand and press against the walls of the mold. Drying of the frozen part is accomplished through the use of absorbent sand.

freeze drying. As used in ceramic production, a process that rapidly freezes the water in clay particles, encasing drops of salt solution that hold the desired metal ions. Also known as *cryochemical drying.*

frit. Particles of glass that have been milled to the desired size. When added to a glaze, frit lowers the necessary firing temperature.

full annealing. Heating a workpiece to a temperature just above the upper transformation temperature, then cooling it slowly in a furnace to a point just below the lower transformation temperature, followed by further cooling in still air to room temperature.

full-cell impregnation. A type of pressure treating used to apply fire retardant and moisture protectant chemicals. In this system, the preservative slowly flows into the cell.

fusing. Gas welding which is accomplished by mixing two gases, oxygen and acetylene, in a welding torch. The heat that is generated is sufficient to melt and weld steel.

fusion sealing method. A process that uses special glass compositions to make direct glass-to-metal seals in an oxidizing environment.

G

gage number. A unit of measure used to determine thickness of metallic stock. The smaller the gage number, the thicker the metal.

gang cutting. A form of traveling wire electrodischarge machining that uses multiple heads to simultaneously cut single or multiple workpieces.

gas metal arc welding (GMAW). An electric arc welding method that uses an inert shielding gas, supplied through a nozzle, to prevent oxidation of the weld. Also referred to as *MIG welding.*

gas tungsten arc welding (GTAW). An electric arc welding method that uses a nonconsumable tungsten electrode. Like GMAW, it makes use of an inert shielding gas supplied through a nozzle.

gel-coat. A specially pigmented resin mixed with a catalyst and sprayed into an open mold to form the smooth finished surface of the product.

geneva mechanism. A popular method for achieving rotary motion by using a continuously rotating driver to index the dial.

gilding. Also known as *banding;* applying gold gilt on the edges of china cups and plates.

gilt. Decorative gold material applied to fine china after glazing.

glass seals. Pieces of glass joined to each other by melting and fusing the two pieces of glass together.

glass solder. A bonding agent that allows glass parts to be joined and later separated without destroying them.

glassy state. State in which there is little or no movement of molecules past one another, and the material is stiff or even brittle.

glass-to-ceramic seals. A bond used to join unlike materials in products such as spark plugs.

glass transition temperature. Temperature below which a thermoplastic material is stiff or brittle, and above which the material becomes plastic and even elastic.

glaze. A specially formulated glass that melts on the surface of a ceramic workpiece, and adheres to the body after cooling.

glost firing. The second firing of a ceramic that is done to sinter the clay body and develop the glaze.

gob. Mass of molten glass used in centrifugal casting.

gouge. A round-nosed, cupped chisel used for roughing and making cove cuts on a lathe.

grade. The classification of wood according to its quality, based on appearance, strength, and lack of defects in the stock.

grafting. A condition that occurs under radiation processing, when two or more elastomers attach to a molecular chain without the use of a catalyst.

grain. The appearance of annual rings and fibers in wood when viewed longitudinally.

grain size. The unit of measure used to grade the coarseness or fineness of an abrasive finishing material.

granulator. A type of grinding machine that is used when the manufacturing process is automated or semiautomated to recycle scrap material back into the process.

gray iron. Type of cast iron that is easily cast and inexpensive.

green body. A dense and durable part produced by the compression of ceramic powders.

green compact. In powder metallurgy, the compressed part that ejected from the die but not yet fired. The part can be handled, but is extremely fragile at this stage.

green machining. The process of machining a dry pressed ceramic part while it is still green (before it has been sintered or fired).

green tape. A ceramic/glass frit held together with a binder and pressed into a sheet carried on a delivery roll.

green veneers. Veneers that have not undergone much drying and aging.

green wood. Stock that has just been cut from a log and contains a great deal of moisture, or sap.

greenware. The delicate state of a molded clay product before firing.

grinding media. Small pieces of rock, rubber, ceramic, or other material used in a ball mill to pulverize the material being ground.

grinding. A cutting process that uses abrasive particles to perform the cutting action.

grog mix. Clay that is mixed with fine particles of previously fired clay and used to make a mold for glass produced by the sagging process.

groundplanes. Conductors screened on the layers of green tape to form horizontal interconnects.

gun drilling. A process used to drill deep, accurate holes. It was originally used to drill gun barrels.

H

half-life. The gradual process of exponential decay where an element exhibits only half of its initial value.

hand forging. A metal forming process dating back to about 4000 B.C. Forging was used to shape primitive metal tools.

hand layup. A fabricating process that essentially consists of applying a layer of roving to a mold, then saturating the roving with resin until desired thickness and strength is reached.

hand-held router. A tool with a spindle-mounted cutter that is held securely against a template or guide to produce the desired pattern.

hard automation. A system in which the machines and tooling are designed to produce a specific part or a family of similar parts or assemblies.

hard machining. Final machining done on parts that have been fired. Hard machining requires the use of diamond tooling and is very slow and time consuming.

hardboard. A continuous mat of wood pulp that is pressed into a sheet of strong, hard material.

hardinge conical mill. A grinding device that turns around a horizontal axis, while the mix is continuously fed in at one end and automatically discharged from the other.

hardwood. A wood produced from a deciduous tree.

Hazardous Materials Transportation Act (HMTA). The Hazardous Materials Transportation Act of 1975 (HMTA) was instituted to assist the DOE (Department of Energy) in improving the regulatory and enforcement authority of the Secretary of Transportation to protect the United States adequately against risks to life and property which are inherent in the transportation of hazardous materials in commerce.

hazwoper. OSHA standard, 1910.120 CFR that requires employees to complete a minimum 40 hours off-site training course with at least three days of on-the-job training before they are permitted to handle hazardous wastes. Supervisors of these workers must also receive 40 hours of off-site training.

headers. The machines used in cold heading to form a head on the end of an unheated metal rod by compressing its length in a die cavity.

heat treating. Processes that use controlled heating and cooling of materials in their solid state to change mechanical or physical properties without altering their chemical composition.

hellerbond induction bonding. A ceramic powder is placed in the joint between the two surfaces to be bonded. The powdered adhesive or bonding agent contains ferromagnetic particles dispersed throughout a polymer. The powder heats up in response to the magnetic field generated by an induction coil.

hemicellulose. A low molecular weight polymer formed from glucose that represents about 25 percent of the composition of wood.

hermetic seal. A method of closure that prevents movement of gases or liquids into or out of a product.

high-energy-rate forming (HERF). Forming process that uses a spark-generated shock wave to force the sheet stock against a forming die.

high-volume, low-pressure (HVLP) paint guns. Gun which uses a turbine to deliver a large volume of low-pressure, low-velocity air, which moves the atomized particles slowly toward the part.

high temperature superconductors. Materials that superconduct at significantly higher temperatures.

hog. A large chipper used to break wood materials into chips.

honeycomb. Very lightweight core material that can be made from a wide range of materials, including paper, aluminum, and glass reinforced, phenolic. Honeycombs create very stiff lightweight laminates that can have very high crush strengths.

horizontal milling machine. This type of milling machine has the cutting tool carried on an arbor or spindle that travels along an axis parallel to the worktable.

horn. In ultrasonic machining, the structure that holds the tool and conveys mechanical motion to the workpiece.

hot extrusion. Extrusion process used to apply a plastic insulation coating to copper electrical wire.

hot forging. A variation of dry pressing that involves heating a pressed part to high temperature, then pressing it in a cold steel die.

hot isostatic pressing. A compacting process involving the use of both heat and pressure to form powdered metal or ceramic parts; also called *gas pressure bonding*.

hot marking. A stamping process used to impress a heated, raised image into the surface of a workpiece.

hot melt adhesives. Thermoplastics used as adhesives. They become fluid when heated, then solidify on cooling. They will not soften on reheating.

hot melts. See *hot melt adhesives*.

hot press compression molding. A molding process in which a plastic mixture of resin, reinforcement, filler, and additives is compressed between heated die halves.

hot wire cutting. A simple process that utilizes a thin wire heated by high electrical resistance. The hot wire slices through foam plastic by melting the material.

hot wire welding. A simple process in which a wire is placed between two thermoplastic surfaces to be joined and heated by electrical resistance.

hot-rolled steel. Steel that is squeezed between rollers while it is hot, and can be identified by its bluish surface coating.

hydrapulping. Blending and agitating a package for about 40 minutes until the paper, plastic, and foil layers separate from each other.

hydration. The chemical process that causes cement to harden when mixed with water.

hydroforming. A draw forming process that employs hydraulic fluid in place of the rubber pad used in marforming.

hydrostatic extrusion. A form of impact extrusion that uses a fluid instead of a mechanical ram to extrude the metal.

hygroscopic. The property of being water-absorbing.

hysteresis. The sensor's ability to provide the same output signal whether the stimulus is increasing or decreasing.

I

impact extrusion. Extrusion process commonly used to make collapsible tubes and cans. Usually a lubricated slug is placed in a die cavity that is then struck by a punch, forcing the metal to flow back around the punch.

impreg. A process that involves saturating the cells of wood with a fiber-penetrating thermoset resin, then curing it without using compression. A 25 to 35 percent gain in weight (when compared to the dry weight of the wood) is achievable.

impregnation. The process of immersing porous P/M parts in heated oil or resin after they are removed from the sintering oven. See *impreg*.

independent variable. Factors to be varied in tests to determine their effect on the output.

indirect two-sheet blow molding. A process in which an air tube is placed between two heated sheets of stock. When the mold closes, air is blown through the tube, expanding the sheets against the walls of the mold.

induction hardening. A fast-acting hardening process that uses electromagnetic currents to generate the necessary heat in the part.

induction welding. Welding method in which a radio-frequency magnetic field is responsible for creating the heat at the joint. It can only be used with nonmetals.

industry productivity program. A unit of the U.S. Department of Commerce that publishes annual measures for output per hour and unit labor costs for over 500, three and four-digit Standard Industrial Classification (SIC) industries (forerunner of NAICS) in the U.S.

inert. There is no excess oxygen or carbon present to contaminate the weld.

inertia friction welding. One workpiece is clamped in a spindle chuck attached to a flywheel and another is held in a stationary holding device. The chuck is accelerated to a prescribed rotational speed. When the desired speed is reached, power is cut and the part held in the chuck is forced against the stationary piece, causing the pieces to heat up and weld together.

infiltration. In powder metallurgy, the practice of placing a lower-melting-point metal on or below the green compact before sintering. The molten metal will be drawn into the porous compact to improve its mechanical properties.

infusion. Molding process that uses atmospheric pressure to squeeze the resin into the reinforcement fibers, similar to the vacuum bagging process. It reduces styrene emissions by wetting out and curing the laminate in a closed system.

ingot. The large brick of stock (iron or other metal) that is melted into a liquid state for casting at a foundry.

initiator. A chemical starting agent mixed with a prepreg so that the composite manufacturer only has to position the material in a mold and heat it to complete the cure.

injection blow molding. A two-step molding process in which a heated parison (plastic tube) is formed in the mold cavity, then forced against the mold's walls by injecting air.

injection molding. A process in which a material, usually a resin or ceramic, is heated until it reaches a plastic state, then forced under pressure into a closed mold.

in-line transfer systems. Chain or belt conveyors that convey a pallet containing a parts-holding nest or fixture between operations.

in-mold decorating. Finishing process accomplished by applying foils to compression-molded and transfer-molded products.

in-mold decorating with foils. Applying foils to compression-molded and transfer-molded products made with melamine, urea, or ammonia-free phenolic thermoset and thermoplastic materials.

input. The part of an open-loop control system representing a value for a characteristic, such as time, temperature, or pressure.

input values. In a control system, arbitrary standards or references that reflected circumstances at the time they were established.

inserting. An application of ultrasonics in which a metal component is embedded into a preformed hole in a plastic part.

intermetallic. A compound made of ceramic and metallic materials.

Institute of Electrical and Electronics Engineers (IEEE). Organization known for developing standards for the computer and electronics industry.

insulation board. A type of composition board that has high resistance to heat transfer, is low in density, and is often used as an acoustical barrier.

interaction. The influence one variable has on another variable as the first variable's level changes.

intercellular layer. The thin adhesive material found between the cell units of wood.

interchangeability. A concept that makes automation possible by allowing the substitution of standardized parts, assemblies, or materials in a manufactured lot. It makes modifications or adjustments to individual parts unnecessary.

interference fits. One of the methods used in joining wood parts, in which the parts are designed to lock together.

intermediate annealing. Heating a part to a temperature close to or just below the lower limit of the critical temperature, then allowing the part to cool. This process softens and stress relieves the part so it can undergo further cold working.

intermittent extrusion blow molding. A process in which the parison, immediately after being ejected from the mold, is quickly formed into a product.

intermolecular attraction. The strength of attraction between like molecules.

internal cylindrical grinding. A form of grinding that is done when the surface of a hole must be accurately smoothed. A tool post grinder attached to a lathe or a specialty grinding machine is used.

internal threads. Threads that are on the inside of a hole.

International Standards Organization (ISO). The agency responsible for establishing and publishing manufacturing and other types of standards worldwide.

intrinsic semiconductor. *Intrinsic* semiconductors are made from pure elements such as silicon.

investment casting. A precision casting method in which the cavity created by the pattern is destroyed when the casting is removed. Also known as *lost wax casting*.

ion implantation. A metal hardening process that improves the material's ability to withstand friction and corrosion.

ionizing radiation processes. Those that produce high-level radiation, either gamma rays generated from radioactive materials such as Cobalt 60, or high-energy electron rays created by electron accelerators.

ions. Electrons that have been drawn free of their atoms in the electrochemical machining or electroplating processes.

irradiated wood. See *wood plastic composition (WPC)*.

irradiation. Exposure of a material to energy from a radioactive source.

ISO certification. A recognized method of determining quality.

isothermal annealing. Heating a ferrous alloy part near the critical temperature (upper transformation temperature) to produce a structure that is partly or wholly austenitic. The part is then cooled and held to an intermediate temperature, causing a transformation of the austenite to a relatively soft ferrite-carbide aggregate.

isostatic pressing. Dry forming process that is conducted by applying pressure to the powder from three directions. Isostatic pressing is also known as *uniaxial pressing, hydrostatic pressing,* or *hydrostatic molding*.

isotropic. Possessing equal strength in every direction.

J

jet mill. Trost fluid energy mill used for dry milling.

jets. Fittings that increase the speed of the water by forcing the water through small holes.

jidoka. See *operator self-control*.

jiggering. A ceramic production process that uses a template or forming tool to force a slice of clay over a revolving convex mold.

JIT supplier. A vendor that supplies orders that relate to the size and number of the manufacturing lots their customer was processing in a day.

job lot. See *batch*.

john mill. A grinding machine in which agitator pegs of tungsten carbide are set into the walls of the grinding cylinder. The cylinder is turned at high speed, producing a uniformly ground product.

jointer. A machine that is used to true the edges of stock. Usually used to complement the work of the planer and circular saw.

jollying. A ceramic production process in which a plaster mold forms the outside surface of the product, and a forming template shapes the inside.

just-in-time manufacturing. A philosophy that eliminates excess inventory by placing the burden directly on suppliers to deliver parts or materials of an agreed-upon quality at a specified time and in the desired quantity.

K

kanban. The trigger (usually a card attached to a bin) used to start production in a pull system. Kanban literally means "card" in Japanese.

kaolin clay. A clay material with a high melting point, used for making refractory products.

kaolinite. A white mineral, primarily aluminum oxide and silica dioxide, found in clay.

kasenit. One type of carbon-hardening powder used in carburizing.

kerf. A cut made in metal or other material by a saw or cutting torch.

kickback. Machine hazard encountered in woodworking when the workpiece is thrown out of the machine back toward the operator.

kiln. A furnace that fires ceramic materials at high temperatures to fuse them.

kiln drying. A method of drying wood, in temperature- and humidity-controlled ovens, to reduce its moisture content to about seven percent.

kiln furniture. Struts or other devices used to support a part with an unusual shape in the kiln.

kiss cutting. A die cutting method that is used for cutting self-adhesive materials without penetrating the paper or film backing.

KLD Nasdaq Social Index. First introduced in 2002, the KLD-NS Index was the first benchmark for socially screened securities traded on the Nasdaq Stock Market. It is a market-value weighted index reflecting the performance of some 280 of the largest U.S. corporations in technology, financial, and telecommunications sectors.

knurling. A technique used to press diamond-shaped indentations on the circumference of a workpiece. Knurling is used to provide a gripping surface on a part or to expand the diameter of a part.

L

lacquer. A finishing material made of nitrocellulose, acetone, varnish resins, a thinner, and plasticizers. Lacquer provides a smooth, hard finish.

laminate. Layers sandwiched together, as in plywood.

laminate bulkers. Products that build laminate thickness quickly or create a barrier to prevent print-out.

laminating. A process that involves sandwiching layers of plastics or composite materials into sheets or parts and applying pressure to bond the layers together.

lancing. A punching process used to make a tab without removing any material.

laser. An active electron device that converts input power into a very narrow, intense beam of coherent visible or infrared light; used for cutting, drilling, welding, heat treating, soldering, and wire stripping.

laser assisted machining. Use of a laser for precision accuracy in final finishing, particularly for drilling, scribing, and burr removal.

lasing medium. A mixture of carbon dioxide and nitrogen used in a laser.

lasing tube. A cylinder filled with a gas, such as CO_2 or argon, that is energized by an electric current and releases energy in the form of coherent light. The output of the tube is focused into a beam to do work.

latex paint. A water-based type of paint that is widely used for finishing.

lathe. The most common machine used to perform machining operations on round parts.

lay-up. Application of woven roving and resin, typically used in open molding of boat hulls.

lay-up schedule. Description of the type and weight of the reinforcement that is needed for an FRP composite to impart the unique properties and performance characteristics required by the designer.

lead. On a bandsaw, the tendency to pull slightly to the left or right of the desired cutting line, usually caused by improper tracking of the blade on the wheels.

lean manufacturing organization. A company that successfully addresses the problem of excess WIP.

lean production. A process of production in which a company goes through continual improvement to the extent necessary to assure that better quality products are being consistently produced at less cost.

lignin. A natural adhesive that bonds adjacent layers of cells together. It makes up about 25 percent of the total volume of wood.

linear friction welding. Welding method in which the heat needed to melt the thermoplastic is generated by pressing one of the parts against the other and rapidly vibrating it.

linear PVC foam. Core material produced mainly for the marine industry. Its unique mechanical properties are a result of a non-crosslinked molecular structure, which allows significant deflection before failure.

linerboard. See *liners*.

liners. Container board component that consists of two pieces of flat paper.

liquid-phase sintering. Wetting ceramic particles with a viscous liquid before heating the part to the sintering temperature.

liquidus line. On a temperature graph, the point at which a molten alloy begins to change from a liquid to a mushy solid.

listed waste. Waste in a fluid or semi-fluid state.

logic. The principles of reasoning that are designed into an automated system.

longitudinally. Shrinkage of wood lengthwise, or in the direction of the grain.

lost motion. The expending of energy without adding any value to the product.

lost wax process. See *investment casting*.

lower transformation temperature. On an equilibrium chart, the temperature where a material begins to change from one phase to another.

M

machine planing. A process that mills wood to cut it to uniform thickness and produce a smooth surface.

magazines. A chamber that keeps its contents secure and ready to be dispensed.

magazine systems. Material handling systems that contain and convey pre-oriented parts.

magnetic induction bonding. An adhesive or bonding agent is placed in the joint between the two surfaces to be bonded. The powdered adhesive or bonding agent contains ferromagnetic particles dispersed throughout a polymer. The powder heats up in response to a magnetic field generated by an induction coil.

major sector productivity and costs program. Unit of the U.S. Department of Commerce that produces quarterly and annual output per hour and unit labor costs for the U.S. business, nonfarm business, and manufacturing sectors.

malleable iron. Type of cast iron made by heating white cast iron to a specific temperature, and then cooling it slowly.

mandrel. A bar, rod, or similar shape that serves as a core around which other materials are wound, cast, forged, or extruded.

manifest. A one-page form used by haulers transporting waste that lists EPA identification numbers, type and quantity of waste, the generator it originated from, the transporter that shipped it, and the storage or disposal facility to which it is being shipped.

manual control. The simplest type of control system, in which an operator is required to start, stop, or adjust the process by operating control devices on the machine.

manufacturability. Designing a product for assembly.

manufacturing productivity. The output of product per unit of effort; a measure of how effectively a nation's manufacturers use labor and equipment.

maquilas. The factories located in maquiladoras.

maquiladoras. Industrial parks along the U.S.-Mexico border that pay no duty on materials imported from the U.S., and pay only duty on value added when finished goods are shipped back into the U.S.

marforming. A flexible draw-forming process that is often a cost-effective alternative to stretch draw forming.

martensite. The metallic structure that is created when steel is rapidly quenched or cooled, usually by plunging it into a liquid bath.

maskant. A type of resist that is not photo-sensitive. It is applied to a surface, then mechanically removed from areas that are to be etched.

mass production. A system involving continuous production of one type of product, such as an automobile, typically in large quantity.

master unit die method. This proprietary quick-change die system uses inserts in a frame or fixture that remains in the machine.

matched mold forming. A molding process involving a two-part metal die used to compact a softened thermoplastic sheet.

material feeding. The process of conveying components to the point of use. Feeding may consist strictly of component movement, but can also include orienting and sorting.

material requirements planning (MRP). The use of software programs to track, purchase, and release materials.

material safety data sheets (MSDS). The "Right to Know Law" requires manufacturers to prepare and maintain access to MSDSs on chemicals that are being used and stored in the workplace. Worksheets provide information on how to handle hazardous chemicals and products, and information on factors such as melting point, boiling point, flash point, toxicity, health effects, first aid treatment, storage and disposal, protective equipment, and spill-proof leak-proof containers.

mechanical attachment. Type of ceramic fabrication process used to join materials, such as shrink fit, and the use of bolts, nuts, and screws.

mechanical etching. Processes, such as sandblasting or bombarding with an abrasive and lead shot mixture, that are used to frost glass, or make it opaque.

mechanical fastener. Wide range of different types of removable fasteners that are used to join two or more materials together. The most common types would be screws and bolts.

mechanical properties. Attributes used to distinguish one material from another, including hardness, tensile strength, wearability, and toughness.

mechanical stretch forming. A process similar to plug-assist thermoforming, but without the use of a vacuum. The mechanical stretch forming process uses only a plug to depress and stretch the stock into the female mold.

mechanization. Making or assembling things with machines and tooling to increase the rate at which work is being done. Mechanization also serves to reduce variability.

memory. The capability of a material to return to its original shape after it is bent, stretched, or otherwise distorted.

metallizing. A process used to apply metal coatings to ceramic, glass, or other types of surfaces. A typical application is to provide a transmission path for electrical current on electrical devices and components.

methanol. Methyl alcohol. A solvent used as a vehicle in non-grain-raising stains.

microfibrils. Crystallites linked together in bundles.

micro-machining. Ultra-fine quality final finishing processes.

MIG. Abbreviation for "metal inert gas" welding. See *gas metal arc welding*.

milling. A machining process that uses a multi-toothed cutter to produce slots, grooves, contoured surfaces, threads, spirals, and many other configurations. Also, a grinding process in which hard clay is broken down into fine particles by passing through a series of rollers.

mistake proofing. Formally and continually work to eliminate the errors and omissions that cause repair and rework.

miter saw. A saw that has the capability of cutting stock at an angle from the vertical, as well as moving through a 90° arc horizontally. It is used primarily when doing interior trim work.

mix muller. A pulverizing machine used for clay. It consists of a large circular pan in which two large steel-tired wheels revolve.

molding. The process in which a material, usually in a softened (plastic) but not liquid state, is formed into the desired shape using pressure and sometimes heat.

Molinex mill. A grinding device in which eccentric grinding disks, mounted on a rotating shaft, are staggered to function as an auger. This mill makes it possible to obtain a superfine grind.

molten ceramic casting. Casting done by pouring the melt into cooled metal molds to accomplish rapid cooling, resulting in a very hard product.

momentary switch. A switch that is only active while pressed.

monomers. Simple molecules that are linked together by the thousands to make up the long chain that is a polymer.

morillonite. A type of clay used to absorb radioactive waste.

mortise. A groove or slot that is cut into a piece of wood to accept a tenon.

mortise-and-tenon joint. A joint that is used to secure drawer components or other pieces of fine furniture.

mortising chisel. Hollow chisel that contains the mortising bit, used to cut mortises for wood joints.

mud frame. See *master unit die method*.

multiple-spindle shapers. Wood-shaping machines that have a number of spindles arranged side by side.

N

nail set. A small cylindrical steel tool, usually tapered at one end, that is used to drive a nail head below or flush with a wooden surface.

National Pollution and Discharge Elimination System (NPDES). National Pollutant Discharge Elimination System (NPDES) permit program, authorized by the Clean Water Act, controls water pollution by regulating point sources such as pipes and man-made ditches that discharge pollutants into waters of the United States. In most cases, the NPDES permit program is administered by authorized states.

natural aging. The process by which some aluminum alloys harden and become stronger over a period of time at room temperature.

Nd.YAG laser. A type of laser that emits either continuous or pulsed beams. It is often used for welding delicate components.

nearly inert. Type of ceramic devices can be implanted in the body without causing toxic reactions.

near-net-shape. Parts are produced close to the desired final size, requiring machining only if the part needs threads, holes, or unusual design features.

necking. A die-reduction process that stretches the metal part at the same time that it reduces its cross-sectional area.

nesting. The positioning of blanks so that multiple parts can be cut from one workpiece with a minimum amount of waste.

nesting fixture. Ultrasonic welding machine attachment that securely holds the part.

neutralize. To stop the action of bleaching once the desired degree of lightening of wood color has taken place.

nibbling. A process that uses a small round or triangular punch to rapidly take small "bites" out of sheet metal, allowing the cutting out of limited numbers of flat parts with complex shapes.

nitriding. A case hardening method that uses heated gaseous ammonia to form a very hard shell on the metal.

nonconsumable products. Goods that are durable in nature and intended for long-term use, such as plastic containers, vehicles, machinery, or furniture.

nonferrous metal. Any metal other than iron and its alloys. Nonferrous metals include copper, aluminum, lead, nickel, zinc, tin, and brass.

non-grain-raising (NGR) stains. A stain consisting of a dye in a methanol solvent, widely used by furniture manufacturers, because they have a short drying time and do not cause swelling (raising) of the wood grain.

normalizing. Annealing process that involves cooling the heated metal to room temperature in still air to produce a more uniform and fine-grained structure.

normally closed. A switch that is operated to close and complete an electrical circuit.

normally open. A switch that is operated to open and break a circuit.

normal state. The position of a switch during no activity.

North American Economic Community (NEC). Canada, the United States, and Mexico.

North American Free Trade Agreement (NAFTA). An agreement made between the U.S., Mexico, and Canada in the 1990s. The agreement was intended to encourage the development of new manufacturing partnerships involving companies in the three member countries.

North American Industry Classification System (NAICS). Introduced in 1997 to replace the Standard Industrial Classification System as the organizing structure for classifying types of manufacturing industries.

North American Product Classification System (NAPCS). System introduced in 1999 by Canada, Mexico, and the United States to develop a market-oriented classification system for products linked to the NAICS industry structure. NAPCS gives special attention to service products, new products, and advanced technology products.

nosing. A die-reduction process that is used to taper or round the end of tubing.

notching. This process, which is similar to nibbling, is typically done using a punch press.

numerical control (NC). A method of automating manufacturing processes by describing tool paths and machine movements in terms of numerical increments. This movement might be along a machine axis, or a vector that is comprised of the resultants of all the axes being controlled on the machine tool.

numerical control program. The method used to store and convey precise instructions for each move along each machine axis; each instruction is sent by the machine control unit to activate the specified output.

O

oil stains. Wood stains with pigments suspended in oil. After being applied, they are wiped off after varying periods of time to achieve the desired depth of color.

oil-base paint. A type of paint that uses a petroleum-based vehicle. It must be thinned or cleaned up with a petroleum-based solvent.

oligomers. Acrylic polymers made up of two, three, or four monomer units that become solid polymers when exposed to light. One basic use of these materials is with the stereolithography process.

open molding. A process used to form many polymeric-matrix composite products.

open-die forging. A method of forging in which the workpiece is formed between flat dies that compact, but do not completely enclose, the heated metal part.

open-grain woods. Hardwoods that have large pores, such as oak.

open-loop control system. A basic control system that attempts to meet a preset standard without monitoring the output or taking corrective action.

operator self-control. Each man and woman in the manufacturing process is responsible for creating and maintaining defect-free product flow.

orbital friction welding. Welding method typically used when butt welding cylindrical or tubular parts end to end. One of the parts is held stationary in a nesting fixture, while the mating part is spun rapidly against it. Friction created at the point of contact generates heat and brings the two pieces to their melting point.

orientation. Properly positioning a component for assembly or processing by a machine.

oriented strandboard. A type of particleboard that is made from large, irregularly shaped wood fibers, rather than particles or flakes.

overaging. A problem that can occur when too high a heat level is used in artificially aging aluminum alloys. Weakening of the part results.

overdamped response. A slow response in a closed-loop control system.

overglaze. A decoration applied over the glaze on a ceramic product. It must melt at a temperature lower than the melting point of the glaze.

overspray. Sprayed finishing material that does not settle on the workpiece.

oxyacetylene rod gun. A finishing tool that operates in a manner similar to that of the flame spray gun, except that a rod of coating material is fed into the oxyacetylene flame, instead of ceramic powder.

oxyacetylene welding. A joining method that uses heat from mixing and burning two gases, oxygen and acetylene, in a welding torch. The heat generated is sufficient to melt and weld many soft metal alloys.

P

P/M injection molding. In this process, very finely ground powders are coated with thermoplastic resin, then parts are injection molded in the conventional manner. After molding, the part is sintered.

pad printing. A stamping process that uses a design in relief to transfer ink from a pad to the surface of the workpiece.

paint. A mixture of pigment suspended in a vehicle, such as water or oil.

pakkawood™. Pakkawood is a plasticized hard wood made from very thin layers of wood that is dyed and then pressure treated with resin.

pallet. The support base, frame, or tray used on an in-line material transfer system.

panel saw. A circular saw attached to steel rails so that it slides up and down easily to cut large sheets of plywood or other panels. The saw can be rotated to cut across panels, as well.

paraffin. A white, solid wax derived from petroleum.

parison. A hollow tube of heat-softened resin that serves as the stock in blow molding.

particleboard. A panel material composed of small discrete pieces of wood bonded together with a resin in the presence of heat and pressure. It is widely used for underlayment and as core material for furniture.

parting line. The seam between closed forging dies, where a thin wing of flash is generated.

parting tool. A lathe tool (chisel) that is used to separate material and to cut off stock.

part-transfer mechanisms. Devices that take a part from the escapement pick-up point and place it in a workholding fixture or nest.

paste filler. A thickened form of wood filler, containing powdered quartz, linseed oil, turpentine, and drying agents, used to close pores and enhance the beauty of open-grain hardwoods before any finish is applied.

PAW. See *plasma arc welding*.

pearlite. The soft and relatively ductile state that occurs when steel reaches 0.8 percent carbon.

penny. The basic unit of size classification for nails.

perforating. A punching process that produces closely and regularly-spaced holes in a straight line across sheet metal, usually to facilitate bending.

peripheral speed. See *cutting speed*.

permanent mold casting. A casting method that uses a two part mold again and again. The mold halves open after the part solidifies, allowing part removal.

perovskites. A particular class of metal oxide that superconducts at significantly higher temperatures.

phase diagram. A graphic representation of the phase changes that takes place in a metal at various temperatures. Also called an equilibrium diagram.

photoinitiator. A component of photopolymeric resin that absorbs laser energy and initiates polymerization.

photopolymer. A liquid that cures and hardens where the light strikes it.

photoresist. A light-sensitive coating that is applied to a substrate, then exposed and developed to provide a mask during chemical etching.

physical frothing. A process accomplished by aggressively whipping air into the resin, and then rapidly cooling the polymer.

physical properties. Attributes used to distinguish one material from another, such as weight, color, electrical conductivity, and reaction to heat.

pick-and-place robots. Programmable mechanical devices that simply pick up a part and move it to another location. These robots are used primarily for material transfer, but are also used to do some simple assembly tasks.

pickling. A chemical process using acid for cleaning.

pig iron. Refined wrought iron, which is mixed with scrap iron or steel to help control the carbon content in cast iron.

pigment. The finely ground material that gives opacity and color to paint.

pilot hole. One of the two holes drilled when fastening wood with a screw. It extends through one piece of stock and into the other, and is slightly smaller in diameter than the screw.

pinch rolls. Cylinders positioned close together and used to collapse a tube of blown film.

pinless moisture meter. Handheld moisture meter that does not use pins, and therefore does not create damage to the sample. It relies on surface contact with a flat electrode that does not penetrate the wood.

plain weave. The simplest fabric, in which the fill yarn alternately crosses over and under each warp yarn.

planing. An operation that is used to remove large amounts of material from horizontal, vertical, or angular flat surfaces.

plasma arc spraying. A metallizing process in which an electric arc heats a high-speed stream of argon or another inert gas. A powdered metal coating material is fed into the stream to be melted and blown onto the part.

plasma arc welding. Welding method that uses a non-consumable electrode recessed in a nozzle that supplies a jet of shielding gas.

plasma-arc gun. See *plasma arc spraying.*

plasma assisted machining. Thermally assisted machining process sometimes used for final finishing.

plastic flow. When metal is forced to move into a new shape.

plasticity. The ability of the material to change shape or size as a result of force being applied.

plasticized. Wood that has been made supple by exposing it to ammonia so it can be more easily shaped or formed.

plasticizers. Additives used to provide plastic materials with increased flexibility.

plastics. Synthetic materials capable of being formed and molded to produce finished products.

plastics welding. A cohesion process using plastic filler rod and a low temperature air stream to join plastic components of relatively heavy thickness.

plastisol. A pourable, viscous liquid consisting of a vinyl resin dissolved in a plasticizer.

platform oscillation. Method of providing mechanical agitation for a washing system. The part is immersed and oscillated with an up-and-down motion that swirls solution over the part.

plating. The deposition of metal on a cathode (negative electrode).

plies. Thin layers of wood are glued together to form plywood.

plug-assist thermoforming. A forming process in which a vacuum, in conjunction with a cylinder-activated plug, is used to depress a heated thermoplastic sheet into the mold.

plug-and-play sensor. Sensors and their associated interface circuitry that are easier to integrate into modern computer-based control systems.

plywood. A laminated wood sheet made by gluing a number of layers, or plies, together at right angles to each other.

PLZT. A unique type of ferroelectric material (lead zirconate-titanate doped with lanthanum) that shifts polarity under the influence of an electric field, acting as a solid-state ceramic switch.

point-to-point system. A type of NC control system that moves the working spindle to a specific location on the workpiece, and then performs a programmed sequence of events. Also known as positioning.

poka-yoke. Formally and continually work to eliminate the errors and omissions that cause repair and rework.

polishing. A process that uses both a fine abrasive and a wax compound to produce a highly polished surface on plastic.

polyethylene. A polymer that consists entirely of ethylene monomers. It is used for many types of containers.

polyethylene glycol (PEG). A white, waxy chemical treatment that results in improved dimensional stability for wood. Also called carbowax.

polyethelene terephthalate (PET). A platice used to make clear blow-molded containers such as soft drink bottles, some plastic sheeting products, carpet yarns, fiberfill (pillow stuffing), and manufacturing geotextiles (fabrics used in road construction).

polymer. A long chain molecule made up of thousands of smaller molecules linked together.

polymerization. The chemical reaction, triggered by heat, pressure, or a catalyst, that causes monomers to blend chemically, linking their molecular chains to form polymers.

polystyrene. A synthetic resin used for injection molding, extrusion, or casting for molding plastic objects.

polyurethane foam. Available in either sheet stock or liquid that can be foamed in-place. This material is often used in the cavities of boat hulls to add stiffness and provide buoyancy and as a core material in the wall panels of refrigeration units. Because of its relatively low sheer strength, this foam is generally not used in critical structural applications.

polyvinyl chloride (PVC). A polymer of vinyl chloride that is widely used for both molded rigid product, such as plumbing pipe, and pliable films.

polyvinyl chloride (PVC) foam. A combination of polyvinyl copolymer with stabilizers, plasticizers, cross-linking compounds, and blowing agents, offering a good combination of strength and low weight.

pores. The open ends of vessels in a hardwood. They appear as holes on cut ends of logs.

porosity. The moisture-absorbing capacity of a material.

post process. Generates the program for the machine tool.

post processed. The process of generating machine code without any further involvement from the designer.

pot broaching. A type of broaching used to economically produce items such as precision external spur gears and automotive front-wheel-drive transmission gears.

potting. Encapsulating method in which polyester thermosetting resin is poured into the housing of a product. The resin and housing form an integral part of the product.

powder metallurgy (P/M). A process that involves compacting metallic powders in a permanent (reusable) mold. Heat is applied to the molded powder to melt and fuse the powder into a finished metal part.

pre-automation program. The steps to eliminate the five forms of waste in an organization.

precipitates. In precipitation hardening of nonferrous metals, small particles of a different phase that are added to the original matrix to achieve hardening.

precipitation hardening. The process in which small particles of a different phase are added into the original matrix to harden a nonferrous metal. Also called *age-hardening*, or *aging*.

precision grinding. Process often used on materials that are too hard to cut with conventional tools.

preforms. Pellets or tablets compressed from powdered polymers, used as stock in some closed molding processes.

prepregs. Cores of partially cured resin with reinforcement, used as stock in some closed molding processes. Also, sheets of polymeric matrix material having fibers that are already saturated with resin.

preservatives. Treatments that impregnate the cells of the wood under pressure to make it fire-resistant, moisture-repellent, colorless, and odorless.

presintering. A heating process that causes drying, decomposes organic binders, vaporizes water from the surface of particles, changes some ions, and then decomposes additives.

press and blow machine. A machine used primarily for container production. The neck of the container is made by pressing or blowing, then the rest of the container is formed by blowing.

pressure-assisted sintering. See *vapor-phase sintering.*

pressure-feed gun. A type of gun used for atomized spraying of finishing materials. In this gun, finishing material is fed into the nozzle under pressure, then mixed with air.

pressure forming. Thermoforming process in which the sheet is softened with heat and placed in the pressure chamber. Air is blown in, providing the pressure to force the stock against the walls of the female mold.

preventive maintenance. A maintenance approach based on following the manufacturers' recommendations about the care that should be given to production equipment.

primary manufacturers. The production of industrial stock.

primary packaging. Primary packaging conveys information about the product.

primary processing. Primary processing industries are those, such as refineries and mines, that make raw materials.

print-out. When the gel coat and matrix resin shrink during curing, causing the pattern of the fiberglass reinforcement to show.

process action. The description of what happens when a process changes the internal structure or the outward appearance of a material. Forming, separating, fabricating, conditioning, and finishing are examples of process actions.

production molder. A type of abrasive planer used to shape long strips of molding.

productive maintenance. Maintenance approach that uses engineers and technicians to quantify the reliability of the key elements in the equipment and determining the element's mode of failure.

productivity/quality improvement program. The steps to eliminate the five forms of waste in an organization, without having automation as an end objective.

programmable logic controllers (PLCs). Control devices widely used in industry to guide machine tools through a programmed sequence of operations. Programs can easily be changed to produce different components or reflect changes in dimensions.

proportional control. A system that is extensively used to control automated processes when an operation requires variation in such factors as speed, tension, or temperature.

prototypes. Physical representations of the final product for visualization and testing.

pugging. A shredding and vacuum process used to remove air bubbles from dewatered clay filter cake.

pugs. Cylinders of extruded clay. See *blanks*.

pull broach. A tapered cutting tool used to shape holes in a workpiece.

pull system. System that tries to ship an order when it is received, creating a demand that moves up the production line to replace what has just been shipped.

pultrusion. A process that involves pulling continuous roving impregnated with resin through a forming die. The resulting shaped material is cured in an oven.

pulverizer. Machine used to break dry clay down further into a fine, dusty powder, and stored until needed for use in dust pressing.

punch. The part of a closed die arrangement that automatically pushes wire stock in precut lengths through the hole in the die block.

punch press. The machinery used in stamping sheet metal parts. Also known as a *stamping press*.

punch-and-die shearing. A production shearing process using a shaped punch and matching die.

punching. A process similar to blanking, except that material stamped out of the sheet metal is usually scrap. The part is what is left.

pure metal. A metal that contains only one kind of atoms, such as iron or copper. No other metallic elements are mixed in.

push stick. A safety device used to hold down the wood stock and push it on through a saw or similar machine.

push system. When an order is received, materials are ordered and then a start date is established to begin production based on when the materials will be available.

pyrometric cones. A set of wedge-shaped pieces of clay, each with a specific melting point, used to determine the proper firing range for ceramics.

Q

quenching. Rapid cooling of metal, usually by plunging into a liquid such as water or oil.

R

radial arm saw. A saw with its circular blade mounted on a horizontal arbor suspended from an arm above the worktable. This type of saw is used primarily for crosscutting.

radial forging. See *rotary swaging*.

radially. In a direction across the rings of the tree.

radiation processing. A type of polymerization process that produces internal change through exposing the stock to high-energy ionizing radiation.

radio-frequency (RF) dielectric heating. A process used to reduce the moisture content of wood. A high-frequency current (microwaves) uniformly distributes heat throughout the wood, vaporizing the moisture.

radio-frequency (RF) heat sealing. The process of using high-frequency radio waves, combined with the energy and pressure exerted by pressing the parts together, to cause the molecules to oscillate and create heat.

rake angle. The angle on the end of each row of teeth of a tap. Wider flutes at the bottom of the tap help to remove chips when the tap is backed out of the hole.

ram pressing. A form of wet pressing in which material is formed under pressure in a die.

ratchet and pawl. The simplest, but least accurate, of the basic driving mechanisms used to provide intermittent rotary motion.

reaction bonding. A process that uses a nitrogen-rich atmosphere during firing. It yields a part with less shrinkage than conventional sintering processes.

reactive maintenance. A maintenance approach in which maintenance departments are ready to respond to unplanned interruptions of work by stocking spare parts and creating "safety stocks" (excess WIP) of critical production materials.

reactive-phase sintering. A variation of liquid-phase sintering, except that the liquid breaks down or disappears, rather than remaining present during sintering.

reaming. A final finishing process that improves the dimensional accuracy and surface finish of a drilled hole.

recessed hex. Opening in a screw head that has more contact with the driving tool and is desired in applications where there is heavy loading.

reel feeding. A variation of magazine feeding in which parts or materials are linked together to provide easier handling and control.

refractories. Ceramics that can withstand continued exposure to high temperatures.

refractory. Nonmetallic ceramic-like materials that remain solid and intact when they are subjected to high temperatures.

refractory metals. High-temperature metals that can withstand heat and maintain strength. Refractory metals include niobium (Nb), tungsten (W), and molybdenum (Mo).

refractory slurry. A substance sprayed on or used as mortar in joints between bricks in a furnace lining.

regrind. The finished product of plastics grinding operations.

relay logic. An older type of machine control system, based on the opening and closing of electro-mechanical relays.

rephosphorized steel. High strength steels with addition of phosphorus.

residual plot. The average weight of the production sample minus the average weight predicted by the guide (10.01 grams). The difference is the residual and that is what is plotted on the chart.

resin. In the plastics industry, all processed material up to the point where industrial stock is created.

resin transfer molding (RTM). A molding process in which catalyzed resin is transferred from a separate chamber into the cavity of a two-part matched mold.

resistance type moisture meters. Moisture meters that require the user to drive a pin-type probe into the wood. These meters measure resistance between the pins inserted into the wood.

resistance welding. A welding method in which the materials to be joined are clamped between two opposing electrodes and a current supplied. The resistance that is produced creates heat, bonding the metal layers together.

resolution. A measure of fineness or detail.

resonance point. The specific frequency at which an ultrasonic cutting tool must vibrate to be effective.

resorbable. Type of bioceramics that have the ability to dissolve and be assimilated into the blood-stream.

Resource Conservation and Recovery Act (RCRA). This act controls the generation, storage, transportation, management, and disposal of hazardous wastes.

response criteria. Methods used to evaluate a closed-loop system's ability to achieve a desired output.

retort. Container used to hold preservative used in impregnation.

reverse roller process. A finishing process in which the rollers turn in one direction and the stock moves in the other. It provides better coverage of finishing material on the workpiece.

ribbon blow molding process. A blow molding process used to make incandescent lightbulb blanks, vacuum bottles, and clock domes from a ribbon of heated glass.

riddle. A screened sifter that is used in the sand casting process to shake sand over the pattern until it is well covered.

right angle butt joint. Two pieces of wood are pressed together in a 90° angle, glued, and nailed.

right-to-know law. The federal *Hazard Communication Rule (CFR 1910.1200)*, also known as the "Hazcom Standard," or "Right-to-Know Law," was first introduced by the Occupational Safety and Health Administration (OSHA) in 1983.

ring broach. A cutting tool, consisting of a series of high-speed steel rings, that is used for pot broaching.

ring-and-stick broaches. Cutting tools that are used to broach teeth on gear blanks at the same time that other operations are being performed.

rip sawing. Saw cuts made in the direction of the grain of the wood.

robocasting. This ceramic fabricating process was developed by Joe Cesarano of Sandia National Laboratories. *Robocasting* requires no molds or machining, and relies on robotics for computer controlled deposition of ceramic slurry through a syringe.

robustness. The property of a product or process that makes it "insensitive" to the fluctuations in its working environment.

roll bender. Machine used to bend circular, curved, and cylindrical shapes from bar, rod, tube, angle, and channel stock.

rolled glass. A translucent flat glass that is poured and rolled to the desired thickness.

roller coating. A high-speed coating process in which stock is squeezed between rubber rollers carrying a film of finishing material.

roller forming. A metal-forming method that uses rollers to progressively squeeze continuous strips of metal to form straight lengths of various crosssections.

roll-on marking. Coining done with flat dies contacting round stock.

rotary coining. Coining done using round dies turning against round stock.

rotary shearing. A hand process that is similar to straight shearing, except that the cutting blades are rotary wheels used to cut either straight or circular shapes.

rotary swaging. A process that takes a solid rod, wire, or tube and progressively reduces its cross-sectional shape through repeated impacts from two or four opposing dies. It is also referred to as *radial forging*.

rotary wheel system. Provides as many as twenty clamping stations. This system enables a parison to be captured from the extruder while some parts are being removed and others are being cooled.

rotational molding. A method used to make hollow plastic articles from plastisols, using a rotating hollow mold. The hot mold fuses the plastisol into a layer of gel, which is then chilled and the product stripped out of the mold.

rotomolding. See *rotational molding*.

rough grinding. A process used for rapid material removal on castings, forgings, and welded parts.

roughing. The initial turning operation performed on a workpiece in a lathe. Also, a type of sanding that is done to remove the maximum amount of stock; it requires coarse-grit abrasive.

rough-sawn lumber. Wood that requires subsequent planing or surfacing to smooth boards prior to use.

round nose tool. A chisel used to produce specialized contours on a lathe.

roving. A woven fiberglass mat used for reinforcement in plastics.

runners. In injection molding, the pathways that carry plastic to the mold cavity.

S

S-glass. The type of glass added to a composite matrix when the product will undergo exposure to excessive temperatures.

sagger or saggar. Refractory box that protects the workpieces from contamination in the firing operation after the glaze is applied.

sand casting. In this process, the mold is expendable; it is destroyed when the part is removed after casting.

sandblasting. A mechanical etching process in which alumina oxide abrasive is propelled against glass by compressed air. It is used to create designs on the glass.

scavengers. Scavengers combine with other impurities that are present in or on the workpiece to prevent the impurities from contaminating the weld.

scribing. Incising the surface of the material by use of a laser.

scrim. An open weave fabric that is glued to the back of core material.

scroll saws. Saws used to produce intricate cuts within the inside dimensions of a workpiece. Also referred to as *jigsaws*.

sealant. Paste-like substance placed along the joint when mechanical fasteners are used in order to make the assembly dust, water, or gas tight.

sealing. A cohesion process that is typically used for joining plastic films.

seam welding. A longitudinal weld used to join sheet metal parts or to make tubing.

secondary bonding. In composite fabrication, the procedure of joining parts together using glues or other bonding materials.

secondary manufacturing. Firms concerned with the production of discrete products.

secondary packaging. Secondary packaging, or unit packaging, is used primarily to transport several primary boxes or packages. The most common example of this is the cardboard, or corrugated paperboard box.

secondary processing. Firms that use the raw material generated in the form of industrial stock, and use this to make hard good manufactured products.

select hardwood. Grade of hardwood that is below FAS and above Commons.

selection management. Individual trees are marked and cut in rows, creating small clearings or openings that allow for growth of new trees through natural reseeding from remaining trees.

select lumber. Wood that is graded from A to D, with A presenting the best quality surface appearance.

semicrystalline. Type of material displaying crystalline regions within an amorphous matrix.

sensitivity. In a control system, the ratio of change in output to the change in input.

sensors. Electrical devices that receive different types of information and transmit it (provide feedback) to a computer or other control device.

separating. Removing material or volume from a workpiece. Some separating processes are chip-producing; others are not.

setters. Solid masses of clay placed under certain areas of a ceramic workpiece to reduce warpage during firing.

shading. A process that involves spraying a tinted lacquer on wood to improve its color uniformity after bleaching.

shank hole. One of the two holes drilled when fastening wood with a screw. It enlarges the pilot hole to the same diameter as the screw shank.

shape memory. A property of some metal and plastic materials that allows them to be heated and formed into a desired shape, then return to their original shape when heated again.

shaper. A machine tool for cutting flat-on-flat, contoured surfaces by reciprocating a single-point tool across the workpiece.

shearing. A mechanical separating process often used to cut flat stock from sheet or plate.

shear line. The path that the shearing follows.

sheet molding compounds (SMC). Sheets of resin that are formed by closed molding.

shell broach. A type of internal cutting tool that is often used to cut helical splines, such as the rifling used inside gun barrels.

shielded metal arc welding (SMAW). An electric arc welding method in which vaporization of a coating on the electrode provides a gas that shields the weld joint from atmospheric contamination.

shot. Small spheres made of cast steel, glass, or ceramic. They are used in the shot peening process to bombard the surface of a part.

shot peening. A cold working process that is accomplished by bombarding the surface of a part with small spheres of cast steel, glass, or ceramic particles called shot.

shrinkage. A reduction in overall size that can occur during curing of plastics, the firing of ceramics, the cooling of a metal casting, or the drying of wood.

shrink fit. When two parts are pressed together, heated and cooled, the parts contract creating a stronger interface between the parts.

shrink wrap. Packaging material that is wrapped around a product and then heated, causing the material to shrink, providing increased protection for parts and components.

shuttle clamping system. Once the parison reaches the desired length, the clamping mechanism shuttles from the blowing station to a position under the die head, where it surrounds the parison, trims it, and returns to the blowing station.

silanes. Widely used coupling agents for increasing adhesion between two different materials.

silicon carbide. The compound of silicon and carbon, widely used for manufacturing grinding wheels. Also, a nonporous composite that is used to form complex, near-net shapes.

silicon nitride. A material, produced by use of reaction bonding, that is resistant to thermal shock and to chemical reagents. It has considerable manufacturing potential.

single plane die cutting. With this type of die cutting machine, a flat die set with multiple patterns is pressed into the stock. Normally the stock to be cut is placed on top of the press bed (a heavy shelf), and the die travels down on top of the stock.

single planer. This tool is a knife-blade planer with a cutterhead above the stock.

single-plane die cutting. A process involving a machine that uses a flat die set with multiple patterns to press into the stock.

single-spindle shaper. A wood-shaping machine with a cutter mounted on a vertical spindle (shaft) that projects through an opening in a horizontal metal worktable.

sintering. The process of holding parts for a short length of time at a temperature just below the base metal's melting point in order to create an internal metallurgical bond between the particles of powder.

siphon gun. A type of gun used for atomized spraying of finishing materials. In this gun, material is drawn upward to a fluid cap by suction, then mixed with air.

sizing. A method of coining that is used with forged parts to improve surface finish.

skew. A flat chisel that is used to smooth cylinders and cut shoulders on a lathe.

skin packaging. Type of packaging used to package components, tools, and other products held to display cards.

slab broach. A tool with cutting teeth on its flat face, used to shape the outside of a workpiece.

slip casting. A process that involves depositing a layer of thin water-clay mixture ("slip") on the inner surfaces of a mold. The process results in a hollow product, such as a figurine. Also called *drain casting.*

slitting. A variation of shearing that creates a slit in metal where another part or device can be inserted.

slotter. A machine tool used for making a mortise or shaping the sides of an aperture.

slumping. This is a technique used for shaping glass by placing sheet glass over a mold and applying heat until the glass softens and drapes downward to conform to the shape of the mold. With slumping, stretching is minimized and a consistent crosssectional thickness is maintained.

slump mold. Mold used to drape flat glass during the slumping process.

slurry. A liquefied ceramic material. Also, an abrasive suspended in a liquid medium.

slush casting. A process that involves inverting the mold when the part has just begun to solidify, so that molten metal can be poured out. The result is a hollow shell casting.

soaking. Used when wet bending wood. Soaking will soften (plasticize) the wood so that it can be more easily formed.

soaking method. Bonding acrylics by immersing one of the pieces of acrylic in solvent or cement, producing a softened cushion of acrylic material that will harden when the two pieces are pressed together and cured.

soft annealing. Heating metal to a temperature above its critical range and then cooling it appropriately to achieve the greatest softness or ductility possible.

soft automation. A system in which the machines and tooling are designed to produce a wide variety of different shapes and material characteristics.

softwood. A wood produced from a coniferous tree.

solder. Filler metal used to form a bond between parts; made from an alloy of tin and lead, or antimony if used in plumbing or food handling.

solid-phase sintering. Using diffusion to move surface material and fusing it where the particles come in contact with each other.

solidus line. On a temperature graph, the point at which a molten alloy changes from a mushy solid to a hardened solid.

solution treating. The first step in precipitation hardening, which involves heating the alloy, which exists as a two-phase solid solution at room temperature, to a temperature that causes it to become a single-phase solid solution.

solution turbulation. To spray under immersion; this process moves solution over, under, around, and through a part.

solvent cements. See *solvents.*

solvent polishing. Cellulosic and acrylic parts can be polished by dipping or spraying them with solvents.

solvent-releasing adhesives. Adhesives that give off a liquid as they cure.

solvents. Chemicals that soften thermoplastic materials, allowing them to be joined by cohesive bonding.

span (input span). The portion of the dynamic range that can be converted by a sensor without causing an unacceptable amount of error.

specific heat. The amount of energy necessary to produce a one-degree change in the temperature of the material.

speed. The revolutions per minute (rpm) that the workpiece or cutting tool is turning.

spin welding. A welding process in which one of the parts is held stationary, while the mating part is spun rapidly against it. Friction created at the point of contact generates heat and brings the two pieces to their melting point.

spindle turning. The process of turning a workpiece that is held between centers on a lathe.

spinning. A process that involves stretching sheet stock over a rotating male or female mold.

spirit stains. Stains consisting of an aniline dye suspended in alcohol. Because they are very fast-drying, these stains are used primarily for repair or touch-up work.

split pattern. A two-piece pattern used in the sand casting process.

spot load. A load placed in a relatively small area.

spray drying. A process used to produce uniformly sized particles of clay powder by spraying slip into a column of heated air.

sprayup. A process that uses a chopper gun to spray strands of roving and catalyzed resin into an open mold.

springback. An increase in the dimensions of the compacted part upon ejection, produced by stored elastic energy.

sputtercoating. A process using a high level of kinetic energy and a cathode that emits electrons to bombard argon gas molecules and ionize them. The sputtering action this creates transfers metal to the part.

sputtering. See *sputtercoating.*

square nose tool. A chisel used to produce specialized contours on a lathe.

stability. The ability of the control system to maintain steady-state conditions.

stabilization. A step in the production of carbon fiber that causes the material to oxidize, changing its internal structure to prevent melting.

stagflex. Flexible plywood that is suitable for cold forming small radii, manufactured and distributed by Clarks Wood Company Ltd.

stain. A finishing material consisting of finely ground pigments, suspended in a vehicle, used to change the overall wood color or to emphasize its grain pattern.

stainless steel. A highly corrosion-resistant material made by alloying iron with nickel and chromium.

staking. Process that uses ultrasonic vibrations to melt thermoplastic studs and mechanically lock dissimilar materials in place.

stamping. A cold-forming process that uses a set of matched molds in a stamping press to compact stock under pressure.

Standard Industrial Classification (SIC). A U.S. Government system of assigning standard numerical codes to industries to separate them into classifications and subclassifications.

staple fiber. A basic type of glass reinforcing fiber, usually between 6″ and 15″ in length.

states of operation. The action or condition of an output device, such as "on" or "off."

stationary mandrel method. Extrusion method used to make hollow tubing.

stationary routers. Tools that use a pin guide mounted in the table and a template that runs against the pin to make grooves and cut irregular shapes.

statistical process control (SPC). Technique that provides manufacturers with a way to confirm that an activity is under control. It consists of three steps: establish control, monitor the activity, and provide for problem solving.

staypak. A modified wood that is less brittle than those made by resin impregnation. Staypak is made by using pressure and heat to cause the natural lignin between the wood fibers to liquefy, and the entire mass to realign itself.

steady-state error. The difference between the actual output and the desired output required by the control system.

steaming. Hot steam used to add moisture to wood prior to forming.

steatite. A type of ceramic material made from the natural crystalline form of magnesium silicate or talc, mixed with clay and barium carbonate flux.

steel. An alloy of iron and carbon, or of iron and other alloying elements.

stereolithography. Prototyping process that builds an object, layer by layer, by exposing a liquid photopolymer to ultraviolet laser light. The photopolymer is cured and hardened as the light strikes it.

stick broach. A cutting tool used for pot broaching.

stickered. The stacking of lumber using spacers between boards.

stock. Raw material that can be processed by secondary industries into manufactured products.

straight shearing. The action of cutting sheet metal to size in rectangular pieces.

straight turning. The simplest form of turning operation, in which the tool is fed into the stock, then moved to the left or right to remove stock and reduce the workpiece diameter. Also called *peripheral turning.*

strand. A collection of more than one continuous glass filaments.

stretch draw forming. This process uses mating dies that must match perfectly.

stretch forming. A process used to shape sheet stock. The sheet is stretched to its yield point, then wrapped around a form die or mandrel. A PLC or CNC system is used to control the machines.

stretch wrap. Material used to secure heavy loads on pallets.

strip feeding. A feeding system that is a variation of magazine feeding. It consists of parts that are formed into, or held in, a long strip to be fed to the point of use.

striplines. See *groundplanes.*

structural lumber. Material purchased by the construction industry for uses ranging from light wall framing members to beams, stringers, posts, and timbers for heavy structural applications.

stud welding. A form of arc welding that is used to fasten a threaded metal stud to another part.

sublimation. A process in which a substance passes directly from the solid state to the gaseous state, without becoming liquid.

substrate failure. A bonding failure that results in the destruction of the substrate (material being joined).

subtractive. Process in which material is removed to create the shape of the final product.

successive transformation. The process of production in stages, one after the other, which may extend across a number of industries.

superalloy. A thermally resistant alloy for use at elevated temperatures where high stresses and oxidation are encountered.

superconductor. A material that exhibits a total absence of electrical resistance below a critical temperature.

Superfund Amendment and Reauthorization Act. This act mandated OSHA to development training requirements for employees handling hazardous wastes.

surface-active. Type of bioceramics that form a chemical bond with the surrounding tissue and encourage growth.

surface energy. The surface tension of a liquid material, which affects how well that liquid will wet a surface.

surface grinding. A form of precision grinding done on flat workpieces.

surplus. Materials that are purchased for manufacture but not used.

sweat soldering. A type of soldering that uses capillary action to draw melted solder into a close-fitting joint.

swing. The maximum diameter of a workpiece that can be turned on a lathe.

synergism. A phenomenon in which the whole is greater than the sum of the parts.

syntactic foams. A form of laminate bulker made by mixing hollow microspheres with resin.

syntactic polymer. Microspheres or microballoons added to resin.

synthetic. Anything that is human-made.

synthetic metal research. A field of research concerning conductive polymers and related topics.

T

tablet pressing. A variation of dry pressing.

tack welding. Applying welds in small beads that are spaced.

tangentially. Shrinkage of wood around a tree (in the direction of the annual rings).

tap. A shanked tool with rows of cutting teeth, separated by flutes, that are used to cut threads inside a hole.

tape casting. See *band casting.*

tapping. Process that is used to cut threads inside a hole.

tapping a heat. The process used to release molten metal from a furnace into a heavy metal container called a ladle.

target. A flat plate of coating material that is bombarded by a beam of electrons during sputtering.

teeth per inch. The measuring unit used to identify many types of saw blades: the greater the number of teeth per inch, the finer the cut.

tempered. Reheating a part to just below the lower transformation point, then cooling it.

tempered hardboard. A type of hardboard impregnated with a resin/oil blend to give it better strength, stiffness, and resistance to water and abrasion.

tempering. The process of reheating a hardened metal, then cooling it to improve toughness and make it less brittle. Also, a heat-treating process used with glass products to increase their annealed strength. In ceramic production, the process of mixing and kneading liquids into a dry clay material to produce stock that is pliable enough for forming.

tenon. A projection from the end of a framing member that is made to fit into a mortise.

tensile strength. The limit of stress and strain a material is capable of bearing.

tertiary packaging. Used for protection during transportation, shipping, and distribution.

theoretical lead time. The minimum time that will elapse from the moment the order for a single part is received to the time it is completed and ready to ship.

thermal conductivity. The ability of a material to transfer heat.

thermal energy. The high-velocity stream of electrons that is generated by the electron beam gun.

thermal expansion. The dimensional changes that take place as a metal is heated or cooled.

thermal expansion resin transfer molding (TERTM). A process in which a preformed foam core is wound with reinforcement material and impregnated with epoxy resin. When the assembly is heated inside a mold, the foam expands to squeeze the reinforcement and matrix materials together.

thermal heat sealing. Using a heated tool or die at a constant temperature to join layers of plastic film. Widely used for packaging applications.

thermal impulse heat sealing. A variation of thermal heat sealing used to provide better joint appearance with thin films.

thermal shock resistance. The ability of a material to withstand sudden changes of temperature.

thermochemical. A process involving both heat and a chemical reaction.

thermocouple. A type of transducer that generates a current proportional to its temperature.

thermoplastics. Plastics that can be melted and reformed as desired into new products. They are produced by addition polymerization.

thermosets. Plastics that are formed by condensation polymerization, a chemical reaction that causes the polymer chains to become cross-linked. Thermosets cannot be melted and reformed like thermoplastics.

thixotrope. An additive that modifies the viscosity and flow characteristics of a resin.

thixotropic. Remaining in a solid condition at rest but becoming fluid when agitated.

thixotropic agent. Gel coat additive that inhibits dripping and sagging.

thread rolling. A chipless cold-forming process that can be used to produce either straight or tapered threads.

three-dimensional printing. A process that requires successive layers of powder and adhesive to build a functional part.

three-plate molds. Molds with two parting planes between each of two plates. The runner system can be located on one parting plane, and the part on the other.

three-roll double-pitch machine. Roll bending arrangement in which the top roll of the three-roll machine turns freely, and the bottom two rolls are powered and adjustable.

through coating. A process that is economical for high-speed coating of flat stock such as wood, metal, plywood, and particleboard. The process consists of squeezing the stock to be coated through rubber rollers carrying a film of finishing material.

throughput. The amount of product that moves through a production operation in a given period of time.

throwing. A foundational process in ceramics that involves hurling a body of clay that is in a plastic state onto a revolving potter's wheel. The clay is then shaped by hand into a symmetrical form.

tie coat. In transfer coating, a second coat that is rolled over the skin coat.

TIG. Abbreviation for *tungsten inert gas* welding. See *gas tungsten arc welding*.

time-and-motion study. Determining the necessary motions or steps of a task and the time it takes a worker to complete it.

time study. The recording of how long it takes to complete an operation.

tin plate. A mild steel that is coated with tin.

tin plate base box. A container in which tin plate is sold by the pound. It is measured in terms of pounds of sheets (112 sheets 14″ × 20″).

tip angle. The included angle formed when measuring from one side of the tip of a drill to the other.

tool steel. A high-carbon steel that is hard and difficult to bend. It is used to make tools, such as forging dies, screwdriver shafts, chisels, and milling cutters.

torch cutting. The use of an oxyacetylene torch to separate metallic materials.

toughness. Resistance to breaking; a physical property used to describe a material.

tracheids. The cells through which sap is transferred vertically in softwoods. In addition to transporting sap, the tracheids provide strength to the wood.

traditional ceramics. Products made from glass and clay.

transducer. A device used to convert one type of energy to another, such as changing an electrical signal into a mechanical motion.

transducer electronic data sheet (TEDS). Specification of the configuration and identifying parameters of the sensor. It is contained in an EEPROM located on the sensor.

transfer coating. A process involving deposit of a surface skin coat on coated release paper, then transferring it to the product being finished.

transformation toughening. A process used to increase the strength and toughness of ceramics in products that must exhibit high wear resistance.

transient response. The response of the output device under control in comparison to an input before steady-state conditions are achieved.

transient sintering. See *reactive-phase sintering.*

traveling extrusion saw. Attachment to the extruder used for cutting long lengths of extruded plastic.

traveling head press. A machine that is designed to make multiple die cuts in wide sheets or roll goods.

traveling wire EDM. A type of EDM in which the cutting is done by using a round wire that travels through the workpiece.

trost fluid energy mill. A dry milling process in which grinding is achieved by collision between the particles being ground and the energy supplied by a compressed fluid entering the grinding chamber at high speed. Also known as a *jet mill.*

true-up. Eliminating bow in a piece of wood by planing.

tumbling. A finishing method in which plastic, wood, or metal parts are placed inside a rotating drum to rub against each other to remove flash and achieve a smooth finish.

tumbling compounds. Polishing materials, such as abrasive particles, waxes, sawdust, or wood plugs, that are rotated with parts in a drum.

turning. Separating method used with a rotating workpiece on a lathe to reduce its diameter, change its profile, or produce a taper on it.

turning center. Machine in which a stationary cutting edge is placed against a rotating bar of metal to reduce the diameter of the bar.

turning processes. Processes that are used to machine rotating parts.

turnkey system. An automated system package that is standardized and ready to be put into operation.

twist drill. The most common type of drill for metals, normally made of high-speed steel or carbon steel.

twist warp. Warpage that involves twisting of the board from one end to the other.

two-roll machine. A widely used roll bending configuration which can produce bends of any diameter using only two rolls and a slip-on tube.

U

ultra cold. Near absolute zero.

ultrasonic casting. Slip-casting process in which low-amplitude audio frequency vibrations of 15–20 kilohertz are transmitted to the plaster mold during casting. Some of the heat generated in the plaster mold by the ultrasonic vibrations is transferred to the workpiece.

ultrasonic cleaning. See *cavitation cleaning.*

ultrasonic machining. A process that removes material by erosion, using vibrations generated by high-frequency sound waves.

ultrasonic vibrations. Moving parts at a minimum of 20,000 hertz. The friction caused by these vibrations is enough to weld and melt metal and plastics parts, spot weld, transform adhesives to a molten state, and melt films and fabrics together.

ultrasonic welding. A welding process that works on the principle of changing sound energy to mechanical movement to generate heat for joining metals or plastics.

uncontrollable factors. Factors that might affect the output being recorded or measured for a test.

underglaze. A decoration that is protected from wear by the overlying glaze, but does not provide the extensive range of colors available with overglaze decoration.

underlayment. Particleboard laid down as a firm base for flooring materials.

unified numbering system (UNS). Classification of steel according to the quantity of carbon in the steel, stated in hundredths of a percent.

unit packaging. See *secondary packaging.*

universal chuck. A three-jawed device used to hold a workpiece in the headstock of a lathe. The three jaws make it self-centering.

universal saws. Saws with two arbors; one with a ripping blade mounted on it; the other with a crosscut blade.

up milling. A process in which the cutter is positioned on top of the workpiece, with the workpiece fed in the direction opposite the rotation of the cutter.

upper transformation temperature. On an equilibrium chart, the temperature where the changing of a material from one phase to another is complete.

upsetting. A type of forging that thickens or bulges the workpiece while also shortening it by compression. Upsetting is actually a combination of forging and extrusion.

V

vacuum bagging. A process in which a vacuum acts to press a plastic bag against resin-impregnated roving to force out excess air.

vacuum metallizing. A process used to apply very thin metal coatings to a plastic substrate by vaporizing the metal in a vacuum.

vapor-phase sintering. Using vapor pressure to cause material to be shifted from the surface of the particles to the point where the particles contact each other to reduce both size and the pore space.

variables. Factors that might affect the output being recorded or measured for a test.

variability. The inconsistency that occurs in manufacturing a particular product.

varnishes. Transparent surface coatings consisting of oil, resin, solvent, and driers that are used to protect the surface of a wood workpiece from water spotting, chipping, and abrasion.

vehicle. The fluid component of paint that acts as the carrier for pigment.

veneer. Thin sheets of hardwood for laminated surfaces.

vertical integration. Manufacturing approach in which the manufacturer creates the parts instead of relying on outside suppliers.

vertical milling machine. A type of milling machine in which the cutter is positioned perpendicularly to the worktable.

very high bond (VHB) tape. An adhesive, in either conventional tape roll form, or as die-cut shapes for specialized applications, that forms extremely strong bonds between dissimilar surfaces.

vessels. Cellular structures found in hardwoods that carry sap vertically.

via holes. Holes punched where vertical interconnects will be required.

vibration welding. A quick method for joining most thermoplastics. It uses vibration to produce frictional heat to make the joints.

vibratory deflashing. A process that uses high-frequency vibrations to improve the abrasive (rubbing) action between the parts.

vibratory finishing. A deburring method in which parts and an abrasive finishing medium are placed in a rotating closed container.

virtual kanbans. Combining kanban push system and pull system, making it possible to quickly change the number of kanbans in use.

viscosity. The thickness or ability of a liquid material to flow.

vitreous glass solder. A low-melting-point glass that can be applied to glass with a higher melting point to create a seal.

vitrification. The process of firing materials at a particular temperature to create a glassy bond and close pores, making the material denser and stronger.

W

waferboard. A type of particleboard that is made using high-quality wood flakes that are about 1.5 in^2.

walking-beam mechanism. A rectangular-motion mechanism that uses a transfer bar to lift work in progress out of a workstation location and move it one position ahead to the next station.

warp yarn. Yarn that runs the length of the fabric, X axis.

waste materials. Materials that are lost or consumed in the manufacturing process, but do not end up as material present in the completed product.

water stains. A wood stain that uses water as a vehicle in which to suspend color pigments.

waterfall coating. See *curtain coating.*

waterjet machining. A process that uses a high-velocity stream of water to cut materials ranging from paper to stone or metals.

wave soldering. A process which places an integrated circuit board over a soldering bath with the connections to be soldered just above the surface of the molten solder. A wave is created that moves across the surface of the bath. The top of the wave briefly touches the connections, making the solder joints on the circuit board.

wear parts. Moving parts exposed to extensive friction.

weft. Another term for *fill yarn.*

weir. A refractory barrier over which molten glass flows out of the furnace.

welding. Process that joins metallic parts by locally heating the metal of two adjacent parts until that area becomes molten, permitting the metal to fuse together.

wet felting. A process that uses water and steam to make hardboard.

wet out. The formation of a composite, made through the integration of a solid reinforcement and a binding material (matrix). The binding material begins as either a liquid or slurry to "wet out" or change the structure of the reinforcement material(s) before it solidifies.

wet pressing. A ceramic forming process that uses higher moisture content and less pressure than dry pressing. However, it can result in dimensional tolerances varying by as much as 2 percent.

wet winding. A process in which the filament is impregnated with resin just before it is wound onto the mandrel.

wettability. A measurement of a liquid's ability to wet the surface of a solid material.

wheel electrodes. Rolling electrical contacts used to resistance-weld a longitudinal seam.

white cast iron. Type of cast iron that is very hard and is used for making parts that must combat fatigue from extreme wear and abrasion conditions.

wiping. Polishing operation performed on plastic using an uncharged wheel.

wire metallizing. See *flame-spraying.*

wood filler. A paste used to close the pores of open-grained wood before a finish is applied.

wood plastic composition (WPC). A process that uses radioactive isotopes to transform wood into a material that is much like plastic. The material is called *irradiated wood.*

woof. Another term for *fill yarn.*

work in process (WIP). Parts or components that are moving through a manufacturing operation. WIP is the stage between raw materials and finished goods.

workholding pallet. A transfer base containing work in process. It is moved from station-to-station by conveyor or other means, then relocated at each station by guide pins.

wrought iron. One of the two basic forms of iron. Wrought iron is tough, because it contains very little carbon.

Y

yarn. Twisted strands, can be made of many different types of materials.

yield. The number of yards of glass fiber roving per pound.

Z

zero drift. The signal level varies from its set zero value even when there are no inputs present to cause a change.

zinc-arc spray metallizing. A process in which zinc wires are melted by an electric arc. The resulting molten zinc is deposited on the plastic substrate by a stream of air.

zirconia. The principal ceramic material in zirconium oxide. It is made more wear-resistant by transformation toughening.

Index